INTERLEUKINS, LYMPHOKINES, AND CYTOKINES

Academic Press Rapid Manuscript Reproduction

INTERLEUKINS, LYMPHOKINES, AND CYTOKINES

PROCEEDINGS OF THE THIRD INTERNATIONAL LYMPHOKINE WORKSHOP

Edited by

Joost J. Oppenheim
National Institutes of Health
Bethesda, Maryland

Stanley Cohen
University of Connecticut
Farmington, Connecticut

Discussion Editor

Maurice Landy
La Jolla, California

1983

ACADEMIC PRESS
A Subsidiary of Harcourt Brace Jovanovich, Publishers
New York London
Paris San Diego San Francisco São Paulo Sydney Tokyo Toronto

ACADEMIC PRESS, INC.
111 Fifth Avenue, New York, New York 10003

United Kingdom Edition published by
ACADEMIC PRESS, INC. (LONDON) LTD.
24/28 Oval Road, London NW1 7DX

Library of Congress Cataloging in Publication Data

International Lymphokine Workshop (3rd : 1982 :)
Interleukins, lymphokines, and cytokines.

Includes index.
1. Lymphokines--Congresses. 2. Interleukins--Congresses. 3. Cytokines--Congresses. I. Oppenheim, Joost J. II. Cohen, Stanley, 1937- . III. Landy, Maurice. IV. Title.
QR185.8.L93I57 1983 599'.029 83-2673
ISBN 0-12-527540-4

PRINTED IN THE UNITED STATES OF AMERICA

83 84 85 86 9 8 7 6 5 4 3 2 1

CONTENTS

SECTION I: CHARACTERISTICS AND EFFECTS OF INTERLEUKIN-2 AND ANTIINTERLEUKIN-2 RECEPTOR ANTIBODIES

SECTION III: B CELL ACTIVATION BY NONSPECIFIC AND SPECIFIC LYMPHOKINES

SECTION IV: LYMPHOKINE-INDUCED DIFFERENTIATION AND GROWTH OF TARGET CELLS

SECTION VI: SOURCES, ACTIVITIES, AND CHARACTERISTICS OF INTERLEUKIN-1 AND ENDOGENOUS PYROGEN

SECTION VII: ACTIONS OF EFFECTOR CYTOKINES

SECTION VIII: PRODUCTION OF LYMPHOKINES BY HYBRIDOMAS, LYMPHOCYTE LINES, CLONES, AND SUBPOPULATIONS

SECTION IX: IMMUNOPHARMACOLOGICAL STUDIES OF LYMPHOKINES

SECTION X: OVERVIEWS AND CONCLUSIONS

CONTRIBUTORS

Numbers in parentheses indicate the pages on which the authors' contributions begin.

Mario Abbud (141), INSERM U25 and Lab.Ass. CNRS 122, Hôpital Necker, 161, rue de Sèvres, 75730 Paris Cedex 15, France

Bharat B. Aggarwal (521), Department of Protein Biochemistry, Genentech, Inc., South San Francisco, California

Andrew Aldridge (123), Department of Experimental Haematology, Paterson Laboratories, Christie Hospital & Holt Radium Institute, Manchester, England

Amnon Altman (599), Department of Immunology, Medical Biology Institute, La Jolla, California

Barbara J. Andersen (401), Microbiology Program, Pennsylvania State University, University Park, Pennsylvania

Paul Anderson (553), Department of Microbiology, New York University School of Medicine, New York, New York

Rene Anderson (325), Department of Immunology, St Thomas' Hospital and Medical School, London SE1, United Kingdom

*Ron N. Apte** (227, 237), Department of Chemical Immunology, The Weizmann Institute of Science, Rehovot, Israel

Douglas L. Archer (51), Food and Drug Administration, Cincinnati, Ohio 45226

Geoffrey Asherson (359, 397), Division of Immunological Medicine, Clinical Research Centre, Harrow, United Kingdom

Elisha Atkins (441), Department of Medicine, Yale University School of Medicine, New Haven, Connecticut

Robert Auerbach (317), Department of Zoology, University of Wisconsin, Madison, Wisconsin

Thomas M. Aune, (383), Department of Pathology and Laboratory Medicine, The Jewish Hospital of St Louis, Department of Pathology, Washington University School of Medicine, St Louis, Missouri

David I. Beller (291, 643), Department of Pathology, Harvard Medical School, Boston, Massachusetts

William R. Benjamin (195, 609, 617), Laboratory of Microbiology and Immunology, National Institute of Allergy and Infectious Diseases, National Institutes of Health, Bethesda, Maryland 20205

*Present address: Department of Microbiology and Immunology, Faculty of Health Sciences, Ben-Gurion University, Beer-Sheva, Israel

Zvi Bentwich (227), Department of Medicine C, Kaplan Hospital, Rehovot, Israel
Florence Bettens (43), Institute of Clinical Immunology, University of Bern, Bern, Switzerland
Michael Blaese (299), Division of Cancer Biology and Diagnosis, National Cancer Institute, National Institutes of Health, Bethesda, Maryland 20205
Jane S. Brandhorst (257), Cancer Metastasis and Treatment Laboratory, NCI-Frederick Cancer Research Facility, Frederick, Maryland
Chaim Brautbar (227), Laboratory of Immunohematology and Immunology, Hadassah Medical School, Jerusalem, Israel
Yael Bromberg (561), Laboratory of Immunopharmacology, Department of Human Microbiology, Sackler School of Medicine, Tel-Aviv University, Tel-Aviv, Israel
Joseph L. Butler (219), Laboratory of Immunoregulation, National Institute of Allergy and Infectious Diseases, National Institutes of Health, Bethesda, Maryland 20205
Graham Carter (723), Department of Chemical Pathology, Charing Cross Hospital, London, United Kingdom
Jacques A. Charon (447), Laboratory of Microbiology and Immunology, National Institute of Dental Research, National Institutes of Health, Bethesda, Maryland 20205
Louis Chedid (465, 699), Service d'Immunothérapie Expérimentale, Institut Pasteur, Paris, France
U. Chen-Bettecken (169), Institut für Virologie und Immunbiologie, Universität Würzburg, Würzburg, Germany
W. Hallowell Churchill (269), Department of Medicine, Harvard Medical School, Division of Rheumatology and Immunology, Brigham and Women's Hospital, Boston, Massachusetts
Ian Clark-Lewis (581), Immunoregulation Laboratory, The Walter and Eliza Hall Institute of Medical Research, Post Office Royal Melbourne Hospital, Melbourne, Australia
Marion C. Cohen (545), Department of Pathology, University of Connecticut Health Center, Farmington, Connecticut 06032
Stanley Cohen, (341, 391, 545, 753), Department of Pathology, University of Connecticut Health Center, Farmington, Connecticut 06032
Vittorio Colizzi (359), Division of Immunological Medicine, Clinical Research Centre, Harrow, United Kingdom
Josephine Cox (325), Department of Immunology, St Thomas' Hospital and Medical School, London SE1, United Kingdom
Pedro Cuatrecasas (511), Molecular Biology Department, Wellcome Research Laboratories, Research Triangle Park, North Carolina
Eric Culbert (245), ICRF Tumour Immunology Unit, Department of Zoology, University College London, London, England
John F. Daley (667), Division of Tumor Immunology, Sidney Farber Cancer Institute, Boston, Massachussetts

C. Damais (465), Immunothérapie Expérimentale, Institut Pasteur, Paris, France

J.-M. Dayer (659), Department of Pathology, University of Geneva, Geneva, Switzerland

Joel M. Depper (19), Metabolism Branch, National Cancer Institute, Bethesda, Maryland 20205

Helen D'Silva (545), Department of Pathology, University of Connecticut Health Center, Farmington, Connecticut 06032

James J. Devlin (527), Department of Molecular Biology and Biochemistry, University of California at Irvine, Irvine, California

T. Michael Dexter (123), Department of Experimental Haematology, Paterson Laboratories, Christie Hospital & Holt Radium Institute, Manchester, England

Tibor Diamantstein (27), Immunology Research Unit, Klinikum-Steglitz, Freie Universität Berlin, Berlin, Federal Republic of Germany

Charles A. Dinarello (495), Department of Medicine, Divisions of Experimental Medicine and Allergy, Tufts University School of Medicine and, New England Medical Center Hospital, Boston, Massachusetts

Rolf Dubs (421), Section of Clinical Immunology, University Hospital, Zurich, Switzerland

Dudley C. Dumonde, (325, 723), Department of Immunology, St Thomas' Hospital and Medical School, London, SE1, United Kingdom

P. A. Dunn (625), Cancer Research Center, Columbia, Missouri 65205

P. L. Durette (691), Department of Membrane and Arthritis Research, Merck, Sharp & Dohme Research Laboratories, Rahway, New Jersey 07065

Richard W. Dutton (179, 687), Department of Biology, University of California, San Diego, La Jolla, California

Michel Dy (141), INSERM U25 and Lab.Ass. CNRS 122, Hôpital Necker, 161, rue de Sèvres, 75730 Paris Cedex 15, France

Susan Edwards (245), ICRF Tumour Immunology Unit, Department of Zoology, University College London, London, England

John M. Ely (643), The University of Chicago, Chicago, Illinois

Karen L. Evans (599), Department of Immunology, Medical Biology Institute, La Jolla, California

J. Fahey (749), Department of Microbiology and Immunology, School of Medicine, University of California at Los Angeles, Los Angeles, California 90024

Werner Falk (109, 287), Institut für Immunologie und Genetik, Deutsches Krebsforschungszentrum, Heidelberg, Federal Republic of Germany

*John J. Farrar** (195, 609, 617, 757), Laboratory of Microbiology and Immunology, National Institute of Dental Research, National Institute of Allergy and Infectious Disease, National Institutes of Health, Bethesda, Maryland 20205

Willam L. Farrar (51), Biological Carcinogenesis Program, Frederick Cancer Research Center, Frederick, Maryland 21701

*Present address: Hoffman-La Roche, Nutley, New Jersey

Anthony S. Fauci (211, 219), Laboratory of Immunoregulation, National Institute of Allergy and Infectious Diseases, National Institutes of Health, Bethesda, Maryland 20205

Marc Feldmann (245, 253), ICRF Tumour Immunology Unit, Department of Zoology, University College London, London, England

E. Fernandez-Cruz (599), Department of Immunology, Medical Biology Institute, La Jolla, California

Isaiah J. Fidler (257), Cancer Metastasis and Treatment Laboratory, NCI-Frederick Cancer Research Facility, Frederick, Maryland

James H. Finke (89, 409), Department of Molecular and Cellular Biology, Cleveland Clinic Foundation, Cleveland, Ohio

Frank W. Fitch (643, 687), The University of Chicago, Chicago, Illinois

Arthur Flynn (409), Department of Molecular and Cellular Biology, Cleveland Clinic Foundation, Cleveland, Ohio

William E. Fogler (257), Cancer Metastasis and Treatment Laboratory, NCI-Frederick Cancer Research Facility, Frederick, Maryland

Adriano Fontana (421), Section of Clinical Immunology, University Hospital, Zurich, Switzerland

Richard J. Ford (203), Department of Pathology, Division of Pathobiology, M. D. Anderson Hospital and Tumor Institute, Houston, Texas

Janet Fuller-Farrar (195), Laboratory of Microbiology and Immunology, National Institute of Dental Research and Laboratory of Immunology, National Institute of Allergy and Infectious Disease, National Institutes of Health, Bethesda, Maryland 20205

Robert C. Gallo (3), Laboratory of Tumor Cell Biology, National Cancer Institute, National Institutes of Health, Bethesda, Maryland 20205

John M. Garland (123), Department of Experimental Haematology, Paterson Laboratories, Christie Hospital & Holt Radium Institute, Manchester, England

Celia L. Gately (335), Laboratory of Microbiology and Immunology, National Institute of Dental Research, National Institutes of Health, Bethesda, Maryland 20205

Diethard Gemsa (589) Institut für Immunologie und Serologie, Universität Heidelberg, Heidelberg, Federal Republic of Germany

J. P. Gerard (63), NCI-FCRF, Frederick, Maryland 21701

Steven Gillis (75), Immunex Corporation, 51 University Building, Seattle, Washington

Markus Glantschnig (421), Section of Clinical Immunology, University Hospital, Zurich, Switzerland

John Goldman (643), The University of Chicago, Chicago, Illinois

Margaret Goldman (643), The University of Chicago, Chicago, Illinois

Eugene Goldwasser (643), The University of Chicago, Chicago, Illinois

Joseph E. Gootenberg (3), Laboratory of Tumor Cell Biology, National Cancer Institute, National Institutes of Health, Bethesda, Maryland, 20205

Julius Gordon (11), Department of Surgery, McGill University, Montreal, Quebec, Canada

E. Gorelik (739), Surgery Branch, National Cancer Institute, Bethesda, Maryland 20205

Fumimasa Goto (427, 435), Department of Immunopathology, Kumamoto University, Medical School, Kumamoto, Japan

Kumiko Goto (435), Department of Immunopathology, Kumamoto University, Medical School, Kumamoto, Japan

B. K. Graham (691), The Department of Immunology, Merck, Sharp & Dohme Research Laboratories, Rahway, New Jersey 07065

Gale A. Granger (527), Department of Molecular Biology and Biochemistry, University of California at Irvine, Irvine, California

Warner C. Greene (19), Metabolism Branch, National Cancer Institute, Bethesda, Maryland 20205

E. A. Grimm (739), Surgery Branch, National Cancer Institute, Bethesda, Maryland 20205

Peter J. Grob (421), Section of Clinical Immunology, University Hospital, Zurich, Switzerland

Petra Gronewold (677), Division of Clinical Immunology, Department of Medicine, Med. Hochschule, D-300 Hannover 61, Federal Republic of Germany

P.-A. Guerne (659), Department of Pathology, University of Geneva, Geneva, Switzerland

Janice K. Gutowski (341), Department of Pathology, University of Connecticut Health Center, Farmington, Connecticut 06032

H. D. Hager (349), Institut für Humangenetik, Universität Heidelberg, 6900 Heidelberg, Federal Republic of Germany

Toshiyuki Hamaoka (161), Department of Oncogenesis, Institute for Cancer Research, Osaka University Medical School, Osaka, Japan

Anne Hamblin (325), Department of Immunology, St Thomas' Hospital and Medical School, London SE1, United Kingdom

G. Hänsch (349), Institut für Immunologie, Universität Heidelberg, 6900 Heidelberg, Federal Republic of Germany

D. F. Hanson (473), Department of Medicine, The Johns Hopkins University, Baltimore, Maryland

Yoshinobu Hara (161), Department of Oncogenesis, Institute for Cancer Research, Osaka University Medical School, Osaka, Japan

Conny Hardt (99), Institut für Medizinische Mikrobiologie, Johannes Gutenberg-Universität Mainz, Mainz, Federal Republic of Germany

Richard N. Harkins (521), Department of Protein Biochemistry, Genentech, Inc., South San Francisco, California

William E. Helsel (335), Laboratory of Microbiology and Immunology, National Institute of Dental Research, National Institutes of Health, Bethesda, Maryland 20205

L. E. Henderson (63), NCI-FCRF, Frederick, Maryland 21701

Mary L. Hilfiker (89, 195, 409), Department of Molecular and Cellular Biology, Cleveland Clinic Foundation, Cleveland, Ohio

Wayne T. Hockmeyer (617), Walter Reed Army Institute of Research, Washington, DC

Michael Hoffman (707), Memorial Sloan-Kettering Cancer Center, New York, New York

Frank den Hollander (723), Organon International B.V., Oss, The Netherlands

Maureen Howard (187, 195), Laboratory of Immunology, National Institute of Allergy and Infectious Diseases, National Institutes of Health, Bethesda, Maryland 20205

Catherine Howarth (123), Department of Experimental Haematology, Paterson Laboratories, Christie Hospital & Holt Radium Institute, Manchester, England

Lothar Hültner (589), Abteilung für Experimentelle Hämatologie, der Gesellschaft für Strahlen- und, Umweltforschung, München, Federal Republic of Germany

James N. Ihle (113, 643), Biological Carcinogenesis Program, NCI- Frederick Cancer Research Facility, Frederick, Maryland 21701

*Peter C. Isakson** (153), Department of Microbiology, University of Texas Health Science Center, Dallas, Texas

Norio Ishii (161), Department of Oncogenesis, Institute for Cancer Research, Osaka University Medical School, Osaka, Japan

Roger James (245), ICRF Tumour Immunology Unit, Department of Zoology, University College London, London, England

Stephanie L. James (617), National Institutes of Health, Bethesda, Maryland

Howard M. Johnson (51), Department of Microbiology, University of Texas Medical Branch, Galveston, Texas 77550

Fransiska Joncourt (43), Institute of Clinical Immunology, University of Bern, Bern, Switzerland

Brigitte Kaltmann (589), Institut für Immunologie und Genetik, Deutsches Krebsforschungszentrum, Heidelberg, Federal Republic of Germany

Ralph F. Kampschmidt (441, 455), Biomedical Division, The Samuel Roberts Noble Foundation, Inc., Ardmore, Oklahoma

Tomohisa Kanatani (161), Department of Oncogenesis, Institute for Cancer Research, Osaka University Medical School, Osaka, Japan

Andrew H. Kang (535), Departments of Medicine and Biochemistry, University of Tennessee Center for the Health Sciences and Veterans Administration Medical Center, Memphis, Tennessee

T. Kasahara, (211), Laboratory of Microbiology and Immunology, National Institute of Dental Research, National Institutes of Health, Bethesda, Maryland 20205

Eva Kasp-Grochowska (325), Department of Immunology, St Thomas' Hospital and Medical School, London SE1, United Kingdom

*Present address: Department of Pharmacology, University School of Medicine, Charlottesville, Virginia

Milena Kataranovski (481), Institute of Microbiology and Immunology, School of Medicine, Belgrade, Yugoslavia

David H. Katz (599), Department of Immunology, Medical Biology Institute, La Jolla, California

Dvora Katz (227), Department of Chemical Immunology, The Weizmann Institute of Science, Rehovot, Israel

Ursula Kees (589), Institut für Immunologie und Genetik, Deutsches Krebsforschungszentrum, Heidelberg, Federal Republic of Germany

Jonathan Keller (113), Biological Carcinogenesis Program, NCI-Frederick Cancer Research Facility, Frederick, Maryland 21701

Anne Keslo (651), Department of Immunology, Swiss Institute for Experimental Cancer Research, Epalinges, Switzerland

Allison Kingsbury (245), ICRF Tumour Immunology Unit, Department of Zoology, University College London, London, England

DeAnna Kirchmayer (317), Department of Zoology, University of Wisconsin, Madison, Wisconsin

Holger Kirchner (677), Institute of Virology, German Cancer Center, D-6900 Heidelberg, Federal Republic of Germany

Jim Klostergaard (527), Department of Molecular Biology and Biochemistry, University of California at Irvine, Irvine, California

Sirkka Kontiainen (245), Department of Bacteriology and Immunology, University of Helsinki, Finland

Nicola Kouttab (203), Department of Pathology, Division of Pathobiology, M. D. Anderson Hospital and Tumor Institute, Houston, Texas

Peter H. Krammer (153, 589), Institut für Immunologie und Genetik, Deutsches Krebsforschungszentrum, Heidelberg, Federal Republic of Germany

Flemming Kristensen (43, 421), Institute for Clinical Immunology, Inselspital, Bern, Switzerland

Frederick C. Kull, Jr. (511), Molecular Biology Department, Wellcome Research Laboratories, Research Triangle Park, North Carolina

Lawrence B. Lachman (441), Immunex Corporation, 51 University Bldg., Seattle, Washington

Michel Lanotte (123), Department of Experimental Haematology, Paterson Laboratories, Christie Hospital & Holt Radium Institute, Manchester, England

Bernard Lebel (141), INSERM U25 and Lab.Ass. CNRS 122, Hôpital Necker, 161, rue de Sèvres, 75730 Paris Cedex 15, France

Claude Leclerc (699), Service d'Immunothérapie Expérimentale, Institut Pasteur, Paris, France

Ruth Lenz (83), Department of Life Sciences, Bar-Ilan University, Ramat Gan, Israel

Warren J. Leonard (19), Metabolism Branch, National Cancer Institute, Bethesda, Maryland 20205

Chungyee Leung (11), Department of Surgery, McGill University, Montreal, Quebec, Canada

Jürg Liechti (635), Institute of Clinical Immunology, University of Bern, Bern, Switzerland

Ruth Lifshitz (237), Department of Chemical Immunology, The Weizmann Institute of Science, Rehovot, Israel

Peter Lindsay (69), BIOTEST-Serum-Institut GmbH, Frankfurt, Federal Republic of Germany

Maria-Luise Lohmann-Matthes (569), Max-Planck-Institut für Immunbiologie, D-7800 Freiburg, Federal Republic of Germany

David H. Lovett (415), Department Internal Medicine, West Haven VAMC and Yale University School of Medicine, New Haven, Connecticut

*Ilana Löwy** (237), Department of Chemical Immunology, The Weizmann Institute of Science, Rehovot, Israel

Richard A. Luben (307), Division of Biomedical Sciences and Department of Biochemistry, University of California, Riverside, California

Thomas A. Luger† (447), Laboratory of Microbiology and Immunology, National Institute of Dental Research, National Institutes of Health, Bethesda, Maryland 20205

H. Robson MacDonald (651), Ludwig Institute for Cancer Research, Lausanne Branch, Epalinges, Switzerland

J. E. McEntire (625), Cancer Research Center, Columbia, Missouri 65205

Jennifer McKimm-Breschkin (581), Immunoregulation Laboratory, The Walter and Eliza Hall Institute of Medical Research, Post Office Royal Melbourne Hospital, Melbourne, Australia

Abby L. Maizel (203), Department of Pathology, Division of Pathobiology, M. D. Anderson Hospital and Tumor Institute, Houston, Texas

Thomas R. Malek (35), Laboratory of Immunology, National Institute of Allergy and Infectious Diseases, National Institutes of Health, Bethesda, Maryland 20205

Miroslav Malkovsky (359), Division of Immunological Medicine, Clinical Research Centre, Harrow, United Kingdom

Dean Mann (299), Division of Cancer Biology and Diagnosis, National Cancer Institute, National Institutes of Health, Bethesda, Maryland 20205

Daniela N. Männel (109, 287), Institut für Immunologie und Genetik, Deutsches Krebsforschungszentrum, Heidelberg, Federal Republic of Germany

Fabrizio Marcucci (589), Institut für Virologie, Deutsches Krebsforschungszentrum, Heidelberg, Federal Republic of Germany

Scott O. Marnoy (495), Department of Medicine, Divisions of Experimental Medicine and Allergy, Tufts University School of Medicine, and New England Medical Center Hospital, Boston, Massachusetts

Otoniel Martinez-Maza (375), Department of Immunobiology, Wallenberg Laboratory, Karolinska Institute, Stockholm, Sweden

Tohru Masuda (367), Department of Immunobiology, Institute for Immunology, Kyoto University, Sakyo, Kyoto, Japan

*Present address: Institut Pasteur, Paris, France

†Present address: Second University Hautklinik Wien, 4 Alstrasse A-1090, Vienna

John Maudsley (245), ICRF Tumour Immunology Unit, Department of Zoology, University College London, London, England

Alma Mednis, (269), Department of Medicine, Harvard Medical School, Division of Rheumatology and Immunology, Brigham and Women's Hospital, Boston, Massachusetts

Shashikant R. Mehta (203), Department of Pathology, Division of Pathobiology, M. D. Anderson Hospital and Tumor Institute, Houston, Texas

*Monte S. Meltzer** (617), National Institutes of Health, Bethesda, Maryland

M. V. Melvin III (691), The Department of Immunology, Merck, Sharp & Dohme Research Laboratories, Rahway, New Jersey 07065

Stefan C. Meuer (667), Division of Tumor Immunology, Sidney Farber Cancer Institute, Boston, Massachusetts

Roger Mitchell (455), Biomedical Division, The Samuel Roberts Noble Foundation, Inc., Ardmore, Oklahoma

Muneo Miyama-Inaba (367), Department of Immunobiology, Institute for Immunology, Kyoto University, Sakyo, Kyoto, Japan

Steven B. Mizel (149, 401), Microbiology Program, Pennsylvania State University, University Park, Pennsylvania

Barbara Moffat (521), Department of Protein Biochemistry, Genentech, Inc., South San Francisco, California

Marjorie A. Mohler (307), Division of Biomedical Sciences and Department of Biochemistry, University of California, Riverside, California

Robert N. Moore (89), Department of Microbiology, University of Tennessee, Knoxville, Tennessee

Edna Mozes (227, 237), Department of Chemical Immunology, The Weizmann Institute of Science, Rehovot, Israel

Bruce Muller (723), Department of Chemical Pathology, Charing Cross Hospital, London, United Kingdom

Atsushi Muraguchi (211, 219), Laboratory of Immunoregulation, National Institute of Allergy and Infectious Diseases, National Institutes of Health, Bethesda, Maryland 20205

P. A. Murphy (473), Department of Medicine, The Johns Hopkins University, Baltimore, Maryland

Carol A. Nacy (617), Walter Reed Army Institute of Research, Washington, D.C.

Shuji Nakamura (427, 435), Department of Immunopathology, Kumamoto University, Medical School, Kumamoto, Japan

Kenji Nakanishi (187), Laboratory of Immunology, National Institute of Allergy and Infectious Diseases, National Institutes of Health, Bethesda, Maryland 20205

Shigeki Nakayama (427), Department of Immunopathology, Kumamoto University, Medical School, Kumamoto, Japan

Glenn E. Nedwin (307), Division of Biomedical Sciences and Department of Biochemistry, University of California, Riverside, California

*Present address: Department of Dermatology, Walter Reed Army Medical Center, Washington, DC

Ruth Neta (715), Department of Microbiology, University of Notre Dame, Notre Dame, Indiana 46556

R. H. Neubauer (63), NCI-FCRF, Frederick, Maryland 21701

Hinnak Northoff (349, 677), Institute of Virology, German Cancer Center, D-6900 Heidelberg, Federal Republic of Germany

Herbert F. Oettgen (731), Memorial Sloan-Kettering Cancer Center, New York, New York

Susumu Ohkawara (427), Department of Immunopathology, Kumamoto University, Medical School, Kumamoto, Japan

Robert K. Oldham (277), Biological Response Modifiers Program, Division of Cancer Treatment, National Cancer Institute – FCRF, Frederick, Maryland 21701

*Joost J. Oppenheim** (211, 299, 447, 769), Cellular Immunology Section, National Institute of Dental Research, National Institute of Health, Bethesda, Maryland 20205

Sally L. Orr (527), Department of Molecular Biology and Biochemistry, University of California, Irvine, Irvine, California

J. R. Ortaldo (63), NCI-FCRF, Frederick, Maryland 21701

Hisao Osawa (27), Immunology Research Unit, Klinikum-Steglitz, Freie Universität Berlin, Berlin, Federal Republic of Germany

V. Paetkau (749), Department of Biochemistry, University of Alberta, Edmonton, Alberta, Canada T6G 2H7

Ronald Palacios (375), Department of Immunobiology, Wallenberg Laboratory, Karolinska Institute, Stockholm, Sweden

Edmund Palaszynski (113), Biological Carcinogenesis Program, NCI-Frederick Cancer Research Facility, Frederick, Maryland 21701

Michael A. Palladino (731), Memorial Sloan-Kettering Cancer Center, New York, New York

B. W. Papermaster (625), Cancer Research Center, Columbia, Missouri 65205

M. Parant (465), Immunothérapie Expérimentale, Institut Pasteur, Paris, France

William E. Paul (187, 195), Laboratory of Immunology, National Institute of Allergy and Infectious Diseases, National Institutes of Health, Bethesda, Maryland 20205

Hans H. Peter (677), Division of Clinical Immunology, Department of Medicine, Med. Hochschule, D-300 Hannover 61, Federal Republic of Germany

Klaus Pfizenmaier (99), Institut für Medizinische Mikrobiologie, Johannes Gutenberg-Universität Mainz, Mainz, Federal Republic of Germany

Edgar Pick (561, 577, 765), Laboratory of Immunopharmacology, Department of Human Microbiology, Sackler School of Medicine, Tel-Aviv University, Tel-Aviv, Israel

Carl W. Pierce (383), Department of Pathology and Laboratory Medicine, The Jewish Hospital of St Louis, Department of Pathology, Washington University School of Medicine, St Louis, Missouri

*Present address: Laboratory of Molecular Immunoregulation, Biological Response Modifiers Program, NCI – FCRF, NIH, Frederick, Maryland 21701

P.-F. Piguet (659), Department of Pathology, University of Geneva, Geneva, Switzerland

Dov H. Pluznik (83), Department of Life Sciences, Bar-Ilan University, Ramat Gan, Israel

Simon Pollack (707), Memorial Sloan-Kettering Cancer Center, New York, New York

Arnold E. Postlethwaite (535), Departments of Medicine and Biochemistry, University of Tennessee Center for the Health Sciences and Veterans Administration Medical Center, Memphis, Tennessee

Michael B. Prystowsky (643), The University of Chicago, Chicago, Illinois

Melanie Pulley (325, 723), Department of Immunology, St Thomas' Hospital and Medical School, London SE1, United Kingdom

Larry Pulliam (455), Biomedical Division, The Samuel Roberts Noble Foundation, Inc., Ardmore, Oklahoma

*Ellen Pure** (153), Department of Microbiology, University of Texas Health Science Center, Dallas, Texas

Jose Quintans (643), The University of Chicago, Chicago, Illinois

H. Rabin (63), NCI-FCRF, Frederick, Maryland 21701

Ellis L. Reinherz (667), Division of Tumor Immunology, Sidney Farber Cancer Institute, Boston, Massachusetts

Heinz Remold (269, 355, 643), Department of Medicine, Harvard Medical School, Division of Rheumatology and Immunology, Brigham and Women's Hospital, Boston, Massachusetts

G. Riveau (465), Immunothérapie Expérimentale, Institut Pasteur, Paris, France

Harold Rode (11), Department of Surgery, McGill University, Montreal, Quebec, Canada

Martin Röllinghoff (99), Institut für Medizinische Mikrobiologie, Johannes Gutenberg-Universität Mainz, Mainz, Federal Republic of Germany

S. A. Rosenberg (739), Surgery Branch, National Center Institute, Bethesda, Maryland 20205

M. M. Rosenstein (739), Surgery Branch, National Cancer Institute, Bethesda, Maryland 20205

Lanny J. Rosenwasser (495), Department of Medicine, Divisions of Experimental Medicine and Allergy, Tufts University School of Medicine, and New England Medical Center Hospital, Boston, Massachusetts

Susan A. Rothmann (89), Department of Laboratory Hematology, Cleveland Clinic Foundation, Cleveland, Ohio

Francis W. Ruscetti (3), Laboratory of Tumor Cell Biology, National Cancer Institute, National Institutes of Health, Bethesda, Maryland 20205

John L. Ryan (415), Department Internal Medicine, West Haven VAMC and Yale University School of Medicine, New Haven, Connecticut

C. G. Sahasrabudde (203), Department of Pathology, Division of Pathobiology, M. D. Anderson Hospital and Tumor Institute, Houston, Texas

*Present address: Department of Cellular Immunology, Rockefeller University, New York, New York

Samuel B. Salvin (715), Department of Microbiology, University of Pittsburgh School of Medicine, Pittsburgh, Pennsylvania 15261

Margrit P. Scheid (731), Memorial Sloan-Kettering Cancer Center, New York, New York

Peter Scheurich (99), Institut für Medizinische Mikrobiologie, Johannes Gutenberg-Universität Mainz, Mainz, Federal Republic of Germany

Anneliése Schimpl (169, 253), Institut für Virologie und Immunbiologie, Universität Würzburg, Würzburg, Germany

Stuart F. Schlossman (667), Division of Tumor Immunology, Sidney Farber Cancer Institute, Boston, Massachusetts

John W. Schrader (581), Immunoregulation Laboratory, The Walter and Eliza Hall Institute of Medical Research, Post Office Royal Melbourne Hospital, Melbourne, Australia

Udo Schwulera (69), BIOTEST-Serum-Institut GmbH, Frankfurt, Federal Republic of Germany

Jeffrey W. Scott (89), Department of Molecular and Cellular Biology, Cleveland Clinic Foundation, Cleveland, Ohio

Ethan M. Shevach (35), Laboratory of Immunology, National Institute of Allergy and Infectious Diseases, National Institutes of Health, Bethesda, Maryland 20205

W. Schmidt (349), Universitätsfrauenklinik, Universität Heidelberg, 6900 Heidelberg, Federal Republic of Germany

Miroslav M. Simić (481), Institute of Microbiology and Immunology, School of Medicine, Belgrade, Yugoslavia

Philip L. Simon (487), Departments of Ophthalmology and Microbiology, University of Pittsburgh School of Medicine, Pittsburgh, Pennsylvania

Kendall A. Smith (19), Dartmouth Medical School, Hanover, New Hampshire

Hans-H. Sonneborn (69), BIOTEST-Serum-Institut GmbH, Frankfurt, Federal Republic of Germany

Barbara Southcott (723), Department of Radiotherapy, Charing Cross Hospital, London, United Kingdom

Beda M. Stadler (131, 635), Institute of Clinical Immunology, University of Bern, Bern, Switzerland

M. J. Staruch (691), The Department of Immunology, Merck, Sharp & Dohme Research Laboratories, Rahway, New Jersey 07065

Patricia Steeg (299, 609), Laboratory of Microbiology and Immunology, National Institute of Dental Research, National Institutes of Health, Bethesda, Maryland 20205

R. Bernd Sterzel (415), Department Internal Medicine, West Haven VAMC and Yale University School of Medicine, New Haven, Connecticut

Richard Stiehm (299), Department of Pediatrics, University of California at Los Angeles, Los Angeles, California

Stanislava Stošić-Grujičić (481), Institute of Microbiology and Immunology, School of Medicine, Belgrade, Yugoslavia

Harold B. Stull (277), Biological Response Modifiers Program, Division of Cancer Treatment, National Cancer Institute-FCRF, Frederick, Maryland 21701

Osias Stutman (731), Memorial Sloan-Kettering Cancer Center, New York, New York

Daniel Suez (227), Department of Chemical Immunology, The Weizmann Institute of Science, Rehovot, Israel

Matsunobu Suko (391), Department of Pathology, University of Connecticut Health Center, Farmington, Connecticut 06032

Deming Sun (569), Max-Planck-Institut für Immunbiologie, D-7800 Freiburg, Federal Republic of Germany

Geoff Sunshine (245), ICRF Tumour Immunology Unit, Department of Zoology, University College London, London, England

Takayo Suzuki (367), Department of Immunobiology, Institute for Immunology, and The First Department of Internal Medicine, Faculty of Medicine, Kyoto University, Sakyo, Kyoto, Japan

Susan L. Swain (179), Department of Biology, University of California, San Diego, La Jolla, California

Marcelo Sztein (299, 447), Laboratory of Microbiology and Immunology, National Institute of Dental Research, National Institutes of Health, Bethesda, Maryland 20205

Kenji Tadokoro (131), Institute of Clinical Immunology, University of Bern, Bern, Switzerland

Kiyoshi Takatsu (161), Department of Oncogenesis, Institute for Cancer Research, Osaka University Medical School, Osaka, Japan

Gary B. Thurman (277), Biological Response Modifiers Program, Division of Cancer Treatment, National Cancer Institute-FCRF, Frederick, Maryland 21701

Shohken Tomita (161), Department of Oncogenesis, Institute for Cancer Research, Osaka University Medical School, Osaka, Japan

Barbara A. Torres (51), Department of Microbiology, University of Texas Medical Branch, Galveston, Texas 77550

Herbert F. Upchurch (455), Biomedical Division, The Samuel Roberts Noble Foundation, Inc., Ardmore, Oklahoma

P. Vassalli (659), Department of Pathology, University of Geneva, Geneva, Switzerland

Jan Vilček (553, 761), Department of Microbiology, New York University School of Medicine, New York, New York

Ellen S. Vitetta (153), Department of Microbiology, University of Texas Health Science Center, Dallas, Texas

Stefanie N. Vogel (643), Uniformed Services University of Health Sciences, Bethesda, Maryland

Hermann Wagner (99), Institut für Medizinische Mikrobiologie, Johannes Gutenberg-Universität Mainz, Mainz, Federal Republic of Germany

Sharon M. Wahl (335, 355), Laboratory of Microbiology and Immunology, National Institute of Dental Research, National Institutes of Health, Bethesda, Maryland 20205

Thomas A. Waldmann (19), Metabolism Branch, National Cancer Institute, Bethesda, Maryland 20205

Christoph Walker (43), Institute of Clinical Immunology, University of Bern, Bern, Switzerland

Sandra L. Watt (317), Department of Zoology, University of Wisconsin, Madison, Wisconsin

David R. Webb (149, 383), Roche Institute of Molecular Biology, Nutley, New Jersey

Elisabeth Weber (421), Section of Clinical Immunology, University Hospital, Zurich, Switzerland

Alain L. de Weck (43, 131, 421, 635), Institute of Clinical Immunology, University of Bern, Bern, Switzerland

E. Wecker (169), Institut für Virologie und Immunbiologie, Universität Würzburg, Würzburg, Germany

Robin G. Weiner (599), Department of Immunology, Medical Biology Institute, La Jolla, California

David Wilde (643), The University of Chicago, Chicago Illinois

Judith B. Willoughby (487), Department of Pathology, The Johns Hopkins University School of Medicine, Baltimore, Maryland

William F. Willoughby (487), Department of Pathology, The Johns Hopkins University School of Medicine, Baltimore, Maryland

Robin A. Wolstencroft (325), Department of Immunology, St. Thomas' Hospital, and Medical School, London SE1, United Kingdom

Grace H. W. Wong (581), Immunoregulation Laboratory, The Walter and Eliza Hall Institute of Medical Research, Post Office Royal Melbourne Hospital, Melbourne, Australia

D. D. Wood (691), The Department of Immunology, Merck, Sharp & Dohme Research Laboratories, Rahway, New Jersey 07065

Peter V. Wussow (731), Memorial Sloan-Kettering Cancer Center, New York, New York

Robert S. Yamamoto (527), Department of Molecular Biology and Biochemistry, University of California at Irvine, Irvine, California

Y. K. Yip (553), Department of Microbiology, New York University School of Medicine, New York, New York

Takeshi Yoshida (391, 397, 545), Department of Pathology, University of Connecticut Health Center, Farmington, Connecticut 06032

Masaru Yoshinaga (427, 435), Department of Immunopathology, Kumamoto University, Medical School, Kumamoto, Japan

Marek Zembala (359), Division of Immunological Medicine, Clinical Research Centre, Harrow, United Kingdom

PREFACE

It is probably important, especially for the sake of the uninitiated, to attempt to define the subject matter of the workshop that was the basis for this volume. It is simplest to define the issues in the context of the rather brief history of this research area. Studies of lymphocyte-derived mediators followed the concomitant discovery by David (1966) and Bloom and Bennett (1966) that the *in vitro* activation of lymphocytes leads to the production of nonimmunoglobulin glycoprotein mediators that inhibited the migration of macrophages. Following the discovery of additional lymphocyte-derived mediators that could regulate the growth or mobility of a variety of leukocytes, these activities were termed "lymphokines" by Dumonde *et al.* (1969). It was subsequently established that factors similar to lymphokines are also produced by nonlymphocytic cells such as normal macrophages, fibroblasts, keratinocytes, and a variety of transformed cell lines. It was, therefore, proposed by Cohen and co-workers (1977) that this entire class of mediators be termed "cytokines." Over the past 18 years, more than a hundred distinct biological activities that are produced by a variety of cell types have been described. Cytokines, in addition to acting on inflammatory leukocytes, such as macrophages, neutrophils, eosinophils, and mast cells, can also act on noninflammatory cells including osteoclasts, fibroblasts, and endothelial cells. Thus, cytokines not only participate in regulating immunological and inflammatory processes, but can also contribute to repair processes and to the regulation of normal cell growth and differentiation. Although cytokines, unlike hormones, are not produced by glands and are not normally present in serum, they are hormone-like in their intercellular regulatory functions and can be recovered from inflamed tissues.

This obviously important area of investigation has justifiably attracted a diversity of scientists including immunologists, pathologists, cell biologists, and biochemists. Despite this, progress in cytokine research has been impeded by a number of technological problems. Cytokines, as exemplified by the interferons, are active at concentrations as low as 10^{-15} *M*. This problem is compounded by the fact that multiple factors are often simultaneously produced. Consequently, these regulatory factors have proven difficult to separate and isolate in quantity, and have had to be identified and quantitated by various *in vitro* and, occasionally, *in vivo* bioassays. The study of the cytokines has been further complicated by the fact that a single factor may exert a multiplicity of effects, and conversely that distinct mediators can have the same effect in a given bioassay. In addition, cytokines frequently are interdependent and can act on cells in an additive, synergistic, or suppressive fashion, which severely limits the capacity of bioassays to define these activities. Finally, although most cytokines function nonspecifically and are not genetically restricted, there is a subset of lymphokines that can bind antigens and function in a genetically restricted manner. These "specific" mediators probably consist of secreted membrane components of T lymphocytes that contain both receptors for antigen and major histocompatibility antigens. However, since interactions between nonspecific cytokines and lymphocytes can influence immunologically specific reactions, the nonspecific and specific factors are also interdependent.

In order to improve communication and interaction of investigators concerned with studying cytokines, the editors convened the First International Workshop in Bethesda, Maryland, in 1976, and to continue to fulfill this need, the workshop has since been reconvened at three-year intervals. At the Second International Lymphokine Workshop, which was held at Ermatingen, Switzerland in 1979, a number of investigators agreed that certain named biological activities produced by macrophages and lymphocytes could probably be attributed to the same molecule and should therefore be renamed interleukins 1 and 2, respectively. This consensus led to a commendable, albeit small, reduction in the profusion of named factors. In addition, the recognition that factors such as the interleukins could have a broad array of crucial biological effects has galvanized increased research effort in this area. Therefore, the theme of the Third International Lymphokine Workshop held at Haverford College, Pennsylvania and the title of this book were chosen to emphasize both the conceptual significance and scope of investigative concern of this subject.

This volume was prepared to make available to all interested scientists the current research approaches and concerns of scientists engaged in studying cytokines. More than 250 scientists met from August 1–5, 1982, to exchange concepts, opinions, and information in the form of oral presentations, discussions, and poster sessions. The chapters in this volume include short versions of the oral presentations and selected poster presentations. In addition, an edited version of the entire discussion following each chapter, as initially transcribed by each of the discussants, was also included in the volume. Rather than reviewing the outline of the volume, suffice it to say that the contributions represent the views of investigators from a broad spectrum of disciplines, including immunology, pathology, biochemistry, cell biology, hematology, molecular genetics, pharmacology, and physiology, present at the meeting who are engaged in the study of cytokines. Although phenomenological reports of factors persist, this volume also reflects the progress that is being made through the application of new techniques of cell hybridization, generation of monoclonal antibodies, affinity purification, and gene cloning. The study of interleukins, lymphokines, and cytokines has already resulted in considerable progress in our understanding of the mechanisms of host defense and wound healing. It is hoped that this volume will promote greated advances in this research area by providing a timely collection of relevant papers as a reference guide for investigators engaged in cytokine research and by recruiting additional scientists to study this provocative subject.

ACKNOWLEDGMENTS

We thank Ms. S. Dougherty, J. Fuller-Farrar, F. Hakim, and Mr. C. Carter for their efficient and invaluable assistance in collecting and recording the discussions. We are grateful to Dr. Dorothy Steffens and her staff at the Haverford Conference Center for providing the local organization and ambience conducive to a successful meeting. We acknowledge the invaluable organizational activities of Ronald Cohen, Dr. Marion Cohen, and Ms. Ann Bernardo.

We owe a special debt of gratitude to Dr. Maurice Landy. He was one of the individuals who provided the stimulus for the First Lymphokine Workshop, and it was the success of that venture that led to his continuing series of meetings. He has given unstintingly of his time and advice. For this meeting, he instigated, coordinated, organized, and edited the written record of the discussions following each presentation. This material provides perspective and helps draw attention to the various themes that emerged during the meeting.

We are also grateful for the organizing efforts of the European co-chairman, Dr. Alain de Weck, and for the able assistance of the other members of the program committee: Drs. John David, Edgar Pick, Anneliese Schimple, Clemens Sorg, James Watson, and David R. Webb, Jr.

Last, but not least, we are indebted for the generous financial support of the NIAID, NCI, and NIH Fogarty Center; and of American Cyanamid; Boehringer Ingelheim, Burroughs Wellcome, E. I. Du Pont de Nemours, Eli Lilly and Company, Genentech, Inc., Genex Corporation, Hoffman-La Roche, Inteferon Sciences, Monsanto, Organon, Pfizer, Inc., Schering Corporation, Searle, Smith, Kline and French, Wyeth Laboratories, Searle Research, Ortho Pharmaceuticals, Cetus Corporation, Hoechst, Medizinisch Pharmazeutische, Studiengellschaft e.v., and Upjohn Company.

SECTION I:
CHARACTERISTICS AND EFFECTS OF INTERLEUKIN-2 AND ANTIINTERLEUKIN-2 RECEPTOR ANTIBODIES

A BIOCHEMICAL VARIANT OF HUMAN T-CELL GROWTH FACTOR PRODUCED BY A CUTANEOUS T-CELL LYMPHOMA CELL LINE

Joseph E. Gootenberg
Francis W. Ruscetti
Robert C. Gallo

Laboratory of Tumor Cell Biology
National Cancer Institute
National Institutes of Health
Bethesda, Maryland

The proliferation of normal human T-lymphocytes requires an initial activation with antigen or lectin followed by a growth stimulus provided by a lymphokine designated T-cell growth factor (TCGF) or Interleukin 2 (1-3). Human TCGF, purified from conditioned media of pooled PHA-stimulated normal lymphocytes, is a protein of molecular weight approximately 13,000 by SDS-PAGE which possesses an isoelectric point of about 6.5 and displays a degree of microheterogeneity attributed to variable sialylation and glycosylation (4-6). In contrast to normal T-lymphocytes, neoplastic cells of mature T-lymphocyte origin (E-rosette receptor-positive, terminal deoxynucleotidyl transferase [TdT]-negative) obtained from patients with cutaneous T-cell lymphoma and leukemia (CTCL) respond directly to lectin-free TCGF (7). Through exploitation of this property, factor-dependent cell lines have been established from several patients with T-cell lymphoma. Two of these cell lines, HUT102 and CTCL-2, have on serial passage lost their requirement for exogenously supplied TCGF (7,8) and are now constitutive producers of a TCGF which can be detected in conditioned media and eluted from their plasma membrane (9). This TCGF has been designated L-TCGF (for lymphoma-leukemia TCGF) to distinguish it from the TCGF obtained from normal lectin-stimulated lymphocytes, termed N-TCGF (for normal TCGF). In this report, we summarize evidence that L- and N-TCGF are biochemically distinct.

ISBN 0-12-527540-4

MATERIALS AND METHODS

N-TCGF was prepared from CM or acid eluates of pooled PHA-stimulated nylon wool-depleted peripheral blood leukocytes (10,11). L-TCGF was obtained from HUT102 cells by the same methods. TCGF activity was quantitated by comparison of dilution curves of sample and standard TCGF preparations in a thymidine incorporation assay utilizing TCGF-dependent long-term cultured human T-lymphocytes as indicator cells (9). Cell surface proteins were eluted by treatment with 0.2 M glycine, pH 2.8 for 15 min. at 4°C (12). Enzyme treatment was accomplished by incubating samples with insolubilized enzyme (neuraminidase, type X-A or alkaline phosphatase; Sigma Chem. Co., St. Louis, MO) for 1 hr. at 37°C. Sialic acid concentrations were determined using a modification of the thiobarbituric acid procedure of Warren (13) and phosphate release was quantitated with a modification of the ascorbic acid-molybdate assay of Ames (14).

RESULTS

Demonstration of TCGF Activity in Acid Eluates (Figure 1)

Conditioned media and acid glycine eluates from both HUT-102 and PHA-stimulated normal lymphocytes did not promote ^{3}H-TdR incorporation by unstimulated normal human lymphocytes.

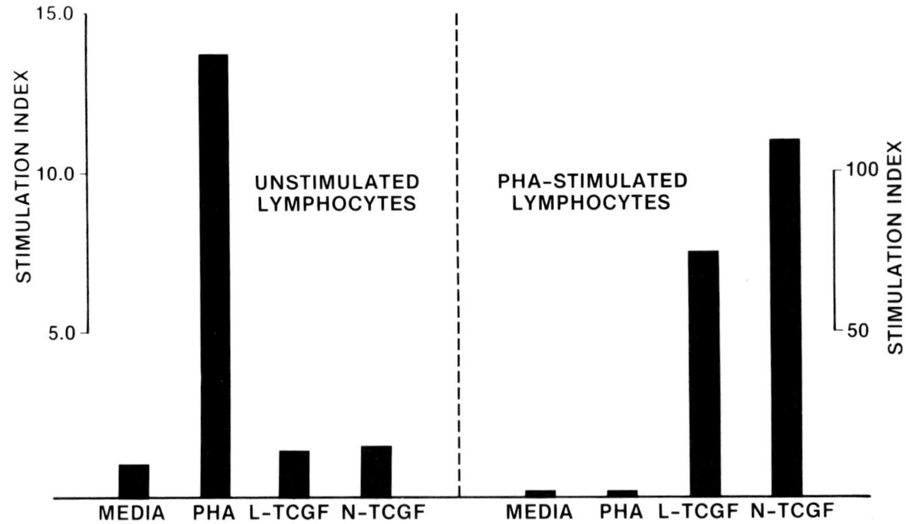

FIGURE 1. Biological Activity of N- and L-TCGF.

In contrast, PHA alone resulted in a 15-fold increase. With PHA-stimulated normal lymphocytes, the supernatants and eluates from HUT102 and normal PHA-stimulated lymphocytes caused 50 to 100-fold increases in ^{3}H-TdR incorporation. In addition, these samples were able to support long-term growth of activated T-cells. Eluates obtained from Molt 4, an immature T-lymphoblastoid line (E-rosette-positive; TdT-positive) and SC-1, a B-lymphoblastoid line, had no activity.

Comparison of Biochemical Properties of N-and L-TCGF

Flatbed isoelectric focusing of N-TCGF prepared from acid eluates or conditioned media revealed activity in the range pH 6.5 to 8.0. In contrast, L-TCGF obtained by acid elution of HUT102 cells yielded a single sharp peak of activity at pH 4.5 (Figure 2). This pattern was conserved during focusing over a range of pH 4 to 6 and in focusing conducted in the presence of 4M urea.

On anion exchange chromatography, N-TCGF, whether obtained from conditioned media or acid eluates, dissociated from DEAE-Sepharose (pH 8.0) at 70mM NaCl while L-TCGF remained bound until 200mM NaCl.

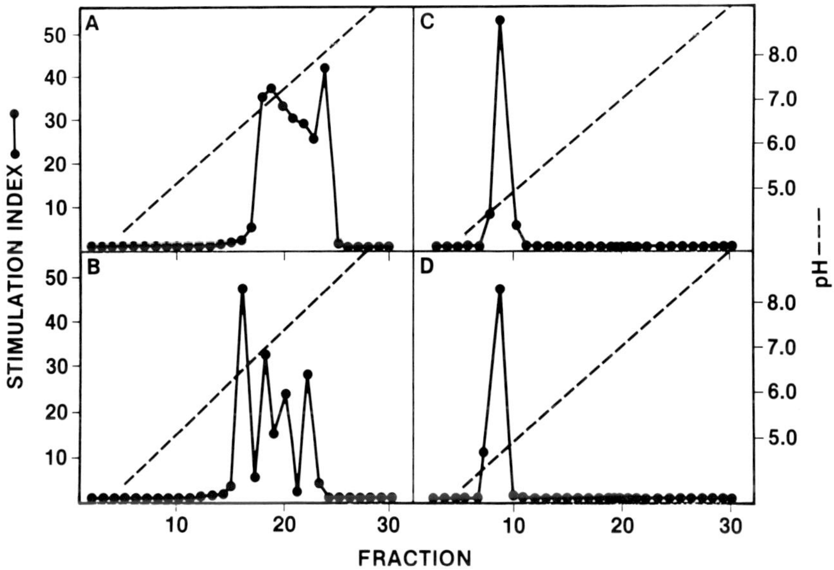

FIGURE 2. Isoelectric Focusing of TCGFs.
A) N-TCGF from Acid Eluate B) N-TCGF from Conditioned Media C) L-TCGF D) Neuraminidase Treated L-TCGF

During Biogel P-60 gel filtration, N-TCGF migrated after and L-TCGF before the chymotrypsinogen standard yielding respective estimated molecular weights of 24,000 and 27,000 (Table 1).

Enzymatic Treatment of TCGF

As shown in Figure 2, neuraminidase treatment did not affect the isoelectric focusing profile of L-TCGF. Specifically, there was no appearance of activity in the pH 6.5-8.0 range characteristic of N-TCGF. Mobility during gel filtration was also unchanged. Eighty percent of the L-TCGF activity remained after neuraminidase treatment and the treated preparations contained levels of sialic acid approaching 100% release.

Treatment with alkaline phosphatase did not change the isoelectric focusing or gel filtration properties of L-TCGF. Phosphate release approximating 25% of total was obtained and prolonged incubation with enzyme (24 hrs.) did not result in further liberation of phosphate. During this treatment, the L-TCGF preparations suffered no loss of activity.

DISCUSSION

The presence of TCGF activity in conditioned media and on plasma membranes from cell lines derived from cutaneous T-cell lymphoma-leukemia has been previously demonstrated (9). This TCGF (designated L-TCGF for leukemic TCGF) displays biological activity similar to the TCGF obtained from lectin-stimulated

TABLE I. Biochemical Characteristics of N- and L-TCGF

Property	*N-TCGF*	*L-TCGF*
Molecular size (gel filtration)	*24,000*	*27,000*
[Na+] of elution from DEAE-Sepharose	*70mM*	*200mM*
Isoelectric point	*6.5-8.0* *4 peaks*	*4.5* *single peak*

normal T-cells (N-TCGF, normal TCGF). Both materials are active only on T-lymphocytes which have undergone prior lectin stimulation, and both support the long-term growth of activated normal T-lymphocytes. However, they possess different biochemical properties. On isoelectric focusing N-TCGF yields a broad range of activity between pH 6.5 and 8.0 while L-TCGF produces a single sharp peak at pH 4.5. This marked acidic character of L-TCGF is reflected in its strong binding to anion exchange resin. Estimated molecular weights obtained by gel filtration differ by 3,000; this could represent a true difference in the molecular weight of N- and L-TCGF, or differences in molecular shape or hydration affecting mobility in the gel matrix (15). N-TCGF derived from conditioned media or from acid eluates displayed identical biochemical properties confirming that these findings are not a result of the preparation of L-TCGF by acid elution.

If L-TCGF were a molecular modification of N-TCGF, the detected shift in pI could be due to the addition of acidic side groups, such as sialic acids of phosphates. Alternatively, L-TCGF could be a precursor protein converted to N-TCGF under normal circumstances by post-translational cleavage of a polypeptide chain or could represent a non-covalent association of a membrane protein with N-TCGF. Finally, there could exist two or more separate TCGF molecules with disparate amino acid sequences. Neuraminidase treatment of L-TCGF does not affect its isoelectric point or molecular weight, implying that sialylation is not involved in the biochemical difference between N- and L-TCGF. This contrasts with the isoelectric point microheterogeneity of N-TCGF which is due to differences in sialic acid content (7,16). In a similar manner, treatment with alkaline phosphatase shows that increased phosphorylation is not the source of L-TCGF's acidic nature. Finally, the presence of 4M urea has no effect on the pI of L-TCGF, indicating that noncovalent protein-protein interactions are not the source of this variation.

The production of this variant TCGF by a cell line which carries a putative human leukemia virus, HTLV (17), raises questions regarding the connection between viral infection, growth factors and malignant transformation. Could this factor be analogous to the sarcoma growth factor which is produced by and supports the growth of simian sarcoma virus transformed cells? (18). Or is L-TCGF a newly detected normal factor found in the subpopulation of cells from which the HUT102 cell line's precursor arose? Further studies characterizing L-TCGF and its interaction with normal and malignant T-lymphocytes will clarify these questions.

ACKNOWLEDGMENTS

We wish to thank Drs. M.G. Sarngadharan and V.S. Kalyanaraman for their helpful discussions and Ms. Anna Iacangelo and J. Blalock for their excellent technical assistance. J.E.G. wishes to thank R. Michael Blaese for his warm encouragement and support.

REFERENCES

1. Morgan, D.A., F.W. Ruscetti, and R.C. Gallo, *Science 193,* 1007 (1976).
2. Ruscetti, F.W., and R.C. Gallo, *Blood 57,* 379 (1981).
3. Smith, K.A., *Immunol. Rev. 51,* 337 (1980).
4. Mier, J.W., and R.C. Gallo, *Proc. Natl. Acad.Sci. (USA) 77,* 6134 (1980).
5. Mier, J.W., and R.C. Gallo, *J. Immunol. 128,* 1122 (1982).
6. Robb, R., and K.A. Smith, *Mol.Immunol. 18,* 1087 (1981).
7. Poiesz, B.J., F.W. Ruscetti, J.W. Mier, A.M. Woods, and R.C. Gallo, *Proc. Natl. Acad. Sci. (USA) 77,* 6815 (1980).
8. Gazdar, A.F., D.N. Carney, P.A. Bunn, E.F. Russell, E.S. Jaffee, G.P. Schechter, and J.G. Guccion, *Blood 55,* 409 (1980).
9. Gootenberg, J.E., F.W. Ruscetti, J.W. Mier, A.F. Gazdar, and R.C. Gallo, *J. Exp. Med. 154,* 1403 (1981).
10. Prival, J.T., M. Paran, R.C. Gallo, and A. Wu, *J. Natl. Cancer Inst. 53,* 1583 (1974).
11. Ruscetti, F.W., J.W. Mier, and R.C. Gallo, *J. Supramol. Str. 13,* 229 (1980).
12. Ehrich, R., and I.P. Witz, *J. Immunol. Methods 26,* 343 (1979).
13. Warren L., *J. Biol. Chem. 134,* 1971 (1959).
14. Ames, B.N. *in* "Methods in Enzymol. Vol. VIII" (S.P. Colowick and N.O. Kaplan, eds.) p. 115. Academic Press, New York (1966).
15. Tanford, C., *Adv. Protein Chem. 23,* 121 (1968).
16. Clark-Lewis, I., and Shrader, J.N., *J. Immunol. 128,* 175 (1982).
17. Poiesz, B.J., F.W. Ruscetti, A.F. Gazdar, P.A. Bunn, J.D. Minna, and R.C. Gallo, *Proc. Natl. Acad. Sci. (USA) 77,* 7415 (1980).
18. Sporn, M.B., and G.J. Todaro, *N. Engl. J. Med. 303,* 878 (1980).

DISCUSSION

GILLIS: How much TCGF is obtained by acid elution of HUT102 cells? In other words, to what extent can acid elution (supernates) be diluted and still express biological activity?

GOOTENBERG: I believe what Gillis is getting at is how much activity is present per cell, because the volume of supernatant is in many respects rather arbitrary; the answer is-not very much. From 100 gm of HUT102 cells we generally obtain 20 ml of an active supernatant, one that will show detectable activity out to a 1:128 or 1:256 dilution. The amount of this material produced is very small-it is detectable in conditioned medium only after a 20-fold concentration. This compels us to use large amounts of cells to obtain sufficient factor for biochemical investigations.

GARLAND: Does L-TCGF-X react with monoclonal antibody to N-TCGF?

GOOTENBERG: I would like very much to explore the possibility of immunological cross-reactivity between N- and L-TCGF, but at present we have no antibody available for this purpose. I look forward to conducting such experiments just as soon as possible.

OPPENHEIM: I would suggest to Gootenberg that it is possible that the L-TCGF represents an intracellular precursor of N-TCGF. Has Gootenberg examined extracts of the normal and transformed cells to ascertain the nature of intracellular TCGF moieties?

GOOTENBERG: One possibility that would be quite exciting would be if this L-TCGF was a protein precursor to the N-TCGF processed in the *non*-malignant cell by post-translational cleavage. However, acid elution of normal PHA-stimulated lymphocytes yields N- not L-TCGF. Moreover, the L-TCGF found in HUT102 cells is associated with the cell membrane, not the cytoplasmic fraction (Gootenberg et al., *J. Exp. Med. 154:*1403 (1981)).

FAHEY: Do other human cell lines with T-helper property and Leu 3 antigen in addition to HUT102 make this factor, L-TCGF? Does Gootenberg know of any human cell lines, of other types, that fail to make L-TCGF?

GOOTENBERG: Some other human cell lines of mature helper phenotype (Rt neg., OKT4+) do produce a low level of TCGF. However, we have not been able to accumulate sufficient TCGF to do biochemical characterization.

Many other lines make *no* TCGF. If we restrict ourselves to lines of T cell origin, constitutive production of TCGF is found *only* with lines of mature phenotype-mostly those raised from cutaneous T cell lymphoma or, speculatively, from the Japanese Acute T cell leukemia. Amongst lines of immature T phenotype ($CTDT^{+}$), such as MOLT 4, Jurkat, HSB2, etc., - only a few will produce TCGF, - and then only after induction with agents such as lectin and PMA. The majority do *not* produce TCGF.

FARRAR: Has Gootenberg tested these two kinds of TCGF on different sub-populations of T cells? This might be a way of getting at the possibility of distinct TCGF's for helper and for cytotoxic T cell lines. Such separate functions for the two distinct TCGF's might well explain his data.

GOOTENBERG: Farrar's idea is an excellent one, one that we are currently exploring. We have the earliest preliminary data about the possibility of such biological differences which have come from known CTLL lines. It is not yet clear whether these differences represent *vertical* differences in response (mouse *vs* human) or whether they reflect *horizontal* differences (helper *vs* cytotoxic cells).

CHARACTERIZATION OF A MONOCLONAL ANTIBODY RAISED TO THE HUMAN LYMPHOKINE BLASTOGENIC FACTOR[1]

Harold Rode
Chungyee Leung
Julius Gordon

Department of Surgery
McGill University
Montreal, Quebec

INTRODUCTION

Blastogenic factor (BF) belongs to the family of mediators designated as IL-2. BF is recovered from the supernatant of a 48-72 hour mixed leukocyte culture (1), and it has been shown to restimulate cells primed in allogeneic mixed cultures independently of the original stimulator cells (2). In addition to eliciting the proliferation of human cytotoxic lymphocytes, BF can also cooperate in a mouse thymocyte co-stimulator assay, as well as sustain both mouse and human IL-2 dependent cell lines. In order to facilitate its study, we have been attempting in vain to raise conventional antibodies to it in several species, as well as hybridomas using spleen cells of Balb/c mice. We have now attempted to produce monoclonal antibodies to BF by utilizing the Biozzi "high antibody producer" mouse as the source of antibody producing cells. The following is a description of the product of a clone established from the fusion of the P3x20 myeloma with spleen cells of a Biozzi mouse immunized with BF.

[1]*Supported by Medical Research Council Grant No. MA-2930*

ISBN 0-12-527540-4

MATERIALS AND METHODS

The antigen used for immunization was a culture supernatant derived from mixed cultures of human peripheral lymphocytes. The crude supernatant was fractionated on a column of G-75 Sephadex, the BF containing fractions pooled and concentrated by ultrafiltration over a UM-10 membrane. One ml of antigen, equivalent to 0.6 l of crude supernatant was injected i.v. followed by an equal dose injected i.p. 8 hours later. Fusion was performed 4 days after immunization and screening begun on the 7th day of culture. The method of screening was to test the ability of culture supernatants to inhibit the proliferative response to BF of 10-14 day old MLC primed cells (2ary cells). The positive cultures were cloned twice by limiting dilution using a (Balb/c x Biozzi)F1 feeder layer. The results reported have been obtained with supernatants derived from one clone, B-5, concentrated 20-fold by

TABLE I. Inhibition by B-5 of the Proliferation of Human 2ary MLC Cells Stimulated by BF

Inhibitor	*Dilution of inhibitor*[a]		
	1:4	*1:8*	*1:16*
Nil	*35,924±3,502*[b]	-	-
B-5	*1,951±661*	*11,538±2,927*	*23,125±2,598*
P3x20	-	*40,915±446*	-
Nil	*26,479±248*	-	-
B-5	*697±187*	*1,190±74*	*13,463±1,280*
αSRBC	-	*19,542±814*	*23,992±296*

[a]*The undiluted B-5 refers to an ammonium sulphate precipitated, 20-fold concentrated culture supernatant. The dilution of the myeloma P3x20 and αSRBC hybridoma tested is equivalent to the dilution of B-5 as listed in each column.*

[b]*CPM±S.E. of H^3T incorporated. $1x10^5$ human 2ary cells, 100 ul BF + 100 ul inhibitor, 72 hr assay, 5 hr pulse.*

precipitation with ammonium sulphate. The immunoglobulin produced by B-5 was shown to be IgG-1 by immunodiffusion; the concentration of IgG-1 in the 20-fold concentrated supernatant was 1 mg/ml by a radioimmunoassay.

RESULTS

The inhibition by B-5 of the proliferation of human 2ary lymphocytes activated by BF is shown in Table I. Routinely, using different responding cells, the endpoint of inhibition was found to vary between B-5 dilutions of 1:16 to 1:64. Accordingly, the amount of IgG-1 required to yield 50% inhibition is in the order of 3 mcg. Control experiments performed with concentrated supernatants of the myeloma cells used for fusion, and of a hybridoma secreting an IgG-1 anti-SRBC-antibody, did not reveal any significant inhibition in the same test system. The antibody was shown not to be directed against the test cells, nor against serum protein components essential for the response. Thus absorption by fresh lymphocytes, or twice by 10^8 2ary cells per ml of B-5 or by glutaraldehyde-insolubilized serum proteins did not abate the inhibitory activity (results not shown).

Inhibition was also assayed in other systems where BF or IL-2 is known to participate. Illustrated in Table II is the inhibition of proliferation of mouse thymocytes in the presence of a low concentration of Con A and IL-2 in a costimulator assay (3). The response of an established murine IL-2

TABLE II. Inhibition by B-5 of the Response of Mouse Thymocytes to Con A in the Presence of IL-2

B-5 Added		*B-5 Added*	
Nil	*16,592±2,561*[a]	*1:32*	*10,987±1,232*
1:2	*793±500*	*1:64*	*11,530±1,181*
1:4	*2,623±1,153*	*1:128*	*10,940±712*
1:8	*7,075±1,188*	*1:256*	*13,103±769*

[a]*CPM±S.E. of H^3T incorporated. $1x10^5$ thymocytes + Con A (2 ug/ml) + IL-2 (100 ul), 72 hr assay, 12 hr pulse.*

TABLE III. Inhibition by B-5 of the Activation of IL-2 Dependent Cell Lines

Stimulant	*No inhibitor*	*B-5 dilution 1:2*	*B-5 dilution 1:4*
Expt. I[a]			
Nil	*73*[b]	*77*	*118*
IL-2	*80,070*	*45,637*	*77,836*
BF	*12,243*	*2,254*	*5,272*
Expt. II[a]			
Nil	*4,376*	*3,217*	-
IL-2	*46,886*	*19,161*	*39,445*
BF	*28,400*	*3,500*	*18,353*

[a]*In Expt. I, 1x10^4 cells/well of a mouse IL-2 dependent cell line were activated by BF or by Con A induced rat IL-2 as indicated. In Expt. II, 5x10^4 human IL-2 expanded cells/well were activated by BF or by PHA induced human IL-2 as indicated.*

[b]*CPM of H^3T incorporated, 48 hr assay, 5 hr pulse.*

dependent cell line stimulated with BF or IL-2 of rat origin was inhibited by B-5 (Table III). Similarly, B-5 also inhibited both thymidine uptake (Table III) and growth of human lymphocytes expanded with IL-2. Both of these target cells were fully IL-2 dependent, neither reacting to lectins at the time of testing. Finally, the response of fresh human lymphocytes to allogeneic cells, to PHA, and to Con A was subject to inhibition by B-5; activation by the B cell mitogen Staph A however was not affected (Table IV).

DISCUSSION

Several factors may be contributory to the difficulty of raising antibodies to BF, such as its low concentration in our preparations, its expected poor antigenicity in mice due to the homology of human and murine material, and its rapid clearance and breakdown (4,5). Having obtained an antibody, which we attribute to having used the Biozzi high responder

TABLE IV. Inhibition of Proliferation by B-5 of Human Cells Stimulated by Mitogens and Alloantigens

Stimulant	No inhibitor	B-5 dilution 1:2	1:4	1:8
Staph A[b]	35,151[a]	50,755	38,780	50,858
PHA[c]	31,154	16,535	45,661	46,329
Con A[c]	44,339	1,996	14,020	23,945
Allogeneic Cells[d]	36,690	-	-	8,173

[a]Values expressed as CPM of H^3T incorporated.

[b]Staph A, $1x10^8$ organisms/well, $5x10^5$ fresh PBL/well, 8 day incubation.

[c]PHA, 2 ug/well, $1x10^5$ fresh PBL/well, 3 day incubation. Con A, 2 ug/well, $1x10^5$ fresh PBL/well, 3 day incubation.

[d]$1x10^5$ x-irradiated stimulating cells and $1x10^5$ responding PBL/well, 7 day incubation.

mouse, shown to be superior to other strains of mice in hybridization experiments (6), the assignment of its specificity is fraught with further difficulties. Not only the immunogen, but also the test system is very complex, cell proliferation depending on the cooperation of several cell types, mediators and serum factors, neutralization of any of which would be expected to result in the inhibition of the overall response. We have shown through absorption experiments that B-5 is not directed against cells, nor against components present in serum in high enough concentration. Accordingly, the inhibition is likely to be due to B-5 being directed against a stimulating agent present in the culture supernatant used for the immunization. This conclusion is supported by the results showing inhibition of all the manifestations of BF regardless of the test cells utilized. The fact that the expansion of IL-2 dependent lines was also inhibited would argue against IL-1 being the target of the antibody. Positive identification of the antigen in the BF has been hindered so far by our inability to retain BF on Sepharose-linked B-5. This failure could be due to a putative low affinity of the antibody, also suggested by the requirement of high concentration for inhibition. Accordingly, although the overall results obtained suggest that B-5 is an antibody directed against BF, the positive assignment of its specificity must await further experimentation.

ACKNOWLEDGMENTS

We wish to express our gratitude to Dr. G. Biozzi for supplying us with his mice.

REFERENCES

1. Gordon, J. and MacLean, L.D., *Nature 208*, 795 (1965).
2. Uotila, M., Rode, H.N. and Gordon, J., *Europ. J. Immunol. 8*, 133 (1978).
3. Paetkau, V., Mills, G., Gerhart, S. and Monticone, V., *J. Immunol. 117*, 1320 (1976).
4. Robb, R.J., Munck, A. and Smith, K.A., *J. Exp. Med. 154*, 1455 (1981).
5. E. Gorelik, *personal communication.*
6. Baumsell, B., *J. Immunol. Methods 38*, 225 (1980).

DISCUSSION

GILLIS: I note that Rode used trypsin-treated cells in his studies. What was the purpose of that particular operation? I see that the IL-2 response of such trypsinized cells was higher than control cells. Does Rode find this to be reproducible, or was it within the region of normal responses?

RODE: We used trypsinized cells so that they would be completely freed of IL-2. We have not ascertained whether this is a fully reproducible phenomenon.

DAVID: What kind of control antibody did Rode use in his various inhibition assays?

RODE: We used as controls the supernatant of the parent P3X20 myeloma, and an anti-SRBC hybridoma also secreting monoclonal antibody of the IgG-1 class. These supernatants were ammonium sulphate-precipitated, concentrated, and utilized at dilutions equivalent to the B-5 supernatants.

OPPENHEIM: My question and comment is directed to Rode, Gillis and Robb. Why is such a high concentration of monoclonal antibody, in the microgram range, necessary to inhibit nanogram or even smaller amounts of IL-2? This is especially perplexing in view of the apparently firm binding and difficulty in eluting IL-2 from immuno-affinity columns prepared with the monoclonal antibodies!

ROBB: We have a monoclonal antibody to IL-2 which directly inhibits the binding of radiolabelled IL-2 to the T cell surface receptor. But we need a 10,000-fold molar excess of antibody over IL-2 to attain *complete* inhibition. This should not come as a surprise since affinity of receptor for IL-2 is 10^{-11} to 10^{-12}M. The antibody in the ELISA assay with purified IL-2 appears to express a relatively low affinity compared to that of the receptor. A second antibody has a higher affinity and can be used as an immuno-affinity support; unfortunately, this does not directly inhibit binding.

MONOCLONAL ANTI-TAC BLOCKS THE ACTION AND MEMBRANE BINDING OF HUMAN INTERLEUKIN-2

Joel M. Depper[1]
Warren J. Leonard
Kendall A. Smith*
Thomas A Waldmann
Warner C. Greene

Metabolism Branch, National Cancer Institute
Bethesda, Maryland
and
*Dartmouth Medical School
Hanover, New Hampshire

I. INTRODUCTION

Interleukin 2 (IL-2) is an inducible glycoprotein hormone of molecular weight 15,000, synthesized and secreted by T cells following encounter with antigen or lectin. IL-2 stimulates the growth of T cells by binding to specific, high affinity membrane receptors, whose appearance also requires the presence of antigen or lectin. IL-2 receptors are present on activated peripheral blood T cells, IL-2 dependent continuous T cell (CTC) lines, and some tumors, but cannot be detected on B cells, monocytes, or resting T cells. Thus, initiation of T cell dependent immune responses requires not only the production of IL-2, but also the formation of IL-2 receptors. The biology of IL-2 has been the subject of several recent reviews (1-3).

Monoclonal antibodies to IL-2 have been isolated (4,5). These antibodies block IL-2 binding, as well as mitogen induced proliferation. In contrast, an antibody to the IL-2 receptor has not been identified, nor has this receptor been characterized. While in our laboratory, Dr. Takashi Uchiyama prepared a monoclonal antibody, anti-Tac (for T cell activation). This antibody binds to mitogen or antigen activated T cells, IL-2 dependent CTC lines, and some T cell malig-

1. JMD is the recipient of an Arthritis Foundation Fellowship

ISBN 0-12-527540-4

nancies, but not to monocytes, B cells, resting T cells nor IL-2 independent T cell lines (6,7). On the basis of this pattern of reactivity, we hypothesized that anti-Tac interacts with the receptor for IL-2. We will describe three lines of evidence, which support this proposed interaction (8,9). First, anti-Tac inhibits the proliferation of IL-2 dependent CTC lines. Second, anti-Tac inhibits a wide range of T lymphocyte proliferative responses and specific T cell functions. Third, anti-Tac blocks the membrane binding of radiolabeled IL-2.

II. EFFECT OF ANTI-TAC ON CTC LINE PROLIFERATION

We began our studies by evaluating the effects of anti-Tac on the proliferation of IL-2 dependent human T cell lines. Cells were washed extensively, preincubated in anti-Tac antibody (in ascites or protein A sepharose purified forms), or control antibody for 30 minutes, followed by activation with IL-2 containing conditioned media. ^{3}H-thymidine incorporation was measured 48 hours later. Anti-Tac ascites at 10^{-2} to 10^{-4} dilution, or purified anti-Tac antibody at $\geq$5 mcg/ml, inhibited ^{3}H-thymidine incorporation by 50-90%. Control SPQC-11 ascites (a gift of Michael Potter, NIH), or purified monoclonal anti-Ia, which binds to these cells, (a gift of Peter Knudsen, Harvard), had little or no effect. Encouraged by these findings, we next examined the effects of anti-Tac on lymphocyte function, and characterized its binding properties.

III. FUNCTIONAL EFFECTS ON PERIPHERAL BLOOD LYMPHOCYTES

A. *Proliferative Responses*

A brief summary of the effects of anti-Tac on peripheral blood lymphocyte (PBL) function is presented in Table 1. In these studies, fresh PBL were isolated from heparinized blood, and separated by a single E rosetting into T and non-T cells where indicated. All proliferative responses were carried out for 6 or 7 days (except PHA, 3 days), and proliferation was measured by ^{3}H-Thymidine incorporation during the final 4 hours of culture. Antibodies were added at the initiation of cultures unless otherwise noted.

Anti-Tac inhibited proliferation induced by the soluble antigens tetanus toxoid and Candida albicans by >90% at dilutions of 10^{-2}-10^{-4}, while identical dilutions of control ascites had no inhibitory effect. The autologous mixed lympho-

TABLE 1. Anti-Tac Effects on Human Lymphocyte Function

	% Inhibition
A. Proliferative Responses	
Soluble Antigens	*80-90*
Autologous MLR	*80-90*
Allogeneic MLR	*75-95*
Mitogenic Lectins	*20-90*
Epstein-Barr Virus	*0*
B. Specific T Cell Functions	
CML: generation phase	*95*
CML: effector phase	*0*
B cell differentiation (PWM)	*75-90*
Natural Killing	*0*

cyte reaction was inhibited by approximately 90% with anti-Tac ascites at a dilution of 10^{-4}. Further, this response remained inhibited by >50% at 10^{-6} ascites, or 1 mcg/ml purified protein. While control SPQC 11 ascites had no inhibitory effect, inhibition by purified anti-Ia was comparable to that of anti-Tac, consistent with the known participation of Ia in this response. The allogeneic mixed lymphocyte reaction (MLR) was inhibited by 75-95% with $\geq 10^{-4}$ ascites whether performed with unseparated PBMC or purified T and non-T cell populations.

In contrast, the inhibition by anti-Tac of proliferation induced by the mitogenic lectins phytohemagglutinin (PHA), concanavalin A (Con A), and pokeweed mitogen (PWM) was variable, ranging from 15 to 90%. As both IL-2 production and cell proliferation are related to lectin dose (10,11) although not strictly proportional, we examined the effect of anti-Tac at varying doses of lectin. While the Con A response was inhibited by >75% in the dose range from 10 to 1.25 mcg/ml, the inhibition of PHA and PWM induced proliferation was inversely proportional to the lectin dose. For example, T cells stimulated by 0.5 mcg/ml PHA incorporated 176,000 cpm ^{3}H-thymidine, and were blocked by only 21% in the presence of 10^{-3} anti-Tac ascites. With graded reduction in PHA dose, anti-Tac produced progressively greater inhibition. Cells stimulated by an 8-fold dilution of PHA incorporated 35,000 cpm. In the presence of anti-Tac, this response was reduced by 75% to 9000 cpm. Thus, the explanation for the variable inhibition of mitogen induced responses may be, in part, a function of the amount of IL-2 produced. (Also see antibody affinity, below).

In contrast to inhibition of T cell proliferation, anti-Tac had no effect on B cell proliferation induced by the helper T cell independent activator, Epstein-Barr virus.

B. *Reversibility and Time Course of Anti-Tac Action*

To evaluate the reversibility of anti-Tac action, parallel MLR flask cultures were prepared with and without antibody. Cells were thoroughly washed at either 24 or 72 hours, and placed into microtiter wells both in the presence and absence of antiTac. All cultures were then harvested 7 days after the initiation of the experiment. Cells incubated in media alone incorporated 67,000 cpm of ^{3}H-thymidine, while cells cultured in the continuous presence of anti-Tac incorporated only 5000 cpm. Cells washed after 24 hours in anti-Tac incorporated 64,000 cpm, while cells cultured for 3 days in anti-Tac incorporated 28,000 cpm. These data suggested that: 1) The effect of anti-Tac is reversible after 24 hour incubation, and corroborates other data suggesting that anti-Tac's effect is not mediated by cytotoxicity. 2) While the effect of anti-Tac after 72 hour incubation was only partially reversed, we could not say whether the potential proliferation of these cells was altered, or the time course for a full response simply delayed.

To evaluate the time course of anti-Tac action, we added anti-Tac at the initiation, or on subsequent days of 6 day MLR cultures. These studies demonstrated that anti-Tac (10^{-4}) inhibited proliferation by 95%, even when added up to 3 days after the initiation of cultures. >75% inhibition was still observed when added on day 4, after which there was no significant inhibition. In other experiments, maximal inhibition of PWM responses was also observed as late as 4 days into culture.

C. *Effects of Anti-Tac on Specific T Cell Functions*

Allogeneic cultures were prepared in the presence and absence of anti-Tac to evaluate its effect on the generation of cytotoxic lymphocytes (CTL). At the end of 7 days, killing was tested at various effector:target ratios against PHA blasts autologous to the stimulator cells. Effector cells generated in the presence of anti-Tac ascites at 10^{-3} were about 70 fold less able to kill stimulator targets. In contrast, the addition of anti-Tac only during the killing phase had no effect.

Because of a report suggesting an effect of TCGF on natural killer (NK) function (12), we examined the effect of antiTac on the ability of fresh PBL to kill K562 cells in a standard ^{51}chromium release assay. Anti-Tac did not inhibit natural killer cell function at any concentration evaluated.

We next evaluated the effect of anti-Tac on helper T cell function. We used a reverse hemolytic plaque assay to enumerate immunoglobulin secreting cells (IgSC) induced by PWM in 7 day cultures, a system dependent on T cell help (13).

Anti-Tac ascites at 10^{-4} inhibited the generation of IgSC by 75-90%, while control ascites had no inhibitory effect.

IV. BINDING STUDIES

To directly assess if anti-Tac blocks the interaction of IL-2 with its receptor, we evaluated the ability of anti-Tac to block the binding of purified radiolabeled IL-2 to a human CTC lines. Leukemic JURKAT cells were stimulated with PHA in the presence of ^{3}H-lysine and ^{3}-leucine, and secreted IL-2 purified by ultrafiltration and molecular sieve chromatography. HUT 102 cells were incubated with purified anti-Tac or anti-Ia for 90 minutes at 37° followed by a 20 minute incubation with ^{3}H-IL-2 (Table 2). ^{3}H-IL-2 binding was completely inhibited by 9 mcg/ml anti-Tac, and 70% inhibition persisted at 3.1 mcg/ml. This was similar to the dose response relationship obtained for anti-Tac inhibition of proliferative responses. Comparable inhibition of ^{3}H-IL-2 binding was observed when binding was carried out at 4°, arguing that inhibition was not secondary to receptor capping with subsequent receptor endocytosis or shedding.

We also studied anti-Tac binding to CTC cells employing ^{125}I anti-Tac. Scatchard analysis of anti-Tac binding to HUT 102 indicated 80-100,000 receptors per cell and a K_d of approximately 10^{-9} moles/liter. Preliminary analysis of binding to PHA blasts demonstrated similar characteristics. IL-2 has an affinity constant of approximately 10^{-12} moles/liter for its receptor (14). The different affinities of IL-2 and anti-Tac may explain our inability to completely inhibit lectin induced T cell responses, as the presence of excess local IL-2 would favor IL-2 action.

TABLE II. Inhibition of ^{3}H-TCGF Binding by Anti-Tac

Anti-Tac (mcg/ml)	DPM Bound	% Control	Anti-Ia (mcg/ml)	DPM Bound	% Control
none	669	--			
88.5	27	4			
27.5	31	5	50	690	103
9.25	45	7	16.7	680	102
3.1	194	29	5.5	663	99
1	792	118	1.8	437	65
0.35	710	106	0.6	690	103

V. CONCLUSIONS

We have discussed three lines of evidence supporting the hypothesis that anti-Tac interacts with the human T cell receptor for IL-2. First, anti-Tac inhibits the proliferation of IL-2 dependent T cell lines. Second, anti-Tac inhibits a variety of T lymphocyte mediated proliferative responses in which IL-2 is felt to participate, including stimulation by soluble, autologous and allogeneic antigens, as well as at least partial inhibition of mitogen induced responses. Further, anti-Tac inhibits specific T cell functions, including the generation but not effector phase of cell mediated lympholysis, and ability of T cells to provide help in a PWM activated system of B cell maturation. Last, anti-Tac specifically inhibits the binding of radiolabeled IL-2 to a human T cell line which bears IL-2 receptors. We are currently employing anti-Tac to characterize its receptor, and investigating the regulatory role of the IL-2 receptor in the human immune response.

REFERENCES

1. Smith, K. A., Immunological Rev. 51, 337 (1980)
2. Ruscetti, F. W., Gallo, R. C., Blood 57, 379 (1981)
3. Watson, J., Frank, M. B., Mochizuki, D., Gillis, S., in "Lymphokines Vol. 6" (Mizel, S., ed.) p. 95 Academic Press New York (1982)
4. Gillis, S., Henney, C. S., J. Immunol. 126, 1978 (1981)
5. Stadler, B. M., Oppenheim, J. J., in "Lymphokines Vol.6" (Mizel, S., ed.) p. 117 Academic Press, N.Y. (1982)
6. Uchiyama, T., Broder, S., Waldmann, T. A., J. Immunol. 126, 1393 (1981)
7. Uchiyama, T., Nelson, D., Fleisher, T. A., Waldmann, T. A., J. Immunol. 126, 1398 (1981)
8. Leonard, W. J., Depper, J. M., Uchiyama, T., Smith, K. A., Waldmann, T. A., Greene, W. C., Nature (London) in press
9. Depper, J. M., Leonard, W. J., Uchiyama, T., Waldmann, T. A., Greene, W. C. submitted for publication
10. Smith, K. A., Gillis, S., Baker, P. E., McLenzie, D., Ruscetti, F., Annals N. Y. Acad. Sci. 332, 423 (1979)
11. Alvarez, J., Silva, A., Landazuri, M., J. Immunol. 123, 977 (1979)
12. Henney, C. S., Kuribayashi, K., Kem, D. E., Gillis, S. Nature(London) 291, 335 (1981)
13. Keightley, R. G., Cooper, M. D., Lawton, A. R. J. Immunol. 117, 1538 (1976)
14. Robb, R. J., Munck, A., Smith, K.A., J. Exp. Med 154, 1455 (1981)

DISCUSSION

GILLIS: Does Depper find that the addition of IL-2 blocks the monoclonal anti-Tac responses? How does Depper account for the discrepancy between his data on receptors and that reported by Robb? I believe that Depper finds more than 10 times the number of receptors reported by Robb.

DEPPER: While we have not actually done the experiment that would deal fully with Gillis' query, the correlation of anti-Tac inhibition with lectin dose suggests that this *would* be the case. That is, as IL-2 production is proportional to lectin dose, and in those instances where anti-Tac only partially inhibited proliferation, a reduction in lectin dose resulted in greater inhibition by anti-Tac. We have attributed this to the presence of excess IL-2, and the 2 to 3 log higher affinity of IL-2 for the receptor, compared with anti-Tac.

I believe that Robb et al. used a different clone of HUT102-a mutagenized 6-thioguanine-resistant clone, for their studies. Our studies have been repeated several times with a number of cell lines, as well as PHA-blast cells. In each instance a similar number of receptors was found. It is of interest that with *their* antibody, Diamantstein et al. report about 80,000 receptors on rat T lymphoblasts.

ROBB: I feel we may have used the same HUT102 line in studies with Kendall Smith at Dartmouth for the initial radiolabelled IL-2 binding assays. We calculated about 14,000 receptors per cell and now appreciate that the specific activity for IL-2 which we used was underestimated. The actual number of receptors should have been about 6000/cell. We seem to be dealing with a substantial discrepancy. Could it be that Depper is picking up low affinity receptors that we had not perceived for some reason or other?

DEPPER: I would point out that our calculations were based on the number of *high* affinity receptors.

ALTMAN: Would Depper recapitulate briefly how the anti-Tac antibody was prepared?

DEPPER: The anti-Tac was produced by immunization with a continuous T cell line initiated from the peripheral blood of a patient with cutaneous T cell lymphoma.

KUNG: Could Depper detect low levels of IL-2 receptors on resting T and B lymphocytes?

DEPPER: No, we could not. A low number of counts is obtained for resting cells, which are employed as controls in all experiments. We believe this represents non-specific binding, as we cannot eliminate it by competing with excess, cold, anti-Tac.

MÜHLRADT: Human IL-2 reacts with mouse cells. One might therefore expect antibody directed against the human IL-2 receptor to crossreact with the IL-2 receptor on mouse cells. Does it?

DEPPER: Anti-Tac does *not* react with mouse cells. We have studied two cell lines: one negative by direct binding, another unable to compete for binding. This is not so unexpected, given that mouse IL-2 does not act on human lines or cells.

VITETTA: Does inhibition with anti-Tac require the presence of the Fc portion of the antibody? Do the Fab or Fab'2 fragments stimulate the cells?

DEPPER: We have employed intact antibody in all our experiments thus far. We have examined this indirectly by preincubating cells with anti-Tac, followed by a rabbit anti-mouse Ig. We were unable to stimulate cells, but it should be noted that we have only done this one time.

BERTOGLIO: Does Depper know whether anti-Tac antibody also binds to activated B cells? What is his interpretation of the inhibition by anti-Tac of PWM-induced Ig synthesis?

DEPPER: We have interpreted the inhibition of PWM-driven B cell maturation by anti-Tac as an interference, in some way, with T helper cell function. That is, IL-2 production and interaction with its receptor appears to be necessary for development of this response. Alternatively, anti-Tac may directly act on B cells, although we have no data whatsoever to support such an interpretation. The lack of inhibition of EBV-induced, T cell-independent B cell activation by anti-Tac argues against this. However, we have detected low, but reproducible anti-Tac binding in two B cell lines out of *many* examined. Such binding is on the order of 5% of the binding to T cell lines. As yet, we do not know the significance of this finding.

A MONOCLONAL ANTIBODY WHICH REACTS WITH A RAT-T-LYMPHOBLAST SPECIFIC ANTIGEN INHIBITS IL-2-RECEPTOR FUNCTIONS *IN VITRO*

Hisao Osawa
*Tibor Diamantstein**

Immunology Research Unit
Klinikum-Steglitz
Freie Universität Berlin
Berlin, F.R.G.

I. INTRODUCTION

According to the interleukin-2 (IL-2) concept, T-lymphocyte proliferation is the consequence of antigen (or lectin) induced generation by the cells of receptors for IL-2, and the binding of IL-2 to such receptors (for rev. see 1). One of the prerequisites for studying the biochemical events taking place in the membrane following IL-2 binding is to isolate the receptor and the surrounding membrane area. Antibodies directed to antigenic determinants specific for either the IL-2 receptor itself or for structures linked to the receptor could serve as a tool for such studies. Recently, we reported on the production (2) and characteristics (3) of a rat-specific monoclonal antibody (McAb) ART18-IgG_1 which binds specifically to rat-T-lymphoblasts inhibiting the capacity of such cells to absorb IL-2. In this study we report 1) on attempts to isolate putative IL-2 receptor containing material using this McAb and 2) on the *in vitro* effects of this McAb on the proliferative response of thymocytes to ConA and that of T-lymphoblast to IL-2.

**This work was supported by the Deutsche Forschungsgemeinschaft*

ISBN 0-12-527540-4

II. MATERIALS AND METHODS

A. Preparation of cells

Thymocytes, T-lymphoblasts of rat (R-T-Bl) or murine origin (M-T-Bl) were obtained as described earlier (4).

B. Preparation of semipurified interleukin-2 (IL-2)

IL-2 was prepared from the 48-h supernatants of either rat (Wistar) or murine (BALB/c) spleen cells cultured with ConA by ammonium sulphate precipitation and Sephadex G-100 filtration as described (2,3).

C. The monoclonal antibodies used

The experiments described were performed with isolated Mc IgG_1 derived from the supernatants of the clones ART18 and ART1 respectively. The production and purification of the $McAb_s$ has been desribed earlier (2). The Ab ART18 binds specifically to R-T-Bl and inhibits absorption of IL-2 by the cells. The Ab ART1 used as a control binds specifically to R-T-Bl but does not affect IL-2 absorption.

D. Binding of ^{125}I-labelled ART18-IgG_1 to cells

ART18-IgG_1 was labelled with ^{125}I as described (3). 1×10^7 cells were incubated in 100 µl of PBS containing 0.5 % BSA, 10 mM NaN_3 and 0.25 µg of ^{125}I-labelled ART18-IgG_1 (5×10^5 cpm) for 60 min at 4°C. The cells were washed with the same buffer and the radioactivity bound to the pelleted cells measured. The number of ^{125}I-ART18-IgG_1 molecules bound was calculated as described earlier (3).

E. Inhibition by $McAb_s$ of absorption of IL-2 by cells

5×10^7 packed cells of different origin were suspended in either culture medium or culture medium containing the McAb ART18 (or, as a control, the McAb ART1) for 1 h at 37°C. The cells were spun down, washed twice with phosphate buffered saline, and resuspended in 0.2 ml of IL-2. The suspensions were incubated at 37°C for 1.5 h. The cells were spun down and the supernatants assayed for remaining IL-2 activity. In order to detect IL-2 activity, 0.2 ml of M-T-Bl of BALB/c origin (2×10^5 ml^{-1}) were cultured with 20 µl of the respective supernatants for 3 days in culture medium and ^{3}H-thymidine uptake by the cells determined as described (4).

F. Inhibition of cell proliferation by the McAbs

The cell suspensions (0.2 ml) were incubated for 3 d in the presence of various amounts of the respective $McAb_s$ and ^3H-thymidine uptake by the cells detected as described earlier (4).

G. Detection of the antigen immunoprecipitated by Ab ART18

5 x 10^7 R-T-Bl or M-T-Bl were surface labelled with ^{125}I by lactoperoxidase according to (5). 100 µl of Nonidet P40 lysates derived from 1 x 10^7 iodinated cells were pre-absorbed with fixed *S. aureus* cells (Pansorbin, Calbiochem) centrifuged and the supernatant incubated with or without 7 µg of ART18-IgG_1 for 30 min at room temperature, and then mixed with 20 µl of packed *S. aureus* cells coated with rabbit anti-mouse IgG (Miles). Immunoprecipitated material was suspended in 50 µl of SDS (2 %), centrifuged and the supernatant (25 µl) subjected to SDS-PAGE (12.5 %) in reducing conditions according to (6). The gel was dried and exposed to Ultrofilm (LKB) for 1 month at -20°C.

III. RESULTS AND DISCUSSION

Recently, the existence of IL-2 receptors was demonstrated by showing that internally radiolabelled and purified IL-2 binds specifically (~10^4 binding sites per cell) to T-lymphoblasts or to IL-2-dependent T-cell lines (7). In order to demonstrate the existence of such receptors, to define their function, and to develop a tool, an alternative approach was used, namely, the production of McAb to IL-2 receptors. We recently (2,3) reported on the production of several $McAb_s$ which bind specifically to rat-T-lymphoblast. One of these antibodies, characterized as an IgG_1 antibody, showed several characteristics of a putative anti-IL-2 receptor antibody. As reported previously (3) and as shown in this report (Table 1) by testing ^{125}I-labelled ART18-IgG_1 binding to cells, ART18 recognizes a species specific antigenic determinant present at the surface of rat-T-lymphoblasts. Furthermore, ART18-IgG_1 inhibits specifically the capacity of R-T-Bl to absorb IL-2-activity. Besides R-T-Bl (~8 x 10^4 binding sites per cell) LPS-stimulated rat-splenocytes bind ART18, although to a much smaller extent. As shown in Fig. 1, the antigenic material (marked by the arrow in the autoradiogram) with an apparent molecular weight of 50-60,000 recognized by ART18-IgG_1 can be extracted by non-ionic detergents from the surface of R-T-Bl and can be separated from other materials by SDS-PAGE. Whether this material contains the IL-2 receptor molecules in an active form remains to be elucidated. The findings 1) that absorption of IL-2 activity by T-lymphoblasts is specifically inhibited by the McAb ART18 and 2) that the proliferation of the T-lymphoblasts in response to IL-2 and the proliferation of thymocytes in response to ConA (but not those of rat spleen cells to LPS) are inhibited in a species specific manner (Table 2) indicate the existence of binding sites for IL-2 at the surface of rat T-lymphoblasts which can be blocked by a species-specific antibody. An antibody which blocks the function of IL-2 receptors does not necessarily have to bind to to the receptor itself. It is possible that binding of the

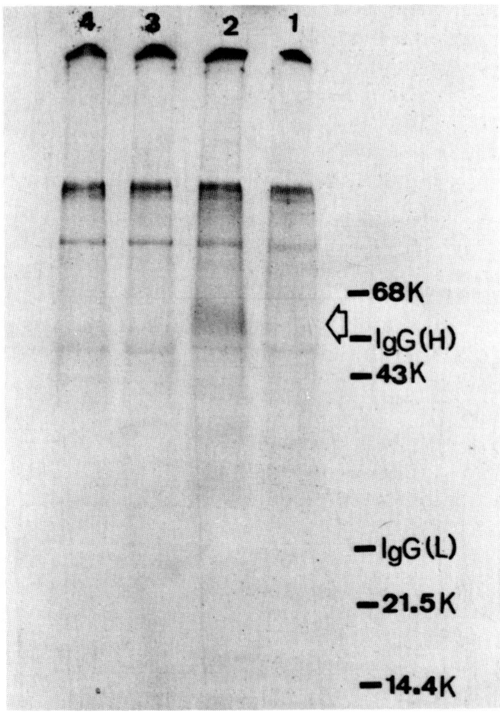

FIGURE 1: Autoradiogram, showing separation by SDS-PAGE of the putative IL-2 receptor containing material

Lysates of R-T-Bl precipitated 1) in the absence or 2) in the presence of ART18-IgG$_1$; lysates of M-T-Bl precipitated 3) in the absence or 4) in the presence of ART18-IgG$_1$. The arrow (line 2) shows the antigenic material, apparent molecular weight 50-60,000, which is only present in the extracts of R-T-Bl precipitated with ART18-IgG$_1$.

TABLE 1: Specificity of Binding of ^{125}I-ART18-IgG$_1$ to Cells and Specificity of Inhibition by ART18-IgG$_1$ of Absorption of IL-2 by Cells

Source of cells	^{125}I-ART18-IgG$_1$ [a] bound (molecules per 10^7 cells)	% inhibition of IL-2 [c] absorption by ART18-IgG$_1$	ART1-IgG$_1$
rat-T-lymphoblasts (Lewis)	5.5×10^{11} b)	92	0
rat-T-lymphoblasts glutaraldehyde-fixed (Lewis)	4.9×10^{11}	64	0
rat-T-lymphoblasts (Wistar)	5.3×10^{11}	88	0
mouse-T-lymphoblasts (BALB/c)	N.S. b)	0	0
rat thymocytes (Lewis)	N.S. b)	-	-

N.S.: No significant ^{125} I-ART18-IgG$_1$ binding detectable.
a) Cold ART18-IgG$_1$ inhibits the binding of ^{125}I-ART18-IgG$_1$ to the cells, whereas ART1-IgG$_1$ does not inhibit the binding.
b) Data taken from Osawa and Diamantstein (3)
c) In this experiment rat derived IL-2 preparation was used.

respective antibody to another T-lymphoblast-specific antigenic determinant may indirectly affect the IL-2 receptors, e.g., by copatching and eliminating the receptors from the cell surface. In order to test this possibility, rat-T-lymphoblasts were fixed with glutaraldehyde, assuming that in this situation such indirect effects could be avoided. As shown in Table 1, the Ab ART18 inhibited IL-2 absorption by the glutaraldehyde-fixed cells, although to a smaller extent. According to these data, we would like to suggest that the Ab ART18 recognizes an antigenic determinant of the IL-2 receptor and that IL-2 receptors of various species are not necessarily identical, even if they can accept and respond to IL-2 from different species. In order to establish conclusively the relationship between IL-2 binding sites and the antigenic determinants detected by the presumably anti-IL-2 receptor antibody ART18, experiments are in progress to isolate the antigenic component of the cell membrane.

TABLE 2: Inhibition by ART18-IgG$_1$ of T-Cell Proliferation

Source of cells (No. of cells/culture)	additions	response to IL-2[c] of rat origin	response to IL-2[c] of murine origin
		cpm x 10^{-3}/culture (%inhibition)	
rat-T-blasts (2×10^4)	none	14.8	12.3
	ART18-IgG$_1$[a]	1.3 (91)	0.8 (93)
	ART1-IgG$_1$[b]	15.1	13.2
mouse-T-blasts (2×10^4)	none	10.8	12.4
	ART18-IgG$_1$	11.2	12.8
	ART1-IgG$_1$	10.6	12.6
		response to ConA[d]	**response to LPS[e]**
		cpm x 10^{-3}/culture (%inhibition)	
rat thymocytes (4×10^4)	none	0.62	-
	ART18-IgG$_1$	0.09 (84)	-
mouse thymocytes (4×10^4)	none	0.48	-
	ART18-IgG$_1$	0.50	-
rat splenocytes (4×10^4)	none	-	0.81
	ART18-IgG$_1$	-	0.79

a) 3 µg/culture; b) 30 µg/culture; c) IL-2 derived from ConA-stimulated rat or murine spleen cells diluted with medium to give 80 % of the maximum proliferative response of the respective lymphoblasts; d) 0.2 µg/culture; e) 6 µg/culture.

ACKNOWLEDGMENTS

We thank Mrs. G. Schulz for excellent technical assistance.

REFERENCES

1. Möller, G., ed., *Immunological Reviews Vol. 63*, (1982)
2. Osawa, H. and Diamantstein, T., *Immunology*, in press (1982)
3. Osawa, H. and Diamantstein, T., *J. Immunol.*, in press (1982)
4. Diamantstein, T., Klos, M. and Reimann, J., *Immunology 43*, 183 (1981)
5. Cone, R.E. and Marchanolis,J.J., *Biochem.J. 104*, 435 (1974)
6. Laemmli, U.K., *Nature (Lond.) 227*, 680 (1970)
7. Robb, R.J., Munck, A. and Smith, K.A., *J. Exp. Med. 154*, 1455 (1981)

DISCUSSION

FARRAR: I believe that Diamantstein mentioned that the isolated receptor competes with the cell receptor for IL-2 in the proliferation assay. Does the isolated receptor also bind IL-2?

DIAMANTSTEIN: We have not examined this point.

ROBB: Instead of an antibody to the IL-2 receptor, I have used purified IL-2 on an affinity support to pull out labelled receptor molecules from receptor-positive murine cells. I also obtained a 50-60,000 MW molecule, which would seem to affirm Diamantstein's results with antibody, and those of Depper *et al.*

GREENE: In our studies with monoclonal anti-Tac which appears to recognize the human IL-2 receptor, ^{35}S precursor-labelling, immuno-precipitation and SDS-PAGE analysis reveals two major proteins with apparent Mr of 113,000 and 50-55,000. In contrast, when these activated human T cells are surface radiolabelled with lactoperoxidase we only detect the 50-55,000 band. In his studies, did Diamantstein likewise detect a higher MW protein which might form an internal component of the IL-2 receptor-effector complex?

DIAMANTSTEIN: We did not perform such experiments with isotopically labelled IL-2.

GERARD: Has Diamantstein done any FACS analysis with his monoclonal antibody? I ask this, as in our laboratory Ortaldo did some FACS analysis with the anti-TAC monoclonal and we were not able to get more than 50% positive binding on an IL-2-dependent T cell line (OH1: monkey line).

DIAMANTSTEIN: We did not perform FACS analysis. However, more than 95% of T lymphoblasts bind rhodamine-labelled ART-18-IgG_1.

OPPENHEIM: In order to crosscheck the pssibility that LPS-activated B lymphocytes may have a receptor for IL-2, has Diamantstein determined whether they absorb IL-2 activity, as well as react with an anti-IL-2 antibody?

DIAMANTSTEIN: In contrast to others we found that LPS-stimulated rat spleen cells (3 day cultures) absorb (or consume) a small but significant amount of IL-2 (possibly due to contamination of such cell populations with T-lymphoblasts?) and, as we showed, they also bind a small amount of ART-18-IgG_1.

IDENTIFICATION AND CHARACTERIZATION OF A MONOCLONAL ANTIBODY THAT INHIBITS INTERLEUKIN-2 DRIVEN T CELL PROLIFERATION BY ACTING ON A CELL SURFACE MOLECULE DISTINCT FROM THE INTERLEUKIN 2 RECEPTOR

Thomas R. Malek[1]
Ethan M. Shevach

Laboratory of Immunology, National Institute of Allergy and Infectious Diseases, National Institutes of Health, Bethesda, Maryland

I. INTRODUCTION

Interleukin 2 (IL-2), is a glycoprotein of 15,000 daltons that appears to function as an obligatory growth hormone for many subsets of T lymphocytes (1, 2). The use of IL-2 has permitted the establishment of long term lines of non-transformed antigen-specific T cells (3). Although the biochemical pathway of IL-2 driven T cell growth has not been completely elucidated, recent studies have demonstrated that IL-2 must interact with specific membrane binding sites (IL-2 receptors) expressed on activated, but not resting, T cells and that following this binding there is an internalization and lysosomal degradation of IL-2 (4).

We have previously prepared and characterized several long term lines of guinea pig alloreactive T cells obtained from individual colonies in soft agar (5, 6). These T cell colonies were specific for distinct epitopes on guinea pig Ia antigens and required stimulation by both allogeneic Ia on macrophages as well as IL-2 containing media for continued growth. We have used these alloreactive T cell colonies as immunogens for the preparation of murine monoclonal antibodies which might be specific for guinea pig T cell differentiatiion antigens present on a very small percentage or at a very low density on unfractionated normal T lymphocytes. In this paper we will describe the initial characterization of one

[1] *Supported by a fellowship from the Arthritis Foundation.*

ISBN 0-12-527540-4

monoclonal antibody, 5C3, produced by this protocol that reacts with a surface molecule expressed preferentially on activated T cells and which totally inhibits IL-2 mediated T cell activation of the immunizing T cell colony as well as IL-2 mediated activation of normal guinea pig T blasts.

II. PRODUCTION OF MONOCLONAL ANTIBODIES TO ALLOREACTIVE T CELL COLONY K5

BALB/c mice were immunized with the strain 13 vs strain 2 guinea pig alloreactive T cell colony, K5, and the spleen cells from these animals were fused to the SP2/0 myeloma line. After approximately 10 days sufficient growth was obtained to begin screening hybridoma supernatant fluids for antibodies cytotoxic for colony K5. In one fusion, 12 of 260 positive wells contained antibodies which had cytotoxic activity for K5. Each of the supernatant fluids from these 12 wells was then screened in a functional assay to assess their ability to enhance or inhibit the proliferative response of colony K5 to allogeneic Ia and IL-2. The supernatant fluid from one well inhibited the proliferative response of K5 by greater than 95% and the cells from this well were cloned by limiting dilution. This IgM monoclonal antibody was subsequently designated 5C3. Additional functional studies demonstrated that 5C3 inhibiited the proliferative response of several other alloreactive T cell colonies that were also IL-2 dependent, but specific for different Ia molecules. Thus, the inhibition by 5C3 appeared not to be related to the fine specificity of these alloreactive T cell colonies.

III. EFFECT OF 5C3 ON ANTIGEN, MITOGEN, AND ALLOANTIGEN INDUCED T CELL PROLIFERATION

As antibody 5C3 was capable of inhibiting T cell proliferative responses of alloreactive T cell colonies, it was of interest to evaluate the effect of 5C3 on the response of T cells to allogeneic Ia antigens in the MLR as well as the response of T cells primed in vivo to soluble protein antigens. The development of the one-way primary MLR of T cells to allogeneic irradiated PEC was consistently inhibited by 5C3 (Table I). In all instances the inhibition detected was greater than 50%, was independent of responder or stimulator cell concentration (data not shown), and generally increased in magnitude as the culture period was increased from 3-7 days. The secondary and tertiary MLR were inhibited to an even greater extent. The antigen-specific proliferative responses of cells primed to PPD, OVA or GL were completely inhibited by 5C3 in all experiments and as little as a 1:100,000 dilution of the

TABLE I. The Effect of Monoclonal Antibody 5C3 on Alloantigen, Antigen, and Mitogen-Induced T Cell Proliferation

Stimulus	*%-Specific Inhibition by 5C3*
Allogeneic Ia (1° MLR)	*64 ± 3*
(2° MLR)	*86 ± 11*
(3° MLR)	*96*
Ovalbumin	*90 ± 4*
L-glutamic acid, L-lysine	*86 ± 7*
PPD	*95 ± 2*
Con A	*19 ± 11*
PHA	*33 ± 9*

ascites was effective. The ability of 5C3 to inhibit Con A or PHA induced proliferation was partial and the magnitude of inhibition varied substantially between experiments.

IV. THE EFFECT OF 5C3 ON IL-2 DRIVEN T CELL PROLIFERATION

Although 5C3 had a variable inhibitory effect on T cell mitogen responses, the observation that this antibody almost completely inhibited antigen and alloantigen induced proliferative responses suggested to us that 5C3 may affect a signal for proliferation common to a variety of T cell subsets. Since IL-2 is required for growth by many T cell subsets, the ability of monoclonal antibody 5C3 to inhibit IL-2 mediated proliferation was tested in assays which directly measure IL-2 activity (Table II). The stimulation of guinea pig Con A blasts by crude guinea pig and human IL-2, as well as a preparation of partially purified human IL-2 was inhibited by a 1:1,000,000 dilution of the ascites or 6 ng of purified antibody in culture. Monoclonal antibodies to other guinea pig lymphocyte surface antigens, including Ia, had no effect on these assays. 5C3 had no effect on the proliferation of human Con A blasts to unfractionated human IL-2 or on the proliferation of the IL-2 dependent murine HT2 cell line to crude rat or guinea pig IL-2. Thus, antibody 5C3 is specific for guinea pig cells. It is unlikely that 5C3 reacts with guinea pig IL-2 since it has no effect on the response of murine HT2 cells to guinea pig IL-2.

TABLE II. Effect of Monoclonal Antibody 5C3 on IL-2 Mediated T Cell Proliferation

Responding Cells	*IL-2 Source*	*Inhibition by 5C3*
Guinea pig Con A blasts	*Guinea Pig*	+++
	Human	+++
	Partially purified human	+++
Human Con A blast	*Human*	-
Murine HT2 line	*Rat*	-
	Guinea pig	-

V. THE RELATIONSHIP BETWEEN THE ANTIGEN DEFINED BY 5C3 AND THE IL-2 RECEPTOR

Cell distribution studies on the FACS demonstrated that the antigen defined by 5C3 was expressed at a low density on 25-50% of purified lymph node T cells and was absent from thymocytes and a B cell leukemic line, BZ-L2C; this surface molecule was present at a significantly greater density on all alloreactive T cell colonies examined as well as on T cells activated by mitogens, alloantigens, or soluble protein antigens. Therefore, one possible explantion to account for the inhibition of IL-2 driven proliferation by 5C3 was that it was directed to the IL-2 receptor and inhibited T cell proliferation by preventing the binding of IL-2 to its receptor. Since human IL-2 was capable of stimulating guinea pig T cell blasts, we were able to directly assess the effect of monoclonal antibody 5C3 on the binding and subsequent degradation of radiolabeled human IL-2 by guinea pig blasts. The addition of 5C3 had no effect on these assays (R. Robb, personal communication). Thus, 5C3 was not directed at or near the IL-2 binding site of the IL-2 receptor. It was still possible, however, that 5C3 reacted with an epitope on the IL-2 receptor distal to the binding site.

Since 5C3 inconsistently affected the T cell response to mitogens, we examined the phenotype on the FACS of Con A blast cells prepared in the presence or absence of 5C3. Greater than 90% of Con A blasts bear the antigen defined by 5C3. In contrast, less than 15% of Con A blasts prepared in

TABLE III. Absorption of Guinea Pig IL-2 Activity by $5C3^+$ and $5C3^-$ Con A Blasts

Absorption	*cpm ^{3}H-TdR Incorporation*[a] IL-2 (1:8)	IL-2 (1:16)
None	72,171	54,055
$5C3^+$ blasts	3,293	1,946
$5C3^-$ blasts	2,986	1,908

[a] *By murine HT2 cell line.*

the presence of 5C3 bear the surface molecule defined by this antibody. This result suggests that during the T cell response to Con A, 5C3 acted to modulate the phenotype of the blasts to a predominantly $5C3^-$ population of cells. The ability to produce $5C3^+$ and $5C3^-$ blasts allowed further examination of the relationship between IL-2 driven growth and the surface molecule defined by this antibody. We compared the ability of $5C3^+$ and $5C3^-$ blasts to absorb guinea pig IL-2 (Table III). $5C3^+$ and $5C3^-$ blasts absorbed equivalent quantities of IL-2 over a range of cell concentrations. This result suggests that 5C3 defines a cell surface molecule distinct from the IL-2 receptor. The $5C3^+$ and $5C3^-$ blasts were also tested for their ability to proliferate to guinea pig IL-2 (Table IV). Surprisingly, the $5C3^-$ blasts responded approximately 10-fold less well than the $5C3^+$ blasts. Thus, these results demonstrated that although $5C3^+$ and $5C3^-$ blast populations absorb equal amounts of IL-2 activity, the expression of the antigen defined by 5C3 was required for IL-2 mediated T cell growth.

TABLE IV. Proliferation of $5C3^+$ and $5C3^-$ Con A Blasts to Guinea Pig IL-2

Responder	*cpm ^{3}H-TdR Incorporation IL-2-Dilution* 1:8	1:32	1:128
$5C3^+$	98,250	81,760	34,320
$5C3^-$	11,112	7,670	4,320

VI. CONCLUDING REMARKS

In this report, we have described the production and initial characterization of a murine IgM monoclonal antibody (5C3) which inhibits IL-2 mediated guinea pig T cell proliferation by reacting with a cell surface molecule expressed preferentially on activated T cells. 5C3 appears to define a cell surface molecule distinct from the IL-2 receptor. The precise function of the antigen defined by 5C3 has not been definitively established in this study. However, the ability of 5C3 to inhibit IL-2 induced proliferation and the proliferative response to a variety of other stimuli as well as the preferential expression of the antigen defined by 5C3 on activated cells suggested that 5C3 might identify a component in a putative IL-2 receptor-effector complex that is involved in modulating the growth signal IL-2 transmits to T cells. Our findings are most consistent with a role of this molecule in the processing or generation of the IL-2 growth signal. We feel that it is likely that the molecule identified by 5C3 may provide a requisite second signal for IL-2 stimulated proliferation and propose that 5C3 inhibits T cell proliferation by blocking this second signal.

REFERENCES

1. Farrar, J.J., Benjamin, W.R., Hilfiker, M.L., Howard, M., Farrar, W.L., Fuller-Farrar, J., *Immunol. Rev. 63,* 129 (1982).
2. Gillis, S., Mochizuki, D.Y., Conlon, P.J., Hefeneider, S.H., Ramthun, C.A., Gillis, A.E., Frank, M.B., Henney, C.S., and Watson, J.D., *Immunol. Rev. 63,* 167 (1982).
3. Paul, W.E., Sredni, B., and Schwartz, R.H., *Nature 294,* 697 (1981).
4. Robb, R.J., Munck, A., and Smith, K.A., *J. Exp. Med. 154,* 1455 (1981).
5. Malek, T.R., Clark, R.B., Shevach, E.M., *J. Immunol. 127,* 616 (1981).
6. Malek, T.R., and Shevach, E.M., *J. Immunol. 127,* 2113 (1981).

DISCUSSION

HOWARD: Many of us believe IL-1 is involved in T cell activation. Is it possible that Malek's antibody is directed against the IL-1 receptor?

MALEK: We cannot directly rule out the possibility posed by Howard, since we are unable to easily measure guinea pig IL-1 activity. However, we believe that this is an unlikely event inasamuch as 5C3 blocked the proliferation of guinea pig Con-A blasts in response to a preparation of partially purified human IL-2, a situation where IL-1 would not be expected to play any role.

FARRAR: Regarding Howard's question about the possibility of an IL-1 receptor, wasn't some of Malek's data based on IL-2-dependent cell lines that do *not* require IL-1?

MALEK: These cell lines are IL-2-dependent, but *also* require stimulation by allogeneic Ia on macrophages. We cannot rule out the possibility that the macrophages also serve to provide IL-1 to the T cell lines.

SWAIN: Could Malek's antibody be like anti-Ly or anti-Leu reagents, which appear to block antigen recognition, but not mitogen-driven induction?

MALEK: The cell distribution of the antigen defined by 5C3 is different when compared to anti-Ly or anti-Leu reagents. Furthermore, 5C3 can inhibit proliferation when added as late as 24 hours before termination of a 3 day antigen-specific culture. This suggests to me that 5C3 is not likely to act during the early events of T cell activation.

KULL: Richard Roth and I have prepared monoclonal antibodies to the insulin receptor, where the structure of the hormone and the receptor are better understood. In that situation, one sees antibodies which inhibit hormone binding, but do not immunoprecipitate subunits characteristic of the receptor.

MALEK: The 5C3 antibody does not affect the binding of IL-2 to the IL-2 receptor. The modulation and absorption experiments, however, suggest that 5C3 recognizes a molecule distinct from the IL-2 receptor, but required for IL-2 growth. It will be necessary to purify this molecule and determine its biological function in order to understand the precise relationships between this surface antigen and IL-2-driven growth.

HOWARD: My question is prompted by Farrar's comment that Malek's antibody blocks proliferation of long term IL-2-dependent T cell lines. My question for Malek is: Are these long term lines propagated on irradiated spleen? If so, they may very well *also* utilize IL-1.

MALEK: Yes, it is entirely possible that the irradiated peritoneal exudate feeder cells may serve as a source of lymphokines in stimulating these T cell lines.

GARLAND: Do any of the anti-TCGF antibodies co-precipitate the mitogen in mitogen-induced responses?

MALEK: Our work would not provide an answer to this query, but according to Diamantstein no such co-precipitation occurs.

HUMAN LYMPHOCYTE PROLIFERATION: RELATIONSHIP BETWEEN CELL CYCLE EVENTS AND INTERLEUKIN-2 (IL-2) PRODUCTION

Flemming Kristensen
Florence Bettens
Christoph Walker
Fransiska Joncourt
Alain L. de Weck

Institute of Clinical Immunology
University of Bern
Bern, Switzerland

Analyses of stimulated T-lymphocytes *in vitro* have revealed that at least two signals are required for proliferation (1,2): (a) some lectins and specific antigens can activate the cells (G_0-G_1 shift)(3,4), a process during which receptors for IL-2 are expressed (1,2), and (b) the interaction between IL-2 and the induced receptors will lead to completion of RNA synthesis (G_{1a}-G_{1b} shift) required for DNA synthesis (5).

This intra-G_1 phase event was demonstrated by cytofluorometric analyses of murine thymocytes. If human peripheral blood lymphocytes (PBL), stimulated with PHA are examined, this shift is not always obvious, because IL-2 is produced in excess and thus allow an unrestricted G_{1a}-G_{1b} transition (6). In the present study, we have inhibited the IL-2 production by adding dexamethason (DEX)(7) or prostaglandin E_2 (PGE_2) (Walker et al., submitted) to the cultures. As it will be shown, such cells are blocked in G_{1a} phase, but addition of IL-2 containing supernatants will reconstitute the G_{1a}-G_{1b} transition and consequently the proliferation. In an attempt to adapt this test system for clinical immunology purposes, kinetics analyses were made to examine the IL-2 production. It was thereby possible to estimate number of activated cells (G_{1a} and G_{1b} cells), IL-2 production and degree of proliferation.

ISBN 0-12-527540-4

RESULTS

A. Effect of Hydroxyurea (HU), DEX and PGE_2 *on Human PBL*

By cytofluorometric analyses of unstimulated human PBL, practically all cells have a uniform low level of RNA and DNA (G_0 cells). Following PHA stimulation, a subpopulation of cells will initiate RNA synthesis (G_1 cells) prior to DNA synthesis. RNA histograms of such cells are illustrated in Fig. 1. As also shown, inhibition of DNA synthesis with HU does not alter the cell distribution, significantly. In contrast, inhibition of IL-2 production by DEX or PGE_2 markedly changes the cell distribution. Accordingly, DNA synthesis is also inhibited (Fig. 2). We use this shift toward less RNA per cell to distinguish between two G_1 subcompartments (G_{1a} and G_{1b}). Finally, this block of lymphocytes in G_{1a} phase can almost completely be reversed by adding an external source of IL-2 to the cultures.

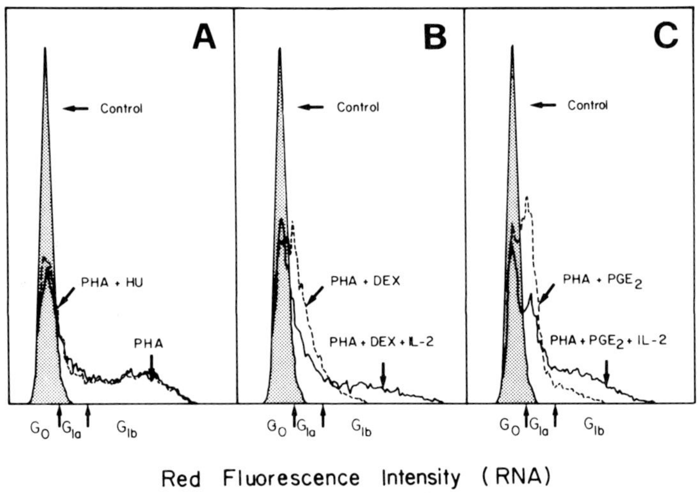

FIGURE 1. Effect of HU, DEX, PGE_2 *and IL-2 on PHA-stimulated PBL. Cells were cultured for 42 hours in RPMI 1640 with 10% AB-serum. The histograms illustrate frequency distribution (100 channels; abscissa) of* 10^5 *acridine orange stained cells, according to red fluorescence (RNA). Cells with elevated DNA content are not included. For the purpose of direct comparisons, histograms shown in A,B and C derive from 3 donors, selected on the basis of similar numbers of PHA-induced* G_1 *(*G_{1a} *+* G_{1b}*) cells.*

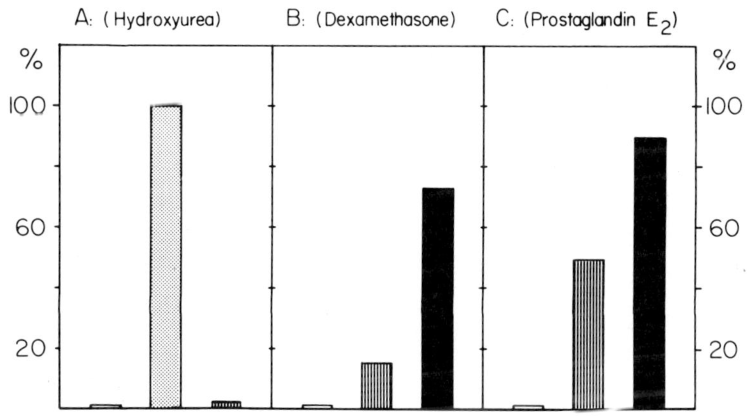

FIGURE 2. PHA stimulation of human PBL. Cells (2 x 10^5 cells/culture) were cultured for 48 hours (6 hours pulsing) in the absence (open bars) or presence of PHA (1:1000)(dotted bar). (A) HU (1.5 mM), (B) DEX (10^{-6} M) or (C) PGE_2 (10^{-6} M) were added to PHA stimulated cultures without (striped bars) or with IL-2 (50 arbitrary units/ml)(black bars). Results are expressed as percent of thymidine incoporation by PBL, stimulated with PHA, alone.

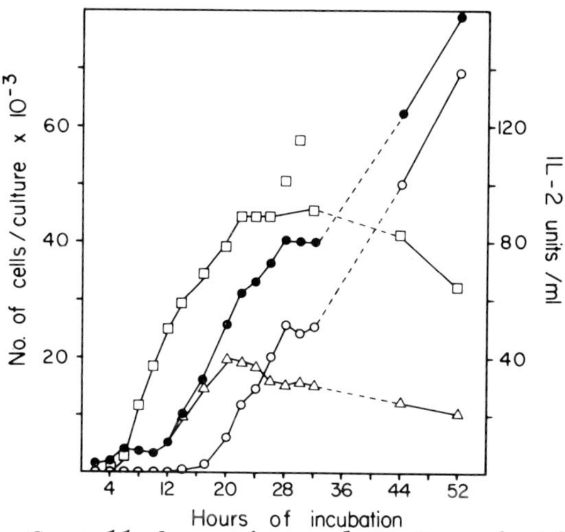

FIGURE 3. G_1 cell formation and IL-2 production of PHA stimulated human PBL. Total number of G_1 (●), G_{1a} (△), G_{1b} cells (○) and corresponding IL-2 titers (□) were measured after various incubation periods. Results are average values (6 donors).

B. Kinetics of the Activation Process and IL-2 Production

If formation of G_{1a} and G_{1b} cells is monitored, it appears that PHA-stimulated PBL require a lag-phase of 12-14 hours before they enter the G_{1a} phase (Fig. 3). Thereafter a rapid increase in number of activated cells occurs and 8-10 hours later cells appear in the G_{1b} phase. After approximately 28 hours of incubation, the total number of G_1 cells reaches a plateau, which lasts for 12-18 hours. Then a second increase occurs.

When IL-2 is quantitated in supernatants from such cultures, significant activities are detected 6-8 hours after stimulation. The IL-2 titers reach a maximum after 22-24 hours of incubation. However, minor differences can be observed from donor to donor. Thus a short lasting peak-titer can be observed in some instances (2 out of 6 donors)(Fig. 3).

C. Correlation between G_1 Cells and ^{3}H-Thymidine Incorporation

Based on analyses of stimulated PBL, we have previously demonstrated the existence of a high correlation (r=0.91) between number of G_1 cells and proliferation (6)(Fig. 4). When the data shown in Fig. 1 and 2 are correlated, it appears that cells stimulated with PHA in the presence of DEX or PGE_2 plus IL-2 supernatants are incorporating the expected amount of ^{3}H-thymidine, based on the number of induced G_1 cells. On the other hand, cultures without IL-2 supernatants will allow cell activation but not proliferation. Finally, cells treated with HU are not responding to IL-2, as expected.

DISCUSSION

Based on cytofluorometric analyses of PHA stimulated human PBL, where the IL-2 production is inhibited by either DEX or PGE_2, it is possible to demonstrate that the G_1 phase can be divided into two subcompartments and that the G_{1a}-G_{1b} transition is IL-2 dependent. Furthermore, results also show that the correlation between number of G_1 cells and ^{3}H-thymidine incorporation is abolished, if IL-2 production is blocked (Fig. 4). This type of analysis will allow to better evaluate some of the causes for an altered proliferative response to lectins, e.g. in patients or selected experimental systems. Thus changes in cell cooperation, in terms of responses to and production of IL-2, can be made.

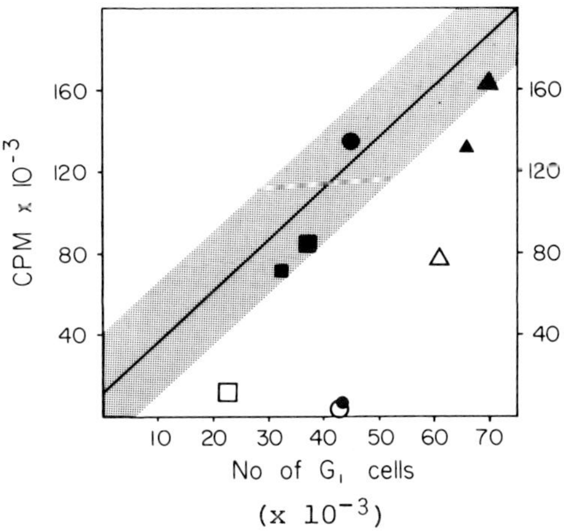

FIGURE 4. Correlation between number of G_1 cells and ^3H-thymidine incorporation (CPM). PBL were cultured in the presence or absence of HU (circles), PGE_2 (squares) or DEX (triangles)(results from Fig. 1 and 2). Control cultures (PHA alone; large filled symbols), cultures without (large open symbols) and with IL-2 supernatants (small filled symbols) are shown. The regression line is calculated on the basis of 36 healthy donors (shaded area represents 95% conficence level).

Obviously, to evaluate the IL-2 production, IL-2 released into culture medium should be examined, as well. But to do so, it is important to know the time at which supernatants should be collected. Therefore, stimulated PBL were monitored, kinetically. As seen in Fig. 3, the IL-2 titer reaches a maximum 22-24 hours after stimulation and a decline takes place thereafter. In all donors, the kinetics of IL-2 production has been rather constant, but significant differences in IL-2 titers have been observed. Furthermore, results show that a decline in IL-2 titers following the plateau takes place, but this decline does not necessarily correlate with the *second* increase in number of activated cells.

The amount of IL-2 released into culture medium is unquestionably dependent on production, consumption and degradation. Therefore, we consider that a combined analysis of free IL-2 after 24 hours of incubation and of number of G_1 cells may give a reasonable estimate of IL-2 production. Consequently, by expanding the standard lymphocyte stimulation assay by cytofluorometric quantitation of resting and activated cells and by estimation of production of excess IL-2, it should be possible to evaluate whether an altered response to lectins is related to a shift in number of responding cells, IL-2 production or ability of cells to respond to IL-2. The experimental data given here illustrate three such situations. In two situations, the lowered response can be corrected by IL-2 enrichment of medium (PGE_2 and DEX), but not in the third situation (HU), and yet, in all three situations the number of cells being activated is normal.

REFERENCES

1. Bonnard, G.D., Yasaka, K. and Jacobson, D., *J. Immunol. 123*, 2704 (1979).
2. Smith, K.A., Gillis, S., Baker, P.E. and McKenzie, D., *N.Y. Acad. Sci. 332*, 423 (1979).
3. Larsson E.-L. and Coutinho, A., *Eur. J. Immunol. 10*, 93 (1980).
4. Bettens, F., Kristensen, F. and de Weck, A.L., *Immunology 45*, 199 (1982).
5. Kristensen, F., Walker, C., Bettens, F., Joncourt, F. and de Weck, A.L., *Cell. Immunol., in press.*
6. Kristensen, F., Walker, C., Joncourt, F., Bettens, F. and de Weck, A.L., *Immunology Letters, in press.*
7. Gillis, S., Crabtree, G.R. and Smith, K.A., *J. Immunol. 123*, 1624 (1979).

DISCUSSION

DIAMANTSTEIN: Do PHA-stimulated, dexamethasone-blocked cells bear IL-2 receptors? At what time after mitogenic stimulation can one detect IL-2 receptors?

KRISTENSEN: We have not yet analyzed dexamethasone-treated, PHA-stimulated cells for the presence of IL-2 receptors. However, the majority of such cells do enter the G_{1b} cycle (and proliferate) if the culture medium is enriched with IL-2. Consequently, Diamentstein's first question can most likely be answered with yes, in that the G_{1a}-G_{1b} cycle transition is IL-2 dependent. His second question cannot be answered on the basis of our data. To do so, it would be necessary to perform IL-2 absorption studies, or to examine PHA-stimulated cells with an anti-IL-2 antibody.

FARRAR: I believe that Kristensen showed IL-2 production ocurring well before the G_1 phase of the cell cycle. How does he reconcile this data with that of Stadler, who showed that IL-2 production is principally associated with the G_1 segment of the cell cycle.

KRISTENSEN: First of all, Stadler's and our own definition of cells in G_1 phase is based on changes in red fluorescence of acridine orange-stained cells. It is to be expected that more accurate criteria will be developed. However, the kinetic analyses as I have given them, show that IL-2 is released into the culture medium well before the major RNA-synthesis is initiated. This has led us to believe that initiation of IL-2 production does not require newly synthesized ribosomal RNA. On the other hand, synthesis of RNA will most likely lead to an increased IL-2 production, as Stadler et al. have shown with lectin-stimulated cells, being blocked in late G_1 (G_{1b} phase according to our definition) by means of hydroxyurea.

VASOPRESSIN AND PHORBOL ESTER REPLACEMENT OF INTERLEUKIN 2 REQUIREMENT IN GAMMA INTERFERON PRODUCTION: CYCLIC GMP AS THE COMMON SECOND MESSENGER[1]

Howard M. Johnson
Barbara A. Torres

Department of Microbiology
University of Texas Medical Branch
Galveston, Texas 77550

Douglas L. Archer

Food and Drug Administration
Cincinnati, Ohio 45226

William L. Farrar

Biological Carcinogenesis Program
Frederick Cancer Research Center
Frederick, Maryland 21701

I. INTRODUCTION

Interferons, including gamma interferon (IFNγ), have been shown to have a broad range of effects on cell function, including immunoregulatory, antiviral, anticellular, and antitumor activities. We will present data here on the cell-to-cell interactions and mediator-cell interactions that

[1]This investigation was supported by PHS Grant Number AI 16428 awarded by the National Institute of Allergy and Infectious Diseases

ISBN 0-12-527540-4

regulate IFNγ production in mouse spleen cell cultures. The cellular interactions are dynamic and involve the regulatory helper and suppressor T cells and the IFNγ-producing T cells (1,2). Data will also be presented which show that helper cells aid IFNγ production by providing interleukin 2 (IL 2), which is apparently required for IFNγ production under normal culture conditions (3,4). The signal provided to the IFNγ-producing cell by IL 2 can be induced or replaced by the neuroendocrine hormones vasopressin and oxytocin and by the tumor promoter phorbol myristic acetate (5,6). These latter observations suggest a possible common second messenger such as cyclic GMP in the mediation of helper cell requirement in IFNγ production.

II. HELPER CELL REQUIREMENT FOR IFNγ PRODUCTION

IFNγ production in C57Bl/6 mice is regulated by cell-to-cell interactions in a manner similar to the interactions that regulate other immune functions (1,2). Treatment of spleen cells from C57Bl/6 female mice with either anti-Lyt 1.2 or anti-Lyt 2.2 monoclonal antibodies resulted in the loss of their ability to produce IFNγ when stimulated with the T-cell mitogen staphylococcal enterotoxin A (SEA) as shown in Table 1 (4). Co-cultivation of the two populations of cells restored the IFNγ response to normal levels. This is evidence that both Lyt $1^+,2^-$ and Lyt $1^-,2^+$ cells are required for IFNγ production.

Table 1. Both Lyt $1^+,2^-$ and Lyt $1^-,2^+$ cells are required for IFNγ production by cultured C57Bl/6 mouse spleen cells

Cultures[a]	SEA-induced IFNγ (units/ml ± SD)
Lyt $1^+,2^-$ cells	<3
Lyt $1^-,2^+$ cells	<3
Lyt $1^+,2^-$ cells + Lyt $1^-,2^+$ cells	478±47
Untreated spleen cells (control)	453±79

[a]Spleen cells, monoclonal anti-Lyt 1.2 or anti-Lyt 2.2 antibody, and guinea pig complement were mixed and incubated at 37°C for 1 h. IFNγ was measured on the second day of culture.

The IFNγ-producing cell was identified as Lyt $1^-,2^+$ (2). This was shown by treating cells *in vitro* with SEA for ongoing IFNγ production 24 h prior to treatment with either anti-Lyt 1.2 or anti-Lyt 2.2 antibody plus complement. The rationale was to reduce the requirement for helper cell activity, and thus direct killing and identification of the IFNγ-producing cell could be assessed. Only the anti-Lyt 2.2 antibody treatment significantly reduced IFNγ production, thus demonstrating that the IFNγ-producing cell was Lyt $1^-,2^+$ and that the helper cell was Lyt $1^+,2^-$ (2). Although the data are not presented, IFNγ production is regulated by a suppressor cell, which is Lyt $1^+,2^+$, and is induced during mitogen activation of cultures for IFNγ production (1,2). The suppressor cell activity was demonstrated in co-cultivation experiments with virgin spleen cells.

III. INTERLEUKIN 2 (IL 2) REPLACEMENT OF HELPER CELL REQUIREMENT IN IFNγ PRODUCTION

The helper cell requirement for IFNγ production can be completely replaced by the Lyt $1^+,2^-$ T-cell product IL 2 (4). This is shown in Table 2 where the addition of IL 2 to Lyt $1^-,2^+$ cells along with SEA resulted in total recovery of IFNγ production, while no effect was seen with similarly treated Lyt $1^+,2^-$ cells. In addition to demonstrating that Lyt $1^+,2^-$ cells provide help for IFNγ production via the lymphokine IL 2, the data also provide further evidence that the Lyt $1^-,2^+$ cell is the source of IFNγ.

The question arose as to whether IL 2 provided help in IFNγ production by stimulation of DNA synthesis and cellular proliferation in Lyt $1^-,2^+$ cells. Accordingly, whole spleen cells and Lyt $1^-,2^+$ cells were treated with mitomycin C and stimulated for IFNγ production in the presence and absence of purified IL 2 (5). Mitomycin C blocked IFNγ production as indicated in Table 3. IL 2 completely reversed the blockage in both whole spleen cells and in Lyt $1^-,2^+$ cultures. The reversal of blockage of IFNγ production was not accompanied by concomitant reversal of blockage of DNA synthesis in the cultures. IL 2 can thus modulate cellular activities such as IFNγ production by a mechanism unrelated to its known proliferative effects on lymphocytes.

Table 2. IL 2 is required for IFNγ production by Lyt $1^-,2^+$ cells

Cultures[a]	IFNγ (units/ml ± SD)
Lyt $1^+,2^-$ cells + SEA	12±4
Lyt $1^+,2^-$ cells + IL 2 + SEA	12±1
Lyt $1^-,2^+$ cells + SEA	<4
Lyt $1^-,2^+$ cells + IL 2 + SEA	759±23
C' control + SEA	443±44
C' control + IL 2 + SEA	685±105
C' control + IL 2	<3

[a]C57Bl/6 mouse spleen cells were treated with monoclonal anti-Lyt 1.2 or anti-Lyt 2.2 antibody and stimulated with SEA (0.5 μg/ml) for IFNγ production for 2 days. Cultures were treated at initiation with approximately 50 units of purified IL 2. Controls are virgin spleen cells treated with complement (C') only.

IV. PHORBOL MYRISTIC ACETATE (PMA) REPLACEMENT OF IL 2 REQUIREMENT FOR IFNγ PRODUCTION

PMA will induce or enhance the induction of a number of lymphocyte and macrophage factors, including the production of IL 1 and IL 2 (7,8). PMA can replace Lyt $1^+,2^-$ or IL 2 requirements for IFNγ production as illustrated in Table 4 where addition of PMA to Lyt $1^-,2^+$ cells along with SEA restored IFNγ production to a normal level (6). Similarly treated Lyt $1^+,2^-$ cells did not produce IFNγ, which was the expected result. PMA replacement of IL 2 requirement for IFNγ production probably occurred through direct interaction with the IFNγ-producing cells, since neither IL 1 nor IL 2 activities were detected in the supernatants from such cultures by a thymocyte assay or IL 2-dependent cell line assay (6). Lyt $1^+,2^-$ cultures did produce IL 2 activity when treated with PMA and SEA under the same conditions, even though IFNγ production did not occur.

Table 3. Effect of IL 2 on IFNγ production by mitomycin C (Mit C) treated mouse C57Bl/6 spleen cells

Cultures[a]	IFNγ (units/ml ± SD)
Spleen cells + SEA	899±93
Spleen cells + SEA + IL 2	1278±167
Spleen cells + IL 2	6±3
Mit C spleen cells + SEA	8±0
Mit C spleen cells + SEA + IL 2	764±62
Lyt $1^-,2^+$ cells + SEA	<3
Lyt $1^-,2^+$ cells + SEA + IL 2	1141±115
Lyt $1^-,2^+$ cells + IL 2	<3
Mit C Lyt $1^-,2^+$ cells + SEA	16±13
Mit C Lyt $1^-,2^+$ cells + SEA + IL 2	760±124

[a]Untreated spleen cells or spleen cells treated with 40 µg Mit C/ml for 30 min. at 37°C, and washed twice with culture media, were incubated for 2 days with SEA (0.5 µg/ml) and IL 2 (50 units/ml) in the indicated combinations for IFNγ production.

Table 4. PMA replacement of IL 2 requirement for IFNγ production by Lyt $1^-,2^+$ cells

Cultures[a]	IFNγ (units/ml ± SD)
Lyt $1^+,2^-$ cells + SEA	<3
Lyt $1^+,2^-$ cells + PMA + SEA	<3
Lyt $1^-,2^+$ cells + SEA	<3
Lyt $1^-,2^+$ cells + PMA + SEA	914±29
Whole spleen cells + SEA	1153±103
Whole spleen cells + PMA + SEA	1557±52
Whole spleen cells + PMA	<3

[a]C57Bl/6 mouse spleen cells were treated with monoclonal anti-Lyt 1.2 or anti-Lyt 2.2 antibody and stimulated with SEA (0.5 µg/ml) for IFNγ production for 3 days. Cultures were treated at initiation with 10 ng PMA/ml.

V. ARGININE VASOPRESSIN (AVP) REPLACEMENT OF IL 2 REQUIREMENT FOR IFNγ PRODUCTION

PMA and the antidiuretic neuroendocrine hormone AVP have been reported to stimulate DNA synthesis in 3T3 cells by a common mechanism (9). For this reason, AVP was examined for IL 2-like properties in IFNγ production (5). AVP at 8 x 10^{-8} M completely restored the ability of Lyt $1^-,2^+$ cells to make IFNγ (Table 5). Significant restoration also was obtained with as little as 10^{-10} M AVP. AVP did not induce IL 2 in cultures; nor did it support the growth of IL 2-dependent cells (data not shown). Thus the AVP help in IFNγ production is related to the nonproliferative properties of IL 2. The structurally related hormone oxytocin was also able to replace the IL 2 requirement for IFNγ production, while insulin was without effect. From the above and other studies (10), it is clear that neuroendocrine polypeptide hormones have lymphokine-like activities and can regulate the immune response. Such findings may have important implications for a functional interaction between the neuroendocrine and immune systems.

Table 5. AVP can replace IL 2 requirement for IFNγ production by Lyt $1^-,2^+$ cells

Cultures[a]	IFNγ (units/ml±SD)	P-value
Lyt $1^-,2^+$ cells + SEA	28±11	
Lyt $1^-,2^+$ cells + AVP + SEA	427±59	<0.01
Whole spleen cells + SEA	220±7	
Whole spleen cells + AVP + SEA	281±19	NS
Whole spleen cells + AVP	<3	

[a] C57Bl/6 mouse spleen cells were treated with monoclonal anti-Lyt 1.2 antibody and complement and stimulated with 0.5 μg SEA/ml for IFNγ production for 3 days. AVP was used at 8 x 10^{-8} M.

VI. CYCLIC GMP AS THE POSSIBLE COMMON SECOND MESSENGER FOR HELPER CELL REQUIREMENTS FOR IFNγ PRODUCTION

The ability of AVP, PMA, and oxytocin to share functions with IL 2 prompted us to examine the IFNγ production system for common second messenger activity. Cyclic guanosine monophosphate (cyclic GMP) was capable of completely replacing the helper cell or IL 2 requirement for IFNγ production by Lyt $1^-,2^+$ cells (Table 6) (11). Activators of guanylate cyclase such as acetylcholine and *Escherichia coli* heat stable toxin also replaced IL 2 requirement for IFNγ production. The cyclic GMP help was independent of DNA synthesis or proliferation in the IFNγ-producing cells, since it reversed mitomycin C blockage of IFNγ production but did not reverse the inhibition of DNA synthesis. Thus the findings presented here are unrelated to the question of the second messenger role of cyclic GMP in the activation of lymphocytes for DNA synthesis and cellular proliferation.

Table 6. Effect of dibutyryl cyclic GMP (dcGMP) on IFNγ production by Lyt $1^-,2^+$ cells

Cultures[a]	IFNγ response (units/ml ± SD)
Lyt $1^-,2^+$ cells + SEA	<10
Lyt $1^-,2^+$ cells + dcGMP + SEA	4857±711
Lyt $1^-,2^+$ cells + dcGMP	<10
Whole spleen cells + SEA (Control)	4748±231
Whole spleen cells + dcGMP + SEA	4524±317
Whole spleen cells + dcGMP	<10

[a] C57Bl/6 female mice spleen cells were cultured in 5% CO_2 with the indicated substances for 3 days. SEA was used at a final concentration of 0.5 μg/ml and dcGMP at 2×10^{-4} M.

VII. SUMMARY

IFNγ production in the mouse appears to be regulated by a dynamic interaction between helper cells (Lyt $1^+,2^-$), suppressor cells (Lyt $1^+,2^+$), and IFNγ-producer cells (Lyt $1^-,2^+$) (Figure 1). Helper cells appear to aid IFNγ production by providing IL 2. The neuroendocrine polypeptide hormones AVP and oxytoxin and the tumor promoter PMA can completely replace the IL 2 requirement for IFNγ production. All of these helper substances may act through a common second messenger, such as cyclic GMP, in a manner independent of DNA synthesis.

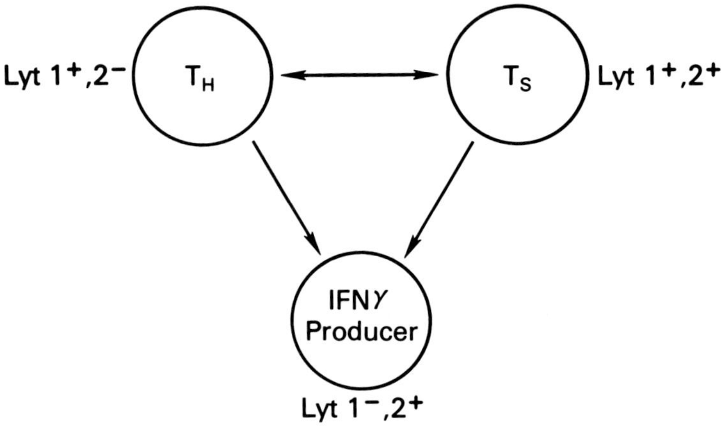

FIGURE 1. IFNγ production is regulated by the dynamic interaction between helper (T_H), suppressor (T_S), and IFNγ producer cells. Helper cells provide IL 2. Suppressor cells may act by absorption of required IL 2 from cultures.

REFERENCES

1. Johnson, H.M., Antiviral Res. 1, 36 (1981).
2. Torres, B.A., Yamamoto, J.K., and Johnson, H.M., Infect. Immun. 35, 770 (1982).
3. Farrar, W.L., Johnson, H.M., and Farrar, J.J., J. Immunol. 126, 1120 (1981).
4. Torres, B.A., Farrar, W.L., and Johnson, H.M., J. Immunol. 128, 2217 (1982).
5. Johnson, H.M., Farrar, W.L., and Torres, B.A., J. Immunol. 129, In press (1982).
6. Johnson, H.M. and Torres, B.A., Infect. Immun. 36, 911 (1982).

7. Farrar, J.J., et al., J. Immunol. 125, 2555 (1980).
8. Fuller-Farrar, J., et al., Cell. Immunol. 58, 156 (1981).
9. Dicker, P. and Rozengurt, E., Nature 287, 607 (1980).
10. Johnson, H.M., et al., Proc. Natl. Acad. Sci. USA 79, 4171 (1982).
11. Johnson, H.M., Archer, D.L., and Torres, B.A., Submitted (1982).

DISCUSSION

SONNENFELD: Are there any other subpopulations of spleen cells, such as adherent cells, required in Johnson's system?

JOHNSON: Yes, there are: macrophages are required.

SONNENFELD: So -- Johnson's cells are only depleted of cells of the appropriate Lyt phenotype?

JOHNSON: Yes. Adherent cells are still present in our cultures.

AGGARWAL: I wonder whether Johnson has explored growth hormone, another neuroendocrine hormone, to look at γ-IFN production.

JOHNSON: No, we have not yet done this.

OPPENHEIM: We are also attracted to the hypothesis that helper Lyt 1^+2^- cells produce IL-2, which in turn helps stimulate Lyt 1^-2^+ cells for IFN production. However, Kasahara and Luger have observed that both FACS-selected human $OKT4^+$ and $OKT8^+$ produce IFN_γ as well as IL-2. It is therefore possible that both these subsets of human lymphocytes, or some even smaller subpopulations of these categories, produce these activities.

JOHNSON: We are not certain that $OKT4^+$ and $OKT8^+$ human T cells are the exact counterparts of mouse $Lyt1^+2^-$ and $Lyt1^-2^+$ T cells, respectively. Both $OKT4^+$ and $OKT8^+$ have been shown, for example, to be capable of cytotoxic activity.

KRAMMER: I call to Johnson's attention that there are some published reports on $_\gamma$IFN production by Lyt 1^- positive T cell lines upon antigen and mitogen stimulation.

JOHNSON: The phenotypic and functional relationships between T cells freshly removed from the individual, and the established T cell lines, needs much further study before we place great reliance on such comparisons.

BENJAMIN: Is it not possible that PMA or vasopressin is stimulating the production of IL-2 by residual Lyt 1^+ T lymphocytes? In that event Johnson's findings could involve such a mechanism, rather than a replacement of IL-2 activity.

JOHNSON: No, I do not think so. We do not find IL-2 activity in the Lyt 1^-2^+ IFN_γ-producing cultures, while we do find it in the Lyt 1^+2^- population after stimulation with SFH and PMA.

FITCH: We have derived several mouse T cell clones which produce interferon and there appears to be no correlation between their Lyt phenotype and interferon production. Clones which are Lyt $1^{+}2^{-}$, Lyt$1^{-}2^{+}$, or which are Lyt "null" -- can all produce interferon when subjected to appropriate stimulation.

JOHNSON: My reply to Fitch would be along the lines of my rebuttal to Krammer. We must be mindful of complexities in the phenotypic and functional relationships between T cells freshly removed from individuals, versus those represented by established T cell lines.

PALACIOS: We have developed a reverse hemolytic plaque assay for detection of $_{\gamma}$IFN secretion at the single cell level. We found that OKT4^{+} but not OKT8^{+} human T cells secrete $_{\gamma}$IFN either spontaneously or after stimulation by OKT3 monoclonal antibody. These are a portion of the data I will be reporting later on during this workshop.

JOHNSON: As I have already replied to a somewhat similar statement by Oppenheim, I again say we are not certain that OKT4^{+} and OKT8^{+} human T cells constitute the precise counterparts of mouse Lyt $1^{+}2^{-}$ and Lyt $1^{-}2^{+}$ T cells, respectively. For example, *both* OKT4^{+} and OKT8^{+} have been shown to be capable of cytotoxic activity.

VILCEK: In agreement with some of the other comments, in connection with Johnson's investigation, my impression is that the human and mouse systems are not really comparable. In unseparated human PBL cultures, PMA causes a marked enhancement in the production of mitogen-stimulated $_{\gamma}$IFN. In contrast, the same combined treatment of unseparated mouse spleen cells does not increase $_{\gamma}$IFN production as compared to stimulation with mitogen alone. In view of Johnson's very interesting data it would seem that the helper-suppressor T cell interactions in the two species (or, at least, in the two culture systems) are not identical.

JOHNSON: I would agree with Vilcek's interpretation.

RESPONSIVENESS OF PURIFIED NATURAL KILLER CELLS TO PURE INTERLEUKIN-2 (IL-2)[1]

J.R. Ortaldo, J.P. Gerard, L.E. Henderson, R.H. Neubauer, and H. Rabin

NCI-FCRF
Frederick, Maryland 21701

I. INTRODUCTION

Natural killer (NK) cells are lymphoid cells possessing spontaneous cytolytic activity toward a variety of malignant and virus transformed target cells (1,2). The lineage of natural killer cells has remained controversial, with evidence for association with both myelomonocytic and T cells (3,4). We recently have identified NK cells as a morphological subpopulation of lymphoid cells, large granular lymphocytes (LGL) which represent 2 to 6% of the peripheral blood white cells (1). LGL can be enriched to greater than 90% purity by discontinuous Percoll gradient centrifugation and subsequent depletion of high affinity E-rosette forming cells (5). There have been a number of reports (6-7) that IL-2 can augment and promote the growth of NK cells. However, all of these studies have been performed with crude or partially purified IL-2 preparations. IL-2 from a gibbon T cell line, MLA-144 (8) was purified to homogeneity by a two-step process (9). In this communication we will show for the first time that purified human LGL can be cultured in the presence of pure IL-2 and that the cultured LGL retain their characteristic NK activity. The immunoregulatory capacity of IL-2 on NK cells will be demonstrated using IL-2 from gibbon and human sources.

II. PURIFICATION OF IL-2

The biochemical and biological properties of the MLA-144 IL-2 are described in Table I. This material which is constitutively produced by MLA-144 is devoid of other detectable

[1] *Supported in part by the NCI, DHHS, under contract No. NO1-CO-75380 with Litton Bionetics, Inc.*

ISBN 0-12-527540-4

lymphokines (Rabin, unpublished results). The IL-2 was purified 8,000-10,000-fold by batch purification on trimethylsilane derivatized controlled pore glass beads followed by reversed phase high performance liquid chromatography with nearly complete recovery of activity. Average purification is shown in Table II. Purified IL-2 had two bands of molecular weight 16,000 and 15,000 daltons by SDS polyacrylamide gel electrophoresis (data not shown) (9). Both bands had biological activity and had essentially the same amino acid composition (9).

TABLE I. Biochemical and Biological Properties of MLA-144 IL-2

Size (G-100)	*peak at 21,500 daltons*
Size (SDS PAGE)	*2 bands at 15 & 16,000 daltons*
Elution from reversed phase support	*55 to 60% Acetonitrile*
Elution from DEAE cellulose	*0.04 to 0.06 M phosphate buffer, pH 7.6*
Isoelectric point	*6.45 (single peak)*
pH and heat stability	*pH 2.0 to 9.0, 60°C*
Responding T cells	*human, simian, rabbit, dog, mouse*
Absence of other lymphokines and cell products	*IL-1, IL-3, CSF, BCGF, MIF, MAF, IF, EGF*

TABLE II. Purification of MLA-144 derived IL-2

Material	*Volume*	*Total protein*	*Total Units of IL-2 Activity*
Serum-free conditioned medium	*15,000 ml*	1.2×10^8 *ng*	5×10^6
TMS-CPG[a] *40-50% Acetonitrile elution*	*500 ml*	4×10^6 *ng*	4×10^6
Elution from reversed phase support	*8 ml*	1.2×10^4 *ng*	4×10^6

[a]*Trimethylsilane-controlled pore glass beads*

III. BIOLOGICAL ACTIVITY OF IL-2 ON LGL

A. Proliferation of LGL

The growth of LGL (10) and highly purified T cells was analyzed by using a limiting dilution assay. This assay provides a minimal estimate of the progenitor frequency present in populations and is accomplished by culturing a range of dilutions of LGL and analyzing the percent of wells which showed proliferation or cytotoxicity in the presence of IL-2. In the present report using crude and pure MLA-144 IL-2, as well as partially purified human IL-2, LGL and T cells were compared for proliferative and cytotoxic progenitors.

As shown with the results of a representative experiment in Table III, both crude and pure MLA-144 as well as partially purified human IL-2 had the ability to promote the growth of LGL and T cells and to support the development of functionally active NK cells. The proliferative capacity of the cells at 200 units of all IL2s was quite similar (LGL 1/55; T cells 1/8). Cytotoxic activity against

TABLE III. Limiting Dilution Assay

Source of IL-2	Units[a] of IL-2	Frequency of Progenitors: Proliferation[b] LGL	Proliferation[b] T	Cytotoxicity (K562)[c] LGL	Cytotoxicity (K562)[c] T
MLA-144 Crude	200	1/58	1/5	1/200	<1/10,000
MLA-144 Pure	200	1/60	1/6	1/165	<1/10,000
	20	1/61	1/7	1/200	<1/10,000
	2	1/46	1/15	1/800	<1/10,000
Human[d]	200	1/54	1/6	1/174	<1/10,000

[a]Units per ml.
[b]Proliferation determined on day 7, after 6 hours pulse with ^{3}H-thymidine.
[c]^{51}Cr-labeled targets (5000/well) were added on day 8 or 9 for 4 hours.
[d]Partially purified human IL-2 (Assoc. Biomed. Systems).

NK-susceptible target K562 was only present in cultures of LGL. The minimum dose of pure IL-2 required for optimum growth of functional NK cells appeared to be about 20 units of IL-2. Similar titrations with the crude and partially purified human material also were seen (data not shown).

B. Augmentation of NK Activity

In addition to their effects on proliferation, IL-2 preparations were assessed for their ability to augment NK cytotoxic activity. Pretreatment of purified LGL with the various IL-2s resulted in significant augmentation of NK cytotoxicity (Table IV). Augmentation was seen at low doses (20 units) of pure as well as partially purified gibbon and human IL-2 at 20 and 42 hrs. The kinetics of the augmentation of cytotoxicity demonstrated that significant enhancement was not seen prior to 20 hours when less than 200 units of IL-2 were used (Figure 1). However, 2000 units of IL-2 resulted in a significant and similar augmentation seen at 6 hours.

TABLE IV. Augmentation of NK Activity by IL-2s

Treatment[a]	*Units IL-2*	*Cytotoxicity (K562)*[b]
Control	*0*	*43.8*
Pure MLA IL-2	*2000*	*89.3*
	200	*70.2*
	20	*53.5*
	2	*49.8*
Human IL-2 (ABS)[c]	*200*	*75.5*
	20	*63.1*

[a]*Pretreatment of LGL for 18 hr, washed 2 times, then assayed in a 4-hour* ^{51}*Cr-release assay.*
[b]*% specific cytotoxicity at effector to target ratio of 3:1.*
[c]*See Table II.*

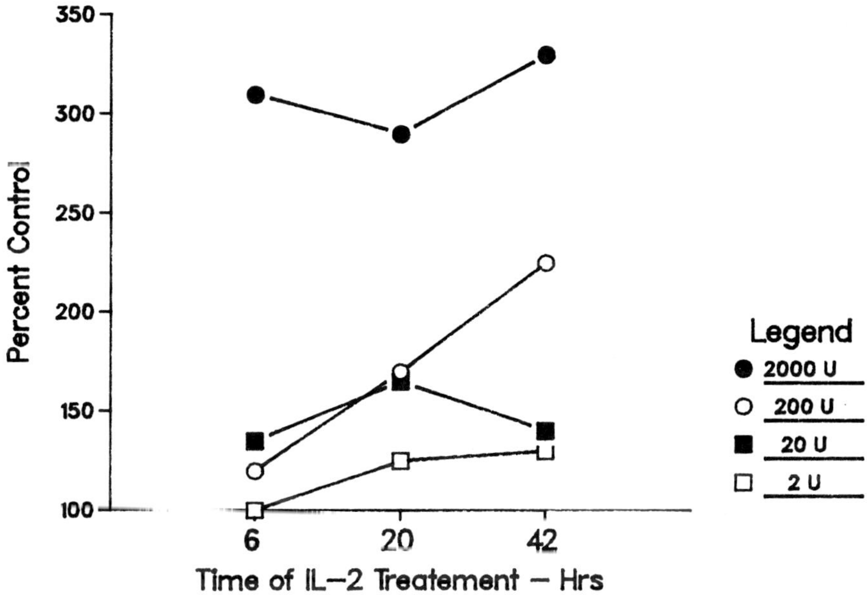

Figure 1. Kinetics of NK boost with pure IL-2. Purified LGL were cultured with pure MLA-144 IL-2 (2000-2 units), washed, and tested in a 4-hr ^{51}Cr release assay. Results are shown relative to control cell, cultured in medium only.

IV. CONCLUSIONS

Using highly purified LGL and pure IL-2, several conclusions can be drawn from the results. 1) Pure IL-2 is able to support the growth of both NK cells and T cells. This refutes previous contentions that contaminating factors other than IL-2 might be responsible for the growth and development of NK cells. 2) Pure IL-2 is sufficient to maintain functional cytotoxic activity of NK cells and as reported previously, T cells were not cytotoxic against NK-susceptible target cells. 3) Pure IL-2 can significantly augment the cytolytic capacity of human NK cells. This augmentation occurred as early as 6 hours after exposure of NK cells to high doses of IL-2; however, low doses of IL-2 required 18 hours for maximal cytotoxic activity to occur. Thus, IL-2 appears to play an important role not only in expanding NK cells but also in regulating their activity.

VI. REFERENCES

1. Herberman, R.B., and Ortaldo, J.R. *Science 214*, 24 (1982).
2. Herberman, R.B. (Editor), "Natural Cell-Mediated Immunity Against Tumors," Academic Press, New York, 1980.
3. Kay, H.D., and Horwitz, D.A. *J. Clin. Invest. 66*, 847 (1980).
4. Lohmann-Matthes, M.-L., Domzig, W., and Roder, J.C. *J. Immunol. 123*, 1883 (1979).
5. Timonen, T., Saksela, E., Ranki, A., and Häyry, P. *Cell. Immunol. 48*, 133 (1979).
6. Ortaldo, J.R., Sharrow, S.A., Timonen, T.T., and Herberman, R.B. *J. Immunol. 127*, 2401 (1981).
7. Timonen, T., Ortaldo, J.R., and Herberman, R.B. *J. Exp. Med. 153*, 569 (1981).
8. Rabin, H., Hopkins, R.F. III, Ruscetti, F.W., Neubauer, R.H., Brown, R.L., and Kawakami, T.G. *J. Immunol. 127*, 1852 (1981).
9. Henderson, L.E., Hewetson, J.F., Hopkins, R.F. III, Sowder, R.C., Neubauer, R.H., Oroszlan, S., and Rabin, H., submitted for publication.
10. Ortaldo, J.R., Timonen, T., Vose, B.M., and Alvarez, J.A. in "The Potential Role of T cell Subpopulations in Cancer Therapy" (A. Fefer, ed.), Raven Press, New York, in press.

REMOVAL OF PHYTOHAEMAGGLUTININ FROM HUMAN INTERLEUKIN-2 CONTAINING SUPERNATANTS BY DYE MATREX LIGAND CHROMATOGRAPHY IN COMPARISON TO OTHER METHODS

Udo Schwulera
Peter Lindsay
and
Hans-H. Sonneborn

BIOTEST-Serum-Institut GmbH
Frankfurt, FRG

INTRODUCTION

Supernatants of human peripheral blood mononuclear cells stimulated with Phytohaemagglutinin (PHA) contain lymphokines, one of which is called interleukin-2 (1, 2). Interleukin-2 (IL-2) is a protein with a molecular weight of 17.000±3000 D. It makes possible the in vitro profagation of T-cells, while retaining their immunological function. Because PHA effects the survival andfunctional capability of cultured T-cells it is important to find methods to remove PHA completly. Several methods are described in the literature e.g. anti-PHA column (3), thyreoglobulin-Sepharose (4), red blood cell adsorption (5), ultrafiltration, PHA-pulse (6), partial purification (7), use of other mitogens (8). Here we present a comparison of these methods in respect to the removal of PHA with results, which could be obtained by affinity chromatography on Amicon Blue A gel.

RESULTS

Several methods have been described for separating PHA from interleukin-2. In table 1 we have compared all these methods in respect to their efficiency in remaining PHA. For these experiments we used the commercially available T-lymphocyte spent medium "Lymphocult T" [R] which was produced as described below (9).

Interleukins, Lymphokines, and Cytokines

ISBN 0-12-527540-4

Table 1: COMPARISON OF VARIOUS PHA REMOVAL METHODS

method	% Removal of PHA from crude extract tested by ^{3}H-Tdr incorporation into PBL and removal of iodinated PHA	Effect on IL-2 activity
anti-PHA-Sepharose 4B (3)	60 - 70%	no effect
Thyroglobulin-Sepharose 4B (4)	40 - 50%	binds measurable amounts of IL-2
Chicken-erythrocyte-adsorption (5)	40 - 50%	no effect, Haemolysis
PHA-puls-stimulation (6)	60 - 80%	lower IL-2 production compared to standard production
Ultrafiltration	0%	IL-2 sticks to membrane and hollow fibers
55% $(NH_4)_2SO_4$-precipitation	20 - 30%	at higher concentration IL-2 is also precipitated
fractioned $(NH_4)_2SO_4$ precipitation and gelfiltration Sephadex G75	100%	no effect

Table 1 clearly shows that all single step methods proved to be inefficient in separating the mitogen. Several adsorption cycles are often necessary in order to remove PHA completly; some of the methods have additional disadvantages such as haemolysis of rat cells and/or loss of IL-2-activity. Only a combination of the above listed methods (10) or a partial purification (7) produced mitogen free preparations, but these are too laborious for routine use.

Other attempts to replace PHA with other mitogens have so far failed. Concanavalin A , lentil-lectin, leucagglutinin (8) or wheat germ agglutinin induce only low IL-2 titers. Soluble staphylococal Protein A, recently shown to be a potent helper T mitogen for human T lymphocytes, produced high titered IL-2 preparation, but only when gamma globulin was present (11).

PHA therefore still seems to be the best-suited mitogen for the induction of Interleukin-2 in the human system.

Here we present a quick and easy method for the quantitative removal of PHA from IL-2 spent medium without affecting the T-cell growth supporting activity. This method, which is known as "Amicon Matrex gel ligand chromatography", is an affinity chromatography carried out using the svthetic textile dye Cibacron [R] Blue 3GA as a ligand bound via an nether linkage to agarose (s. figure 1).

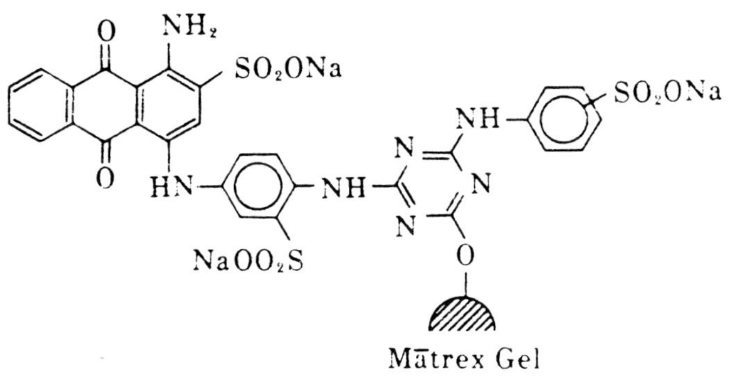

Figure 1: Partial structure of Matrex Gel Blue A (12)

The availability of various dye ligands with different properties (Dyematrix Kit) enables a screening to show which of the five Matrex Gel dye ligand media are most suitable for the particular purification. The results of these experiments are shown in figure 2.

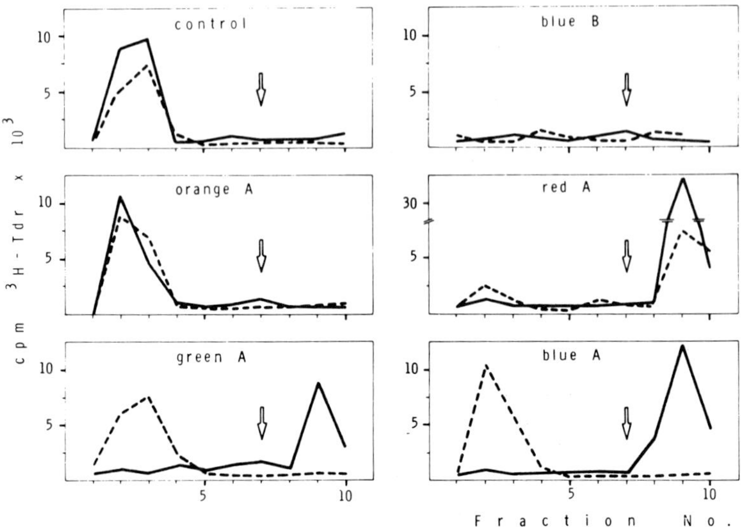

Figure 2: Interaction of Matrex Gel ligand columns with interleukin-2 and/or Phytohaemagglutinin. The arrow indicates the elution of the IL-2 activity with 1M NaCl. PHA activity (---) was measured by ^{3}H-Tdr incorporation into PBL. IL-2 activity (-) was measured by ^{3}H-Tdr incorporation into human MLC- and/or PHA-Blast cells.

As previously shown (13), only the Blue A media allow a complete removal of PHA in a single column run. IL-2 was bound to the dye reversibly and could be eluted in a highly active state by using a salt elution step or a salt gradient, whereas the mitogen passed through unadsorbed. The green A media showed lower IL-2 binding capacity and lower IL-2 recovery rates, while the controls and orange A columns did not bind PHA or IL-2. Red A gels bind both components reversibly, while blue B media show a strong adsorption, and neither PHA nor IL-2 could be eluted under the conditions tested.

In serveral experiments we worked out optimal conditions for the use of the Cibacron Blue A column. According to the manufacturers instructions (12) the column was washed and then equilibrated with 10mMTris-HCL pH 7,4 or 10mM Phosphate Buffer pH 7,4. Lymphocult T ^{R} was then applied directly to 2ml columns using a flow rate of 30ml/hr. The batch-method with an incubation time of 15 - 30 minutes at 4°C is also possible. After washing 5 column volumes with the same buffer used above, the IL-2 activity was eluted by means of 1M NaCl or KCl salt step elution. The eluted IL-2 is more than 95% free of PHA as tested by ^{3}H-Tdr incorporation into PBL and by monitoring this procedure iodinated PHA.

In order to determine the IL-2 binding capacity of a 2ml Cibacron Blue A column we applied increasing amounts of crude lymphocyte supernatant. The results are shown in figure 3.

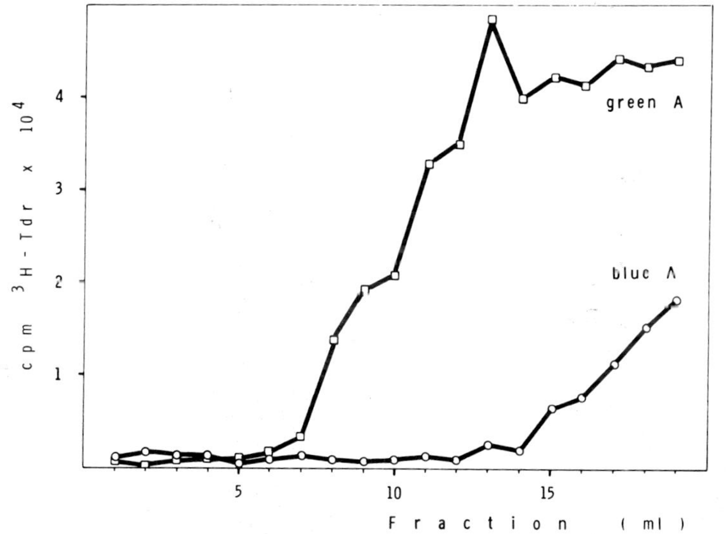

Figure 3: Comparison of the IL-2 binding capacity using Blue A and Green A Matrex Gel column. IL-2 containing conditional medium was applied directly to the columns. The curves show the wash-through of IL-2 activity which was not bound by the column.

As figure 3 shows, a 2ml green A column binds IL-2 from 5 - 6ml conditioned media whereas a 2ml blue A column binds IL-2 from 14 - 15ml. Preliminary immunological studies show that human albumin and gamma-interferon also adheres to the Cibacron Blue A-gel. The capacity of the column could be increased when no albumin was present in the conditioned medium because albumin appears to compete for the dye binding side on the Matrex beads. Further purification of IL-2 could be archieved by applying a salt gradient or a multistep salt elution.

SUMMARY

The Blue A method is an easy and quick method which allows the quantitative removal of the mitogenic component of PHA from interleukin-2 containing lymphocyte supernatants. The amount of PHA removal is, however, limited by the capacity of the column when human albumin is present. The column can be used 20 times or more without a significant decrease in capacity.

Because of the high Il-2 binding capacity when no albumin is present, this method is also the best suited for the purification and concentration of human and mouse IL-2 (results not shown).

REFERENCES

1. Morgan, D. A., Ruscetti, F. W., and Gallo, R. Science 193 1007 (1976)
2. Smith, K. A. Immunological Rev. 51, 337 (1980)
3. Kurnick, J. T., Grönvik, K.O., Kimura, A. K., Lindblom, B. Skoog, J.T., Sjörberg, O., and Wigzell, H. 122, 1255 (1979)
4. Fagnani, R., and Braatz, J.A. J. Immunol. Methods 33, 313 (1980)
5. Alvarez, J., Rodriguez, J., Lopez-Botet, M., Silva, A., and de Landazuri, M.O.J. Immunol Methods 40, 289 (1981)
6. Fabricus, H.A., and Stahn, R. Immunobiol. 156 364 (1979)
7. Mier, J.W., and Gallo, R. C. Proc. Natl. A cad. Sci. 77 6134 (1980)
8. Solbach, W., Barth, S., Röllinghoff, M., and Wagner, H. Immunbiol. 157 197 (1980)
9. Lindsay, P. Schwulera, U., and Sonneborn, H.-H. in "Human Lymphokines" (Khan A., and Hill N. D.) p. 479 Academic Press New York (1982)
10. Rosenberg, S. A., Schwarz, S., Spiess, P.J., and Brown, J. M., J. Immunol. Methods 33 337 (1980)
11. Fleischer, B. J. Immunol. Methods 47 191 (1981)
12. "Dye Ligand Chromatography" Amicon Corporation Lexington, USA
13. Schwulera, U., Lindsay, P., and Sonneborn, H.-H. 12th Workshop on Leukocyte culture. 87, (1982)

ACKNOWLEDGMENTS

This work was supported by the BUNDESMINISTERIUM FÜR FORSCHUNG UND TECKNIK (PTB 03 8114)

INTERLEUKIN-2 RESEARCH: PLATEAU OR PLATFORM?

A Critical Assessment of the Interleukin 2 Symposium at the Third International Lymphokine Workshop

Steven Gillis

Immunex Corporation
51 University Bldg., Suite 600
Seattle, Washington

Research experimentation into the role that Interleukin 2 (IL-2) plays in controlling cellular immune responses has flourished over the past five years. This molecule, (which, despite consensus agreement as to its nomenclature at the Second International Lymphokine Workshop four years ago, continues in some circles to be referred to as T-cell growth factor) is clearly the proliferation-inducing agent which controls the expansion of activated T-cell clones (1). Biochemical characterization of IL-2 has progressed rapidly due to several pivotal accomplishments brought forward by several lymphokine laboratories.

First, unequivocal assays for IL-2 activity have been standardized. IL-2 continues to be the only molecule which allows for the overnight proliferation of dependent T-cell clones. Such proliferation can be quanitated rapidly by the incorporation of tritiated thymidine into replicating cells. Such a rapid readout (which can be conducted in very small volumes) has allowed investigators, worldwide to subject IL-2 containing conditioned media to multiple separation techniques based on its size, proteinaceous nature, and electrophoretic mobility (2,3,4,5,6).

Secondly, many laboratories have identified cloned, cellular sources (of either malignant or normal phenotype) which have the capacity to either constitutively produce large quantities of IL-2 or which upon mitogen and/or

ISBN 0-12-527540-4

phorbol ester induction will synthesize and secrete copious quantities of the lymphokine (7,8,9,10,11,12). This development has further fostered chemical characterization. Monoclonal antibodies have been described by two laboratories which have the capacity to neutralize the biological activity of IL-2 and in some instances precipitate either biosynthetically, radiolabelled or iodine substitution, labelled IL-2 (13,14).

On the molecular front, numerous laboratories have isolated Poly A$^+$ mRNA which when translated in either oocyte, rabbit reticulocyte, or wheat germ in vitro translation systems gives rise to a product which has limited, yet clearly detectable, IL-2 biological activity (15,16,17). Monoclonal antibodies directed against IL-2 have further been used to attempt to immunoprecipitate radiolabelled IL-2 prepared by in vitro translation of IL-2 mRNA containing, message pools (17).

Many of the recent strides in IL-2 research were reviewed and highlighted at the symposium which addressed IL-2 biochemistry and biology, held as part of the Third International Lymphokine Workshop. After digesting what was presented at that particular symposium, it would appear that IL-2 research in general, and in particular in the area of molecular characterization, has reached a plateau. Although critics may view such a plateau as suggestive evidence that IL-2 research has stalled, I would prefer to consider the plateau has a platform from which the next round of hallmark experimentation and observations must be built. It is clear that what remains to be accomplished in the area of IL-2 experimentation involves generation of complete amino acid sequence and the cloning and expression of the IL-2 gene in a sympathetic vector system. It is this symposium chairman's opinion that many investigators outside the direct field are frustrated that these accomplishments have not occured with the rapidity with which initial biochemical characterization and isolation of cellular sources of IL-2 genetic information, have progressed. In looking back over the field, however, one must pause and reflect carefully that it is only within the past two to three years that lymphokine researchers, and to some extent other cellular immunologists, have learned that crude conditioned media cannot be labelled with acronyms and titles suggestive of molecular homogeneity. It is only recently that cellular immunologists as a whole have discovered that precise biochemical separation of lymphokine containing supernates must be conducted in order to ascribe biological activities to discrete molecular entities. Sober, realistic assessment of the field suggests that right or wrong, amino acid sequence information for IL-2 will likely not be available until the gene is cloned and expressed. The intense scientific, political and business pressures on today's lymphokine researchers will more than likely prohibit publication of a sequence in the absence of a protected bacterial clone.

Review of much of the data presented at this year's IL-2 Symposium might suggest that cellular immunologists are not quite as close as we might like to be with respect to final determination of IL-2 sequence information, as well as production of the lymphokine by recombinant technologies. Existing monoclonal antibodies, although useful in neutralizing IL-2 activity, have proven unwieldy with respect to affinity purification of the molecule. Although these serological reagents have been shown to precipitate biologically active IL-2, standardized assays for detecting picogram amounts of the lymphokine on a serological basis have not been forthcoming.

It is relatively naive of cellular immunologists to suspect that existing monoclonal antibodies must work in assays for molecular detection of IL-2. Many monoclonals have been known to give complex and bewildering results in immunoprecipitation trials. Perhaps additional monoclonal antibodies must be generated for molecular purposes, using as a screen, the immunoprecipitation of a single IL-2 polypeptide from message translation systems. Perhaps totally new assays for detection of IL-2 biological activity must be sought in order to have success in the molecular arena.

Recently published data with respect to translation of IL-2 mRNA into a biologically active product are somewhat disappointing in that the highest levels of activity do not approach 1/20, or at best 1/10, the response manifested by positive controls (15,16,17). As a group of lymphokine researchers, we must be wary of such data. Although luck may prevail, and the IL-2 gene may be cloned and expressed in a matter of weeks, one might suspect that we currently do not have all the tools in place with which to make the endeavor a routine exercise. The field has established a platform from which detailed and painstaking molecular experimentation must be conducted. The cloning of the IL-2 gene to many investigators may seem like a trivial exercise in something which is "just around the corner", however, more realistic assessment of the situation would suggest that success, if in the same town, is perhaps miles distant.

Another area of IL-2 research was discussed at great length at this year's meeting. This involved IL-2 receptor display and putative generation of anti-receptor monoclonal antibodies. Published reports of quanitation and enumeration of IL-2 receptors on activated T-cells using biosynthetically radiolabelled human IL-2 have recently appeared (18). These data suggest that activated T-cells contain on the average approximately 8-10,000 IL-2 recepters/cell. Such precise quanitation was made possible by use of "purified" human IL-2 prepared biosynthetically by incorporation of radiolabel into the IL-2 protein during mitogen and phorbol ester induction of human, IL-2 producer, leukemic T-cells. The final criterion of purity, for the radiolabelled reagent used in binding studies was that autoradiographs of the material elaborated a single protein band at an appropriate molecular weight. Although migration of a single band on SDS gels has for many years been heralded as a criterion of purity, the advent of high pressure liquid chromatography (HPLC) has shown (in a number of protein systems) that proteins migrating as single bands on denaturing polyacrylamide gels when subjected to HPLC give rise to multiple protein species. Investigators conducting such binding studies, and even biochemical characterization of unlabelled IL-2, should take considerable warning from these observations. In fact, future biochemical analysis launched from the platform of research activity currently established, may in fact prove that purported claims of molecular homogeneity for IL-2, and other lymphokines, are erroneous, particularly when analyzed in more sophisticated means of biochemical separation. Furthermore, binding data, either previously published or presented at this year's symposium using biosynthetically radiolabelled IL-2, is conducted with a protein of relatively low specific activity. Use of such reagents yields differences between cell populations whose receptors are totally void of ligand (an unbound condition) and totally saturated for ligand (a bound condition) discriminated by only a few hundred counts. It would of course be beneficial to current investigators studying IL-2 receptor display if a stable, iotinated species could be generated.

Several investigators at this years's conference presented data with respect to generation of B-cell hybridomas whose supernates, or antibody purified therefrom, had the capacity to inhibit IL-2 dependent T-cell line proliferation. Although many of the presentations were quite dramatic, (particularly results obtained with an antibody previously shown to identify the TAC antigen), lymphokine researchers involved in this experimentation should take note of three essential observations. The first is that monoclonal antibodies have been shown to block various cellular functions without the antigen to which the antibody was directed, being directly involved in that particular cellular function. For example, anti-Ly^{+2} antibodies have been known for many years to block T-cell mediated (19) cytotoxicity. However, involvement of the anti-Ly^{+2} antigen in the lytic reaction remains to be proven. Similar data are now available for the capacity of anti-human T-cell antigen monoclonal antibodies to block human cell mediated immune responses. Once again, the involvement of these antigens in the lytic process remains conjecture.

Secondly, the capacity of antibody to block IL-2 dependent proliferation may not have anything to do with that antibody's ability to recognize the IL-2 receptor. Antibody binding to a totally different antigen near the receptor may cause receptor masking and inability of IL-2 to bind to a confirmationally altered receptor. Third, it is imperative that appropriate specificity controls be conducted. In other words investigators must show that the antibody has no effect on LPS responses. Or with the aid of a cell sorter or isotopic antiglobulin binding asssay it must be determined that the antibody does not bind to cell types which have previously not been shown to possess functional IL-2 receptors (either by binding radiolabelled IL-2 or by absorption of its biological activity from appropriate supernates).

In summary, research on IL-2, its biological and biochemical characterization has progressed rapidly since the Second International Lymphokine Workshop. Key advancements have allowed the work to progress in such a rapid fashion. Researchers in the field now sit poised, hoping to be able to use the latest techniques proffered by microbiologists and 1980-style biochemistry to allow them to generate enough homogeneous lymphokine to pursue its physiological relevance _in vivo_.

REFERENCES

1. Aarden, L.A., Burnner, T.K., Cerottini, J.C., Dayer, J.M., deWeck, A.L., Dinarello, C.A., Di Sabato, G., Farrar, J.J., Gery, I., Gillis, S., Handschumacher, R.E., Henney, C.S., Hoffmann, M.K., Koopman, W.J., Karane, S.M., Lachman, L.B., Lefkowits, I.,Mishell, R.I., Mizel, S.B., Oppenheim, J.J., Paetkau, V., Plate, J., Rollinghoff, M., Rosenstreich, D., Rosenthal, A.S., Rosenwasser, L.J. Schimpl, A., Shin, H.S., Simon, P.L.., Smith, K.A., Wagner, H., Watson, J.D., Wecker, E., and Wood, D.D., _J. Immunol. 123_, (1979).

2. Farrar, J.J., Simon, P.L., Farrar, W.L., Koopman, J., and Fuller-Bonar, J., Ann. N.Y. Acad. Sci. 332, 303, (1980).
3. Farrar, J.J., Simon, P.L., Koopman, W.J., and Fuller-Bonar, J., J. Immunol. 121, 1353, (1978).
4. Mochizuki, D.Y., Watson, J. and Gillis, S., J. Immunol. Methods 39, 185. (1980).
5. Watson, J., Gillis, S., Marbrook, J., Mochizuki, D.Y., and Smith, K.A., J. Exp. Med. 150, 849, (1979).
6. Caplan, B., Gibbs, C., and Paetkau, V., J. Immunol. 126, 1351, (1981).
7. Gillis, S., Scheid, M., and Watson, J., J. Immunol. 125, 2570, (1980).
8. Gillis, S. and Watson, J., J. Exp. Med. 152, 1709, (1980).
9. Rabin, H., Hopkins, R.F., Ruscetti, F.W., Neubauer, R.H., Brown, R.L. and Kawakami, T.G., J. Immunol. 127, 1852, (1981).
10. Harwell et al., J. Exp. Med. 152, 893, (1980).
11. Shimizu, S., Konaka, Y., and Smith, R.T., J. Exp. Med. 152, 1436, (1980).
12. Farrar, J.J., Fuller-Farrar, J., Simon, P.L., Hilfiker, M.L., Stadler, B.M. and Farrar, W.L., J. Immunol. 125, 2555, (1980).
13. Stadler, B.M., Berenstein, E.H., Siraganian, R.P. and Oppenheim, J.J., J. Immunol. 128, 1620, (1982).
14. Gillis, S. and Henney, C.S., J. Immunol. 126, 1978, (1981).
15. Lin, Y., Stadler, B.M. and Rabin, H., J. Bio. Chem. 257, 1587, (1982).
16. Bleackley, R.C., Caplan, B., Havele, C.R., Ritzel, G., Mosmann, T.R., Farrar, J.J., Paetkau, V., J. Immunol. 127, 2432, (1982).
17. Gillis, S., Mochizuki, D.Y., Conlon, P.J., Hefeneider, S.H., Ramthun, C.A., Gillis, A.E., Frank, M.B., Henney, C.S., Watson, J.D., Immunological Rev. 63, (1982).
18. Robb, R., Munck, A. and Smith, K.A., J. Exp. Med. 154, 1455, (1981).
19. Sarmiento, M., Glasebrook, A.L., and Fitch, F.W., J. Immunol. 125, 2665, (1980).

SECTION II:
LYMPHOKINE ACTIONS ON PRECURSOR CELLS

ROLE OF GANGLIOSIDES IN THE INTERACTION OF CSF WITH ITS MURINE BONE MARROW TARGET CELLS

Dov H. Pluznik[1]
Ruth Lenz

Department of Life Sciences
Bar-Ilan University
Ramat Gan, Israel

I. INTRODUCTION

Colony stimulating factor (CSF) are glycoproteins which regulate the proliferation and differentiation of committed hemopoietic stem cells to mature granulocytes and/or macrophages in soft agar cultures. The mechanism controlling this regulation is unclear at present. Previous studies suggest that CSF interacts with its target cells via cell membrane receptors (1). Gangliosides are present in the cell membrane and serve as receptors to some bacterial toxins (2)(3), peptide hormones (4)(5) and interferon (6). The present study was undertaken to determine whether CSF initiates its regulating activities via binding to ganglioside receptors present on bone marrow (BM) stem cells. Two experimental approaches were used: (a) competitive inhibition of target cell CSF receptors by substances such as cholera toxin (CT), which is known for its binding to gangliosides (7); (b) inactivation of CSF by exogenous gangliosides. Both approaches would be expected to reduce the proliferation and differentiation effects of CSF on bone marrow cells in soft agar cultures.

II. MATERIALS AND METHODS

Mice: Male ICR and C57Bl/6 mice, obtained from the Weizmann Institute of Science, Rehovot, Israel were used in all experiments.

[1] *Leave of absence from Bar-Ilan University*
Present address: LMI, NIDR, NIH, Bethesda, Maryland

Interleukins, Lymphokines, and Cytokines

ISBN 0-12-527540-4

Reagents: Cholera toxin (CT) was purchased from Schwarz-Mann, NY USA. Cholera Toxoid (choleragenoid) was a gift from Dr. R. A. Finkelstein, University of Texas, Southwestern Medical School, Dallas, Texas. Mixed gangliosides from bovine brain (type II) were purchased from Sigma, St. Louis, MI. Purified gangliosides were obtained from Supelco, Bellefonte, PA. Rabbit antiserum to the GM_1 ganglioside was a gift from Dr. Gitler from the Weizmann Institute, Rehovot, Israel.

Sources for CSF: Two sources were used in all the experiments (a) GM-CSF from mouse lung conditioned medium (MLCM) (8) and (b) M-CSF from L909 fibroblasts (L-CM) (9).

Assay of CSF: Murine bone marrow (BM) cells were cloned in soft agar on top of a hard agar base supplemented with CSF. After 7 days of incubation number of macroscopic colonies is counted (10).

III. RESULTS AND DISCUSSION

BM cells were cultured for 7 days in soft agar supplemented with 1-40% MLCM and increasing concentrations of cholera toxin (CT) (Table 1). As shown, MLCM in the absence of CT stimulated dose dependent clonal growth of BM cells, as measured by colony counts. CT, in doses ranging from 10^{-8}-10^{-10}M, inhibited the CSF dependent clonal growth of BM cells. Similar results were obtained with L-fibroblast CM.

Table 1. Inhibitory Effect of CT on Clonal Growth of BM Cells

	Mean No. of Colonies/10^5 BM Cells			
MLCM (%) \ CT(M)	10^{-10}	10^{-9}	10^{-8}	0
40	299	260	60	306
4	153	65	0	202
2	87	21	0	108
1	20	7	0	36

Cholera toxoid (choleragenoid), which is a competitive non-cytotoxic inhibitor of CT (11) also inhibited the clonal growth of BM cells in soft agar. The inhibitory effect of the toxoid, however, is less effective by a factor of 10-100 when compared to that of CT.

Preincubation of BM cells with increasing concentrations of CT prior to cloning in soft agar also inhibited in a dose dependent manner CSF dependent clonal proliferation (Table 2). Such an inhibitory effect of CT can be reversed by pre-exposure of BM cells to high concentrations of CSF. Results of such an experiment are shown in Table 2.

Table 2. Pretreatment of BM Cells with CT Inhibits CSF-dependent Colony Formation: Reversal by Pre-exposure to CSF

Pretreatment of BM Cells	*% of colonies/10^5 BM cells*
PBS	*100*
PBS (30 min); CT (10^{-10} M, 30 min)	*63*
CSF (50% MLCM)	*97*
CSF (30 min; CT 30 min)	*87*

BM cells can also be protected against the inhibitory effect of CT by preincubating the CT with the monosialo-ganglioside GM_1 (Table 3). This ganglioside is considered to be the cellular receptor to which CT binds (12) (13).

Table 3. Abrogation of CT Inhibitory Effect by Pre-treatment with Gangliosides

Pretreatment of CT	*% of colonies/10^5 BM cells*
PBS	*100*
CT ($5X10^{-10}$M)	*55*
CT & PBS	*63*
CT & GM_1 (25 ng/ml)	*90*

The data suggests that CT may compete with CSF for ganglioside receptors, present on BM cells. We, therefore, examined, whether exogenous gangliosides could compete with membranal receptors and block the active site of the CSF molecule. As a result, an inhibition of interaction of CSF binds to gangliosides, a column of mixed gangliosides covalently linked via a poly L-lysine spacer to sepharose (14) was prepared. When MLCM was passed over such a ganglioside-sepharose column over 90% of the material was retained on the column. CSF was eluted with 6M Guanidine-HCl in PBS (data not shown). Gangliosides can spontaneously bind to polystyrene via their ceramide moiety, which leaves the oligosaccharides portion free to react with specific ligands (15). Polystyrene tubes were, therefore, coated with purified preparations of mono, di and trisialogangliosides and a constant amount of MLCM was applied to the coated tubes for 2 hours at 0º to permit binding of CSF to the gangliosides. When these supernatants were collected and tested for CSF activity, the following results were obtained (Table 4).

Table 4. Absorbance of CSF Activity in Polystyrene Tubes Coated with Various Purified Gangliosides

Treatment of Tubes:	*PBS*	*GM_1*	*GM_2*	*GD1a*	*GT*
% CSF activity remaining in culture supernatant	*100*	*54*	*81*	*98*	*100*

Similar results were obtained when the purified gangliosides were incorporated into soft agar cultures supplemented with MLCM (data not shown). Preincubation of BM cells with antibodies against GM_1 before cloning the cells in soft agar cultures was followed by a decrease in number of colonies as shown in Table 5.

Table 5. Effect of Preincubation of BM Cells with Antibodies to GM_1

Treatment of BM Cells	*PBS*	*NRS*	*Rabbit Anti- GM_1*
% Decrease in No. of Colonies	*0*	*10*	*55*

In summary, the data presented in the report indicate that the CSF interacts with its target cells, present in BM, via ganglioside receptors. Blocking such receptors with CT, which is known to bind to membranal monosialicganglioside GM_1 inhibits CSF activity. Furthermore, gangliosides were shown to bind to CSF and thus reduced its activity. GM_1 was the most efficient, of the monospecific gangliosides tested, however its inhibitory activity did not exceed about 50%. Further studies are required to determine whether GM_1, or a very similar ganglioside are the actual receptors to CSF.

REFERENCES

1. Price, G.B., Senn, J.S., McCulloch, E.A. and Till, J.R. *J. Cell. Physiol.* 84:383, 1974.
2. Cuatrecasas, P., *Biochemistry* 12:3547, 1973.
3. Van Heyningen, W.E., *Nature* 249:415, 1974.
4. Mullin, B.R., Fishman, P.H., Lee, G., Aloj, S.M., Ledley, F.D., Wimund, R.J., Kohn, L.D. and Brady, R.O., *Proc. Natl. Acad. Sci.* 73:842, 1976.
5. Lee, G., Aloj, S.M. and Kohn, L.D., *B.B.R.C.* 77:434, 1977.
6. Besancon, F. and Ankel, H., *Nature* 252:478, 1974.
7. Holmgren, J., Lonnroth, I. and Svennerholm, L., *Scand. J. Infec. Dis.* 5:77, 1973.
8. Burgess, A.W., Camakarias, J. and Metcalf, M., *J. Biol. Chem.* 252:1998, 1977.
9. Stanley, E.R. and Heard, P.M., *J. Biol. Chem.* 252:4305, 1977.
10. Pluznik, D.H. and Sachs. *Exp. Cell Res.* 45:553, 1966.
11. Cuatrecasas, P., *Biochemistry* 12:3577, 1973.
12. Cuatrecasas, P., *Biochemistry* 12:3578, 1973.
13. Hollenberg, M.D., Fishman, P.H., Bennet, V., Cuatrecasas, P., *Pros. Natl. Acad. Sci.* 71:4224, 1974.
14. Cuatrecasas, P., Parikh, I., Hollenberg, M.D., *Biochemistry* 21:4253, 1973.
15. Holmgren, J., Elwing, H., Fredman, P., Svenenholm, L. *Eur. J. Biochem.* 106:371, 1980.

DISCUSSION

MIZEL: How does Pluznik view the specificity of the CSF cell target, since most cells have GM_1 in their membranes?

PLUZNIK: Our results show that the receptor involved is a ganglioside. There are indications that it is GM_1 or a closely related ganglioside.

LINKER-ISRAELI: In Pluznik's inhibition assays, I discern an incomplete reduction in the number of colonies found. Did he check the nature of the developing colonies? Does he think that GM_1 is specific for a certain subset of either the monocytic or granulocytic/CFU-C?

PLUZNIK: It could be that by using higher concentrations of ganglioside, we could attain augmented inhibition of clonal growth. There was no change in the ratio between granulocyte and macrophage colonies in the presence or absence of GM_1. Accordingly, it does not appear that GM_1 is selective for either CFU-C.

OPPENHEIM: Do other agents that react with GM_1, such as cholera toxoid, or anti-GM_1, emulate CSF or enhance CSF effects?

PLUZNIK: Neither cholera toxoid nor anti-GM_1 exerts such an effect. In each instance it took CSF to induce clonal growth.

ZSCHIESCHE: Would Pluznik identify for us the target cell for the CSF in his experiments?

PLUZNIK: The target cell employed in our CSF studies were committed hemoppoietic stem cells derived from bone marrow. These cells differentiate and proliferate to mature granulocytes and/or macrophages.

DUMONDE: Would Pluznik tell us whether anyone has ever infused large amounts of gangliosides into experimental animals and followed the effects on hemoploiesis?

PLUZNIK: Not to my knowledge.

GARLAND: Does Pluznik know of any interactions between agar sugars and cell surfaces? Could such reactions affect his findings?

PLUZNIK: The only information I can offer Garland in this regard is that inclusion of Me-cellulose has little if any effect in our system.

HELPER FACTOR REQUIREMENTS FOR CTL DEVELOPMENT: EVIDENCE FOR A T CELL DERIVED CSF-LIKE FACTOR AND A MØ DERIVED DIFFERENTIATION FACTOR.[1]

James H. Finke
Mary L. Hilfiker
Jeffrey W. Scott

Department of Molecular and Cellular Biology
Cleveland Clinic Foundation
Cleveland, Ohio

Susan A. Rothmann

Department of Laboratory Hematology
Cleveland Clinic Foundation
Cleveland, Ohio

Robert N. Moore

Department of Microbiology
University of Tennessee
Knoxville, Tennessee

Generation of cytotoxic T lymphocytes (CTLs) is regulated in part by soluble mediators that are produced by T cells and macrophages (MØ) in response to antigen. Several factors have been identified that provide signals to pre-CTLs as well as to helper and accessory (MØ) cells involved in CTL development. Interleukin-2 (IL-2) provides the proliferative signal to CTLs whereas Interleukin-1 (IL-1), a MØ derived product, is thought to induce IL-2 production from helper T cells (1-2). Gamma interferon (IFNγ) also is involved in CTL development and may provide a differentiation signal to CTLs (3).

We and others have reported the existence of a T cell derived CTL helper factor distinct from IL-2 and IFN that mediates its helper activity through a MØ (4,5). This factor appears to stimulate MØ to produce factors which act upon pre-CTLs (4,5). Furthermore, studies from our laboratory and others suggest that a differentiation/induction factor separable from IL-2 and IFNγ is required for CTL development (4,6-10).

[1]This work was supported by U.S.P.H.S. grant CA 24474.

ISBN 0-12-527540-4

Additional evidence for a MØ activation signal and a CTL differentiation signal are provided here.

Evidence for a factor that mediates CTL helper activity through adherent cells (MØ) is shown in Table 1. Supernates from activated T cells induced a cytotoxic response from nylon wool nonadherent splenocytes stimulated by heat treated thymocytes in the presence but not the absence of adherent accessory cells. These results agree with earlier work in which this T cell factor induced a cytotoxic response from unfractionated but not nylon wool nonadherent spleen cells (4). The inability of adherent cells to reconstitute a cytotoxic response in the absence of the T cell factor suggested that adherent cells were not activated by heat-treated cells. Further the T cell factor appeared to provide a signal necessary to activate MØ to produce monokine(s) involved in CTL development.

TABLE 1: A Factor from Activated T Cells Mediated CTL Helper Activity through Adherent Cells.

C57Bl/6 Spl T cells plus C3H/He Δ thymocytes were cultured with and without a T cell derived helper factor (30% v/v)[a]	E:T ratios	No. of adherent PECs (C57Bl/6) in the cultures: none	$5x10^5$	$1x10^5$	$5x10^4$	$1x10^4$	$5x10^3$
-	16:1	-1.4	-6.0	-4.2	-2.8	-1.8	-1.5
	4:1	0.0	N.D.	-1.0	-1.4	0.4	2.6
+	16:1	2.9	-4.5	15.3	26.6	24.8	6.1
	4:1	0.7	-0.4	4.8	10.6	5.3	2.9

[a]Nylon wool nonadherent splenocytes plus allogeneic heat-treated (Δ) thymocytes were cultured for 5 days in the presence or absence of T cell helper factor and various concentrations of peritoneal adherent cells. Effector cells were assayed for their ability to lyse C3H Con-A blast in a 3 hr ^{51}Cr-release assay. Helper supernate was derived from T cells primed *in vivo* to *Listeria monocytogenes* and restimulated *in vitro* with heat-killed organism.

The CTL helper factor used in these experiments was produced by activated T cells which had been primed *in vivo* to *Listeria monocytogenes* and then restimulated *in vitro* with heat-killed organism. This supernatant fluid was analyzed for lymphokine activities and was found to contain colony stimulating factor(s) (CSF) but no detectable levels of IL-2 or IFN (Table 2). Others have reported that under specific conditions activated T cells will produce CSF in the absence of detectable levels of IL-2 (11). These results suggested that the CTL helper activity was distinct from IL-2 and IFN but may be associated with CSF.

TABLE 2: Biological Activities Present in Supernatant Fluid from Cultures of Activated T Cells.

Supernate tested for the following activities[a]	Assay employed	Presence or absence of activity
CTL help	MLC (splenocytes plus allogeneic Δ thymocytes)	+
Interleukin-2	Proliferation of cloned T cell line. Activity absorbed by CT6 cells.	-
Interferon	Virus yield reduction plaque assays.	-
Colony stimulating factor(s)	Colony growth in agar bone marrow proliferation (uptake ^{3}H-TdR)	+

[a]Induction of CTLs from splenocytes stimulated with allogeneic heat-treated thymocytes (4). Proliferation of clone T cell line (CT6) (4). Helper activity was not removed by incubation with CT6 cells which did remove 87% of the IL-2 activity from supernates known to contain IL-2. Interferon was assayed by two methods; MuLV yield reduction plaque assay and encephalomycarditis virus yield reduction assay (15). Supernate was tested for its ability to induce colonies in soft agar from bone marrow cells and to induce bone marrow cells to proliferate in liquid culture (manuscript in preparation, R. Moore).

Recent experiments support the existence of an association between CSF and CTL helper activity. We found that the EL4 thymoma (kindly provided by J. Farrar NIDR, Bethesda, MD) produced a CTL helper factor that was separable from IL-2 (and IFN) but coeluted with CSF through several purification procedures (Table 3). Supernates from EL 4 cells stimulated

TABLE 3: Biochemical Properties of a CTL Helper Activity Derived from EL4 Thymoma Cell Line that Coelutes with CSF Activity

Biochemical procedures:	Characteristics of:	
	CTL helper activity:	CSF activity:
Hydrophobic chromatography phenyl-Sepharose	weakly hydrophobic elutes in early portion of the gradient (<10% ethanediol)	weakly hydrophobic elutes in early portion of the gradient.
Gel filtration	40,000 - 30,000 daltons	≃30,000 daltons
Chromatofocusing	pI 4.8 - 4.2	pI 4.8 - 4.2

[a]Supernate from PMA stimulated EL4 cells was precipitated with ammonium sulfate (40 - 90%) and applied to phenyl Sepharose column (12). Fractions containing CSF activity were pooled and subjected to either gel filtration or chromatofocusing. Fractions were tested for CSF activity and for CTL helper activity.

with phorbol-12-myristate-13-acetate (PMA) were precipitated with ammonium sulfate and fractionated on phenyl Sepharose as previously described (12). Both the CTL helper and CSF activity eluted from the phenyl Sepharose column in the early portion of the gradient, with considerable overlap between the two activities. Fractions from phenyl Sepharose columns that contained CSF activity were pooled and fractioned by chromatofocusing or gel filtration. In both cases, the fractions containing CSF activity also contained CTL helper activity. This helper activity does not appear to be Interleukin 3 (IL-3) (13). Although IL-3 is produced by EL4 cells and coelutes with CSF on phenyl Sepharose (J. Ihle, personal communication), purified IL-3 (kindly provided by J. Ihle, Frederick Cancer Research Facility, Frederick, MD) did not generate a cytotoxic response in our bioassay (unpublished data). Furthermore, the helper activity from EL4 supernates is not attributable to residual PMA since various concentrates of PMA (10^{-5} to 10^{-9}M) were unable to generate a cytotoxic response in our bioassay.

As seen in Table 4, the helper factor produced by EL4 cells mediated CTL helper activity only in the presence of adherent cells. Pooled fractions from a phenyl Sepharose column that contained both CSF and helper activity generated

TABLE 4: A CTL Helper Factor from EL4 Thymoma Requires Adherent Cells to Express its Activity.

The following supernates were added to the bioassay[a]	% Volume added	% Specific lysis Responder cells: Unfractionated spleen cells, E:T ratio 4:1	8:1	16:1	32:1	Nylon wool nonadherent spleen cells, E:T ratio 4:1	8:1	16:1	32:1
none	none	1.8	2.8	4.0	ND	ND	ND	-1.3	ND
pooled fractions (CSF activity)	10%	ND	6.8	11.0	20.3	ND	ND	ND	ND
	40%	ND	22.2	39.2	59.6	ND	1.6	3.1	1.2
positive control	30%	59.9	76.1	81.0	ND	28.0	47.8	64.5	ND

[a]Supernates were incubated with C58Bl/6 responder cells and DBA/2 Δ-thymocytes for 5 days after which effector cells were assayed for their ability to lyse P815 target cells. EL4 supernate was subjected to phenyl Sepharose chromatography and the fractions that contained CSF activity were pooled and tested for helper activity. Supernate from co-culture of T cells and MØ from Listeria primed animals served as positive control.

a cytotoxic response from unfractionated but not from nylon wool nonadherent splenocytes.

Results presented here indicate that a T cell derived factor involved in the generation of CTLs is associated with CSF and may induce CTL development indirectly by stimulating MØ. That CSF may stimulate MØ to produce factors which are involved in CTL development is supported by the work of Moore, et al. who showed that CSF would induce MØ to produce IL-1 (14).

Although our results are consistent with the hypothesis that CSF can function as a CTL helper signal, further purification of the EL4 produced helper factor and CSF will be required to determine whether both activities are mediated by a single molecule.

Generation of CTLs appears to require antigen activated pre-CTLs to receive induction/differentiation signal in addition to a proliferation signal. Evidence for such a differentiation factor is presented in Table 5. Partially purified IL-2 (see Legend, Table 5) from the EL4 cell line was not a sufficient signal to generate CTLs from nylon wool nonadherent splenocytes stimulated by heat-treated allogeneic thymocytes (Table 5). Similar results were obtained with IL-2 purified from the LBRM-33 cell line (kindly provided by S. Gillis, Immunex, Seattle, WA; manuscript in preparation). However, supernates from cultures of activated MØ (see Legend, Table 5) that did not contain significant amounts of IL-2 induced a cytotoxic response. This MØ derived differentiation factor appears to synergize with IL-2 to produce a potent cytotoxic response. The cytotoxic response generated in the presence of the differentiation factor and IL-2 was much greater than the additive responses of cells exposed to each factor separately. The differentiation factor appears to be produced by MØ although from our present evidence we cannot rule out the possibility that T cells play a role in the production of this factor. The differentiation signal does not appear to be IL-1 since the differentiation factor has an approximate molecular weight of 38,000 daltons (4). The differentiation factor is also distinct from IFN since elimination of IFN activity with anti-IFN α/β and pH-2 treatment did not affect the CTL activity (15). However, we cannot exclude the possibility that the differentiation factor induces IFN production in the mixed lymphocytes cultures (MLCs).

TABLE 5: Synergy between IL-2 and Supernates Containing MDF in the Generation of CTLs

C57Bl/6 splenic T cells plus DBA/2 Δ-thymocytes were cultured with the following[a]:	% volume added to cultures	% specific lysis effector:target ratio		
		16:1	8:1	4:1
none	none	0.5	3.4	ND[b]
EL4 - IL-2	10%	3.5	5.1	0.8
MDF	10%	20.8	13.6	10.5
IL-2 + MDF	10%+10%	67.4	54.7	38.3

[a]Nylon wool nonadherent splenocytes (C57Bl/6) were cultured with DBA/2 Δ-thymocytes in the presence of EL4-IL2 (10% vol. 11.9 U/ml) MDF (10% vol., <.1 U/ml of IL2) or both. Pooled fractions from phenyl Sepharose column of EL4 supernatant fluid that contained IL-2 activity. MDF refers to supernatant fluid from cultures of activated MØ, primed in vivo to L. monocytogenes and restimulated in vitro with heat-killed organism. After 5 days of culture, responder cells were assayed for cytotoxic activity against P815.

[b]Not done.

These results suggest that a MØ derived factor distinct from IL-1 and IFN provides an induction/differentiation signal to CTLs which synergizes with IL-2 to produce a strong cytotoxic response. The ability of MØ factors to act directly on CTLs is supported by the work of Lutz, et al. who showed that MØ or their products in conjunction with IL-2 were necessary for a cloned T cell line to express cytolytic activity (16). We also provided evidence that a T cell derived factor which is associated with CSF activity is involved in CTL development and may stimulate MØ to produce monokines that provide additional CTL helper signals. Further studies are required to delineate the biochemical nature of these factors and to clarify the role they play in CTL development.

ACKNOWLEDGEMENTS

We thank Dr. Mike Caulfield for critical evaluation of the manuscript as well as Katrina Sturkie and Tom Oluic for their expert technical assistance. We also thank Catherine Reinhard for secretarial assistance.

REFERENCES

1. Gillis, S. and Smith, K. A., *Nature*. 268, 154. (1977).
2. Larsson, E. L., Iscove, N. N. and Coutinho, A., *Nature*. 283, 664. (1980).
3. Farrar, W., Johnson, H. M. and Farrar, J. J., *J. Immunol.* 126, 1120. (1981).
4. Finke, J. H., Sharma, S. D. and Scott, J. W., *J. Immunol.* 127, 125. (1981).
5. Wagner, H., Hardt, C., Stockinger, H., Pfizenmaier, K., Bartlett, R. and Rollinghoff, M., *Immunol. Rev.* 58, 95. (1981).
6. MacDonald, H. R. and Lees, R. K., *J. Immunol.* 124, 1308. (1980).
7. Reddehuse, M., Svessmuth, W., Moyers, G., Falk, W. and Dorege, W., *J. Immunol.* 128, 61. (1982).
8. Zuberi, R. I. and Altman, A., *J. Immunol.* 128, 817. (1982).
9. Raulet, D. H. and Bevan, M. J., *Nature*. 296, 754. (1982).
10. Wagner, H., Hardt, C., Rouse, B. T., Rollinghoff, M., Schevrich, P. and Pfizenmaier, K., *J. Exp. Med.* 155, 187 (1982).
11. Moore, R. N., Hoffeld, J. T., Farrar, J. J., Mergenhager, S. E., Oppenheim, J. J. and Shadduck, R. K., *Lymphokines*. 3, 119. (1981).

12. Hilfiker, M. L., Moore, R. N. and Farrar, J. J., *J. Immunol.* 127, 1983. (1981).
13. Ihle, J. N., Rebar, L., Keller, J., Lee, J. and Hapel, A. J., *Immunol. Rev.* 63, 5. (1982).
14. Moore, R. N., Oppenheim, J. J., Farrar, J. J., Carter, C. S., Waheed, A. and Shadduck, R., *J. Immunol.* 125, 1302 (1980).
15. Hsu, L. Finke, J. H., Scott, J. W. and Proffitt, M. R., *J. Inf. Res.* 1, 271 (1981).
16. Lutz, C. T., Glasebrook, A. L. and Fitch, F. W., *J. Immunol.* 126, 1404. (1981).

DISCUSSION

GRIMM: Can CSF be absorbed by IL-2-dependent T cells?

FINKE: We found that the helper factor from activated T cells was not removed by absorption with CT6 cells which had been shown by Hilfiker and Farrar to absorb IL-2 activity.

FARRAR: The resistance to pH2 of the macrophage differentiation factor may rule out γIFN, but does it rule out the possibility of macrophage type 1 IFN?

FINKE: We have added anti-IFN-α/β antibody to this bioassay and did not discern any reduction in the cytotoxic response generated by the macrophage factor. This antibody was obtained from Dunnick, and is directed against L cell interferon.

PLUZNIK: Did Finke try to remove CSF from his preparations by absorbing with spleen or with bone marrow cells?

FINKE: We have not done such experiments.

VADAS: Did Finke's EL-4 produce biological activity constitutively? What was the morphology of the clones elicited by his EL-4 supernatant?

FINKE: The helper activity was not produced constitutively; EL-4 cells had to be stimulated with PMA. The CSF from EL-4 induced both macrophage and granulocyte colonies from bone marrow cells.

WEBB: Did Finke utilize purified IL-1 in his helper assay?

FINKE: No, we did not. But we did use fractionated material from activated macrophages. This product had an approximate molecular weight of 18,000 daltons, it possessed IL-1 activity but did not generate any cytotoxic responses in our bioassay.

MIZEL: Since Robert Moore et al. have found that CSF induces IL-1 production, is it possible that CSF is active in Finke's system via the *induction* of IL-1? Furthermore, it is known that there are synergistic effects between IL-1 and IL-2 in a variety of immune responses. If the CSF-induced product is IL-1 then Finke's data are certainly consistent with previous studies on the mediators involved in the generation of cytotoxic T cells.

FINKE: The macrophage factor described has an approximate molecular weight of 38,000 daltons and does not appear to be IL-1. If it were IL-1 it would have to be doing more than merely inducing IL-2 production, because IL-2 alone does not work in our bioassay.

DISSECTION OF THE PROLIFERATIVE AND DIFFERENTIATIVE SIGNALS CONTROLLING MURINE CTL RESPONSES

Peter Scheurich
Conny Hardt
Klaus Pfizenmaier
Martin Röllinghoff
Hermann Wagner

Institut für Medizinische Mikrobiologie
Johannes Gutenberg-Universität Mainz
Mainz, Federal Republic of Germany

I. INTRODUCTION

It is widely accepted that distinct cell types are involved in the development of a primary cytotoxic T cell response in vitro (1-3). Both direct cell to cell contact as well as the transmission of signals via soluble mediators may occur. To explain primary activation, growth and differentiation of cytotoxic T cells (CTL) the interleukin 2 model has been proposed as minimal scheme (4). Accordingly, interaction of antigen with receptors of clonally committed CTLp induces sensitivity to interleukin 2 (Il-2). Il-2 as produced by helper T cells in turn drives the activated CTLp into proliferation and differentiation into CTL (5).

To date most conclusions on T-T cell interactions are based on experiments using crude or semipurified supernatants of activated T cells which may contain an array of lymphokines and other biological active substances. The Il-2 content of such supernatants usually is determined by measuring its ability to keep cytotoxic T lymphocyte lines (CTLL) in continuous proliferation. However, by using refined sources of Il-2 it soon became apparent that there existed a poor correlation between the Il-2 content of the factors used, as measured on CTLL, and their ability to provide help for the in vitro development of primary CTL responses. In this communication we provide arguments that the interleukin 2 model has to be extended

ISBN 0-12-527540-4

to include a cytotoxic T cell differentiation factor (CTDF) and a soluble mediator inducing sensitivity for Il-2.

II. MATERIALS AND METHODS

Animals: C57BL/6 and CBA/Ca mice were obtained from OLAC Ltd. Shaw's farm, Blackthorn, England. Wistar rats were bred in our own colony.

Il-2 batches: The Il-2 producing EL4 variant (6) was kindly provided by Dr. Farrar (NIH). The Il-2 producing T cell lymphoma LBRM-33 (7) was a kind gift of Dr.G.Cudkowicz, Buffalo, N.Y. The preparation of semipurified Il-2 from these sources has been described (8). Briefly, the EL4 cells were stimulated with 10 ng/ml PMA for 36 h, while the Il-2 batches from the clone LBRM-33 were obtained after stimulation with 1 % PHA. Mouse and rat factors were obtained by stimulation with Con A (1 µg/ml), under serum free conditions. The thymoma derived supernatants were enriched in Il-2 and freed from mitogenic activity by either $(NH_4)_2SO_4$ precipitation followed by an Sephadex G-100 chromatography or, alternatively, by chromatography over an column filled with Phenyl-Sepharose followed by an ion exchange chromatography on DEAE-Sephacel.

Assay for Il-2: Il-2 activity was quantitated by its ability to sustain in vitro growth of Il-2 addicted, cloned Cl-3 cells (kindly provided by Dr.Th. Hünig, Institut für Virologie, Würzburg, FRG).

Selective depletion of Il-2 from crude supernatant: Cl-3 cells were washed free of growth medium and incubated in fresh medium for 2 h at 37°C. Three times washed cells were suspended at a density of 5×10^7/ml in crude supernatants at 4°C for 30 min, then pelleted, and reincubated at 37°C for 2 h in fresh medium. The cells were used then for absorption a further 3 to 5 times, after which the supernatant was tested for Il-2 content.

Preparation of T cell subpopulations: Negatively selected Lyt 1^-23^+ cells depleted for macrophages were prepared by passage of the spleen cells over nylon-wool and Sephadex G-10 followed by treatment with a monoclonal anti Lyt 1.1 antibody (NEN-Chemicals, Dreieich, FRG) and complement. Positively selected Lyt 2^+ cells were produced by panning the cells on petri dishes coated with goat anti mouse IgG antibodies (9,10).

Polyclonal activation of CTLp: Replicates of responder cells were seeded at various numbers into microplates containing

200 µl of medium supplemented with 2.5 µg Con A/ml. The proliferative response was assayed by pulsing the cultures for 6 h with ^{3}H-Tdr. The induction of CTL was measured in a lectin mediated readout (10 µg PHA/ml, or 20 µg Con A/ml) using 10^3 tumor cells as targets.

III. RESULTS AND DISCUSSION

In a first approach negatively selected Lyt 1^-23^+ cells depleted for macrophages were used as source of CTLp. Four different batches of Il-2 with identical activity on Cl-2 cells were compared in their ability to induce proliferation as well as differentiation of the CTLp into CTL. As shown in Table I in the presence of Con A these factors caused comparative proliferative responses independent whether or not Cl-3 cell absorbed Con A supernatant devoid of Il-2 was added. Note that only the Con A derived supernatant was able to induce alone both proliferative responses as well as high CTL responses in the lectin dependent assay system (Table I). However, complementation of semipurified Il-2 batches plus Il-2 deficient Con A supernatant allowed both proliferative and CTL responses to occur. Thus, under the conditions used the over-all response of negatively selected Lyt 1^-23^+ spleen cells could be dissected into an Il-2 driven proliferative response as opposed to a CTDF controlled differentiative response.

Table I. Proliferation and cytolytic response of polyclonally activated Lyt 1^-2^+ T responder cells

10^5 Lyt 1^-23^+ responders: Supernatant added — Il-2 (10 % v/v)	Absorbed SN (5 % v/v)	^{3}H-Tdr incorp. cpm (x 10^3)	% Lysis Effector : target 50 : 1
-	-	3.9 ± 0.3	0
-	+	5.3 ± 0.2	18
crude Con A SN	-	16.2 ± 0.6	68
LBRM (semipurified)	-	13.5 ± 0.3	8
LBRM (semipurified)	+	10.7 ± 0.3	42
EL4 (semipurified)	-	18.1 ± 0.6	14
EL4 (semipurified)	+	17.9 ± 1.7	70
Con A SN (semipurified)	-	10.8 ± 0.4	12
Con A SN (semipurified)	+	10.4 ± 0.6	64
-	Il-1	7.2 ± 0.7	9

In a second approach positively selected Lyt 2^+ cells (i.e. Lyt 123^+ and Lyt 1^-23^+ cells) of CBA mice were set up to respond towards Con A. After three days in the presence of different factors the microcultures were split to measure proliferation as well as lytic activity. As shown in Table II no positive results were obtained in the absence of factors. Addition of a rat Con A supernatant derived semipurified Il-2 resulted in proliferation, but low lytic activity. The lytic activity but not the proliferative response could be amplified by the addition of a mouse spleen cell derived Con A supernatant with negligable Il-2 content (about 5 % of the rat factor). Again, these results support our contention that factor(s) other than Il-2 are required for differentiation of proliferating CTLp into CTL.

Table II. Enhancement of lytic activity by a mediator distinct of Il-2

10^4 Lyt 2^+ responders supernatants added	^{3}H-Tdr incorporation (cpm x 10^3)	% Lysis
-	0.6	0
semipurified Il-2 (8 %)	27.3	19.5
" plus 3 % SN II[a]	23.4	24.5
" plus 10 % SN II	26.5	30.4
" plus 25 % SN II	21.3	55.4

[a] *Mouse spleen derived Con A supernatant SN II is almost devoid of Il-2 (<5 units/ml) which alone neither induce specific lysis nor ^{3}H-Tdr incorporation. 10 units/ml of Il-2 results in 30 % maximal proliferation of the CTLL Cl-3.*

In a third set of experiments the number of positively selected Lyt 2^+ responder cells was decreased to 200-500 cells per microculture. As shown in Table III in the presence of Con A optimal concentrations of semipurified Il-2 from the thymoma EL4 was unable to even induce cell proliferation. If, however, unselected spleen cells had been preactivated for 12 h in a bulk culture with 2.5 μg Con A/ml (10^7 cells/ml), prior to the positive selection procedure, selected Lyt 2^+ cells were now able to proliferate in the presence of semipurified EL4-derived Il-2. Note that in the "deficient" microcultures containing non-preactivated, positively selected Lyt 2^+ cells plus semipurified Il-2 both the proliferative response and the development of CTL could be reconstituted, provided, a rat spleen cell derived Con A supernatant devoid of Il-2 was added at the initiation of the culture. The reconstituting element in this supernatant was not dialysable and was stable to pH 2 treatment for 24 h, the latter finding arguing against a role of γ-interferon.

Table III. Either preactivation of unselected T cells by Con A or a mediator distinct of Il-2 is necessary to induce Il-2 sensitivity.

350 Lyt 2^+ responders supernatants added[a]	Preactivation in bulk culture	^{3}H-Tdr incorp. (cpm x 10^3)	% Lysis
-	-	0.27	2.8
-	+	0.32	5.5
Il-2	-	0.52	3.3
Il-2	+	5.55	37.2
Il-2 + 3 % SN III	-	3.88	17.3
Il-2 + 3 % SN III	+	6.31	34.0
Il-2 + 5 % SN III	-	2.67	28.5
Il-2 + 5 % SN III	+	5.16	31.5

[a]Il-2 semipurified from the thymoma EL4 was used at 7 units/ml. SN III is a rat Con A supernatant containing <5 units/ml of Il-2. It induces alone neither lytic activity nor ^{3}H-Tdr incorporation (compare Table IV).

Table IV shows the effect of different factors on positively selected Lyt 2^+ cells activated in a bulk culture system. After three days of culture the effectors were purified by a ficoll gradient, counted and tested for their lytic capability in the lectin dependent assay system. As can be seen even under bulk culture conditions large amounts of EL4 thymoma derived Il-2 induced only marginal proliferation. On the other hand good proliferative responses were obtained after addition of Con A derived supernatant almost devoid of Il-2.

Table IV: Il-2 sensitivity is induced by soluble mediator(s)

$4x10^5$ Lyt 2^+ responders supernatant added[a]	Il-2 content (units/ml)	Cells recovered	% Lysis 100:1	20:1[b]	Lytic units
1 % rat Con A SN	1	$< 10^4$	-	-	< 1
10 % rat Con A SN	10	5.4×10^5	31.2	12.1	5.4
5 % SN III	< 0.3	$< 10^4$	-	-	< 1
10 % Il-2	50	1.1×10^5	-	16.7	~ 1
10 % Il-2 + 5% SNIII	50	1.2×10^6	65.0	37.4	>60

[a] *Rat Con A SN is a crude supernatant. For further details see Table III.*

[b] *effector : target ratio*

In conclusion the data presented provide strong evidence that Il-2 per se is insufficient to drive a mitogen exposed CTLp into proliferation and subsequent differentiation into CTL. The role of Il-2 appears to be confined to induce cell proliferation of Il-2 sensitive CTLp. In contrast to previous views (4,11,12) Il-2 sensitivity appears not to be induced by contact of CTLp with mitogen or antigen per se but requires an additional soluble mediator distinct from Il-2 which is also contained in culture supernatants of Con A induced spleen cells. Our results further argue for the existence of a third class of factor(s) involved in the primary activation of CTLp which we have termed CTDF(8). This factor is involved in the differentiation of preactivated proliferating CTLp into functional CTL. On the basis of experiments presented here and elsewhere (8, 13 - 16) we have summarized in Fig. 1 our present view on the role of the various signals involved in the primary activation of CTL. According to this scheme an "inducing factor" is essential to render a mitogen/antigen binding CTLp sensitive to Il-2. Subsequently Il-2 drives such CTLp into cell proliferation (clonal growth). Finally differentiation into CTL is controlled by a "differentiation factor", CTDF.

Fig. 1: Minimal scheme of primary CTL activation

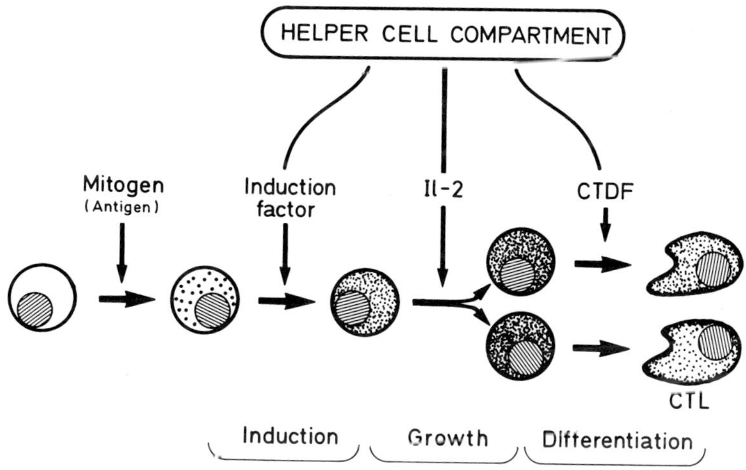

REFERENCES

1. Bach,F.H. et al., *Science 180*, 403 (1973).
2. Cantor,H., and Simpson,E.,*Eur.J.Immunol. 5*, 330 (1975).
3. Pilarski,L.M., *J. exp. Med. 145*, 709 (1977).
4. Wagner,H. et al., *Immunological Rev. 51*,215 (1980).
5. Wagner,H., and Röllinghoff,M.,*J.exp.Med.148*,1523 (1978).
6. Farrar,J.J. et al., *J. Immunol. 125*, 2555 (1980).
7. Gillis, S. et al., *J. Immunol. 125*, 2570 (1980).
8. Wagner, H. et al., *J. exp. Med. 155*, 1876 (1982).
9. Mage, M. et al., *Eur. J. Immunol. 11*, 228 (1981).
10. Krönke, M. et al., *J. exp. Med. 156*,41 (1982).
11. Larsson, E.-L. et al., *Nature 283*, 664 (1980).
12. Smith, K.A., and Ruscetti,F.W.,*Adv.Immunol.31*, 137 (1981).
13. Finke, J.H. et al., *J. Immunol. 127*,2354 (1981).
14. Reddehase, M. et al., *J. Immunol. 128*, 61 (1982).
15. Raulet, D.H., and Bevan, M.J., *Nature 296*, 754 (1982).
16. Conzelmann, A. et al., *Nature 298*, 170 (1982).

DISCUSSION

DIAMANTSTEIN: We have recently found that IL-2 receptor generation (IL-2-R) requires at least two signals. The first signal is delivered to the resting T cell by mitogens, such as PHA or Con-A. The second signal is delivered to these "pre-activated" cells by a factor we designate IL-2R, an "inducing" factor which is not IL-1, not CSF, not interferon and different from IL-2. This factor, IL-2-RiF, is produced by T-cells and induces IL-2 receptors by itself or together with other known lymphokines, e.g. interferon, when added to antigen-or mitogen-pre-activated T-cells.

MANNEL: Did Pfizenmaier do kinetic studies, in order to determine whether one needs the differentiation factor from the beginning of the cell culture and during the entire culture period? I ask this as we reported earlier during this workshop that for T cell activation, two more mediators - both different from IL-2 - are necessary. One of the two mediators can be added at a later stage of culture, as late as day 3 of a 5 day culture period.

PFIZENMAIER: Yes, we know that the "inducing factor", which we assume causes IL-2 receptor expression on CTL-P, has to be provided in the culture from the beginning. We further assume that the "differentiation factor" can act also at a later point in time during culture. To test this, we are currently separating these two biological activities, both of which are present in IL-2 deficient Con-A supernatants.

GARLAND: Were the Con-A supernatants alluded to by Pfizenmaier of syngeneic origin?

PFZIENMAIER: The Con-A supernatants were *not* from syngeneic cells.

GARLAND: Is there any relationship of the differentiation factors to a peripheral phase of restriction?

PFIZENMAIER: There was no evidence of restriction.

OPPENHEIM: We are being overwhelmed by the number of factors being proposed. I cannot help wondering whether some of Pfizenmaier's observations can be attributed to known cytokines. Can he rule out the possibilities that, as proposed by Dutton, IL-1 may have induced the IL-2 receptor or, as documented by Farrar, that interferon could promote the expression of the IL-2 receptor? Is it not conceivable that one of these is actually the "induction" molecule he proposes?

PFIZENMAIER: Admixing an IL-1 preparation with our IL-2 has not thus far resulted in reconstitution either of proliferative or cytotoxic activity.

OPPENHEIM: Since it has been observed by Farrar and coworkers that interferon can promote the effector functions of cytotoxic T lymphocytes, can Pfizenmaier really rule out the possibility that this CTDF activity could be due to interferon?

PFIZENMAIER: The biological activity detected in IL-2 deficient Con-A supernatants appears to be resistant to treatment at pH-2 for 24 hours, arguing against it being IFN.

FALK: When Pfizenmaier was testing selected cells which contained 2/3 CTL precursor cells why did he have to wait 48 hours for cytotoxicity to be manifested - assuming that he was dealing with a differentiation factor?

PFIZENMAIER: The most likely explanation is, that because of the low number of responding cells, we need clonal expansion in order to get significant levels of cytotoxic activity. Lytic activity is low after the first day of culture.

INDUCTION OF T CELL CYTOTOXICITY IS REGULATED BY A MINIMUM OF 3 DIFFERENT LYMPHOKINES

Werner Falk
Daniela N. Männel

Institut für Immunologie und Genetik
Deutsches Krebsforschungszentrum
Heidelberg, F.R.G.

The induction of cytotoxic T lymphocytes (CTL) requires the interaction of different cell types (1). This requirement can be bypassed by factors which are present in supernatants of ConA-stimulated spleen cell cultures (2). The involvement of IL-1 and IL-2 has been shown (3). Recently, participation of differentiation factors has been described (4,5,6). In this report, we present evidence for the necessity of 2 factors different from IL-2 for induction of cytotoxicity.

In microcultures, spleen cells or nylon-wool purified T cells show cytotoxic responses after 5 days of culture with UV-killed TNP-modified syngeneic spleen cells only when ConA supernatants were added. However, IL-2 preparations which were obtained from phorbol myristic acid stimulated EL4 thymoma cells after ammonium sulfate precipitation, Phenylsepharose- and Sephadex G100 chromatography (7) failed to induce any significant response even at IL-2 titers which were 10-fold higher than in a standard ConA supernatant. No inhibitory effect was found when this IL-2 preparation was added to a standard ConA supernatant, indicating that there must be factors present in ConA supernatants different from IL-2 which are required for induction of cytotoxicity. ConA supernatants devoid of IL-2 were prepared by absorption with an IL-2 dependent T cell line. These preparations did not inhibit the response to a standard ConA supernatant. However, as shown in Table I a strong synergistic effect was observed when the IL-2 preparation and the IL-2 free supernatant were mixed. This synergism was seen with spleen cells and with purified T cells. The factor present in the absorbed supernatants was termed T cell cytotoxicity factor (TCF-1).

ISBN 0-12-527540-4

TABLE I. Synergistic Effect of IL-2 and TCF-1 Containing Preparations

Responder Cells	Lymphokine (ul/well) IL-2	TCF-1	Cytotoxicity % spec. ^{51}Cr-release
Nylon wool	60[a]	0	0
purified	0	60	0
T-cells	60	20	5
10^5	60	40	31
	60	60	39
standard supernatant (80 U/ml)			25

[a]IL-2 preparation with 950 U/ml

To test the pH-stability of this lymphokine, a standard supernatant was kept overnight in pH 2 buffer and the cytotoxic response was compared with the same preparation treated at pH 7. The IL-2 titer was unchanged after this procedure but induction of cytotoxic activity of purified T cells was reduced by about 75% (Table II). Also, this pH 2-treated lymphokine did not inhibit the response to a normal supernatant when it was added to it. If the pH 2-sensitive material were TCF-1, it should be possible to show a synergism between TCF-1 and the pH 2-treated preparation. Surprisingly, a synergism between these 2 preparations was not detected (Table III). From these observations, namely a) no change in IL-2 titer but strongly reduced cytotoxicity after pH 2-treatment and b) no enhancing effect when TCF-1 was added to the pH 2-treated material, one has to conclude that there must be a third, pH 2-sensitive factor in ConA supernatants which is necessary

TABLE II. Loss of Cytotoxic Activity after pH 2-treatment

ul of lymphokine per well	% specific ^{51}Cr-release pH 7-treated 108 U/ml of IL-2	pH 2-treated 108 U/ml of IL-2
0	1	0
20	13	1.5
40	36.5	6
80	42.5	18

[a]Nylon wool purified thymus cells were used as responders (10^5/well)

TABLE III. Lack of Synergism between TCF-1 and pH 2-treated Supernatant

ul per well of TCF-1	pH 2-treated	% specific ^{51}Cr-release[a]
40	0	0
0	20	2
0	40	18
20	20	4
20	40	17

[a]Responders: 10^5 nylon-wool purified thymus cells

for induction of cytotoxicity. This factor was termed TCF-2. The IL-2 preparation used in these studies was found to be a convenient source for TCF-2. As shown in Table IV a strong synergism was found when the IL-2 preparation and the pH 2-treated lymphokine were mixed.

Combining these data, one can draw the following conclusion: a) the IL-2 preparation contains IL-2 and TCF-2. b) The absorbed preparation contains TCF-1. c) The pH 2-treated preparation contains IL-2 and TCF-1. Only when all 3 factors were present a cytotoxic response was measured.

To show that TCF-1 was indeed not pH 2-sensitive, as has to be concluded from the above data, a TCF-1 preparation was subjected to pH 2-treatment overnight and tested for synergism with the IL-2 preparation. Table V shows that the pH 2-treatment had no influence on the synergistic effect.

TABLE IV. Demonstration of Synergism between IL-2 and pH 2-treated Supernatant

ul per well of IL-2 (950 U/ml)	pH 2-treated	% specific ^{51}Cr-release (10^5 responders)
60	0	4
0	40	9
0	60	25
20	20	46
20	40	48
20	60	50

TABLE V. Resistance of the Synergistic Effect of TCF-1 to pH 2-treatment

ul of lymphokines per well			% specific ^{51}Cr-release
IL-2 (950 U/ml)	TCF-1	TCF-1 pH 2	
60	0	0	0
0	60	0	0
0	0	60	0
20	60	0	28
20	0	60	24

Thus, T cell activation for cytotoxicity seems to be a highly complex process involving a variety of different lymphokines. The isolation, biochemistry and biological role of these two new lymphokines with respect to their influence on growth and differentiation of CTL-precursors remains to be elucidated.

REFERENCES

1. Larsson, W.L., Coutinho, A., and Martinez, A.C., Immunol. Rev. 51, 61 (1980).
2. Farrar, W.L., Johnson, H.M., and Farrar, J.J., J. Immunol. 126, 1120 (1981).
3. Farrar, J.J., Mizel, S.B., Fuller-Farrar, J., Farrar, W.L., and Hilfiker, M.L., J. Immunol. 125, 793 (1980).
4. Reddehase, M., Suessmuth, W., Moyers, C., Falk, W., and Droege, W., J. Immunol. 128, 61 (1982).
5. Raulet, D.H., and Bevan, M.J., Nature 296, 754 (1982).
6. Wagner, H., Hardt, C., Rouse, B.T., Röllinghoff, M., Scheurich, P., and Pfizenmaier, K., J. Exp. Med. 155, 1826 (1982).
7. Hilfiker, M.L., Moore, R.N., and Farrar, J.J., J. Immunol. 127, 1983 (1981).

INTERLEUKIN-3 REGULATION OF A LINEAGE OF LYMPHOCYTES EXPRESSING 20 ALPHA HYDROXYSTEROID DEHYDROGENASE

James N. Ihle,[1] Jonathan Keller, and Edmund Palaszynski

Biological Carcinogenesis Program
NCI-Frederick Cancer Research Facility
Frederick, Maryland 21701

The enzyme 20 alpha hydroxysteroid dehydrogenase (20αSDH) was initially demonstrated to be associated with a mature T cell subpopulation in hematopoietic tissues (1-3). In particular, 20αSDH activity is associated with a Thy 1+ splenic lymphocyte population and with the glucocorticoid resistant medullary thymocyte population. This enzyme is not present in the cortical, immunologically immature, terminal transferase (TdT) positive thymus population. In addition, to tissue distribution studies, the observation that splenic lymphocytes from young nu/nu mice have extremely low levels of 20αSDH relative to normal splenic lymphocytes suggested that 20αSDH is restricted to a T cell related lineage. To further explore the lineage specificity of 20αSDH expression, we have examined a variety of characterized cell lines for the enzyme. As demonstrated in Table 1, most of the cell lines had no detectable 20αSDH including B cell lines, erythroleukemias, macrophages or macrophage lines, fibroblasts or a melanoma line. Of the T cell lines examined, TdT positive lymphoma lines have not been found to express 20αSDH. In contrast a cloned, IL 2 dependent cytotoxic T cell line had readily detectable levels of 20αSDH. These results, therefore, further demonstrate that the expression of 20αSDH is highly restricted and may be uniquely associated with a T cell lineage.

To initially investigate the regulation of the lineage characterized by the expression of 20αSDH, we examined the ability of a variety of sources of factors for the presence of a factor capable of inducing the differentiation of 20αSDH positive cells in cultures of nu/nu splenic lymphocytes. These studies demonstrated that conditioned media from mitogen or antigen-activated T cells had a readily detectable factor capable of inducing 15- to 20-fold increases in 20αSDH levels

[1]Research sponsored by the National Cancer Institute under contract no. 75380 with Litton Bionetics, Inc.

ISBN 0-12-527540-4

Table 1. Expression of 20αSDH in Various Cell Lines

Cell Line	Characteristics	pmoles/hr/10^8 cells
NP-3	B-cell lymphoma	<50
5F4-B	B-cell lymphoma	<50
GM86	Erythroleukemia	<50
RL-12	TdT^+ lymphoma	<50
PEM	Normal peritoneal macrophages	<50
PU5-1R	Macrophage tumor line	<50
RAW 264	Macrophage tumor line	<50
3T3	Fibroblast	<50
F10	Melanoma	<50
CT6	IL 2 dependent cytotoxic T cell line	2070

over a 24 hr culture period. Because the factor responsible for 20αSDH induction was biochemically distinct from interleukin 1 (IL 1) or interleukin 2 (IL 2), the term interleukin 3 (IL 3) was proposed. IL 3 is predominantly the product of activated T cells as indicated by a variety of studies including the observation that mitogen stimulation of normal but not nu/nu splenic lymphocytes induces IL 3 (4,5). In addition, IL 3 is constitutively produced by the WEHI-3 cell line at high titers which is the most convenient source of IL 3 for large scale purification (6). The production of IL 3 by the WEHI-3 line was not anticipated since this line has been classified as a myelomonocytic and not a T cell line. However, in examining a variety of additional myelomonocytic and macrophage lines, it is clear that WEHI-3 is unique in its ability to produce IL 3 as well as the expression of the Thy 1 antigen.

Using the WEHI-3 cell line as a source of IL 3, a purification protocol has been developed for the routine purification of IL 3 to homogeneity (7). Generally starting volumes of 150 liters of conditioned media are used which will yield from 10-100 μg of homogeneous IL 3. The purification scheme involves ammonium sulfate fractionation, DEAE cellulose columns, hydroxylapatite chromatography, G-75 Sephadex sizing columns and chromatography on C18 hydrophobic support resins as a final step in purification. The fold purification required to obtain homogeneous IL 3 is generally 5.0×10^5 - 2.0×10^6. Homogeneous IL 3 has a specific activity of approximately 0.05

ng/ml for 50% of maximal induction of 20αSDH. Multiple forms of IL 3 can be isolated which vary in molecular weight, but the major component of WEHI-3 CM has an apparent molecular weight of 28,000 daltons by SDS-PAGE analysis. The amino acid composition of IL 3 indicates a relatively large percentage of acidic residues. In addition, there is approximately 12 μg of glucosamine per 100 μg of IL 3 demonstrating the existence of a substantial carbohydrate moiety. The pI of IL 3 is heterogeneous in the acidic pH range apparently due to the carbohydrate. The amino terminal sequence through the first 29 residues has been established and a possible glycosylation site at residue 10 has been identified (Ihle, Henderson and Oroszlan, unpublished data). Homogeneous IL 3 can be iodinated with chloramine T to high specific activities without loss of biological activity (6) and can be fluoresceinated. In both cases derivatized IL 3 can be used to characterize cell surface receptors as described below.

The biological activities associated with homogeneous IL 3 have been examined and found to include a number of previously defined "factor" activities. As initially defined, IL 3 is a factor capable of inducing 20αSDH in cultures of nu/nu splenic lymphocytes. In addition, IL 3 is equivalent to WEHI-3 growth factor, a special form of colony stimulating factor, mast cell growth factor, P-cell stimulating factor, Thy 1 inducing factor and NC (natural cytotoxicity) cell growth factor. Some of the properties of each of these activities are described below.

The WEHI-3 growth factor was initially defined as a factor in WEHI-3-conditioned media required for the establishment and maintenance of continuous cell lines from cultures of bone marrow cells (7). The properties of these cell lines have been examined and all the lines uniquely require IL 3 for viability and proliferation (8,9). This requirement provides a simple, highly specific assay for IL 3 which distinguishes IL 3 from a number of potential mitogenic factors in Con A-conditioned media. All the IL 3 dependent lines are lymphoblastoid in appearance and are characterized by the expression of varying levels of 20αSDH activity. Using iodinated homogeneous IL 3 it also has been possible to demonstrate the unique presence of receptors for IL 3 on the cells. A typical analysis of one of the cell lines is demonstrated in Figure 1 which shows the existence of a saturable receptor with an apparent Kd of 5.4×10^{-11} M and a receptor density of approximately 5000/cell. In spite of the common presence of 20αSDH, IL 3 receptors and an absolute requirement for IL 3 for growth, the cell surface phenotypes of the cells varies considerably. Most cell lines express Ly 5 and H-11 and several express Thy 1. Although no truely constant pattern of expression is seen, the distribution suggests a sequential acquisition of markers from none to Ly 5^+, H-11^+ and then to Thy 1^+, Ly 5^+, H-11^+ (8).

In collaboration with Drs. M. Prystowsky and E. Goldwasser (University of Chicago), homogeneous IL 3 has also been shown to have activity in colony stimulation factor (CSF) assays. The characteristics of the colonies has not been extensively examined, although based on the results described below, this activity is probably related to the proliferation and differentiation of the mast cell-like lineage. Although IL 3 has CSF activity it does not constitute the major CSF activity associated with conditioned media from mitogen-

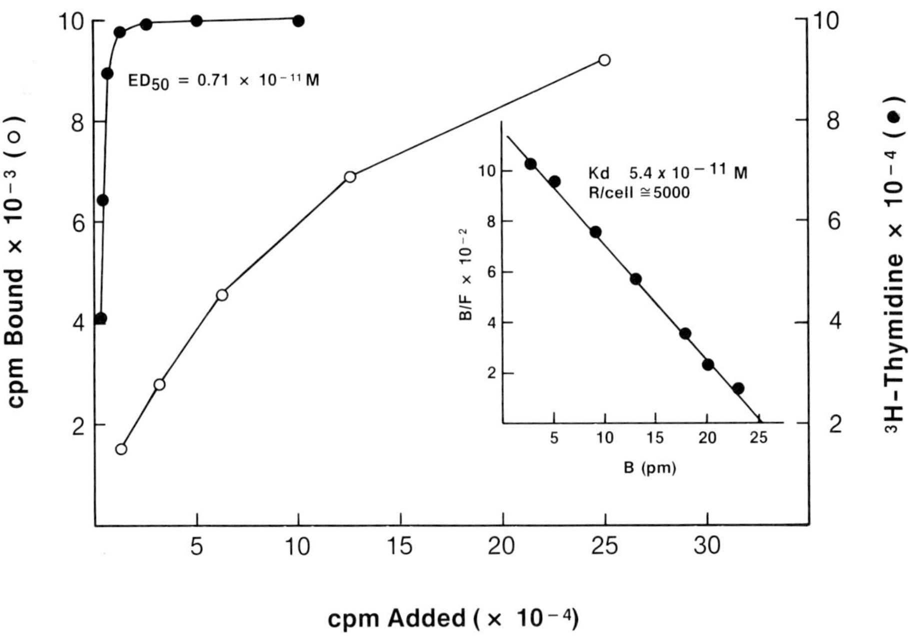

Figure 1. Characterization of IL 3 receptors on the 32D cl-23 line. Homogenous IL 3 was iodinated with chloramine T and ^{125}I. The titration of biological activity of the iodinated preparation (●——●) was determined in a standard proliferation assay. The binding curve (O——O) is the specific binding derived from subtracting background binding in the presence of excess unlabeled IL 3 from the total binding. These data are plotted according to Scatchard in the insert.

stimulated lymphocytes. In particular, DEAE cellulose chromatography resolves the IL 3, which elutes in the run through, from the major CSF activity which elutes at 0.1 M NaCL. The major CSF component does not induce 20αSDH nor does it support the growth of IL 3 dependent cell lines and, therefore, clearly represents a distinct factor. Since the main CSF activity is thought to be a factor regulating myeloid differentiation, the implication is that the IL 3 regulated lineage is clearly distinct from the myeloid series.

The presence of a factor in conditioned media from mitogen activated lymphocytes which is capable of promoting the *in vitro* growth of mast cells or a related cell, the P-cell, has been described by several investigators (10-13). Homogeneous IL 3 has been shown to function as a mast cell growth factor or P-cell stimulating factor. The IL 3 effect on mast cells *in vitro* is more complex, however, than simply a growth factor. In particular, we have analyzed the sequence of events which occur when splenic lymphocytes from nu/nu mice are cultured in IL 3. This type of experiment is illustrative since nu/nu lymphocytes are blocked in differentiation at a point at which IL 3 is required, thus the sequence observed *in vitro* is more synchronized. Second, nu/nu splenic lymphocytes lack a population of "helper" T cells which potentially could contribute factors which would obscure the effects of the added IL 3. By examining the cell phenotypes with time in such cultures, the sequence of events shown in Figure 2 is observed. The initial event involves the induction of 20αSDH in a cell which is Ly 5^-, H-11^- and Thy 1^-. This cell is characterized by extremely high levels of 20αSDH activity. Subsequently, at 5-7 days in culture, the induced cell becomes Ly 5 and H-11 positive and the majority of the cells express these markers. These cells, however, are still primarily lymphoblastoid in appearance. In addition, there is a consistant increase in Thy 1^+ cells. By 2 weeks an increasingly larger percentage of the population is a heavily granulated cell with the morphology typical of mast cells and which are Thy 1^-, Lyt 1^-,2^-, Ly 5^+, H-11^+. In long term cultures these cells persist and will eventually die off only after 2-4 months while maintaining an absolute requirement for IL 3.

In addition to the above activities, IL 3 is also capable of inducing Thy 1 positive cells in cultures of anti-Thy 1 treated bone marrow or spleen cells or as demonstrated above in cultures of nu/nu splenic lymphocytes. This type of activity is generally associated with a Thy 1 inducing factor.

Last, when cultures of IL 3 grown cell populations have been examined for cytotoxic activity, they have been shown to exhibit an activity normally associated with NC (natural cytotoxic) cells (14). The ability to specifically expand NC cells in vitro indicates an NC cell growth factor activity.

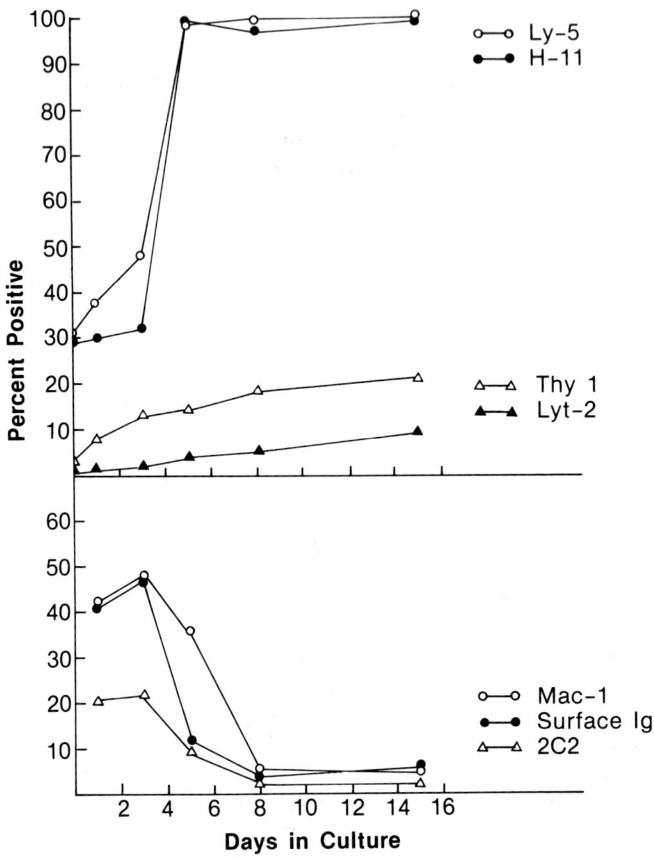

Figure 2. Cell surface phenotypes observed in cultures of nu/nu splenic lymphocytes in IL 3. Splenic lymphocytes were obtained from nu/nu mice and partially purified by passage through a nylon wool column. The cells were subsequently cultured with IL 3 and the percentage of the population expressing the indicated phenotypes established by standard techniques.

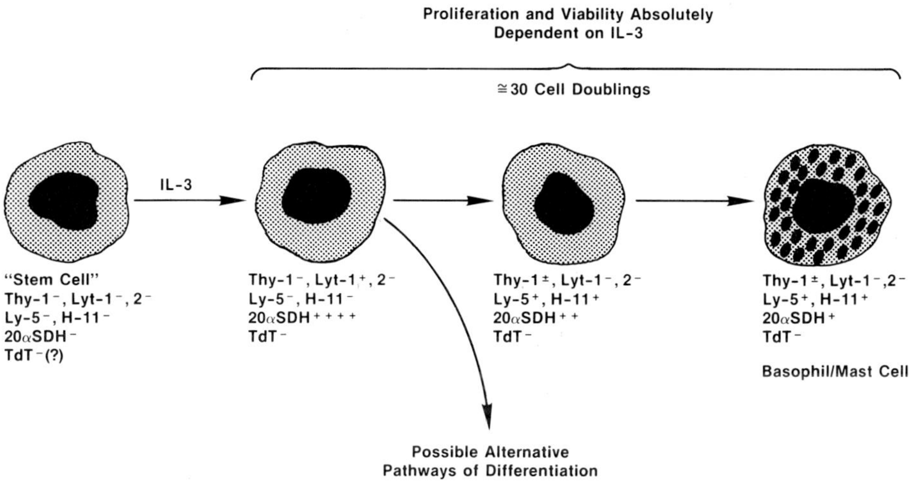

Figure 3. Proposed sequence of differentiation regulated by IL 3.

The relationship between these various factor activities can best be reconciled by proposing a cell lineage with the sequence of differentiation indicated in Figure 3. This lineage is in part characterized by the presence of the enzyme 20αSDH but perhaps more characteristically is a lineage uniquely expressing receptors for IL 3 and dependent on IL 3 for growth. This lineage has a functional "end product" cells with the properties of mast cells. However, whether all "mast cell type" lineages are regulated by IL 3 is unresolved. We would further propose that the IL 3 dependent cell lines are "transformed" intermediates in this lineage and are unable to differentiate to the mast cell phenotype. Last, whether the mast cell-like phenotype is the sole possible progeny of this lineage is unknown and in particular the expression of 20αSDH in cloned cytotoxic T cells raises the possibility that alternative pathways, perhaps unique to the thymus, may exist.

ACKNOWLEDGMENTS

We would like to thank Allan Scott, Debbie Gilbert and Linda Rebar for their expert technical assistance. We also would like to thank Linda Fawley for help in preparation of the manuscript.

REFERENCES

1. Weinstein, Y., Linder, H. R., and Eckstein, B. (1977). Thymus metabolizes progesterone, a possible enzymatic marker for T lymphocytes. Nature 266, 632.

2. Weinstein, Y. (1977). Twenty-alpha-hydroxysteroid dehydrogenase: a T lymphocyte associated enzyme. J. Immunol. 119, 1223.

3. Pepersack, L., Lee, J. C., McEwan, R., and Ihle, J. N. (1980). Phenotypic heterogeneity of Moloney leukemia virus-induced T cell lymphomas. J. Immunol. 124, 279.

4. Ihle, J. N., Pepersack, L., and Rebar, L. (1981). Regulation of T cell differentiation: *In vitro* induction of 20 alpha hydroxysteroid dehydrogenase in splenic lymphocytes is mediated by a unique lymphokine. J. Immunol. 126, 2184.

5. Ihle, J. N., Lee, J. C., and Rebar, L. (1981). T cell recognition of Moloney leukemia virus proteins. III. T cell proliferative responses against gp70 are associated with the production of a lymphokine inducing 20 alpha hydroxysteroid dehydrogenase in splenic lymphocytes. J. Immunol. 127, 2565.

6. Ihle, J. N., Keller, J., Henderson, L., Klein, F., and Palaszynski, E. (1982). Procedures for the purification of interleukin 3 to homogeneity. J. Immunol., submitted.

7. Greenberger, J. S., Newburger, P. E., Karpas, A., and Moloney, W. C. (1978). Constitutive and inducible granulocyte-macrophage functions in mouse, rat and human leukemia-derived continuous tissue culture lines. Cancer Res. 38, 3340.

8. Ihle, J. N., Keller, J., Greenberg, J. S., Henderson, L., Yetter, R. A., and Morse, III, H. C. (1982). Phenotypic characteristics of cell lines requiring interleukein 3 for growth. J. Immunol. In press.

9. Ihle, J. N., Rebar, L., Keller, J., Lee, J.C., and Hapel, A. J. (1982). Interleukin 3: Possible roles in the regulation of lymphocyte differentiation and growth. Immunol. Rev. 63.

10. Nabel, G., Galli, S. J., Dvorak, A. M., Dvorak, H. F., and Cantor, H. (1981). Inducer T lymphocytes synthesize a factor that stimulates proliferation of cloned mast cells. Nature 291, 332.

11. Nagao, K., Yokova, K., and Aaronson, S. A. (1981). Continuous lines of basophil/mast cells derived from normal mouse bone marrow. Science 212, 333.

12. Schrader, J. W., Lewis, S. J., Clark-Lewis, I., and Culvenor, J. G. (1981). The persisting (P) cell: Histamine content, regulation by a T cell-derived factor, origin from a bone marrow precursor, and relationship to mast cells. Proc. Natl. Acad. Sci. USA 78, 323.

13. Yung, Y. P., Eger, R., Tertian, G., and Moore, M. A. S. (1981). Long-termin *in vitro* culture of murine mast cells. II. Purification of a mast cell growth factor and its dissociation from TCGF. J. Immunol. 127, 794.

14. Lanza, E., Hapel, A.J., Ihle, J.N., and Djeu, Y. (1982). Growth of natural cytotoxic (NC) effector cells in interleukin-3. Nature. Submitted.

DISCUSSION

SCHRADER: The best assay for growth factors requires the use of homogeneous (cloned) populations of target cells. We can therefore be sure only then that IL-3 acts directly on P cell/mast cell lines. It will be important to determine whether lines giving rise to megakaryocytes, granulocytes, etc., can also respond. IL-3/PSF is active in assays involving heterogeneous populations of bone marrow cells in producing a variety of myeloid/erythroid colonies and in supporting pleuripotential stem cells and therefore may act on earlier cells not yet committed to the mast cell lineage.

IHLE: No argument with Schrader's dictum. It is however also important to do experiments addressing questions such as the following: do the non-mast cell differentiated cells in colonies possess IL-3 receptors? Will the addition of other factors take stem cells along other pathways?

PLUZNIK: The last protocol presented by Ihle can be interpreted as a process of selection of mast cells with the help of IL-3, not necessarily as a process in which cells change their surface markers. Did Ihle do any of his experiments with cloned cells?

IHLE: In replying to Pluznik I would point out that if one takes cultures that have been in IL-3, clones out these cells and allows the emerging individual clones to grow out, (because only then does one know what proportion are granulocytic and non-granulocytic) and then examines the morphology of those clones, it turns out that a month later-when one finally has sufficient cells to work with-they all prove to be granulocytic.

RELATIONSHIP OF WEHI GROWTH FACTOR (WGF), INTERLEUKIN-3 (IL-3) AND 20α-STEROID DEHYDROGENASE (20αSDH) TO HAEMOPOIETIC PROGENITOR CELLS

John M. Garland
T. Michael Dexter
Michel Lanotte
Catherine Howarth
Andrew Aldridge

Department of Experimental Haematology
Paterson Laboratories
Christie Hospital & Holt Radium Institute
Manchester

INTRODUCTION

The term "IL-3" has been given to a growth factor isolated from mitogen-stimulated lymphocyte conditioned media and supernates from myelomonocytic leukaemic WEHI-3b cell line cultures (1). It promotes a number of cell lines *in vitro*, and induces the enzyme 20αSDH in nu/nu spleen cells (1). Whilst "IL-3" promoted lines have certain characteristics of lymphoid cells (2), we have identified a very similar factor in WEHI-3b supernates (WGF) that promotes myeloid-like lines *in vitro*(3). We have attempted to relate these lines, called FDCP, and WGF to 20αSDH and "IL-3", and to progenitor cells from which the cell lines may arise.

Factor-dependent Cell Lines (FDCP) - Phenotypic Studies

Our original studies reported the isolation of WGF-dependent lines from murine long-term marrow cultures without leukaemic virus infection in contrast to the myeloid lines described by Greenberger *et al* (4). Phenotyping of FDCP showed that nearly all the lines expressed various amounts of Thy 1,

ISBN 0-12-527540-4

and some lines had small numbers of Ly-positive cells (5). Thy antigen expression was independent of both clonogenic potential in agar and Fc receptor expression. In agar supplemented with WGF, significant numbers of immature and mature granulocytes were developed within isolated colonies, and cytochemistry of liquid-culture cells showed the presence of myeloid esterase. The original FDCP were derived from cultures at over eight weeks; other WGF-dependent FDCP have since been derived from murine marrow cultures at earlier times (4 weeks) and these lines are all Thy 1 negative. However, they all express Mac-1 and F4.80 antigens characteristic of myeloid cells (Table 1). FACS sorting has

TABLE I. Phenotypes of FDCP (derived from DBA long-term cultured cells at 4 weeks)*

Sub. no.	*Clone no.*	*Thy 1.2*	*ThB*	*Mac 1*	*F4.80*	*FcR*
A	*1*	*0*	*0*	*9*	*12*	*32*
	2	*0*	*0*	*3*	*17*	*49*
A1	*2*	*0*	*0*	*7*	*9*	*0*
A1.1.1	*2*	*0*	*0*	*16*	*7*	*36*
A2	*1*	*0*	*0*	*10*	*12*	*32*
	2	*0*	*0*	*3*	*11*	*22*

* *Percent fluorescent*

shown that the clonogenic cells are null for both antigens, and therefore their expression appears to be related to differentiation and loss of clonogenic potential.

Response Of FDCP To "IL-3"

All the FDCP lines so far tested have optimally responded to purified "IL-3" supplied by Dr. J. Ihle, as assayed by ^{3}HTdR proliferation. Growth in "IL-3" does not result in any morphological changes, and thus WGF and "IL-3" appear to be very similar if not identical molecules in promoting the cell lines.

Progenitor Cells Responding To "IL-3" And WGF

Autoradiography of nu/nu spleen cultures has shown that purified "IL-3" increases the number of labelled immature myeloid cells and blast cells without any increase in the labelling of lymphoid cells (Table 2). Similar results were observed for a semipurified WGF fraction. Further, "IL-3" increased the number of progenitors responding to G/M-CSF. Thus, both "IL-3" and WGF appear to promote normal myeloid cells and progenitors, although it has not been determined yet that the effect is a direct one. We have also investigated the effect of purified "IL-3" on the proliferation of multipotential stem cells (CFU-S) using ^{3}HTdR suiciding. Incubating normal marrow in "IL-3" for 3 hours in vitro increases firstly the number of spleen colonies developing in a syngeneic recipient and secondly the number of CFU-S killed by 'S' phase cytotoxic doses of ^{3}HTdR *in vitro* (Table 3). As similar effects were seen in fractionated marrow cells devoid of endogenous stimulator producing cells (6) we conclude that the "IL-3" preparation also promotes multipotential stem cells.

TABLE II. Promotion of cell lineages by IL-3 at 48 hrs. in nu/nu marrow and spleen

Population	*% labelled*	*L*	*P-M/MM*	*G*	*BI*
Spleen	*2.3*	*24*	*17*	*0*	*58*
Spleen + IL3	*7.3*	*5*	*46*	*0*	*49*
Marrow	*2.9*	*3*	*13.5*	*0*	*83.5*
Marrow + IL3	*10.4*	*3*	*32*	*0*	*65*

20αSDH Expression In Cell Lines And Marrow Cultures

Because of previous reports that 20αSDH, a T-lineage restricted enzyme in haemopoietic cells (7,8), was expressed in "IL-3" dependent lines (2,9) we investigated FDCP, stromal cell lines and murine marrow cultures for 20αSDH (10). All FDCP tested expressed low levels of 20αSDH irrespective of phenotype, but much greater expression occurred in cloned marrow stromal cells (10). Further, similar levels were found in the adherent cells of recently-initiated long-term marrow cultures (where "IL-3" or WGF was absent) and the lipid products identified by TLC were very similar to those found in the stromal lines (11). The strict confinement of 20αSDH to T-

TABLE III. Effect of IL-3 on Thymidine Suiciding

Experiment no.			IL-3 (0.25%)
1	(BDF_1)	9 ± 6[1]	34 ± 7
2	(nu/nu)[2]	10 ± 5	29 ± 6
3	(BDF_1)	18 ± 7	38 ± 5
4	High dens. BDF_1	26.6 ± 7.2	46.8 ± 9.9

[1] *Figures are given for mean ± S.E.*

[2] *Normal Balb/c recipient*

lineage cells in haemopoietic tissue is thus questionable, and expression may occur in myeloid-lineage cells.

WGF And Factor-dependent Lines In Vivo

One key question is whether WGF or "IL-3" are normal regulators. To determine this, we investigated two situations *in vivo* where the factor might be expected, namely in acute GvH mice (representing a pan-lymphocyte stimulation analogous to *in vitro* mitogen stimulation) and in mice made leukaemic by injection of WEHI-3b cells which are constitutive producers of the factor(s). In neither situation was the factor demonstrable in serum by the standard ^{3}HTdR assay on a dependent line or in sonicates particularly of grossly leukaemic spleens containing some 85% leukaemic cells. Co-culture experiments showed that soluble inhibitors of the factor were not present, nor did normal cells actively suppress production of the factor *in vitro*. Finally, the cells from both GvH and leukaemic mice were capable of generating the factor *in vitro*, and the conclusion was that WGF (and "IL-3") production *in vivo* is either absent or severely curtailed to undetectable levels. Further experiments showed that the half-life of WGF in the serum *in vivo* is extremely short (approx 20 min). Of some interest also is that to date it has not been possible to "rescue" FDCP after *in vivo* injection, although studies using Inoxine have demonstrated that the cells do recirculate for nearly 24 hours after injection. Thus, the *in vivo* relevance of the factor(s) and lines remain unclear.

DISCUSSION

Our studies suggest that WGF promotes myeloid-like cells *in vitro*, as demonstrated by morphology and expression of certain antigens. WGF appears very similar to "IL-3", and the expression of lymphoid antigens appears to be "aberrant". 20αSDH is induced by both "IL-3" and WGF, but it may occur without addition of the growth factors as shown by its expression in stromal lines and early long-term marrow cultures. We have found evidence that "IL-3" and WGF both promote myeloid cells in short-term incubation, but no evidence that lymphoid cells are promoted. Our findings therefore seem to be in direct conflict with those of Ihle *et al.* However, WEHI-3b conditioned medium contains many factors, and at least one promotes mast cells or P-cells (12), and another mixed colonies (13); the question therefore is whether the various activities are due to a single factor or to a number of very similar ones. Our results also suggest that a variety of phenotypes are promoted *in vitro*, and factor-dependency is dominant to phenotype. The fundamental questions therefore are the mechanism by which the factor(s) promotes cell division and whether the factor-promoted cell lines retain function *in vivo*. These are being pursued vigorously.

ACKNOWLEDGMENTS

We thank Stella Crompton for excellent technical assistance. This work was supported by the Cancer Research Campaign.

REFERENCES

1. Ihle, J., Pepersack, C. and Rebar, L. J. Immunol. 126 2184-2189 (1981).
2. Hapel, A., Lee, C.J., Farrar, W.L. and Ihle, J. Cell 25 179-186 (1981).
3. Dexter, T.M., Garland, J.M., Scott, D., Scolnick, E. and Metcalf, D. J. Exp. Med. 152 1036-1047 (1980).
4. Greenberger, J.S., Gans, P.J., Davisson, P.R. and Moloney, W.C. Blood 53 987-993 (1979).
5. Garland, J.M. and Dexter, T.M. *in* "In vivo Immunology" Proceedings 7th Germinal Centre Conference, p.127-136. Plenum Press, New York, (1982).
6. Lord, B.I. Cell Tiss. Kinet. 14 425-431 (1981).
7. Weinstein, Y. J. Immunol. 119 1223-1229 (1977).
8. Weinstein, Y. Thymus 2 305-320 (1981).
9. Ihle, J. Immunology Rev. 63 5-32 (1982).
10. Garland, J. and Dexter, T.M. Eur. J. Immunol. 12 332-336 (1982).

11. Garland, J. and Dexter, T.M. Eur. J. Immunol. 1982 submitted.
12. Schrader, J., Lewis, S.J., Clark-Lewis, I. and Culvenor, J.G. Proc. Nat. Aca. Sci. 78 323-326 (1981).
13. Iscove, N., Roitsch, C.A., Williams, N. and Guilbert, L.J J. Cell Physiol. Supp.1 65-78 (1982).

DISCUSSION

SCHRADER: By stimulating mice with antigens (e.g. skin painting) we have obtained evidence consistent with the *in vivo* production and action of IL-3/PSF. Thus, the frequency of P cell precursors (i.e. mast cells which continue to proliferate *in vitro* in the presence of PSF) in the antigen-activated draining lymph node is greatly increased. Mast cells can be found in increased numbers and, when dissociated lymph node cells are cultured overnight in serum-free medium, they release large amounts of PSF.

GARLAND: *In vivo* the mast cells and P cell precursors are not necessarily being generated by PSF. *In vitro*, PSF/IL-3 could merely be taking over from this other process.

GILLIS: Is there any evidence that 20αSDH is expressed on human hemopoietic cells or is there any evidence to support an IL-3 growth/differentiation system for the human model?

GARLAND: Thus far, we have only found 20aSDH in human bone marrow-derived fibroblast colonies (CFU_f). With regard to human "IL-3", we need a factor-dependent cell line to test supernates or to find a cross-reacting factor. So far, no human conditioned medium-even FDCP, and WEH1-3b conditioned medium - appears to be toxic for human cells.

FACTOR-DEPENDENT DEVELOPMENT OF NORMAL BONE MARROW-DERIVED HUMAN BASOPHILS WITH HISTAMINE-RELEASING ACTIVITY[1]

Kenji Tadokoro
Beda M. Stadler
Alain L. de Weck

Institute of Clinical Immunology
University of Bern
Bern, Switzerland

INTRODUCTION

Basophils have an important role in many immunological reactions mainly through their specific receptors for IgE and their release of chemical mediators including histamine. On the other hand, basophilopoiesis also seems to be influenced by immunological reactions, because the number of basophils is elevated in tissue transplantation and immunological disorders such as atopy and ulcerative colitis. In experimental animals, repeated injections of antigen have increased basophil production (1). Denburg and his co-workers have shown that in guinea pigs, conditioned medium from antigen- as well as lectin-stimulated mononuclear cells enhanced normal basophilopoiesis in vitro (2).

In human, several reports have demonstrated that basophils were produced in vitro without the addition of exogeneous conditioned medium, when culturing bone marrow cells from patients with chronic myelogeneous leukemia (3-5).

[1]*This work was supported in part by the Swiss National Science Foundation. Grant NF 3.165-0.81*

ISBN 0-12-527540-4

Aglietta et al. have shown the production of small number of basophils from normal human bone marrow in semi-solid agar cultures with leukocyte feeder layers (6). Recently, Rasin and his co-workers have reported the spontaneous and selective in vitro growth of basophils from human fetal liver cells (7).

Here we report on the in vitro growth of normal bone marrow-derived basophils, dependent on lectin-stimulated mononuclear cell conditioned medium. Furthermore we show that these cultured basophils resemble morphologically and functionally normal human basophils.

MATERIAL AND METHODS

A. Lectins and other Agents

Lectins were purchased as follows: Concanavalin A (Con A) from Calbiochem (San Diego, USA), phytohemagglutinin (PHA) from Wellcome Reagent Limited (Beckham, England) and pockweed-mitogen (PWM) from Pharmacia (Uppsala, Sweden). Hydroxyurea and phorbol myristate acetate (PMA) were purchased from Sigma Chemical Co. (St. Louis, USA).

B. Conditioned Medium

Peripheral blood mononuclear cells were isolated from residual cell pellets after thrombopheresis by Ficoll-Hypaque density sedimentation. Cultures were set up at 5×10^6 cells/ml in RPMI 1640 with 1% fetal calf serum (FCS) and either 10 μg/ml Con A, PHA or PWM. Supernatans were harvested after 2, 3 and 4 days of culture respectively.

Supernatants from the HSB-2 T cell line were prepared by stimulating 1×10^6 cells/ml in RPMI 1640 containing 1% FCS, 10 μg/ml Con A and 5 ng/ml PMA as described recently (8).

Rat cell supernatant was harvested from rat spleen cells cultured at 1×10^6 cells/ml in RPMI 1640 containing 1% FCS and 1 μg/ml Con A.

C. Bone Marrow Cell Culture

Nucleated cells from normal bone marrow were cultured at 3.3×10^5 cells/ml in McCoy's 5A modified medium supplemented with 20% FCS, 1.5×10^{-5} M 2-mercaptoethanol, and 33% conditioned medium in upright 25-cm^2 tissue culture flasks (Falcon). Cultures were split in half once or twice a week.

D. *Staining and Counting of Cells*

Viable cells as well as the percentage of cells with metachromatic granules stained with toluidine-blue were counted at the same time in order to calculate the absolute number of metachromatically-staining cells (MSC).

Smears made by cytocentrifuge were stained with 0,1% toluidine-blue in 50% ethanol, pH 3.0 (9).

E. *Histamine Assay and Histamine Release*

Measurement of histamine was performed using a Technicon Autoanalyzer II (Technicon Instruments Co., Tarrytown, N.Y. USA) as described by Syraganian (10).

Histamine release with synthetic releasers was performed as follows: Cultured cells were washed two times with HEPES-buffered balanced salt solution and resuspended in the same buffer containing additional 0.6 mM of $CaCl_2$ and 1 mM of $MgCl_2$. After the incubation of 37°C, for 40-60 min with the calcium ionophore A23187 (Sigma) and Compound 48/80 (Sigma), the supernatants were obtained by centrifugation at 300 x g, 4°C for histamine determination.

Passive sensitization with IgE and histamine-release with anti-IgE was performed as follows: washed cells were incubated with 10 µg/ml of purified myeloma IgE for 2 hours. After three washings, cells resuspended in HEPES-buffered balanced solution with Ca and Mg, were incubated with different dilutions of anti-IgE (Hoechst Behringer, West Germany) for 1 hour. The supernatant was obtained as described above.

RESULTS

A. *Development of Metachromatically Staining Cells*

Con A-conditioned medium (Con A-CM) increased the number of MSC in liquid bone marrow cultures up to 100 times within a culture period of 2-3 weeks. At this time 40 - 60% of the total cells were metachromatically staining with toluidine-blue. The number of MSC remained constant for additional 3 - 4 weeks and then decreased (Figure 1, upper panel). In order to study whether the development of MSC was due to proliferation of MSC or due to proliferation-independent differentiation of preexisting precursor cells, bone marrow cultures were treated with hydroxyurea or x-irradiation. As shown in Figure 2, blocking of bone marrow cells at the G_1/S interphase with hydroxyurea, or inhibition of proliferation by x-irradiation, inhibited the development of MSC. Thus, our data presented in Figures 1 and 2 suggest that the development of MSC was dependent on an actual proliferation of either MSC precursor cells or MSC. The highest number of MSC was usually found at the 33% Con A-CM concentration. Significant MSC development was observed with as little as 4% of CM. Conditioned medium from un-

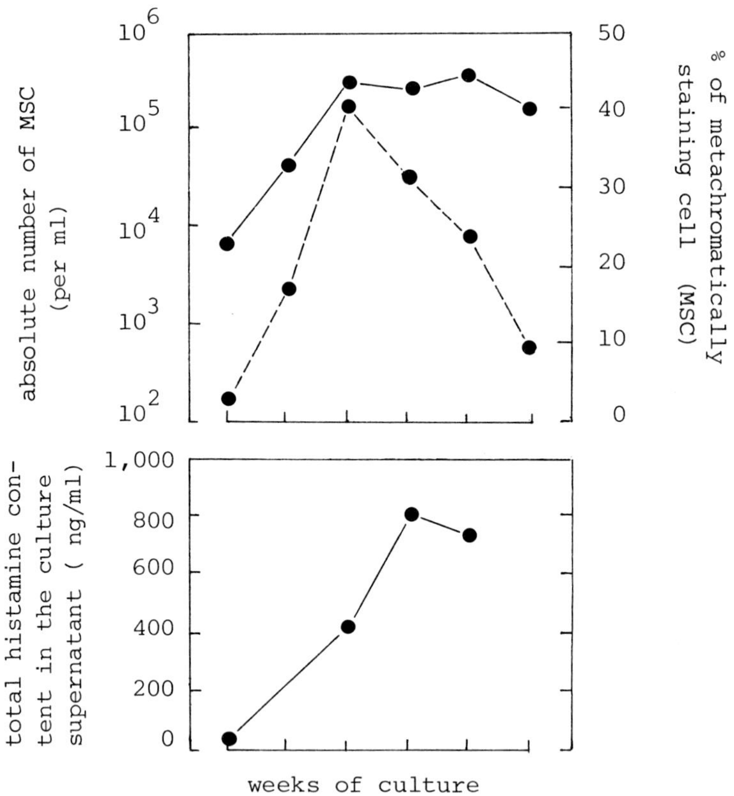

FIGURE 1. Development of metachromatically staining cells (upper panel; absolute number of MCS ●——●; percentage of MCS ●---●) and its correlation to histamine levels in the supernatant (lower panel).

stimulated mononuclear cells also contained some MSC promoting activity, but this was significantly less than the activity of 4% of Con A-CM.

B. MSC-promoting Activity

Highest MSC-promoting activity was usually detected in Con A-CM harvested after 2 days. PHA-CM harvested after 3 days and PWM-CM harvested after 4 days contained lower levels of MSC-promoting activity. Con A-CM contained only marginal levels of Interleukin 2 (IL 2), determined as previously described (11). Interestingly, heat treatment at 65°C for 30 min, which abrogated IL 2 activity, increased the MSC-promoting activity. Supernatants from the HSB-2 cell line, which contained high levels of IL 2 (8), had no MSC-promoting activity. Therefore, our data suggest that the MSC-promoting activity is different from the IL 2 activity. Con A-CM prepared from rat

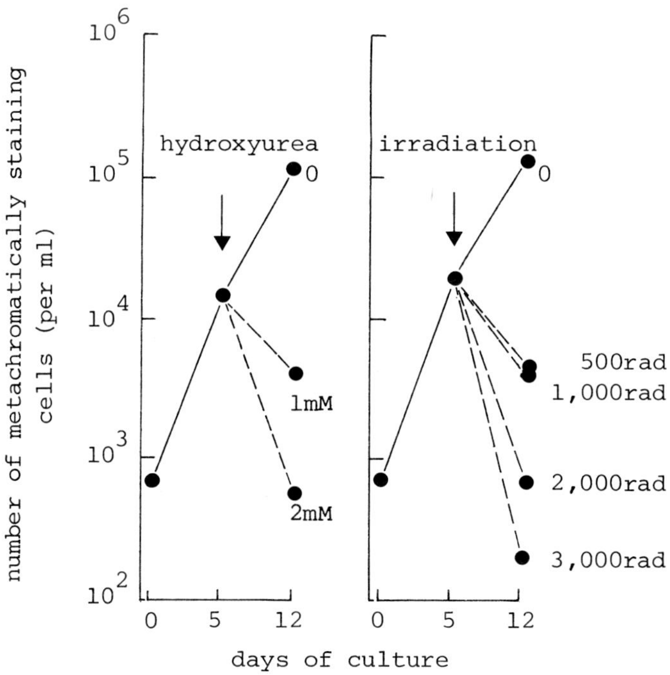

FIGURE 2. Effect of hydroxyurea and x-irradiation treatment on the development of MSC.

spleen cells contained MSC-promoting activity. This finding suggests that, in contrast to rat IL 2 which has had no effect on human lymphoid cells, the rat MSC-promoting factor was active on human bone marrow cells, Thus, it seems that the MSC-promoting activity is not species restricted.

C. Morphological and Functional Characterization of MSC

Cultured MSC had round to segmented nuclei. The intracytoplasmic granules stained metachromatically with toluidine-blue. These were fewer and finer in MSC with round nuclei, and denser and larger in MSC with segmented nuclei. However, transi-

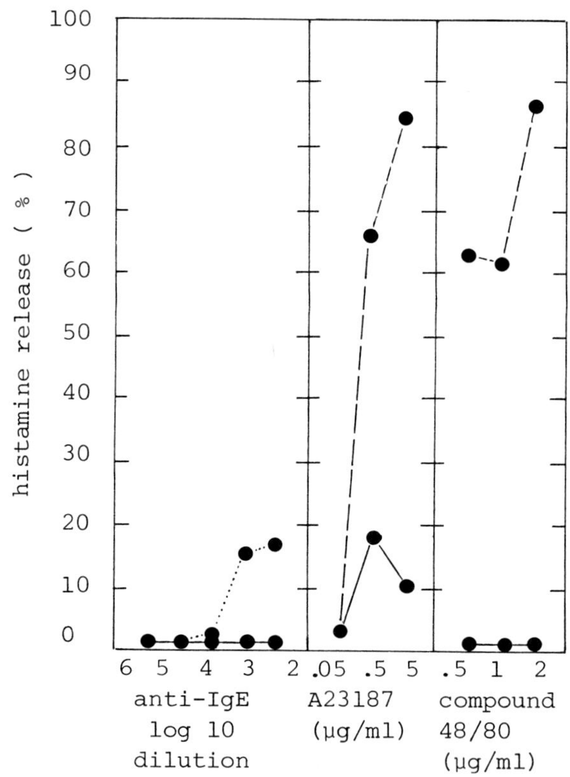

FIGURE 3. Histamine release from cultured bone marrow cells. Normal bone marrow cells cultured with Con A-CM for 15 days were used: ●——● cultured cells, ●·······● cultured cells passively sensitized with 10 μg/ml purified myeloma IgE, ●---● rat peritoneal mast cells.

tional forms between cells with round and segmented nuclei were also found without clear discrimination. Consequently, these cells seemed to be one lineage of cells in different stages of differatiation, rather than two different cell lineages.

As is shown Figure 1 (lower panel), histamine levels in the culture supernatants paralleled the increase of MSC. In preliminary experiment we estimated the histamine-content as 0.8 pg/cell on day 15 of culture.

As shown in Figure 3, histamine-release was induced with anti-IgE from cells passively sensitized with IgE, but not from unsensitized cells. Histamine-release was also induced with the calcium ionophore but not with Compound 48/80, while both agents induced histamine-release from rat peritoneal mast cells.

These data suggested that MSC resemble functionally and morphologically normal human basophils rather than mast cells. Our data show that the human basophil-promoting activity was different from the IL-2 activity. Some characteristics of the basophil-promoting activity such as heat stability and cross-species stimulating activity were similar to those of colony stimulating activity. Further biochemical characterization of this factor is now underway.

DISCUSSION

Metachromatic granules and IgE receptor-linked histamine-release are main characteristics of human basophils and mast cells. However, round to segmented nuclei and histamine-release with calcium ionophore but not with Compound 48/80 favour the assumption that MSC enriched in our liquid culture system were actually basophils at different stages of differentiation.

We have tested various kinds of conditioned medium from human mononuclear cells stimulated with Con A, PHA and PWM. However, these conditioned medium never promoted the appearance of typical human mast cells. We are presently investigating other sources of conditioned medium in order to eventually obtain human mast cells, as described in mice and rats (9, 12-16).

ACKNOWLEDGMENTS

We would like to thank Dr. G. Brun del Re and Miss K. Bütikofer for their kind supply with bone marrow cells; Dr. K. Deubelbeiss for the donation of thrombophoresis cell pellets; and Dr. G. D. Bonnard for his critical review of this manuscript . We are also grateful to Dr. Odavic for his technical and critical advice.

REFERENCES

1. Winquist, G., *Exp. Cell Res. 19,* 7 (1960).
2. Denburg, J. A., Davison, M., and Bienstock, J., *J. Clin. Invest. 65,* 390 (1980).
3. Miyoshi, I., Uchida, H., Tsubota, T., Hiraki, S., and Kitajima, K., *Scand. J. Hematol. 19,* 321 (1977).
4. Denegri, J. F., Naiman, S. C., Gillen, J., and Thomas, J. W., *Brit. J. Hematol. 40,* 351 (1978).
5. Denburg, J. A., Wilson, W. E. C., Goodacre, R., and Bienenstock, J., *Brit. J. Hematol. 45,* 13 (1980).
6. Algietta, M., Camussi, G., and Piacibello, W., *Exp. Hematol. 9,* 99 (1981).
7. Rasin, E., Rifkind, A. B., Cordon-Cardo, C., and Good, R. A., *Proc. Natl. Acad. Sci. 78,* 5793 (1981).
8. Stadler, B. M., Dougherty, S. F., Farrar, J. J., and Oppenheim, J. J., *J. Immunol. 127,* 1936 (1981).
9. Ishizaka, T., Okudaira, H., Mauser, L., and Ishizaka, K., *J. Immunol. 116,* 747 (1976).
10. Siraganian, R. P., *J. Immunol. Methods, 7,* 283 (1975).
11. Stadler, B. M., and Oppenheim, J. J., *Lymphokines 6,* 117 (1982).
12. Hasthrope, S., *J. Cellular. Physiol. 105,* 379 (1980).
13. Tertian, G., Yung, Y., Tuy-Grand, D., and Moore, M. A. S., *J. Immunol. 127,* 788 (1981).
14. Schrader, J. W., *J. Immunol. 126,* 452 (1981).
15. Nabel, G., Galli, S. J., Dvorak, A. M., Dvorak, H. F., and Cantor, H., *Nature, 291,* 332 (1981).
16. Nagao, K., Yokoro, K., and Aaronson, S. T., *Science 212,* 17 (1981).

DISCUSSION

SCHRADER: With regard to the failure to find mast cells, rather than basophils, in the human, there is also a lack of agreement on whether typical "mast cells" are present in the mouse. We would do well to remember that there are probably distinct subsets of mast cells and it is possible that the cells (e.g. P cells) that require a T cell-dependent factor for *in vitro* growth, are related not to typical peritoneal cavity mast cells, but rather to the T cell-dependent "atypical" mast cells found in association with mucosal surfaces.

STADLER: Yes, I agree. The point I sought to make was that based on the morphology and function of the MSC, we obviously culture a more "basophil-like" cell rather than a "mast cell-like" cell.

MIZEL: Would Stadler comment further on the biochemical properties of their factor promoting mast cell activities?

STADLER: The MSC-promoting activity is non-dialysable, suggesting a molecular weight >10,000; it is stable at 65°C for 30 min. Further biochemical characterization is presently underway. However, the present bioassay which is based on microscopy is not very well suited for this purpose; we hope to develop an assay based on proliferation, somewhat analogous to the microassay presently used for IL-2 determination.

HISTAMINE-PRODUCING CELL STIMULATING FACTOR (HCSF) PRODUCTION DURING ALLOGENIC OR MITOGENIC LYMPHOCYTE RESPONSE

Michel Dy
Bernard Lebel
Mario Abbud

INSERM U25 and Lab.Ass. CNRS 122
Hôpital Necker
161, rue de Sèvres
75730 Paris Cedex 15, France

I. INTRODUCTION

We have recently shown that supernatants of mixed leukocyte cultures between donor and recipient of a mouse skin allograft contain a factor which can strongly increase histamine synthesis by normal spleen or bone marrow cells (1). We have termed this factor: Histamine-producing Cell Stimulating Factor (HCSF). HCSF-containing supernatants also increase the proliferation of mast-like cells from spleen or bone marrow cells (2,3). These results suggest that the increase in histamine production could represent a consequence of a mast cell proliferation and that HCSF could be similar to other factors recently described acting on mast cell proliferation (4,5,6,7,8).

The same results have been found following Con A stimulation of normal spleen cells.

This paper briefly presents our recent findings concerning this phenomenon.

II. RESULTS AND DISCUSSION

A. *Nature of the HCSF-producing Cell*

HCSF-producing cells are probably T lymphocytes since: a) concanavalin A (con A) stimulation of normal spleen cells induces HCSF production, b) con A does not induce HCSF

Interleukins, Lymphokines, and Cytokines

ISBN 0-12-527540-4

production from nude mice, c) previous treatment of spleen cells by monoclonal anti-mouse Thy 1-2 antibody abolishes HCSF production following secondary MLC or con A stimulation (table I), d) T cell clones established from secondary alloreactive cells maintained in culture with TCGF produce large amounts of HCSF in response to con A (9) and e) preliminary results show that T lymphoma cells (EL4) can produce HCSF in response to con A and PMA.

TABLE I. Nature of HCSF producing cells

pretreatment	*HCSF units[a] produced during* con A stimulation of normal spleen cells	secondary MLC[b]
complement alone	5.2 ± 1.4	69.7 ± 12.4
α *Thy 1-2 + complement*	0	1.3 ± 0.6 ($p<0.001$)
α *Lyt 1-2 + complement*	0	26.3 ± 11.2 ($p<0.05$)
α *Lyt 2-2 + complement*	6.3 ± 1.4	21.8 ± 11.9 ($p<0.02$)

[a] *HCSF unit = quantity of HCSF contained in 1 ml of culture supernatant which can induce a 100% increase in histamine production by normal bone-marrow cells in our test conditions (1).*

[b] *Secondary MLC = skin allograft recipient spleen cells + donor spleen cells.*

The phenotype of the HCSF-producing cells has been determined by pretreating spleen cells by monoclonal anti-mouse Lyt 1-2 or Lyt 2-2 antibody. As shown in table I, HCSF production in response to con A stimulation is not modified by previous treatment with anti-Lyt 2 antibody and complement but is abolished by anti-Lyt 1 antibody and complement treatment. In contrast, HCSF production by spleen cells from allografted mice in presence of donor cells is strongly diminished by previous treatment of recipient spleen cells by either anti-Lyt 1 or Lyt 2 antibody and complement. Similar to what was observed with other lymphokines (10), this would indicate that both Lyt 1+ and Lyt 2+ cells can produce HCSF under appropriate conditions and that cells involved in HCSF production depend on the stimulation.

HCSF production by recipient spleen cells requires specific stimulation by donor cells or with cells bearing at least partly the H-2 antigens of the donor cells. Partial homology at the H-2 complex between donor and stimulating cells is sufficient to induce HCSF production but such homology must be coded by K and D loci (1).

HCSF production is not modified by previous thymectomy at weaning (3 weeks) while it has been demonstrated that such a treatment causes a long-lasting diminution of T cell proliferation (11).

However, although our results clearly demonstrate that T cells are HCSF producing cells in response to con A or secondary MLC, they do not exclude the possibility that under appropriate conditions HCSF could be produced by other cells. For example, preliminary results show that HCSF could be produced by B cells in response to LPS or anti-IgG serum (M. Abbud et al: manuscript in preparation).

B. Physicochemical characteristics of HCSF

HCSF obtained from secondary MLC or con A stimulation of normal spleen cells possess the same characteristics. It is a molecule of 30,000 daltons (as determined following gel filtration chromatography). It is thermostable, destroyed by trypsin and chymotrypsin but not affected by ribonuclease. Following isoelectrofocusing its isoelectric point is around 4.7 (B. Lebel et al, manuscript in preparation). All these biochemical procedures, applied to secondary MLC supernatants, allow us to obtain a 1000-fold purification with a recovery of 35% (table II).

Studies of HCSF production by different T cell clones demonstrate that HCSF differs from Interferon, CSF-GM, MAF or TCGF (10). The main question now raised is to know whether we can discriminate between HCSF and the factors acting on mast cell proliferation such as MCGF (6), PSF (5) or IL3 (8).

TABLE II. Summary of the partial purification of HCSF

Purification step	Activity (units)	Protein (mg)	Specific Activity (units/mg)	Purification (- fold)	Cumulative Per cent Recovery
Starting MLC II supernatant	32,000	324	99	1	100
Negative Pressure Dialysis Concentration (x 100)	32,000	36	890	9	100
Heating 80°C/10 mn	32,000	3	10667	108	100
Ultrogel AcA54	27,200	0,64	42500	429	85
Isoelectrofocusing (1% agarose)	9792	0,09	108800	1099	36

C. Nature of HCSF-target cells and mechanism of action

HCSF target cells are clearly much more abundant in bone marrow than in other sites and are located in the less dense layers following discontinuous Ficoll gradient (1).

Although it has been clearly demonstrated that HCSF acts by increasing histamine synthesis (1), at least two possibilities remain: a) HCSF acts by increasing histidine decarboxylase activity (HDC) in its target cells or b) HCSF increases the number of cells which can synthetize histamine. We found that long-term culture of bone-marrow cells in the presence of HCSF-containing supernatant leads to a pure cell population having all the characteristics of mast cell (2,3), suggesting that HCSF could act by increasing the number of cells which can synthetize histamine; however, inhibition of cell proliferation by mitomycin C or irradiation does not abolish the increase in histamine production found in 48h of culture (2).

On the whole, the above results have led us to suggest that HCSF could act on pre-mast cells (stem cells) by first inducing a phase of maturation (or differentiation) with low or no proliferation during which immature cells with low HDC activity acquire higher HDC activity. This phase would be followed by a phase of intensive proliferation leading to a pure population of mast-like cells.

If this hypothesis is correct, the increase in histamine production observed during the first days of culture would reflect an early event in mast cell differentiation. However, we cannot yet exclude the possibility that histamine production and mast cell proliferation in response to HCSF-containing supernatant are two distinct phenomena.

III. CONCLUSIONS

In the allograft situation the production of HCSF could explain the increase in HDC found in skin allografts at the time of rejection (12). The increase in local production of histamine could play a role in the alteration of capillary permeability. It could also represent a local feedback immunoregulatory mechanism (13) by stimulating T lymphocytes bearing H2 receptor for histamine (14).

REFERENCES

1 Dy, M., Lebel, B., Kamoun, P., and Hamburger, J., *J. Exp. Med. 153,* 293 (1981).

2 Dy, M., Guy-Grand, D., and Nabarra, B., *in* "Mechanism of lymphocyte activation" (K. Resch, H. Kirchner, eds.), p. 554 Elsevier/North Holland, (1981).

3 Tertian, G., Dy, M., Yung, Y.P., Guy-Grand, D., Moore, M.A.S., and Hamburger, J., *C.R. Acad. Sci. (Paris), 291,* 1041, (1980).

4 Denburg, J., Davison, M., Bienenstock, J., *J. Clin. Invest. 65,* 3900 (1980).

5 Schrader, J.W., *J. Immunol. 126,* 452 (1981).

6 Tertian, G., Yung, Y.P., Guy-Grand, D., and Moore, M.A.S., *J. Immunol. 127,* 788 (1981).

7 Nabel, G., Galli, S.J., Dvorak, A.M., Dvorak, M.F., and Cantor, H., Nature, 291, 332 (1981).

8 Ihle, J.N., Rebar, L., Keller, J., Lee, J.C., and Hapel, A.J., *Immunological Rev. 63*, 5 (1982).

9 Krammer, P.H., Dy, M., Hultner, L., Isakson, P., Kees, U., Lohmann-Matthes, M.L., Marcucci, F., Michnay, A., Pure, E., Schimpl, A., Staber, F., Vitetta, E., and Waller, M., *in* "Isolation, characterization and utilization of T lymphocytes" (G. Fathman, F. Fitch, eds.), Academic Press, New York (In press).

10 Henney, C.H.S., Okada, M., and Gillis, S. , *in* "Behring Institute Mitteilungen: Interleukin 2" p. 26, (1980).

11 Kaiserlian, D., Dardenne, D., and Bach, J.F., *Cell. Immunol. 64*, 93 (1981).

12 Moore, T.C., and Schayer, R.W., Transplantation (Baltimore), 7, 99 (1969).

13 Lewis, G.P., Nouri, A.M.E., Fan, T.P., and Gordon, D., *in* "Advances in pharmacology and therapeutics II" Vol. 4 (H. Yoshida, Y. Hagimara and S. Ebashi, eds.), Pergamon Press, p. 159 (1982).

14 Rocklin, R.E., Greineder, D.K., and Melmon, K.L., *Cell. Immunol. 44*, 404 (1979).

DISCUSSION

MIZEL: Does HCSF, as present in homogenates, function efficiently to induce histamine synthesis? Does HCSF operate to increase enzyme activity, or is it rather a determinant of the amount of enzyme elaborated.

DY: Thus far, we have not succeeded in causing our factor to increase histamine synthesis in cell homogenates. Up to now our experiments have not been designed so as to distinguish whether HCSF increases enzyme activity or the quantity of enzyme produced.

SCHRADER: There is one experiment which suggests that the activation of bone marrow progenitors which do not give rise to continuous lines of histamine-storing cells could also be associated with stimulation of histidine decarboxylase. This is an experiment in which we collaborated andy Dy found that a supernatant of the hybridoma T190 which contains only GM-CSF and does not support (except perhaps at very high concentrations) the growth of P cells, nevertheless had considerable activity in Dy's assay. There are other experiments suggesting that high levels of histidine carboxylase activity (but *not* of stored histamine) are associated with growing tissues, (e.g. bone marrow and fetal tissues).

DY: Yes, that could be right. Another explanation could be that one needs relatively large amounts of factor to support the growth of mast-like cells since 100% of the cells used in this test depend on the presence of this factor. In our test, if our hypothesis is correct, a small amount of the factor is sufficient to induce a detectable increase in histamine production since only a small percentage of the bone marrow cells (i.e. precursors of mast cells) require the presence of this factor. Accordingly, the discrepancies could as well be explained by a lack of test sensitivity.

PLUZNIK: Dy's last protocol resembles one given by Ihle. I can't help wondering whether Dy tried to induce 20αSDH with HCSF and whether Ihle tried to induce histidine decarboxylase via IL-3.

DY: I have never had occasion to attempt induction of 20αSDH with HCSF, nor do I know whether Ihle sought to induce histidine decarboxylase with IL-3. However, I hope we can effect an exchange of reagents to answer this point in the near future.

VADAS: Does Dy's factor contain GM or G-CSF and, if so, how does he explain the lack of mature granulocytes in his 12 day cultures?

DY: I know that our partially purified factor still contains CSF for macrophages, but these cells are lost during subculture at day 3 and 7, since we used *only* non-adherent cells to continue the culture with our factor. As regards granulocytes, I do not know if we have CSF for these cells. If that is indeed the case, it may be that in our situation the intensive development of mast-like cells prevents their proliferation.

CHAIRMEN'S COMMENTS

Session II dealt with lymphokines that affect precursor cells; in particular, mast cell, basophil, granulocyte, and macrophage precursors. Lenz and Pluznik presented their results on the binding of CSF to gangliosides, e.g., GM_1. The suggestion was made that gangliosides may be part of the CSF receptor. These findings on the interaction on CSF with gangliosides indicate that caution is required when studying lymphokine receptors since gangliosides are ubiquitious cell surface components. Finke and co-workers reported that CSF possessed the ability to enhance the generation of alloreactive cytotoxic T cells. One pathway for this CSF effect may involve the induction of IL1 by CSF. The IL1 would then be able to stimulate IL2 synthesis, an event that would lead to the expansion of specific clones of cytotoxic T cells.

Several speakers discussed their findings on lymphokines that stimulate mast cell/basophil growth and/or differentiation. Ihle described his studies on a lymphokine, termed interleuken 3 (IL3), that stimulates the synthesis of an enzyme, 20α hydroxysteroid dehydrogenase (20α-SDH), by spleen cells. Ihle has previously reported that IL3 also promotes the *in vitro* growth of splenic lymphocytes. However, more recent evidence suggests that this lymphokine may act on mast cell/basophil precursors. From his purification studies Dr. Ihle was able to report a high degree of purity but very low recovery, 3 μg from 150 liters. At this rate one would need 1500 liters of cell supernatant in order to obtain the sequence data Dr. Ihle showed. Based on Dr. Ihle's work and the work presented by Dr. Garland (see below) we believe that the name IL-3 is inappropriate for this factor. Unlike IL-1 and IL-2 there is no consensus regarding the biological properties of this factor or agreement on its biochemical nature. Central to the success of future studies is to establish a solid biological assay with which all who study this factor can agree. Otherwise "IL-3" will remain a factor in search of a function. The studies of Garland et al. certainly put to rest the idea that 20α SDH is a T cell marker. These investigators demonstrated that both lymphoid and granulocyte/macrophage antigens are present on 20α-SDH positive cells from long term bone marrow cultures. This is in direct conflict with Ihle's earlier reported study and lends further confusion as to what "IL-3" is supposed to do. Stadler et al. reported on a lymphokine that possesses the ability to stimulate basophil or mast cell proliferation or differentiation. This mediator appears to be distinct from IL2. Dy et al. described a lymphokine that stimulates 1) histamine synthesis by spleen or bone marrow cells, and 2) the appearance of a mast cell-like population in long term bone marrow cultures. Is it possible that all of these studies involve the same lymphokine? To us it appears that much additional work needs to be accomplished in order to clarify the nature of the mast cell growth factor. Perhaps an informal exchange of reagents between researchers could help to straighten out what is at this juncture, a very confusing story.

S. Mizel
D. Webb

SECTION III:
B CELL ACTIVATION
BY NONSPECIFIC AND SPECIFIC LYMPHOKINES

IDENTIFICATION AND CHARACTERIZATION OF A LYMPHOKINE WHICH INDUCES MURINE B CELLS TO SECRETE IgG[1]

Ellen S. Vitetta

Peter C. Isakson[2]

Ellen Pure[3]

Department of Microbiology
University of Texas Health Science Center
Dallas, Texas

Peter H. Krammer

Institute for Immunology and Genetics
German Cancer Research Center
Heidelberg, Federal Republic of Germany

A. DEFINITION OF A T CELL-DERIVED B CELL DIFFERENTIATION FACTOR WHICH INDUCES IgG SECRETION ($BCDF_{\gamma}$)

Small $IgM^{+}IgD^{+}$ B cells can be induced to secrete IgM and to switch to the production of other classes of immunoglobulins (Ig) (1-3). This switch can be induced by either antigenic (4,5) or mitogenic stimulation and is relatively T cell-dependent (6-8). T cells can influence

1 Supported by NIH grants AI-12789 and AI-11851

2 Present address: Department of Pharmacology, University School of Medicine, Charlottesville, Virginia

3 Present address: Department of Cellular Immunology, Rockefeller University, New York, New York

Interleukins, Lymphokines, and Cytokines

ISBN 0-12-527540-4

the subclasses of IgG secreted by B cells (9,10), although in low density cell cultures in the presence of lipopolysaccharide (LPS) IgG secretion by murine B cells may not require T cells (11-14).

We have used murine B cells stimulated by LPS to investigate the effects of several monoclonal sources of T cell-derived supernatants (SN) on isotype switching in normal splenic B cells. Two potent sources of $BCDF_\gamma$ are SN from PMA-stimulated EL-4 thymoma cells (15) and the Conconavalin A-stimulated alloreactive long-term T cell line, 7.1 (16). Both SN induce IgM secretion by B cells. In the presence of LPS, they also enhance IgG secretion. The induction of IgG secretion is particularly evident when the SN are added 1-2 days after the initiation of the LPS culture (17). As shown in Table I, $BCDF_\gamma$-containing SN induce IgG secretion in adult B cells: similar results (not shown) were obtained when B cells from neonatal mice were used (18).

TABLE I: EFFECT OF $BCDF_\gamma$ ON LPS-TREATED B CELLS[a]

Source of Supernatant	Day of addition	IgG per Culture PFC	IgG per Culture ng
7.1	-	657 ± 211	23 ± 5
	0	2033 ± 172	142 ± 26
	1	3867 ± 1069	164 ± 26
	2	1780 ± 320	126 ± 4
	3	1200 ± 327	52 ± 4
EL-4	0	1117 ± 422	40 ± 20
	1	2387 ± 555	134 ± 14
	2	1813 ± 531	84 ± 14
	3	1227 ± 247	46 ± 4

a BALB/c B cells were cultured at 2×10^5 cells/ml (0.2 ml) for 6 days with LPS, then assayed for IgG plaque forming cells and for total secreted IgG in a RIA.

Since the 7.1 SN does not contain IL-2, conventional (19) TRF or IFN-γ (16), $BCDF_\gamma$ appears to be different from these three lymphokines. By molecular weight analysis, $BCDF_\gamma$ is different from a lymphokine which induces IgM secretion

($BCDF_{\mu}$) since $BCDF_{\gamma}$ has a molecular weight of less than 20,000 whereas $BCDF_{\mu}$ is larger (18). $BCDF_{\gamma}$ has the same molecular weight as B cell growth factor (BCGF). However, since some sources of $BCDF_{\gamma}$ lack BCGF, these lymphokines are probably also different from each other (18).

B. $BCDF_{\gamma}$-CONTAINING SN ENHANCE SECRETION OF IgG_1 AND SUPPRESS SECRETION OF IgG_2

As shown in Table II, an analysis of the subclasses of IgG secreted following the addition of $BCDF_{\gamma}$-containing SN to LPS-stimulated B cells indicates a striking increase in IgG_1 with a concomitant decrease in IgG_2 (17).

TABLE II: Suppression of IgG_2 Secretion by PK 7.1 SN[a]

Cells	Experiment No.		% Change from LPS Treatment: Total IgG	IgG_2
	1		650	-100
Unstimulated	2		850	- 90
B Cells	3		1070	-100
	4		700	- 80
	5		710	-100
		Average	796	- 94
	1		320	- 64
	2		300	- 53
LPS Blasts	3		150	-100
	4		650	- 80
	5		670	- 54
		Average	420	- 70

a Data was calculated by comparing the amount of IgG or IgG_2 secreted during 6 days culture in the presence of an optimal concentration of PK 7.1 SN + LPS to cultures containing LPS only.

$$\text{Total IgG} = \frac{\text{ngIg with BCDF}\gamma\text{ and LPS}}{\text{ngIg with LPS only}} \times 100$$

$$\text{IgG2} = \frac{\text{ngIgG}_2\ (\text{LPS} + \text{BCDF}\gamma) - \text{ngIgG}_2\ (\text{LPS})}{\text{ngIgG}_2\ (\text{LPS})} \times 100$$

Extending the culture period did not result in the appearance of other subclasses of IgG, indicating that both $BCDF_\gamma$-containing SN selectively induce IgG_1 secretion. There are several possibilities to explain the finding that $BCDF_\gamma$-containing SN enhance the secretion of one isotype and suppress the secretion of another. 1) $BCDF_\gamma$ acts on a precursor cell, directing it to synthesize and secrete a particular subclass of IgG; 2) $BCDF_\gamma$ suppress IgG_2 secretion in some cells and enhances IgG_1 secretion in others; 3) there are several lymphokines in the $BCDF_\gamma$-containing SN. One may enhance IgG_1 and another may suppress IgG_2. The different lymphokines may act on the same or different target cells.

C. $BCDF_\gamma$ DOES NOT BIND TO Ig

One possibility to explain the subclass specificity of the $BCDF_\gamma$ is that it binds to $sIgG_1$ on LPS-activated B cells. This possibility appears unlikely for two reasons: 1) $BCDF_\gamma$ cannot be absorbed from the SN with Sepharose bound to IgG_1 (17); and 2) When $sIgG^+$ B cells are eliminated from the starting culture by staining with an anti-γ reagent followed by negative selection on the fluorescent activated cell sorter, there is no dimunition in the amount of IgG secreted into the culture (20).

Taken together, these results suggest that LPS induces the appearance of a non-Ig receptor on the surface of the B cells which is responsible for binding $BCDF_\gamma$ and transmitting a signal to the interior of the cell. Since an analysis of the precursor frequency in $BCDF_\gamma$-treated cells indicates that there is significant increase in precursors secreting IgG_1 with little or no increase in their burst size (21), these results support the notion that $BCDF_\gamma$ induces a switch in isotype production by LPS-stimulated B cells rather than expanding cells which have already switched to IgG_1.

D. CONCLUSIONS

Our results suggest that $BCDF_\gamma$ is a new lymphokine which differs from others thus far described and which can induce LPS-stimulated B cells to secrete IgG_1. From other reports in the literature, there are indications that T cells secrete $BCDF_\varepsilon$ (22) and possibly $BCDF_\alpha$ (23). If BCDFs for all classes and subclasses of Ig do exist, our results would suggest that the switch to any particular isotype by an activated B cell might be directed by specific lymphokines secreted by T cells. Furthermore, by analogy with T cell growth factor (24) and conventional hormones it is possible that BCDFs for different classes and subclasses of Ig act *via* specific receptors on the surface of activated B cells.

ACKNOWLEDGMENTS

We thank Mr. Y. Chinn, Mr. P. Genthner, Ms. K. Sill, Ms. P. May, Mr. W. Muller and Ms. A. Wyatt for expert technical assistance and Ms. G. Cheek for secretarial help.

REFERENCES

1. Nossal, G.V.J., Warner, N.L. and Lewis, H. Cell Immunol. 2,41 (1971).
2. Pernis, B. XLI Cold Spring Harbor Symposium on Quantitative Biology. Origins of Lymphocyte Diversity. 41,175 (1976).
3. Wabl, M.R., Forni, L. and Loor, F. Science 199,1978 (1978).
4. Press, J.L. and Klinman, N.R. J. Exp. Med. 138,300 (1973).
5. Gearhart, P.J., Sigal, N.H. and Klinman, N.R. Proc. Natl. Acad. Sci. 72,1707 (1975).
6. Braley-Mullen, H. J. Immunol. 113,1909 (1974).
7. Torrigiani, G. J. Immunol. 108,161 (1972).
8. Taylor, R.B. and Wortis, H.H. Nature 220,927 (1968).
9. Slack, J., der-Balian, G.P., Nahm, M. and Davie, J.M. J. Exp. Med. 151,853 (1980).
10. Mongini, P.K.A., Stein, K.E. and Paul, W.E. J. Exp. Med. 153,1 (1981).
11. Andersson, J., Coutinho, A., and Melchers, F. J. Exp. Med. 147,1744 (1978).
12. Andersson, J., Coutinho, A., and Melchers, F. Eur. J. Immunol. 8,336 (1978).
13. Kearney, J.F., Cooper, M.D., and Lawton, A.R. J. Immunol. 117,1567 (1976).
14. Gronowicz, E.S., Doss, C., Assisi, F., Vitetta, E.S., Coffman, R.L., and Strober, S. J. Immunol. 123,2049 (1979).

15. Farrar, J., Fuller-Farrar, J., Simon, P., Hilfiker, M., Stadler, B. and Farrar, W. J. Immunol. 125,2555 (1980).
16. Krammer, P.H., Dy, M., Hultner, L., Isakson, P., Kees, U., Lohmann-Matthes, M.-L., Marcucci, F., Michnay, A., Pure, E., Schimpl, A., Staber, F., Vitetta, E. and Waller, M. in "Isolation, Characterization and Utilization of T Lymphocytes" (Fathman, G., and Fitch, F., Eds.), pp. 252-262, Academic Press, New York, (1982).
17. Isakson, P., Pure, E., Krammer, P.H., and Vitetta, E.S. Manuscript in preparation, (1982).
18. Pure, E., Isakson, Kappler, J., Marrack, P., P., Krammer, P.H., and Vitetta, E.S. Submitted for publication, (1982).
19. Schimpl, A., and Wecker, E. Transplant. Rev. 23,176 (1975).
20. Isakson, P.C., Pure, E., Vitetta, E.S., and Krammer, P.H. J. Exp. Med. 155,734 (1982).
21. Isakson, P.E., Pure, E., Vitetta, E.S. and Krammer, P.H., Unpublished observations.
22. Ishizaka, K. Adv. Immunol. 23,1 (1976).
23. Elson, C.O., Heik, J.A., and Strober, W. J. Exp. Med. 149,632 (1979).
24. Robb, R.J., Munck, A., and Smith, K.A. J. Exp. Med. 154,1455 (1981).

DISCUSSION

LANDY: I would remind Vitetta that there are a few immunostimulants indeed that can match the potency of LPS in the evocation of IL-1 production by macrophages. How certain is Vitetta that her B cell preparations are completely freed of the very few macrophages that would suffice to release a significant amount of IL-1 -- possibly enough to greatly complicate analysis of the system she is studying.

VITETTA: We have not excluded the possibility that there are still macrophages in our B cell suspensions, *or* that IL-1 plays *some* role in the action of $BCDF_{\gamma}$. Nevertheless, $BCDF_{\gamma}$ *does* (in the presence or absence of other factors) induce B cells to secrete IgG_1.

CLEMENT: How does Vitetta rule out the possibility that $BCDF_{\gamma}$ simply acts in concert with LPS to stimulate another (different) subset of B cells that are pre-committed to γ_1 production rather than truly "inducing" a switch.

VITETTA: We cannot rule out that possibility. Nor do we know which B cell produces IgG_1. We only know that the producer B cell is $sIgG^-$ and that its precursor frequency increases after adding LPS+$BCDF_{\gamma}$. We do not yet know whether in the process it has rearranged its C_{γ} genes.

FELDMANN: Does Vitetta know whether B cell differentiation via factor $_{\gamma}$ is operative *in vivo*?

VITETTA: No, we do not.

FELDMANN: The relevance of $BCDF_{\gamma}$ to adjuvants which modulate Ig class, e.g. alum and pertussis for IgG_1, would be of particular interest.

VITETTA: I agree. We rather expect that $BCDF_{\gamma}$ has a role *in vivo*, but the relevant experiments have not yet been done.

ROLE OF T CELL-REPLACING FACTOR (TRF) PRODUCED BY MONOCLONAL T CELL HYBRID IN B CELL DIFFERENTIATION

Kiyoshi Takatsu
Shohken Tomita
Yoshinobu Hara
Norio Ishii
Tomohisa Kanatani
Toshiyuki Hamaoka

Department of Oncogenesis
Institute for Cancer Research
Osaka University Medical School
Osaka, Japan

Several reports have demonstrated that soluble products of T cells regulate B cell growth and differentiation (1). These T cell factors have been obtained mostly from mixed populations of T cells and their biochemical characterization has remained unclear. Recently, however, the establishment of T cell hybridomas and long-term T cell lines(2-6) have allowed to facilitate studies of T cell-derived lymphokines on proliferation and differentiation of B cells.

We have established a T cell hybrid clone (B151K12) with the ability to produce T cell-replacing factor (TRF)(or B cell differentiation factor) without any mitogenic stimulation. Cell-free supernatant (CFS) derived from the B151K12 clone induced PFC responses in rigorously T cell-depleted antigen-primed B cells and also rendered B cell chronic leukemia (BCL_1) cells to differentiate into immunoglobulin producing cells (7). CFS from a B151K12 cell line exerts only TRF activity without any other activities such as T cell growth factor (TCGF) and B cell growth factor(BCGF). B151K12-TRF was eluted by chromatofocusing in the range of pI 4.9-5.1, and by gel filtration on Sephadex G-150 in the range of 45K-60K daltons. The B151K12-TRF induced the differentiation of B cell responses in a totally antigen non-specific manner beyond the MHC barrier. On the other hand, supernatant from TCGF

ISBN 0-12-527540-4

producing T cell hybrid (FS6-14.13) contains BCGF as measured by anti-IgM antibody-induced normal B cell proliferation. BCGF activity was eluted by chromatofocusing in the range of pI 6.3-6.5.

By utilizing these factors, genetic control of B cell responsiveness to TRF and BCGF was analyzed. DBA/2Ha mouse carries an X-linked recessive defect, reflected in part by the low responsivity of its B cells to TRF. However, the B cells from this strain responded well to BCGF. In contrast, B cells from another X-linked B cell defective mouse, CBA/N, were not stimulated significantly with BCGF, whereas antigen-primed B cells from CBA/N mice responded quite well to TRF. To analyze the acceptor site(s) for TRF in relation to the acceptor for BCGF, the alloantiserum was raised in (DBA/2Ha x BALB/c)(DC)F_1 male mice (TRF low-responder) against antigen-primed B cells from BALB/c mice (TRF high-responder). This antiserum selectively inhibited the in vitro B cell response mediated by TRF when added continuously to the culture, whereas it did not block proliferation of B cells induced by BCGF. These results suggest that the TRF acceptor site(s), being defective in DBA/2Ha mice are distinct from acceptor site(s) for BCGF.

I. CHARACTERIZATION OF TRF IN SUPERNATANT DERIVED FROM B151K12

The cell-free supernatant(CFS) derived from a T cell hybridoma (B151K12), which had been obtained by fusion of Tbc-primed T cells with BW5147 thymoma cells was used as a source of TRF (4). TRF activity was assessed using rigorously T cell-depleted DNP-primed B cells, viz., spleen cells from DNP-KLH-primed and boosted BALB/c mice having received i.p. injection of anti-thymocyte serum (ATS) two days prior to the experiments and then treated with anti-Thy 1.2 antiserum plus rabbit complement (C). These DNP-primed B cells were stimulated in vitro with DNP-OVA or DNP-D-GL (D-glutamic acid:D-lysine copolymer) in the presence of varying dilutions of CFS from B151K12 cells before and after fractionation. After 5 days anti-DNP IgG PFC responses were enumerated. In some experiments, TRF activity was determined using BALB/c B cell chronic leukemia (BCL_1) cells as described previously (7).

A significant anti-DNP IgG PFC response was observed when DNP-primed B cells were stimulated with either DNP-OVA or DNP-D-GL in the presence of CFS from B151K12. Moreover, a significant IgM PFC response was also observed when percoll-purified BCL_1cells were cultured in the presence of B151K12-CFS.

To characterize physicochemical properties of TRF, CFS of B151K12 was salted out with $(NH_4)_2SO_4$ in the range of 40-85% saturation. Most of the TRF activity was recovered in this

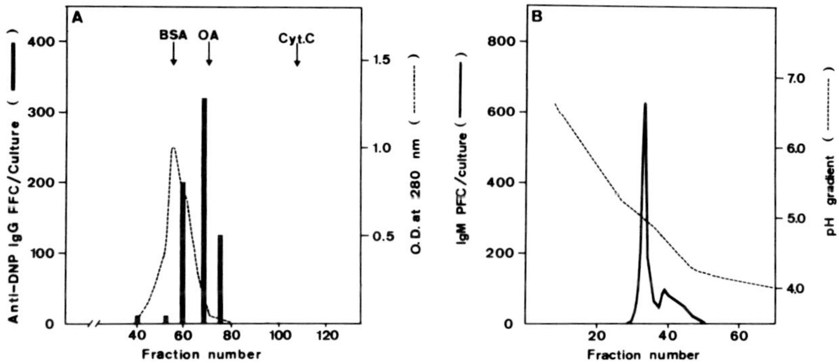

FIGURE 1.(A). Gel filtration profile of B151K12-CFS on Sephadex G-150. Supernatant from B151K12 cells was precipitated with ammonium sulphate (40-85%), subjected to gel filtration on a Sephadex G-150 column. Eluted fractions were assayed for TRF activity using DNP-primed B cells stimulated with DNP-OVA. (B). Chromatofocusing analysis of B151K12-CFS. After removal of polybuffer, each fraction was assayed for TRF activity by adding to the culture of BCL_1 cells.

fraction. After extensive dialysis, samples were applied to either chromatofocusing column (at pH range 7 to 4) or Sephadex G-150 column. TRF activity in each fraction was assayed using DNP-primed B cells or BCL_1 cells. As shown in FIGURE 1, TRF activity was mainly eluted at the fraction of 45K-60K daltons of molecular weight as estimated from gel filtration (A), and between the isoelectric points of 4.9 to 5.1 (B).

II. CHARACTERIZATION OF BCGF DERIVED FROM FS6-14.13-CFS

The CFSs from a battery of T cell clones and hybrids were examined for BCGF activities, and found in CFS from Concanavalin A-stimulated T cell hybridoma (FS6-14.13), which had been established by Harwell et al.(2).

BCGF activity in CFS of FS6-14.13 cells was determined by a standard BCGF assay using purified B cells, which had been prepared according to the method described by Leibson et al.(7). Briefly, spleen cells from BALB/c mice which had been treated with 0.2 ml ATS 2 day before the experiments were passed through Sephadex G-10 column and further treated with anti-Thy 1.2 antibody plus C. The BCGF assay employed involved the culture of $5x10^4$ purified B cells/well with anti-IgM antibodies at a final concentration of 2 μg/ml, in the

TABLE I. Lymphokine Activities in CFS from T Cell Hybridomas

CFS	*Lymphokine*		
from	*TRF*	*BCGF*	*IL-2*[a]
B151K12	*Yes* *(4.9-5.1)*[b]	*No*	*No*
FS6-14.13	*No*	*Yes* *(6.3-6.5)*	*Yes* *(4.6-4.9)*

[a] *Assayed utilizing IL-2 dependent CTLL.*
[b] *Numbers in parenthesis represent isoelectric range.*

presence of varying dilutions of the CFS. Cultures were incubated for 3 days, with a final 16 hr pulse of ^{3}H-thymidine, and the ^{3}H-thymidine incorporated into cells was measured as described previously (5).

Active components possessing BCGF activty in FS6-14.13-CFS was localized at the isoelectric range of 6.3 to 6.5, and separable from IL-2, its isoelectric points being around 4.6 to 4.9.

Interestingly, CFS from B151K12 revealed no BCGF activity. In contrast, CFS from FS6-14.13 did not show any significant TRF activities in our assay. The lymphokine activities in CFS utilized herein were summarized in TABLE I.

III. DISTINCT GENETIC CONTROL OF B CELL RESPONSE TO TRF AND BCGF

To examine responsiveness of B cells (X-linked recessive B cell defective mice) to BCGF, purified B cells from either DBA/2Ha or CBA/N mice were cultured with 2 μg/ml of anti-IgM in the presence of various concentrations of CFS of FS6-14.13. As evident from FIGURE 2, B cells from DBA/2Ha mice responded well to BCGF in CFS of FS6-14.13, whereas B cells from CBA/N mice did not respond significantly, indicating that X-linked B cell defect inherited in DBA/2Ha is different from that expressed in CBA/N mice.

To analyze whether or not acceptor site(s) for TRF are distinct from those for BCGF, we examined blocking activity of DCF_1 male anti-BALB/c B cell antiserum on B cell proliferation. DCF_1 male anti-BALB/c B-cell alloantiserum containing anti-TRF acceptor site antibody was prepared. As reported

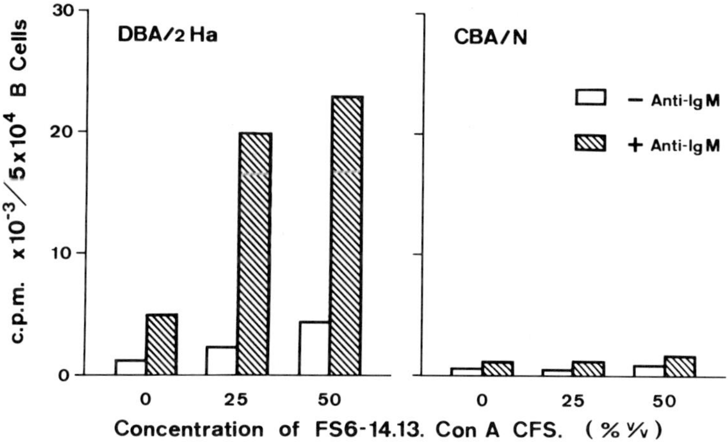

FIGURE 2. Responsiveness of B cells from DBA/2Ha or CBA/N mice to BCGF. Purified B cells from DBA/2Ha mice or CBA/N mice were cultured with anti-IgM antibody in the presence of various concentrations of FS6-14.13-CFS.

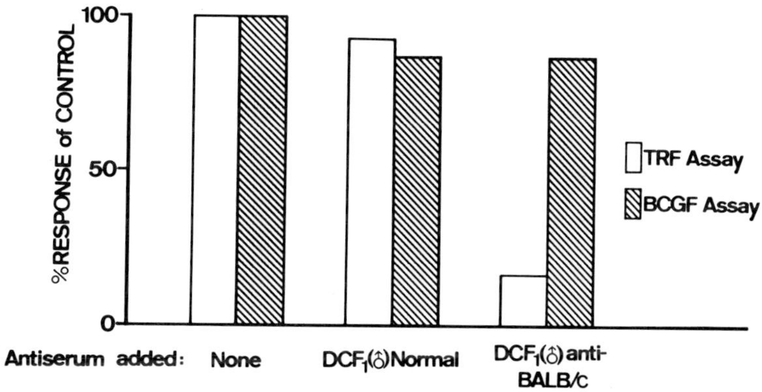

FIGURE 3. Failure of blocking of DCF$_1$ male anti-BALB/c B cell antiserum in expression of BCGF activity. DCF$_1$ male anti-BALB/c B cell antiserum (final 0.25%) was added to the culture of B cells from BALB/c mice plus anti-IgM antibody and CFS from FS6-14.13 (final 25%). To assess blocking activity of the antiserum on TRF-mediated PFC responses, the same lot of the antiserum was added to the culture of BCL$_1$ cells stimulated with CFS from B151K12. Results represented as % response of control. In BCGF assay, anti-IgM plus CFS from FS6-14.13 gave 10600 c.p.m./well of B cell proliferation in control. In TRF assay, control response was 968 PFCs/culture.

previously (10), this antiserum selectively inhibited TRF-mediated antibody response but did not affect cognate T-B cell interaction, when added continuously to the culture.

To ascertain whether or not the same antiserum can block proliferation induced by BCGF, the antiserum was added to the culture continuously during the BCGF assay. As shown in FIGURE 3, addition of the antiserum in BCGF assay hardly inhibited B cell proliferation induced by FS6-14.13-BCGF plus anti-IgM antibody. In contrast, the same antiserum strikingly inhibited IgM PFC responses of BCL_1 cells induced by B151K12-TRF, indicating that antiserum contains antibody against acceptor site(s) for TRF to block TRF action upon the B cells. Thus, blocking of acceptor site(s) for TRF by the antibody did not affect BCGF function. Acceptor site(s) for TRF may be distinct from those for BCGF.

ACKNOWLEDGMENTS

This work was partially supported by a grant-in-aid for scientific research from the Japanese Ministry of Education, Culture, and Science. We wish to express our appreciation to Drs. J. Kappler and P. Marrack for their generous supply of FS6-14.13 used in this study.

REFERENCES

1. *Immunol. Rev. 64,* (1982)
2. Harwell, L., Skidmore, B., Marrack, P. and Kappler, J., *J. Exp. Med. 152,* 893. (1980)
3. Swain, S. L., Dennert, G., Warner, J. F. and Dutton, R. W. *Proc. Natl. Acad. Sci. 78,* 2517. (1980).
4. Takatsu, K., Tanaka, K., Tominaga, A., Kumahara. Y., and Hamaoka, T., *J. Immunol. 125,* 2646. (1980).
5. Howard, M., Farrer, J., Hilfiker, M., Johnson, B., Takatsu, K., Hamaoka, T. and Paul, W. E., *J. Exp. Med. 155,* 914. (1982).
6. Isakson, P., Pure, E., Vitetta, E. S. and Krammer, P., *J. Exp. Med. 155,* 734. (1982).
7. Pure, E., Isakson, P., Takatsu, K., Hamaoka, T., Dennert, G., Swain, S. L., Dutton, R. W., Uhr, J. and Vitetta, E. S., *J. Immunol. 127,* 1953. (1981).
8. Leibson, M., Marrack, P. and Kappler, J., *J. Exp. Med. 154,* 1681. (1981).
9. Takatsu, K., Tominaga, A., and Hamaoka, T., *J. Immunol. 124,* 2414. (1980).
10. Tominaga, A., Takatsu, K. and Hamaoka, T., *J. Immunol. 124,* 2423. (1980)

DISCUSSION

DAVID: Could Takatsu's antisera be acting on the TRF itself? Has Takatsu ascertained whether TRF will bind to antibody on an affinity column?

TAKATSU: We have not done the kind of experiments David alludes to. However, it is a real possibility, for the reason that the antibody reacts only with a population of B cells (up to 20%) and can be absorbed out with B cells. Moreover, treatment of the antigen-primed B cells with this antibody, under appropriate conditions, mimics antibody-forming B cells.

FELDMANN: What evidence does Takatsu have that his antibody does *not* react with TRF, i.e. that it does not absorb it?

TAKATSU: We have not done TRF absorption experiments using antibody-coupled to Sepharose columns. However, the antibody can inhibit the absorbing ability of BCL_1 cells for TRF activity under the conditions where BCL_1 cells had been pretreated with the antibody prior to absorption.

DUTTON: I understood Takatsu to say that BCL_1 absorbs out the TRF activity and that pretreatment of the BCL_1 with the antiserum blocks that absorption. Does the BCL_1 absorb out the BCGF activity?

TAKATSU: We have not done that particular experiment. I do not know the answer.

MOND: Does X-irradiation of B cells decrease the incremental effect B cells have together with T cells in the production of BCGF?

MAIZEL: Up to the present, we have not examined the effect of B cell irradiation on the incremental effect of these cells on T cells in the production of BCGF. Our experiments have been limited to examination of the dose-response of the effect and to the linearity of the phenomenon.

OPPENHEIM: Did Maizel try murine as well as human B cells stimulated by anti-μ to detect BCGF, in order to ascertain the species specificity of BCGF?

MAIZEL: Our experiments have been strictly limited to human B cells in terms of our assay for BCGF. Furthermore, we have been producing BCGF only from human cell populations.

A MONOCLONAL ANTIBODY WHICH INHIBITS MATURATION OF MURINE B CELL TO ANTIBODY-PRODUCING CELLS

A. Schimpl
U. Chen-Bettecken
E. Wecker

Institut für Virologie und Immunbiologie
Universität Würzburg
Würzburg

TRF has been described as a T-helper cell product which acts directly on proliferating B cells and provides them with a maturation signal required for subsequent synthesis and secretion of antibodies (1). TRF assays are, however, not always easy to handle. Firstly, this is due to contaminating T cells in the assay cultures which can be expanded by T cell growth factors contained in the usual source of TRF, i. e. supernatants from mitogen or alloantigen activated T cells. Secondly, B cells at different stages of maturation or responding to different antigens may require the synergistic action of TRF plus other factors (2). A detailed analysis thus certainly depends on homogenous cell preparations or factor preparations devoid of other biological activities. Great progress towards this goal has been made both by biochemical separation of different activities (3) and by the use of T cell hybridomas producing a restricted range of lymphokines (4). Both approaches, however, have so far failed to deliver lymphokine preparations verifiably containing only one biologically active molecular entity.

ISBN 0-12-527540-4

We therefore attempted to prepare monoclonal antibodies directed against such biologically active molecules. In order to favour antibodies which might be directed against TRF, rats were first immunized with crude preparations of ConA induced mouse spleen cell supernatants in CFA, followed by several booster injections with TRF containing preparations obtained after various purification procedures based on different separation criteria which might be expected to contain different contaminants (3). Fusion of the spleen cells of immunized rats with P3X63Ag8 (5) led to the establishment of several hybridomas, presumably clones as judged by limiting dilution analysis. The supernatant of one of the clones, RE 3, was particularly strongly inhibitory in the TRF assay on SRBC activated nu/nu spleen cells (1). RE 3 has been shown to contain rat chromosomes and produce rat Ig at 5-20 μg/ml. The γ globulin fraction (33 % $(NH)_2SO_4$) of RE 3 not only inhibits antibody formation in the nu/nu-TRF assay, but also inhibits plaque formation in normal spleen cells at concentrations corresponding to 10-50 ng rat Ig. The secondary in vitro anti-hapten responses are also inhibited, although higher concentrations of the RE 3 γ globulin fractions have to be used and inhibition is not complete (Fig. 1). Inhibition did not seem to be due to toxicity, since in all cases the number of viable cells was unaffected or even increased. Also, RE 3 was neither directly nor indirectly cytotoxic for ^{51}Cr labelled normal spleen or lymph node cells and thus does not seem to be directed against a structure on the surface of normal lymphocytes or macrophages.

RE 3 γ globulin fractions were then tested for inhibitory activity in the following assay systems: 1) LPS induced ^3H thymidine incorporation and polyclonal Ig-production, 2) late ^3H Thy incorporation by goat anti mouse μ $(Fab)_2$ stimulated nu/nu spleen cells in the presence of Con A induced spleen cell supernatants (presumably a test for BCGF (8)), 3) IL-2 dependent growth of cloned cytotoxic T cell lines (6) (Cl 3 and A 13 were kindly provided by T. Hünig, Würzburg), 4) Ag dependent

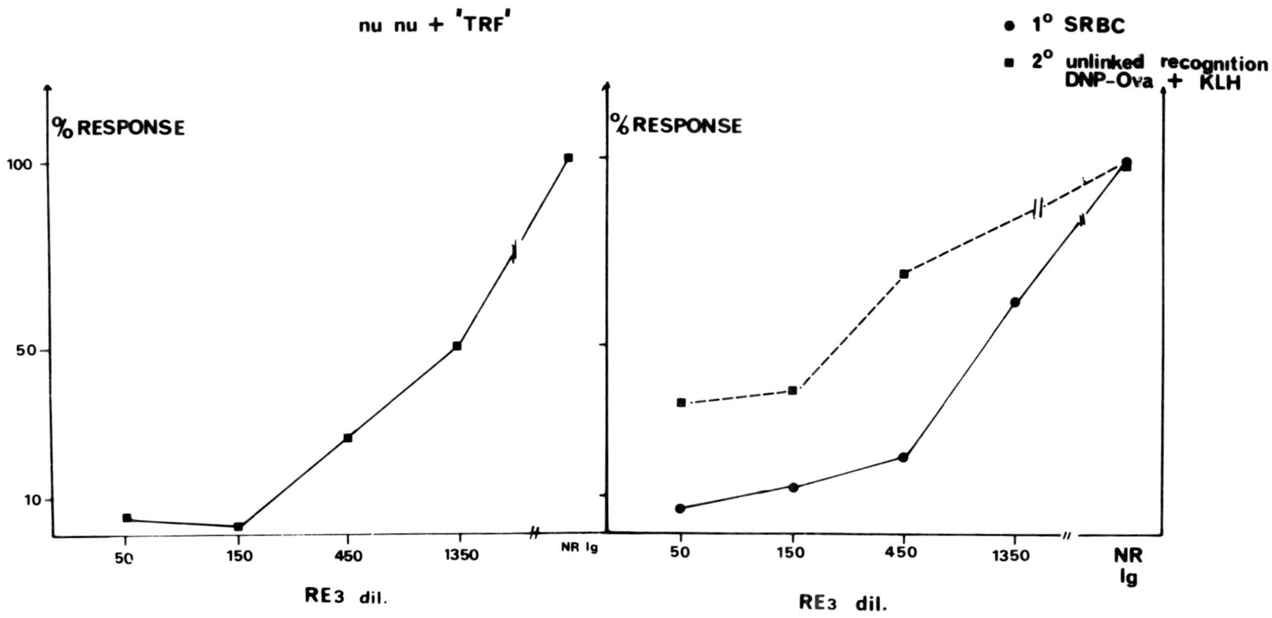

Fig. 1 Inhibition of primary and secondary plaque responses by RE 3. Experimental conditions as given in (13).

^3H Thy incorporation by spleen cells from Ag primed mice (=T helper cell proliferation). Controls received normal rat Ig. Table 1 shows that in none of these assay systems RE 3 exerted a significant dose dependent inhibition. Taking

Table I Lack of inhibition by RE 3 in various assay systems

	% Inhibition Assay system				
Final dil. of RE 3-Glob	Cl 3 (IL-2)	$2^{O}T_{KLH}$	α/u+BCGF	LPS prolif.	LPS PFC
1:50	7 %	18 %	3 %	0 %	13 %
1:150	-	0 %	3 %	0 %	0 %
1:450	20 %	9 %	0 %	-	0 %
	cpm	cpm	cpm	cpm	PFC Ig
pos. Co	13.500	10.900	11.760	7.740	20.600
BG with Med.	82	4.900	1.300	610	4.300

Assay conditions: Cell numbers: Cl 3 CTL: 5×10^4/ml; 200 /ul; 2nd T_{KLH}: $7{,}5 \times 10^5$/ml; 100 /ul; /u + BCGF: $7{,}5 \times 10^5$/ml; 100 /ul; LPS prolif: 8.10^5/ml; 200 /ul; LPS PFC: 2,5 x 10^6/ml; 1 ml. ^3H thymidine incorporation was measured by adding 0,25 /uCi (2Ci/mM) to each well for 16 hr. LPS-PFC were assayed with Rabbit Mouse Ig coupled to SRBC with C_3Cl_3 (12).

the known involvement of lymphokines in the various assay systems (6-8) into account, the data allow the conclusion that the inhibitory activity in RE 3 is not directed against TCGF (IL-2), LAF (Il-1) or BCGF, nor does it significantly affect B cells responding to LPS with increased ^{3}H Thy incorporation and plaque formation. Inhibition via a determinant only present on the surface of B cells which depend on TRF for their maturation can, however,

Table II Absorption of TRF by RE 3-immunosorbents

cells cultured		Additions d 2	PFC/culture
Exp. I	nu/nu	medium	60
		Con A-supernatant	800
		El 4[a]	120
		RE 3 Is unbound	30
		RE 3 Is eluted	60
		RE 3 unbound + El 4[a]	120
		RE 3 eluted + El 4[a]	670
Exp. II	nu/nu	medium	170
		"TRF"	5.150
		RE 3 Is unbound	990
		RE 3 Is eluted	5.110
			cpm/5x10^{3}cells
	CTL	medium	220
		ConA supe	10.120
		RE 3 Is unbound	8.730
		RE 3 Is eluted	180

Culture conditions as described in (13).

a) Supernatants from PMA induced El 4 cells after hydrophobic chromatography on Phenyl-Sepharose

b) Supernatants from ConA activated normal spleen cells after Phenyl-Sepharose.

not be excluded by these data. Such structures might either concern a differentiation marker of such cells (e. g. Lyb 2 (9)) or surface molecules directly concerned with TRF action, e. g. the TRF receptor. Alternatively, RE 3 might indeed be directed against TRF itself and thus prevent its function. To more directly address these alternatives, immunosorbent columns were prepared with RE 3 γ globulin. TRF containing supernatants were passed over the columns and bound material was eluted at pH 3.0, followed by immediate neutralization. Results are shown in Table 2. The total TCGF (IL-2) activity as assayed on CTL clones was contained in the unbound material. The same fractions were, however, greatly reduced in TRF activity as judged by PFC induction in the SRBC/nu/nu assay system. The eluted material showed two different patterns of reactivity, depending on the condition of the assay cultures: as typified by Exp. I, eluates by themselves did not induce plaques but could synergize for PFC production with fractions derived from PMA induced El 4 cells (10). These fractions by themselves only led to a marginal increase of PFC's as compared to controls. Other cultures, however, were able to respond to the material contained in the eluate fractions (see Exp. II). Whether these differences in the responsiveness of cultures is due to their varying contents in different B cell subsets or to varying degrees of in situ production of factors required in addition to TRF is not clear. In any case, the data seem to favour the notion that RE 3 is directed against TRF itself rather than a B cell differentiation marker.

In further experiments we attempted to characterize the molecule interacting with RE 3. Two approaches were taken 1) indirect precipitation with RE 3, Rabbit anti-rat Ig and Staphylococcal protein A using crude ConA induced spleen cell supernatants biosynthetically radiolabelled with ^{3}H Leu or ^{35}S-Meth and SDS-PAGE analysis of the precipitates. 2) Western blotting of SDS-PAGE separated crude cold ConA-supernatants with subsequent treatment of the blots with peroxydase-conjugated rabbit anti-rat Ig or ^{125}I-labelled rabbit anti-rat Ig. Both techniques revealed an interaction of RE 3 with a molecule of an apparent MW of slightly lower than 30.000. The band is not detectable in PMA induced El 4 (10) or AOFS (4) supernatants, not containing TRF activity. An ~ 30.000 MW agrees with the one described for TRF and coincides with a band obtained after precipitation of ^{3}H-leu labelled material with a goat anti-TRF antiserum (11).

In summary the data are consistent with the idea that RE 3 is indeed directed against TRF. However, there is one big caveat. While the inhibitory molecule is contained in 33 % $(NH_4) SO_4$ precipitates of RE 3 hybridoma supernatants and on Sephacryl-300 exhibits a MW consistent with Ig, RE 3-IG has not yet been purified. There is thus still no formal prove that the inhibitory activity of RE 3 is indeed carried by a rat Ig molecule. If this can be firmly established, RE 3 should provide a valuable tool in the further characterization of TRF, the final definition of its biological role and the establishment of its interrelationships with other lymphokines.

Acknowledgments: We thank Mr. A. Zant and Mr. M. Clark for technical assistance. This work was supported by the Deutsche Forschungsgemeinschaft through the Sonderforschungsbereich 105.

References

1) Schimpl, A., Hünig, T. and Wecker, E. Progr. Immunol. II, 135, 1974
2) Swain, S. L., Dennert, G., Warner, J. F. and Dutton, R. W. Proc. Natl. Acad. Sci. U.S.A. 78, 2517 (1981)
3) Schimpl, A., Hübner, L., Wong, C. and Wecker, E. Behring Inst. Mitt. 67, 221 (1980)
4) Harwell, L., Skidmore, B., Marrack, P. and Kappler, J., J. Exp. Med. 152, 893 (1980)
5) Köhler, G. and Milstein, C. Nature (London) 256, 495 (1975)

6) Gillis, S. and Smith, K. A. Nature (London) 268, 154 (1977)
7) Wagner, H., Röllinghoff, M., Kurrh, F. and Seiler, F. R. (eds) Behring Inst. Mitt. 67 (1980)
8) Parker, D. C. Immunol. Rev. 52, 115 (1980)
9) Yakura, H., Shen, F.-W., Bourcet, E. and Boyse, E. A. J. Exp. Med. 155, 1809 (1982)
10) Farrar, J. J., Fuller-Farrar, J., Simon, P. L., Hilfiker, M. L. and Farrar, W. L. Behring Inst. Mitt. 67, 58 (1980)
11) Hübner, L., Schimpl, A. and Wecker, E. Mol. Immunol. 17, 591 (1980)
12) Molinaro, G. A., Maron, E., Eby, W.C. and Dray, S. Eur. J. Immunol. 5, 771 (1975)
13) Hünig, T., Schimpl, A. and Wecker, E. J. Exp. Med. 145, 12165 (1977)

DISCUSSION

KRAMMER: Would Schimpl want to comment on the relationship of the different T cell factors acting on B cells: TRF, BCGF, BCDF, BRMF, etc.?

SCHIMPL: BRMF has recently been described by Melchers et al. as being a mixture of BCGF and TRF. It would thus seem that, generally speaking, in addition to antigen, B cells need growth factors (possibly derived from both macrophages and T cells) and differentiation factors (e.g. TRF or BCDF). As shown by Vitetta, the differentiation factors may be specific for the Ig clones induced. Whether *all* factors are needed in the responses to *all different* antigens is not yet clear.

BERTOGLIO: Did Schimpl ascertain whether her antibody cross-reacts with human TRF?

SCHIMPL: It does not, at least as assayed in a TRF assay which registers the increase in Ig production in an EBV-transformed human B cell line.

BELLER: I believe Schimpl stated that her antibody was neither directly nor indirectly cytotoxic; this is to be expected if the antibody is directed against TRF itself. However, it might be cytotoxic with complement if evaluated in the presence of TRF, its purported ligand. Has Schimpl determined whether activated B cells can absorb TRF activity?

SCHIMPL: Cytotoxicity assays have not been performed using activated B cells preincubated with TRF. Absorption experiments with TRF are very difficult to carry out since it was not clear from such experiments whether TRF is absorbed, inactivated, or simply destroyed.

GARLAND: Has Schimpl investigated the CBA/N system? Is a molecular species of the same molecular weight produced?

SCHIMPL: The experiments Garland refers to have not been done with hybridoma RE3. In previous experiments we have shown, however, that CBA/N mice produce normal or even slightly higher amounts of TRF *and do not exhibit* a TRF receptor defect, at least in B cells derived from DNP-KLH-primed and recently boosted animals.

FELDMANN: What controls did Schimpl utilize to exclude a non-specific effect of rat Ig on her tissue cultures? I ask this especially in view of the fact that Fc fragments can activate B cells.

SCHIMPL: We used normal rat Ig as control reagent.

FELDMANN: I would think that if Schimpl did not know the Ig class involved, the precise, appropriate control would be hard to perform.

SCHIMPL: Feldmann has a point. Since we had no way of knowing the Ig class involved, we used as control a mixture of all rat Ig clones.

ROLE OF DIFFERENT T CELL REPLACING FACTORS (TRFs) IN THE PROLIFERATION AND DIFFERENTIATION OF B CELLS

Susan L. Swain[1,2]
Richard W. Dutton[2]

Department of Biology
University of California, San Diego
La Jolla, California

INTRODUCTION

Recent major advances in lymphokine research can be attributed primarily to the development of T cell lines producing large quantities of particular factor(s) and to assays which unambiguously respond to single factors giving both a reliable assay and revealing function. This is certainly true for IL1 and IL2.

However, the two classes of factors of major concern for their effects on B cells, the factors responsible for B cell proliferation (BCGF) and B cell differentiation to Ig secretion (BCDF or TRF) are considerably less well-defined. We have been studying the activities present in supernatants of C.C3.11.75 (DL)(1-4). Supernatants from this cell line are active in synergizing with IL2 in the development of antibody-producing B cells (1-3). We call this putative differentiation factor (DL)TRF. In addition, the same supernatants have been found to contain an activity(ies) which promote initiation of growth of clones of mitogen-stimulated B cells(2-4). We will call this activity (DL)BCGF, based on the results to be presented here.

We have compared the ability of (DL)BCGF and a variety of other sources of lymphokines to cause proliferation of normal and BCL_1 tumor cells (passaged in vivo). Normal B cells respond with modest proliferation to sources of IL1, IL2 and TRF.

[1]*Recipient of Established Investigatorship Award from the American Heart Association (81173).*

[2]*Supported by NIH Grant AI8795, CA9174 and ACS Grant IM.*

ISBN 0-12-527540-4

Further proliferation is seen when anti μ and/or mixes of factors are added. In contrast, BCL_1 proliferated strongly only in response to (DL)BCGF and to one other putative source of TRF. Anti μ or other factors have no additional effect. In addition, BCL_1 absorbs the BCGF but not the TRF activity present in DL supernatants. Thus we have identified a high titered source of a B cell growth factor which can be assayed by an unambiguous method.

RESULTS AND DISCUSSION

Because DL supernatants have activity in both assays of B cell differentiation (1-3) and proliferation (2-4), we sought assays that might indicate whether the same or different factors within the supernatant were working in different assays. Our assay for TRF uses spleens of ATS pretreated mice which are treated in vitro with two monoclonal anti Thyl.2 reagents and monoclonal anti Lyt2.2 and which are G10 passed. In cultures with 10^6 cells such as these, the development of large numbers of SRBC specific PFCs is dependent on the addition of antigen (SRBC), IL2, and a source of TRF, which is operationally defined by this assay. IL1 and additional IL2 have little or no effect in this assay (1-3). When the same population of B cells (at 10^5 per well), without antigen, is used in a proliferation assay, DL supernatant, sources of IL1 and many sources of IL2 have enhancing effects on proliferation. Addition of anti-Ig reagents or mixtures of factor sources often have synergistic or additive effects (2,3). Thus this kind of assay, similar to that used by others describing B cell proliferation such as Parker (5) and Howard (6) appears very complex. Indeed Howard and coworkers find that several cells are involved which can be mostly but not yet totally replaced by soluble factors (Howard, personal communication). Because of these considerations, the ability of a factor to enhance proliferation in such an assay does not connote that it plays a role as a growth factor.

Recently we noted that BCL_1 tumor cells, removed from BALB/c mice 4-8 weeks after innoculation of mice with 10^6 tumor cells and purified by treatment with anti Thyl.2 monoclone and anti Lyt2.2, proliferated vigorously when cultured at 5 x 10^4 cells per well in the presence of DL supernatants. The extent of proliferation of these B cells in response to DL supernatants and to sources of lymphokines IL1 and IL2 is shown in Figure 1. A summary of the results testing various factors for their effects on proliferation in more than 20 experiments with normal B cells (as described above) and more than 10 experiments with BCL_1 cells, is shown in Table I.

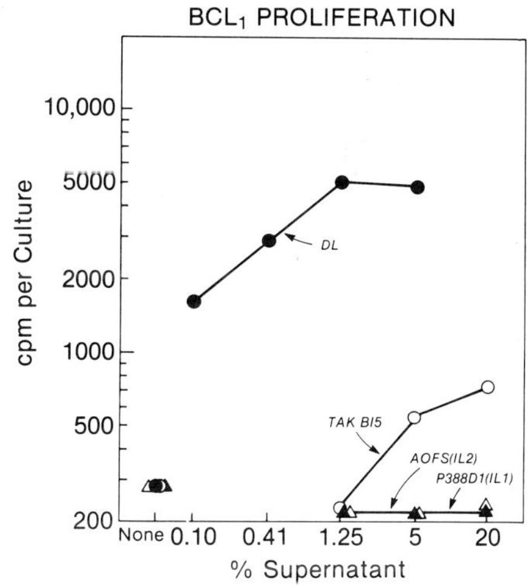

FIGURE 1. 5×10^4 BCL_1 cells per well cultured for 4 days. Proliferation determined by I^{125}UdR incorporation for final 6 hours.

It is absolutely clear that BCL_1 responds to a very much more limited range of factors than normal B cells. A simpler response by BCL_1 is also noted when mixtures of factors and anti-Ig are tested in normal B and BCL_1 proliferation, as seen in Table II. While sources of IL2 and DL supernatant show additive or greater effects when normal B cells are used, addition of IL2 in addition to DL supernatant has no effect on BCL_1 proliferation. Anti μ synergizes markedly with DL in normal B cell proliferation, but if anything slightly suppresses DL-induced BCL_1 proliferation.

This set of findings suggests that many cells of the in vivo BCL_1 tumor may be in a state of differentiation where they are already expressing receptors for a B cell growth factor present in DL supernatants, whereas normal spleens may contain few B cells in this state of differentiation or of the subset responsive to the growth factor present in DL. Experiments absorbing DL supernatants with BCL_1 further substantiate this concept.

Figure 2 shows the effect of absorbtion of 1 ml of DL supernatant with 4×10^8 BCL_1 cells in 3 batches over a 3-hour

TABLE I. Factors Active in the Proliferation of Normal and BCL_1 B Cells

Source of factor		Normal B cells S.I.	BCL_1 cells S.I.
IL_1	$P388D_1$	2.96	0.84
IL_2	FS6.14.13	4.16	1.06
	AOFS	2.76	0.96
TRF	DL	2.75	11.26
	TAKB15	3.77	4.93

TABLE II. Factor Mixes in B Cell Proliferation

	Source of factor	*Normal B cell S.I.*	*BCL_1 cells S.I.*
1.	*None*	*1.00*	*1.00*
	DL 0.5%	*8.14*	*21.94*
	DL 0.02%	*4.74*	*13.02*
	AOFS 10%	*3.83*	*0.94*
	AOFS 10% + DL 0.5%	*12.53*	*21.48*
	AOFS 10% + DL 0.02%	*6.74*	*7.63*
2.	*None*	*1.00*	*1.00*
	DL 1%	*2.43*	*12.54*
	anti μ 5 μg/ml	*2.78*	*0.95*
	anti μ + DL	*13.01*	*5.44*

period at room temperature. Absorbtion rendered such a supernatant completely unable to promote either normal B cell or BCL_1 proliferation but had much less effect on the activity of the supernatant in a TRF assay. Control supernatants were incubated at room temperature and dialyzed like those absorbed on BCL_1. A mixture of absorbed and control supernatants was able to support proliferation indicating that the absorbtion did not result in the release of toxic or inhibitory materials (not shown).

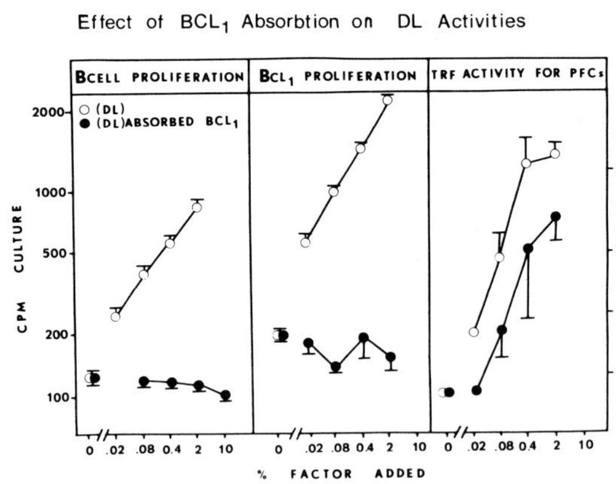

FIGURE 2. Factors were added to 5×10^4 T-depleted BCL_1, 10^5 normal splenic B cells or 10^6 splenic B cells with 30% AOFS (IL2) and 0.01% SRBC (TRF activity).

The absorbtion of BCGF but not TRF activity indicates that DL supernatants contain at least two lymphokines which we will call, for the time being, by their activities: (DL)TRF (active in the TRF assay for B cell antibody formation) and (DL)BCGF (active in the BCL_1 proliferation assay).

Two further kinds of absorbtion experiments, shown in Table III,

TABLE III. Abaorbtion of (DL)BCGF on BCL_1

Absorbtion	BCL_1 proliferation S.I.
1. None	13.65
BCL_1 untreated	2.73
BCL_1 ATS plus anti Thy1.2	2.53
BCL_1 glutaraldehyde-fixed 1%	1.55
2. None	9.44
BCL_1 anti Thy1.2	0.74
BALB/c anti Thy1.2	18.77

confirmed our interpretation. In one, both T-depleted BCL_1 (taken from ATS pretreated mice, anti Thyl.2 treated in vitro) and glutaraldehyde-fixed BCL_1 were effective at removing (DL)BCGF. In the second experiment, BCL_1 but not normal BALB/c B cells (both T-depleted) removed (DL)BCGF activity.

These results suggest that a large number of BCL_1 cells removed from the spleens of tumor-bearing mice have receptors for (DL)BCGF.

What more can we say about (DL)BCGF? Several kinds of experiments indicate that this factor(s) is produced by T cells. Most convincing, several subclones of DL produce high amounts of (DL)BCGF activity in the absence of recent stimulation of recent stimulation of any kind (not shown). Second in a large series of experiments we found that the highest levels of (DL)BCGF are produced when DL cells are stimulated with sources of IL2 rather than with stimulators or mitogens as seen in Figure 3. IL2 sources also induce several subclones of DL to release greater amounts of BCGF activity (data not shown).

Our results are most compatible with (DL)BCGF acting directly on B cells. Earlier experiments on single B cells showed unequivocally that DL supernatants had some activity

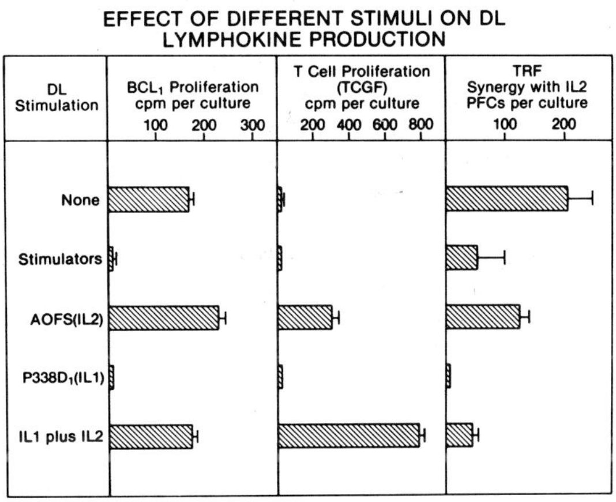

FIGURE 3.

with direct effects on the growth of single B cells (4). We would postulate that the factor responsible was (DL) BCGF and are currently testing that hypothesis.

We have no information as yet on how (DL)BCGF relates to BCGF activities described by Howard and Farrar and colleagues which is present in EL-4 supernatants (6) or that which we have found in other IL2 sources (2,3). Further experiments comparing other sources of BCGF with our own in the BCL_1 assay and biochemical characterization of (DL)BCGF should resolve these issues.

To summarize. We have described a second activity in DL supernatants previously found to have (DL) TRF, an activity which synergizes with IL2 in the development of antibody-forming cell to SRBC. This second activity appears to be a B cell growth factor (DL) BCGF. It causes BCL_1 tumor cells obtained from BALB/c spleens to proliferate strongly. Such BCL_1 cells appear to have receptors for (DL)BCGF since they absorb the growth activity of DL supernatants without absorbing the (DL) TRF activity. What is probably the same activity appears to work on some subpopulation of normal splenic B cells. Optimal induction of (DL)BCGF is found when IL2 is added to the DL line or to subclones of DL, some of which show significant "constitutive" production of (DL)BCGF.

Further studies should allow us to better understand the role of both (DL) TRF and (DL) BCGF in B cell responses.

REFERENCES

1. Swain, S. L., Dennert, G., Warner, J. F., and Dutton, R. W. *Proc. Natl. Acad. Sci. 78*, 2517 (1981).
2. Swain, S. L. and Dutton, R. W. *Lymphokines 6*, 219 (1981).
3. Swain, S. L., Wetzel, G. D., Soubiran, P. and Dutton, R. W. *Immunol. Rev. 63*, 111 (1982).
4. Wetzel, G. D., Swain, S. L. and Dutton, R. W. *J. Exp. Med.* (In press, 1982).
5. Parker, D. C. *Immunol. Rev. 52*, 115 (1980).
6. Howard, M., Farrar, J., Hilfiber, M., Johnson, B., Takatsu, K., Hamaoka, T., and Paul, W. E. *J. Exp. Med. 155*, 914 (1982).

DISCUSSION

ALTMAN: Will BCL_1 cells grow in long-term culture in the absence of exogenous growth factor? What is the minimal number of BCL_1 cells that will respond in Swain's BCGF assay?

SWAIN: Most of the *in vivo* BCL_1 cells will not grow when plated *in vitro* without added growth factor, but it is possible that a small subset of them might do so. As regards the minimal number of BCL_1 cells that can function in our assay, we routinely use $5x10^4$ cells but as few as 10^4 will respond.

MOND: Has Swain tried BCL_1-absorbed BCGF to ascertain whether the residual material synergizes with anti-μ to induce proliferation? Takatsu has reported he has no BCGF activity, yet Swain finds BCGF activity using the BCL_1 tumor line.

SWAIN: Absorbed BCGF does *not* synergize with anti-μ in normal B cell proliferation. The batch of Takatsu's factor in which we found BCGF activity has activity on both normal *and* BCL_1 cells. He also finds that his particular batch has BCGF activity in his own experiments.

IHLE: Would Swain comment on the relationship of her results to those of Irving Weissman who has shown that the BCL_1 immunoglobulin is specific for a retroviral protein, which he feels is the growth factor required for proliferation of the BCL_1 cells?

SWAIN: Assuming our BCL_1 cells are the same, I would have to say that in the absence of virus the tumor responds to BCGF made by T cells. Since the tumor does not grow in nude mice, it is also likely that the B cell growth factor is required *in vivo*.

SCHIMPL: Has Swain characterized the B cell clones growing out after stimulation *in vitro*?

SWAIN: No, we have not.

GARLAND: Is there any possibility that stimulation of BCGF by IL-2 in BCL_1 is simply a re-processing of IL-2?

SWAIN: I would judge Garland's proposition as unlikely. There is a small possibility, but it is really unlikely. Most sources of (DL)BCGF have not been induced with IL-2 and none is found in the supernatant. Furthermore some sublines and clones produce BCGF constitutively.

SOLUBLE FACTORS REQUIRED FOR B CELL PROLIFERATION AND DIFFERENTIATION

Maureen Howard, Kenji Nakanishi and William E. Paul

Laboratory of Immunology
National Institute of Allergy and Infectious Diseases
National Institutes of Health
Bethesda, Maryland

I. INTRODUCTION

Considerable evidence has been accumulated to support a conceptual division of splenic B lymphocytes into two populations which differ from one another in the membrane antigens they express, in the immunogens to which they respond, and in their sensitivity to distinct regulatory mechanisms (1-7). These subsets, often referred to as Lyb 5^+ and Lyb 5^-, are also distinguished in terms of their activation requirements, with Lyb 5^+ B cells being readily responsive to antigen plus soluble factors, and Lyb 5^- B cells requiring an MHC-restricted interaction with helper T cells (6,7). In this report, we describe our own studies on the activation of Lyb 5^+ B cells, and in particular our analysis of the soluble factors required for proliferation of these cells and their differentiation into antibody producing cells. As we have selected anti-IgM antibodies as a polyclonal analog for antigen, we believe our data reflect the behaviour of most if not all Lyb 5^+ B cells.

II. B CELL PROLIFERATION FACTORS

To identify factors involved in B cell proliferation, we have developed a short-term B cell co-stimulator assay in which highly purified B cells are cultured at low cell density (8,9).

Interleukins, Lymphokines, and Cytokines

ISBN 0-12-527540-4

Three signals were found to be required for induction of DNA synthesis by the particular subset of small B lymphocytes investigated: a signal delivered by antibodies specific for the IgM receptor expressed on the B cell membrane; a signal delivered by a T cell derived factor designated with the provisional name B cell growth factor (BCGF); and a signal delivered by the macrophage derived factor interleukin 1 (IL1). BCGF and IL1 act in a non-antigen specific, synergistic manner. Their requirement is not observed at high B cell numbers, presumably reflecting accessory cell contamination and endogenous factor production at such cell densities.

In collaborative experiments with J. Farrar, we have examined in detail the properties of BCGF (8,10,11). It has an approximate Mr of 18,000 and an isoelectric point in the range of 6.5-7.5. It can be separated from the T cell growth factor interleukin 2 (IL2) by gel filtration, phenylsepharose chromatography, sodium dodecyl sulfate-polyacrylamide gel electrophoresis, and isoelectricfocusing (8,10,11). It is not absorbed by cells which bear IL2 receptors (8). It can be distinguished both functionally and by molecular weight from other previously described T cell factors, namely T cell replacing factor (TRF) (12), colony stimulating factor (13), interleukin 3 (14), B cell replication and maturation factor (15), and B cell differentiation factor (16). It does not produce proliferation of PHA stimulated thymocytes, and so is distinct from the monokine interleukin 1 (IL1) (17). BCGF has no apparent action on resting B cells, and its effect on anti-Ig activated B cells is polyclonal, non-H2 restricted, and seemingly restricted to DNA synthesis without the production of antibody forming cells (AFC) (8).

Identification of the monokine involved in anti-Ig induced B cell proliferation being IL1 is based on correlation of B cell and T cell co-stimulating activities following an extensive series of biochemical purification procedures (9). Murine IL1 purified to apparent homogeneity by S. Mizel, and human IL1 purified 50,000-fold by L. Lachman, both showed excellent B cell co-stimulating activity in our assay systems. Like BCGF, IL1 has no apparent action on resting B cells and its action on activated B cells is genetically unrestricted and does not induce the development of AFC (9).

The provocative concept that IgM-specific induction of B cell proliferation requires three stimuli prompted a detailed kinetic analysis to explore the relative roles of the three signals. DeFranco *et al.* (18) have shown elsewhere that anti-Ig acts as expected on the G_o (resting) B lymphocyte. These studies also show that in the first round of anti-Ig induced B cell activation few cells enter S phase before 30 hr. By culturing purified B cells with anti-Ig and one co-factor, while adding the second co-factor at different times of the culture period, we were able to demonstrate that BCGF and IL1 acted at independent times of the cell cycle (9). BCGF was required in the first 6 hr of a 42 hr culture period, suggesting it acted on G_o or early G_1 B cells. In contrast, addition of IL1 could be delayed 16-20 hr after initiation of the culture without any significant decrease in the response, suggesting it governed entry of the B cell into late G_1 phase of the cycle. These data have led us to propose the model of B cell activation illustrated in Fig. 1. Verification of these assigned roles will obviously require reagents which precisely identify membrane bound factor-acceptor-sites and specific binding to these.

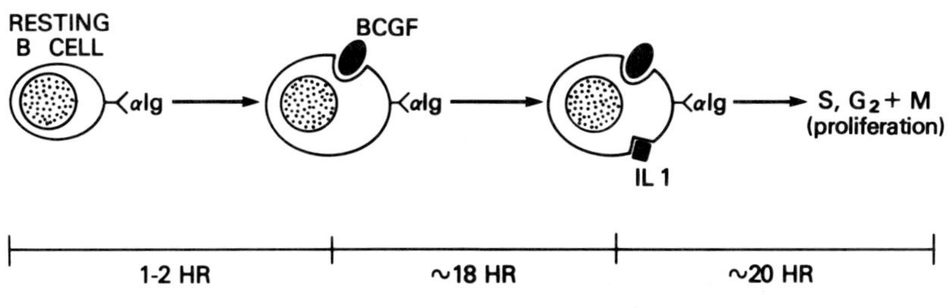

FIGURE 1. Activation of Lyb 5^+ B cells

III. B CELL DIFFERENTIATION FACTORS

B cells stimulated with anti-IgM and BCGF proliferated but failed to develop into Ig-secreting cells (19) as monitored by staining cells for cytoplasmic (c) IgM with a fluoresceinated monoclonal anti-IgM antibody. Others have shown that antigen and anti-Ig driven differentiation into antibody secretion requires additional factors, such as "T cell replacing factor" (TRF) and/or IL-2 (20-24). We found that the development of specific plaque-forming cells by B cells stimulated with sheep erythrocytes (8) and the appearance of $cIgM^+$ cells in cultures of anti-IgM stimulated B cells (19) required

TABLE 1. Two Distinct B Cell Differentiation Factors

Factor	*Properties*
B15-TRF	1. *In supernatants of B15K12 T hybridoma* 2. *Acts early in AFC response* 3. *Distinct from IL2*
EL-TRF	1. *In induced supernatants of EL4* 2. *Acts late in AFC response* 3. *Distinct from IL2*

the addition of three distinct T-cell derived factors: BCGF, purified from induced EL-4 supernatants by either IEF or phenylsepharose chromatography; a supernatant obtained from the T cell hybridoma B151K12 (22) (designated B15-TRF); and the pI-4.5 fraction from the IEF on an induced EL-4 supernatant (designated EL-TRF). The addition of any one or two of these factors to cultures of B cells stimulated with anti-IgM produced few, if any, $cIgM^+$ cells. Interestingly while neither B15-TRF nor EL-TRF caused proliferation of anti-IgM activated B cells, B15-TRF substantially enhanced cell yields in cultures also containing anti-IgM and BCGF. The relative roles of these three co-factors in B cell development were investigated in kinetic analyses. The results showed that BCGF and B15-TRF are required early in the culture while EL-TRF appears to be a late-acting differentiation factor. Although EL-TRF preparations contain IL-2, we believe the B cell differentiation factor is a distinct entity since other preparations rich in IL-2 fail to replace EL-TRF in these assays. EL-TRF may be identical to the recently described factor BCDFμ (25).

These data demonstrate there are two distinct differentiation factors involved in B cell development, both of which fit the general description "T cell replacing factor." The properties of these two factors are summarized in Table 1.

IV. CONCLUSIONS

The studies outlined here can be primarily regarded as forming a foundation for the fuller understanding of the cellular and molecular biology of the activation of Lyb 5^+ B cells.

They also allow us to propose a model, illustrated in Fig. 2, for the action of various factors which regulate the activation, proliferation and differentiation of these cells. We suggest that agents which can cross-link membrane Ig receptors in an appropriate manner, such as anti-Ig antibodies and certain antigens, cause resting B cells to become activated and to enter the G_1 phase of the cell cycle. As a result of this activation, these G_1 B cells become sensitive to certain lymphokines and interaction with these agents induces the cells to enter S phase. The sensitivity of G_1 B cells to such growth factors is most likely determined by the expression of growth factor receptors on the cell membrane. Lyb 5^+ B cells stimulated to enter G_1 by anti-IgM are sensitive to the action of BCGF, a T cell derived factor, and also to the macrophage-derived factor IL1. The differentiation of these cells into antibody-secreting cells requires two additional factors: B15-TRF, which apparently acts as a growth factor, and a differentiation factor in EL-4 supernatant (EL-TRF).

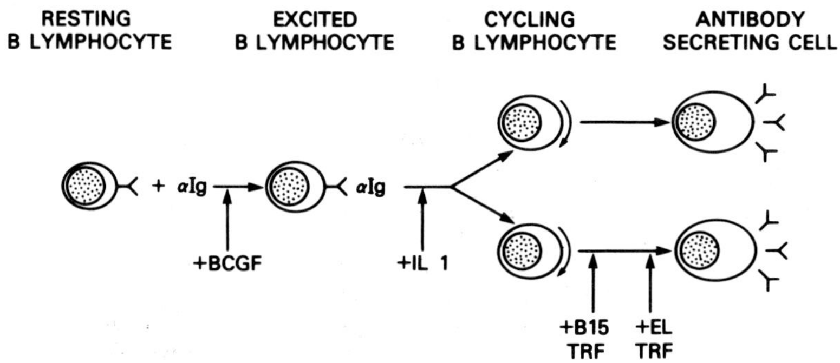

FIGURE 2. Non-specific factors involved in B cell development

REFERENCES

1. Mosier, D.E., Zitron, I.M., Mond, J.J., Ahmed, A., Scher, I, and Paul, W.E., *Immunol. Rev. 37,* 89 (1977).
2. Ahmed, A., Scher, I., Sharrow, S.O., Smith, A.H., Paul, W.E., Sachs, D.H., and Sell, K.W., *J. Exp. Med., 145,* 101 (1977).

3. Huber, B., Gershon, R.K., and Cantor, H., *J. Exp. Med. 145,* 10 (1977).
4. Sieckmann, D.G., Scher, I., Asofsky, R., Mosier, D.E., and Paul, W.E., *J. Exp. Med. 148,* 1628 (1978).
5. Perlmutter, R.M., Nahm, M., Stein, K.E., Slack, J., Zitron, I., Paul, W.E., and Davie, J.M., *J. Exp. Med. 149,* 993 (1979).
6. Singer, A., Morrissey, P., Hathcock, K., Ahmed, A., Scher, I., and Hodes, R., *J. Exp. Med. 154,* 501 (1981).
7. Asano, Y., Singer, A., and Hodes, R.J., *J. Exp. Med. 154,* 1100 (1981).
8. Howard, M., Farrer, J., Hilfiker, M., Johnson, B., Takatsu, K., and Paul, W.E., *J. Exp. Med. 155,* 914 (1982).
9. Howard, M., Mizel, S., Lachman, L., Johnson, B., and Paul, W.E., manuscript submitted (1982).
10. Farrar, J.J., Benjamin, W.R., Hilfiker, M.L., Howard, M., Farrar, W.L., and Fuller-Farrar, J., *Immunol. Rev. 63,* 129 (1982).
11. Farrar, J., Howard, M., Fuller-Farrar, J., and Paul, W.E. manuscript in preparation (1982).
12. Schimpl, A., Hubner, L., Wong, C., and Wecker, E., *Behring Inst. Mitt. 67,* 221 (1980).
13. Nicola, N., Burgess, A., and Metcalf, D., *J. Biol. Chem. 254,* 5290 (1979).
14. Hapel, A., Lee, J., Farrar, W., and Ihle, J., *Cell 25,* 179 (1981).
15. Andersson, J., and Melchers, F., *Proc. Natl. Acad. Sci. 78,* 2497 (1981).
16. Isakson, P., Pure, E., Vitetta, E., and Krammer, P., *J. Exp. Med. 155,* (1982).
17. Lachmann, L., Hacker, M., Blyden, G., and Handschumacher, R., *Cell. Immunol. 34,* 416 (1977).
18. DeFranco, A.L., Raveche, E.S., Asofsky, R., and Paul, W.E., *J. Exp. Med. 155,* 1523 (1982).
19. Nakanishi, K., Howard, M., Muraguchi, A., Farrar, J., Takatsu, K., Hamaoka, T., and Paul, W.E., manuscript in preparation.
20. Kishimoto, T., and Ishizaka, K., *J. Immunol. 114,* 585 (1975).
21. Parker, D., *Immunol. Rev. 52,* 115 (1980).
22. Takatsu, K., and Hamaoka, T., *Immunol. Rev. 64,* 21 (1982).
23. Leibson, H.J., Marrack, P., and Kappler, J.W., *J. Exp. Med., 154,* 1681 (1981).
24. Swain, S., Dennert, G., Warner, J., and Dutton, R., *Proc. Natl. Acad. Sci. USA 78,* 2517 (1981).
25. Pure, E., Isakson, P.C., Takatsu, K., Hamaoka, T., Swain, S.L., Dutton, D.W., Dennert, G., Uhr, J.W., and Vitetta, E.S., *J. Immunol. 127,* 1953 (1981).

DISCUSSION

BELLER: Howard suggested that both BCGF and IL-1 are required for B cell proliferation, implying distinct roles for these factors. What is the evidence for that statement? Are these factors synergistic? How substantial is the response to these components supplied singly?

HOWARD: BCGF and IL-1 enhance anti-IgM-induced B cell proliferation in a synergistic manner. This is our only evidence that they act on the same B cells rather than on different B cell subsets. Both factors will, by themselves, enhance anti-IgM-induced B cell proliferation, but we believe this reflects endogenous production of sub-optimal amounts of the other factor, with both factors being required absolutely.

STADLER: Did Howard seek to absorb BCGF, IL-1 or TRF on B cells?

HOWARD: We have not yet attempted absorption of IL-1 or TRF on B cells. We have made some limited attempts to absorb BCGF on B cell lymphomas, but obtained either no absorption whatsoever, or non-specific absorption of a number of factors in an EL4 supernatant.

SCHRADER: Nossal has shown that antigen-specific B cells plated singly in culture, without filler cells and stimulated with specific antigen, respond to a factor or factors present in supernatants of Con-A-stimulated EL4 or certain T cell hybridomas with an increased frequency of clones as measured either by direct observation of the number of cells developed per culture or by antibody-forming cells. Ian Clark Lewis has biochemical evidence that this activity is distinct from IL-2. How do these results fit with Howard's model?

HOWARD: The principal difference between Nossal's work and ours seems to be that everything required for complete AFC development in a single cell assay appears to be contained in induced EL4 supernatants, whereas we additionally need to provide IL-1 and B15-TRF. The most likely explanation for this is that EL4 supernatants could contain trace amounts of B15-TRF which are sufficient for the extraordinarily-sensitive single cell assay, but rather insufficient for our higher cell density assay. Regarding the requirement for IL-1, Nossal's culture system is FCS-dependent and there is now growing concern that the "good batches" of FCS we all select for *in vitro* assays, are in fact "good" because they contain variable amounts of IL-1.

Again, trace IL-1 contamination in the FCS may well be sufficient for the super-sensitive single cell assay.

FELDMANN: I speak to Howard's comment on differences between Nossal's single cell assay and her assay with more cells, and consequent requirement for further growth factors. It is *also* possible that the difference in the two investigations is due to activation of B cells during Nossal's purification procedure. This was shown previously when B cell populations were purified by absorption onto anti-Ig columns.

MOND: Can Howard use CBA/N spleen cells as fillers to supply the BCGF and IL-1 activities required for low density cultures to respond to αIg? Will BCGF + αIg induce IL-1 receptors on B cells?

HOWARD: We have not attempted the use of CBA/N cells as a growth factor source. My feeling is that the use of any filler system to generate the growth factors, at a time when we have in hand highly purified factors which work, would be a backward step. As to Mond's second query, we do not yet have reagents capable of addressing this question. IL-1 receptors may be on resting B cells already, or alternatively be induced as a result of stimulation with either αIgM_1, BCGF or both.

SWAIN: Howard's data seem to suggest that her (EL)BCGF and our (DL)BCGF are different and that it would, perhaps, be worthwhile to consider two classes of "BCGF". One, an initiation factor, and the second, one that maintains cycling B cells.

HOWARD: I agree that the BCGF Swain finds in DL supernates which enhances BCL_1 proliferation is distinct from our BCGF. I think we should explore whether Swain's BCGF is either IL-1 or $B15\text{-}TRF_1$ or alternatively something which induces BCL_1 cells to make IL-1 or B15-TRF. Neither Takatsu nor I find B15-TRF to be a proliferation factor in our BCGF assays. However, it may well register this way in the BCL_1 assay.

FORD: At the low concentrations of B cells used, do many cells absorb out any IL-1 activity?

HOWARD: We have not tried this yet. The difficulty in this particular experiment is that in order to activate B cells to the point of IL-1 responsiveness, without endogenous production of IL-1 which would drive them past this point, one needs to culture purified B cells at very low density, i.e. 10^4/microtiter well. Consequently, generating sufficient cells to enable factor absorption studies becomes a major undertaking.

BIOCHEMICAL CHARACTERIZATION OF A MURINE B-CELL GROWTH FACTOR DISTINCT FROM INTERLEUKIN-2

John J. Farrar[1]
Maureen Howard
Mary L. Hilfiker
William R. Benjamin[1]
Janet Fuller-Farrar[1] and
William E. Paul

Laboratory of Microbiology and
Immunology, National Institute
of Dental Research and
Laboratory of Immunology
National Institute of Allergy
and Infectious Disease
National Institutes of Health
Bethesda, Maryland

I. INTRODUCTION

A number of studies have shown that T-cells produce a variety of antigen-nonspecific soluble factors which augment the antibody response of B cell-enriched cultures to the thymic-dependent antigen, sheep erythrocytes. Until recently, only two subclasses of helper factors had been delineated. The first type of factor is closely associated with IL-2 (1,2) and the second with T-cell replacing factor (3). Because the IL-2 associated helper activity has been shown to be early acting, there has been speculation that IL-2 might actually be the factor which induces the B-cell prolifera tive response. It was, therefore, of importance to determine the relationship between IL-2 and the newly described B-cell growth factor (BCGF) as a prerequisite to an examination of the relative roles of BCGF and IL-2 in the induction of antibody synthesis.

In this report, we present further chromatographic data which serve to characterize murine BCGF and differentiate this factor from IL-2.

[1] *Present address: Hoffmann-La Roche, Nutley, New Jersey*

ISBN 0-12-527540-4

II. RESULTS AND DISCUSSION

In a previous report we showed that IL-2 and BCGF produced by phorbol myristate acetate-stimulated EL-4 cells could be distinguished by two criteria (4). Whereas IL-2 eluted during gel filtration with an apparent molecular weight of approximately 30,000, the BCGF exhibited a smaller molecular weight of 15,000 to 20,000. Additionally, the two factors were differentially adsorbed by an IL-2-dependent cloned T-cell line. Whereas greater than 95% of IL-2 was adsorbed by the cloned T-cells, less than 20% of the BCGF was adsorbed. Figure 1 presents the results of a preparative isoelectric focusing experiment in which IL-2 and BCGF are also shown to have distinct pI's. Whereas, IL-2 elutes with a broad pI between 3.5 and 4.5, the BCGF exhibits a more

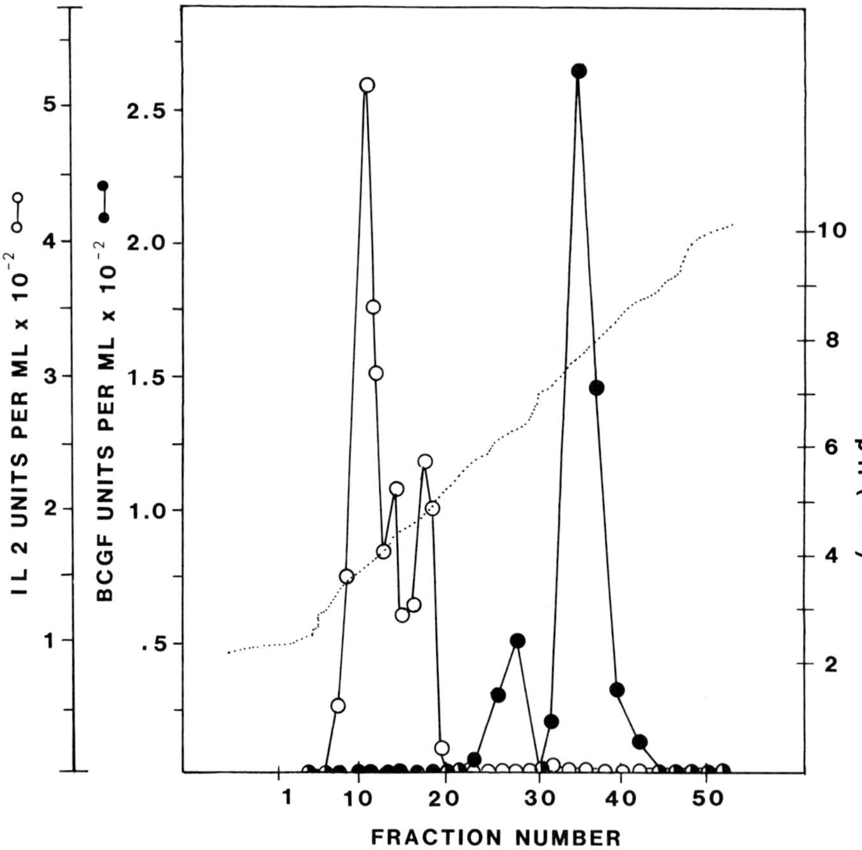

FIGURE 1. *Profile of IL-2 and BCGF activity following separation by preparative isoelectric focusing.*

alkaline pI range with distinct peaks at pH 6.4 and 7.4. Thus, by three criteria; molecular weight, isoelectric point, and differential adsorption to factor-dependent T-cells, IL-2 and BCGF appear to be distinct molecules.

In order to prepare BCGF completely devoid of IL-2, a three step purification scheme was developed. Crude serum-free EL-4 thymoma-derived culture supernatant was precipitated by treatment with $(NH_4)_2SO_4$. The precipitate obtained between 40% and 80% saturation was chromatographed on a phenyl sepharose column to achieve a 90% separation of IL-2 and BCGF. The results of a representative experiment are shown in Figure 2 in which it is seen that BCGF elutes in the early portion of the ethanediol gradient and the IL-2 elutes in the latter portion of the gradient. A pool of fractions (40-58) containing BCGF and the leading edge of

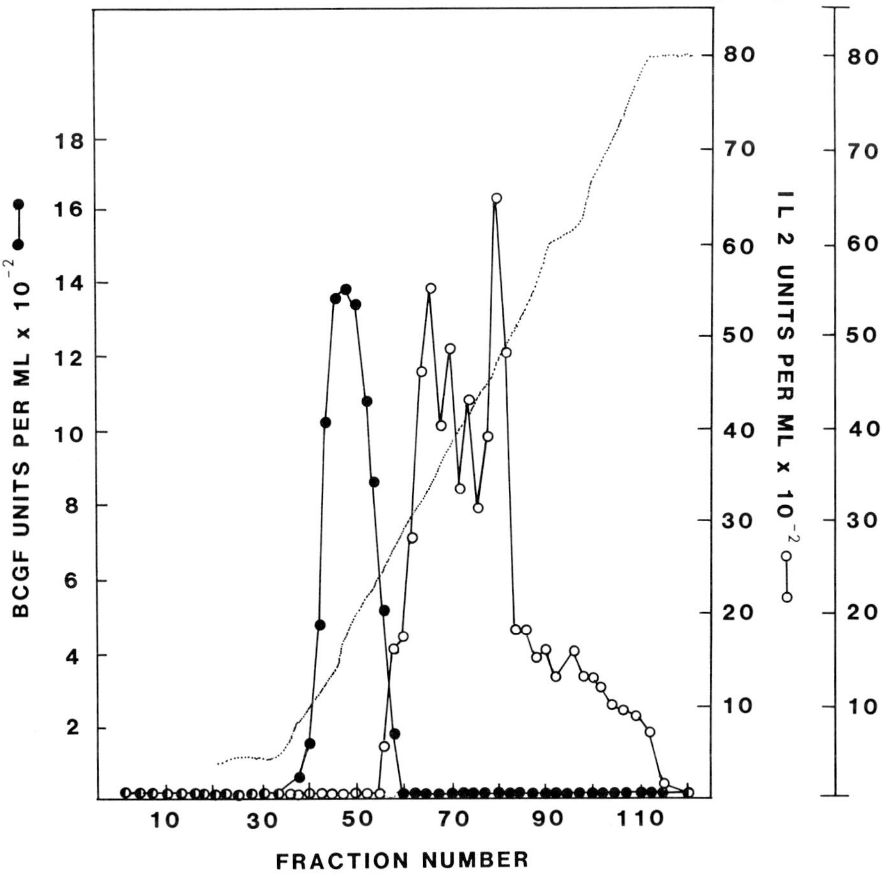

FIGURE 2. *Phenyl sepharose hydrophobic chromatography of ammonium sulfate precipitated EL-4 thymoma-derived culture supernatant containing IL-2 and BCGF.*

the IL-2 peak were prepared and further fractionated by electrophoresis in a pH 8.9 tris glycine 7.5 % polyacrylamide gel. The gel was then sliced and each individual gel slice eluted and assayed for both IL-2 and BCGF. The results of this experiment are shown in Figure 3. Despite the lower molecular weight of BCGF, the factor migrates through the gel with a much slower Rf of .326 than IL-2 with an Rf of .558 presumably due to the alkaline pI of the factor. Thus, through the use of ammonium sulfate precipitation, phenyl sepharose chromatography, and polyacrylamide gel electrophore sis, IL-2 and BCGF can be completely separated.

In order to begin to examine the chemical nature of BCGF, we tested the sensitivity of BCGF to proteolytic enzymes. Table 1 shows that both BCGF and IL-2 were sensitive to trypsin inactivation but not to RNase. Control experiments

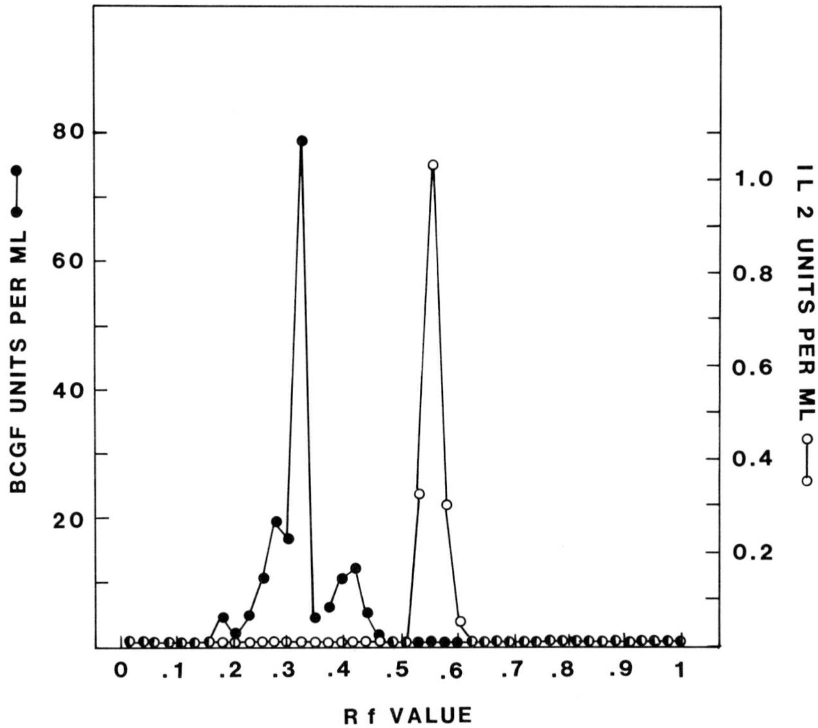

FIGURE 3. *Tris-glycine PAGE of phenyl sepharose purified BCGF.*

Table 1. Sensitivity of BCGF and IL-2 to Trypsin and RNase treatment

Lymphokine	Treatment	Activity (Units)
IL-2	None	5048.6
	Trypsin	0.4
	RNase	2232.1
BCGF	None	99.6
	Trypsin	0.1
	RNase	88.2

showed that a trypsin inactivated preparation of BCGF did not inhibit the response to untreated BCGF. Therefore, it is unlikely that treatment with the proteolytic enzyme is activating a suppressor molecule rather that digesting BCGF. Recent results indicate that treatment of BCGF with neuramindase results in a shift of the pI of the factor to a more alkaline and homogenous pI of 9.3 (Data not shown). These results are similar to what has previously been observed for EL-4 thymoma-derived IL-2 in which the desialated IL-2 exhibited a single pI of 5.0.

The results in this report indicate that IL-2 and BCGF are distinct glycoprotiens with molecular weights of 21,000 and 12,000, and pI's of 5.0 and 9.3 respectively. Experiments are currently in progress to determine the relative mechanisms-of-action of the two factors in the induction of antibody synthesis.

References

1. Farrar, J.J., P.L. Simon, W.J. Koopman, and J. Fuller-Bonar. 1978. Biochemical relationship of thymocyte mitogenic factor and factors enhancing humoral and cell-mediated immune responses. J. Immunol. 121:1353.

2. Watson, J., L.A. Aarden, J. Shaw and V. Paetkau. 1979. Molecular and quantitative analysis of helper T-cell-replacing factors on the induction of antigen sensitive B and T-lymphocytes. J. Immunol. 122:1633.

3. Schimpl, A., L. Hubner, C.A. Wong, and E. Wecker. 1980. Distinction between T-helper cell replacing factor (TRF) and T-cell growth factor (TCGF). Behring Inst. Mitt. 67:221.

4. Howard, M., J. Farrar, M. Hilfiker, B. Johnson, K. Takatsu, T. Hamaoka, and W.E. Paul. 1982. Identification of a T-cell-derived B-cell growth factor distinct from Interleukin-2. J. Exp. Med. 155:914.

DISCUSSION

MOND: Since Farrar has shown that at high cell density BCGF is not required for responsiveness to αIg, why should BCGF be required to obtain a normal *in vitro* SRBC response? The cells should already be proliferating adequately in response to antigen (which is akin to αIg) and to constitutively-elaborated BCGF.

FARRAR: Mond's argument assumes that αIg-induced proliferation is functionally the counterpart of a thymic-dependent antigen-induced proliferation--something that I do not believe is necessarily the case. Furthermore, additional factors besides BCGF may be required to support proliferation in the antigen-driven system, which factors are not essential in the αIg system. Factors such as these might be contributed, in part, by the BCGF preparation itself.

STADLER: Does Farrar have an explanation for the heterogeneity of BCGF in terms of its PI, and MW on SDS-gels?

FARRAR: At this point I do not. We have not yet had an opportunity to examine whether variable glycosylation accounts for this heterogeneity, as has been shown by Kendall Smith for IL-2.

RAPPAPORT: In the course of purification of BCGF by Sepharose or acrylamide gel columns, there is the possibility of Farrar's product becoming contaminated with bacterial endotoxin. Does he have data relative to the avoidance of endotoxin in his purified or partially purified BCGF fractions?

FARRAR: I must admit that we have never quantitated the amount of endotoxin in the preparations. However, in our experience, endotoxin does not synergize with anti-mμ in this assay. Moreover, the response can be obtained as well in B cell cultures from endotoxin-resistant mice. Consequently, I rather doubt that minimal and perhaps almost unavoidable endotoxin contamination constitutes a significant problem in the basic interpretation of these experiments.

CELLULAR ORIGIN OF A HUMAN B CELL MITOGENIC FACTOR[1]

Abby L. Maizel
Shashikant R. Mehta
C. G. Sahasrabuddhe
Richard J. Ford
Nicola Kouttab

Department of Pathology, Division of Pathobiology
M. D. Anderson Hospital and Tumor Institute
Houston, Texas

The current studies have focused upon the examination of the cellular sources responsible for the production of a human B cell mitogenic factor. This proliferative factor was assayed using anti-u bound to an insoluble matrix support for the initial activation step. Lymphocytes derived from peripheral venous blood were purified and either cultured alone, or in combination, for a period of 72 hours. Conditioned media from these sources were biochemically fractionated by ion exchange chromatography and units of activity calculated. Experiments indicated that lectin stimulated T cells, cultured with an obligate number of monocytes, were capable of producing a B cell mitogenic factor. The effect of the monocyte could be substituted by partially purified interleukin-1. Factor production could be significantly augmented when lectin-activated T cells were cultured both with interleukin-1 and syngeneic B lymphocytes.

INTRODUCTION

The modulation of human B cell proliferation, by multiple signals, has been determined to be analogous to the obligate signal requirements for T cell proliferation. Experiments have demonstrated that an activated B lymphocyte may be stimulated to enter the S phase of the cell cycle by factors

[1]*This work was supported by NIH grant CA21927, a grant from the Faisal Foundation and an institutional grant from the Interferon Research Fund.*

ISBN 0-12-527540-4

present in lectin stimulated peripheral blood lymphocyte conditioned media (1). One factor responsible for promoting human B cell proliferation has been shown to be distinct from the host of other cytokines presently known to exist in mononuclear cell conditioned media (2). This soluble factor, known as B cell growth factor (BCGF), is currently thought to be a trypsin sensitive protein of 12,000-13,000 daltons. The protein possesses a mildly acidic isoelectric point (pH 6.3-6.6), is stable from pH 3 - pH 8 is uneffected by reducing agents (i.e. β mercaptoethanol), and is relatively heat unstable at 56°C (3,4). The B cell growth factor has been shown to possess target cell and functional specificity. A proportion of activated human B lymphocytes, stimulated with a partially purified growth factor preparation, may be demonstrated to enter the S phase of the cell cycle in the absence of detectable immunoglobulin secretion (2).

A question of importance that remains to be fully elucidated concerns the cell of origin for the human BCGF. Work in the murine system, using either putatively malignant thymoma lines and/or T hybridoma populations, has suggested a T cell origin for one such factor capable of promoting B cell S phase entry (5,6). In the present report we have explored this question using normal human peripheral blood mononuclear cells. The experiments have indicated that the normal human T cell is capable of producing a B cell specific lymphokine and its production is dependent upon multiple cellular interactions.

RESULTS

For the experiments directed at the determination of the cellular origin of BCGF, an assay system was utilized which depended upon insolubilized anti-μ for an initial activation step. Human B lymphocytes were purified from peripheral venous blood by either negative or positive selection (2). Cells prepared in this manner were greater than 88% sIg positive with less than 1% contamination by E-rosette positive cells and less than 4% contamination by nonspecific esterase positive cells. The human B cells were exposed to the insolubilized anti-μ, together with multiple dilutions of the putative growth factor containing preparations, for a period of 96 hours. 16 hours prior to termination of the culture the cells were labelled with 1uCi [^{3}H]-Tdr (6 Ci/mMol) and subsequently harvested onto glass fiber strips. Thymidine incorporation values of the experimental samples were compared to known standards prepared in the same manner and arbitrary units of activity calculated. The standard preparation of BCGF was derived from conditioned media from pooled peripheral blood lymphocytes (>20 donors) cultured at a concentration of 1 x 10^6/ml for 72 hours. Both experimental and standard samples of the growth factor preparations were biochemically fractionated at least through ion exchange chromatography as previously described (2).

Table 1 presents representative thymidine incorporation values for a standard preparation of partially purified human B cell growth factor. For these experiments human B cells were continuously exposed to anti-μ and growth factor during the course of a 96 hour incubation. The response to these two signals was seen to be both anti-μ dependent and factor dependent. Determination of the number of cells in S phase, as indicated by cytofluorography of propidium iodide stained preparations, revealed that 47% (mean of 3 independent replicate experiments) of the cells may be seen in the DNA synthesis phase 72 hours following anti-μ and factor stimulation. Control incubations using the insoluble matrix support lacking the relevant antibody routinely gave values consistent with factor addition alone. Experiments were next directed at the examination of the effect of the time of addition of growth factor, to an anti-μ stimulated B cell population, on thymidine incorporation. For these studies human B cells were exposed to anti-μ for periods of time ranging from 0-72 hours prior to factor addition or control medium addition. The growth factor preparations were added to a constant number of viable cells and thymidine incorporation was tested during the last 16 hours of culture which was continued for 96 hours subsequent to factor addition. Table 2 presents representative values seen with the delayed addition of growth factor. Of interest is the observation that the response to anti-μ alone and the response to anti-μ plus factor continually diminishes with time although the efficacy of factor addition, as indicated by a stimulation index, continually increases. Experiments were next conducted examining the efficacy of anti-μ stimulation in reference to the cells' subsequent ability to absorb BCGF from a partially purified preparation of the growth factor. Cells that were stimulated with anti-μ for periods of time from 0 through 120 hours were incubated for 2 hours at 4°C with a preparation of BCGF that had been purified by ion exchange chromatography and multiple gel filtration steps (2). The growth factor sample, after absorption, was tested on activated B cells for its ability to promote S phase entry. The results indicated that a population of human B lymphocytes could effectively absorb BCGF from a growth factor preparation subsequent to anti-μ stimulation; and, that the ability of the population to

TABLE I. Effect of Anti-μ and Growth Factor on the B Cell Proliferative Response

	$[^3H]$ - Tdr/2x10^5 cells[a]		
Anti-μ (ug/ml)[c]	10% BCGF	5% BCGF	1% BCGF[b]
30ug/ml	12808	10447	10122
15ug/ml	34980	16448	10520
5ug/ml	20222	15116	8136
1ug/ml	9016	7295	4248
0ug/ml	7253	6488	2176

a. Cells were cultured in microtiter plates at 2x10^5 cells/well. Thymidine incorporation was measured during the last 16 hours of a 96 hour total culture period.

b. The percentages indicate the final amount of factor added to the cultures on a volume per volume basis. The BCGF was partially purified from the conditioned medium by ammonium sulfate precipitation, ion exchange chromatography, p30 and p100 gel chromatography (2).

c. The concentrations listed refer to the actual amount of specific anti-μ antibody, bound to the insoluble matrix support bead, per ml of culture volume.

absorb BCGF was dependent upon the time of exposure of the B lymphocytes to anti-μ. Specifically, a cell population that had been exposed to anti-μ for 96 hours was capable of absorbing at least three fold more factor than cells that were exposed for 24 hours. Control experimentation revealed that non-activated human B cells and/or continuously cultured T cells failed to specifically absorb BCGF from the factor containing samples.

Using the above described assay where anti-μ and growth factor were added simultaneously, experiments were undertaken to determine the cellular origin of the human B cell specific factor. For these studies human T cells, B cells, and monocytes were purified from peripheral blood (7). The cell populations were lectin stimulated either alone, or in combination for a period of 72 hours. At this time, the conditioned media were clarified and fractionated through DEAE sephadex chromatography (2). The amount of activity promoting S phase entry in anti-μ activated B lymphocytes was determined in comparison to a known standard and units of activity were calculated. From Table 3 it may be seen that lectin activated T cells, by themselves, were relatively poor producers of a

TABLE 2. Effect of Delayed Factor Addition on the B Cell Proliferative Response

	$[^3H]$ - Tdr/2x10^5 cells	
Time of anti-μ prestimulation[b]	anti-μ(15ug/ml) -BCGF	anti-μ(15ug/ml) +BCGF (10%)[a]
0	12730	61882
24	4400	53260
48	1950	36265
72	1148	31719

a. Cells were exposed to anti-μ for variable periods of time prior to addition of factor (+BCGF) or the addition of control medium (-BCGF). For the cells receiving either control medium or factor, the material was added at 10%(v/v) of the final 200 ul microtiter culture volume. The BCGF used for these experiments was prepared by ammonium sulfate precipitation and ion exchange chromatography (2).

b. B cells were exposed to anti-μ for the stated time periods prior to the addition either of control medium or factor. For the group referred to as 0, both anti-μ and factor (or control medium) were added together at time zero and cultured for the subsequent 96 hours. For the groups labelled 24, 48, or 72 the cells were first exposed to anti-μ for 24, 48, or 72 hours, respectively. At this time cell viability was determined, cell number adjusted to 2x10^5 cells/well, factor or control medium added, and the cells cultured for an additional 96 hours. Thymidine incorporation was tested during the last 16 hours of the 96 hour period subsequent to factor or control medium addition.

human B cell growth factor. Addition of monocytes at a concentration of 10% of the added T cells, resulted in substantive augmentation of the factor production. Further augmentation of factor production could be observed if the T cells, cultured with monocytes, were further cocultured with syngeneic B cells. Control cultures of lectin stimulated B lymphocytes alone, or in combination with monocytes, without the presence of T cells failed to result in substantial factor generation. Experiments were also undertaken to examine the efficacy of purified IL-1 to substitute for the macrophage in the generation of a T cell derived B cell growth promoting agent. Human IL-1 was prepared by a fractionation sequence inclusive of ion exchange chromatography, gel filtration on a p30 (BioRad) column and gel filtration on a high pressure

TABLE 3. Cellular Origin of the B Cell Mitogenic Factor

	Units of activity[a]
T cells + lectin (PHA)	9.4
T cells + monocytes + lectin[b]	32.1
T cells + IL-1 + lectin[c]	37.6
T cells + B cells + monocytes + lectin	59.5
T cells + B cells + IL-1 + lectin	57.8
B cells + monocytes + lectin	5.8
B cells + IL-1 + lectin	5.3
Standard BCGF preparation	100.0

a. Units of activity in an experimental sample are herein defined in relation to that activity demonstrated by a standard preparation which has been partially purified through ammonium sulfate precipitation and DEAE-sephadex chromatography. All experimental samples, prior to testing for activity have been processed through DEAE chromatography. The standard preparation of growth factor, prepared from a multiple donor pool, was tested in a mitogenic microtiter assay and the degree of thymidine incorporation mediated by the standard preparation at a fixed dilution was assigned a value equivalent to 100 units/ml. The results presented represent the mean values from three experiments.

b. Monocytes were added at a fixed ratio of 1 monocyte per 10 T lymphocytes. Monocytes were purified as indicated in the text. In those experiments where B lymphocytes were added, these cells were added at a final 1:1 ratio of B cells to T cells. PHA was utilized in the factor production phase at a final concentration of 0.75% (v/v).

c. IL-1, purified as described in the text, was utilized at dilutions ranging from 1/20 to 1/400. The values presented are those seen at a fixed dilution of 1/100.

liquid chromatography column (TSK 2000, LKB). IL-1 prepared in this manner could effectively substitute for the monocyte in the generation of the B cell growth promoting factor(s). Further confirmation for the central role of the T cell in the generation of BCGF was also seen in experiments where T cells and non T cells were repurified following an initial preincubation period and subsequently retested for factor production. Lectin exposed T cells, monocytes and B lymphocytes were cultured together for 24 hours. Following this preincubation period, the cells were repurified and placed alone in culture for an additional 72 hours. The conditioned media from this secondary 72 hour cultivation period were fractionated and units of BCGF determined. During the secondary culture period, the activated T lymphocytes that had been initially placed together with either monocytes alone or in combination with B cells continued to produce B cell stimulatory factors while the non-T cells failed to produce significant amounts of these factors.

DISCUSSION

The results presented herein demonstrate that the T lymphocyte plays an important role in the production of that factor specific for human B lymphocytes. Previous experimentation has documented the specificity of the signal both by biochemical and biologic means (4). In the present report we have extended the biologic information to include the observation that anti-μ activated B lymphocytes may specifically absorb the human B cell mitogenic factor from partially purified preparations. The combination of the activation signal, provided by anti-μ, and the growth factor results in a substantial proportion of the population of cultured B lymphocytes going into the S phase of the cell cycle. These lymphocytes may be kept in a proliferative state by continuous supplementation of the culture media with the factor. Knowledge of the cellular source for the production of the factor in addition to the means by which the B lymphocytes may be continually cultured should significantly aid in our basic understanding of lymphoid biology.

ACKNOWLEDGMENTS

We would like to thank Rose Sandler and Jean Glenn for their excellent technical assistance; and Linda Kimbrough for the preparation of the manuscript.

REFERENCES

1. Ford, R. J., Mehta, S. R., Franzini, D., Montagna, R., Lachman, L.G., and Maizel, A. L., Nature *294*, 261 (1981).

2. Maizel, A. L., Sahasrabuddhe, C., Mehta, S., Morgan, J., Lachman, L., and Ford, R. J., *Proc. Natl. Acad. Sci.* (USA) *79*, 5998 (1982).

3. Maizel, A. L., Mehta, S., Kouttab, N., Morgan, J., Ford, R., and Sahasrabuddhe, C., *Fed. Proc.*, in press, (1982).

4. Lachman, L., and Maizel, A.L., *Cont. Topic. Mol. Immunol.*, in press, (1982).

5. Howard, M., Kessler, S., Chused, T., and Paul, W. E., *Proc. Natl. Acad. Sci.* (USA) *78*, 5788 (1981).

6. Howard, M., Farrar, J., Hilfiker, M., Johnson, B., Takatsu, K., Hanaoke, T., and Paul, W. E., *J. Exp. Med.* *155*, 914, (1982).

7. Maizel, A. L., Mehta, S. R., Hauft, S., Franzini, D., Lachman, L. B., and Ford, R. J., *J. Immunol.* *127*, 1058 (1981).

BIOCHEMICAL CHARACTERIZATION OF HUMAN B CELL GROWTH FACTOR (BCGF)

T. Kasahara
J.J. Oppenheim

Laboratory of Microbiology and Immunology
National Institute of Dental Research
National Institutes of Health
Bethesda, Maryland

A. Muraguchi
A.S. Fauci

Laboratory of Immunoregulation
National Institute of Allergy and Infectious Diseases
National Institutes of Health
Bethesda, Maryland

INTRODUCTION

Recent success with the long-term culture of normal mouse and human B cells has suggested the existence of a T-cell derived growth factor, termed B cell growth factor (BCGF), which acts as a second signal to support B cell proliferation following initial B cell activation (1-3). BCGF-dependent B cell lines are not yet widely available, and the consequent lack of an easy assay system for human BCGF has hampered the biochemical characterization of human BCGF. Recently, Muraguchi and Fauci have developed a simple and reproducible assay system for the detection of BCGF using *Staphylococcus aureus* Cowan I bacteria (SAC)-activated B cells (4). We have utilized this assay for the separation and characterization of human BCGF.

ASSAY OF HUMAN BCGF

Human peripheral blood B cells were obtained after depletion of monocytes by adherence to plastic surfaces and elimination of T cells by rosetting with AET-treated sheep red blood cells followed by treatment with anti-Leu 1 plus

Interleukins, Lymphokines, and Cytokines

ISBN 0-12-527540-4

complement. This B cell population consisted of 70-86% SIg^+ cells with less than 5% monocytes and less than 0.5% T cells.

BCGF activity was measured as previously described (4,5). Briefly, purified human B cells were incubated with 0.002% (v/v) SAC for 3 d in 96-well flat-bottom microtiter plates (Costar) at a cell density of 1-2 x 10^5 cells/well. After 3 d of culture, serially diluted test supernatants were added, and the cells were incubated for an additional 3 d. The cultures were pulsed with 3H-thymidine (1 µCi/ml) overnight, and the incorporation of 3H-thymidine was determined.

In order to obtain reliable BCGF assays, reduction of contaminanting T cells to less than 0.5% by anti-leu 1 plus complement was critical. Otherwise IL 2 and/or mitogens in the crude supernatants may act on the residual T cells, which can in turn produce a second factor that activates B cells. Therefore, the presence of contaminating T cells in the B cell preparation should be assessed in each experiment by determining if there is a proliferative response to PHA.

BCGF-containing culture supernatants were prepared by stimulating normal human peripheral blood lymphocytes with 2 µg/ml of PHA or 10 µg/ml of Con A plus 10 ng/ml of PMA in the presence of 1% fetal calf serum for 4 d.

BIOCHEMICAL CHARACTERIZATION OF HUMAN BCGF

Initially, culture supernatant was concentrated to 50 to 100-fold using an Amicon YM-10 membrane and chromatographed using a Phenyl-Sepharose column (Φ1.5 x 25 cm) in the presence of 0.8M ammonium sulfate eluted with an ethanediol gradient of 0 to 50%. Both IL 2 and BCGF activities, however, eluted together at 25 to 40% of ethanediol (Fig. 1). A minor peak of BCGF activity was also observed in the break-through fractions along with concomitant IL 2 activity. The major peak of IL 2 and BCGF activity was then dialysed extensively against 15mM phosphate buffer (pH 7.8) and applied to a DEAE-Sephacel column (35ml). This column was eluted with a gradient of 0-300mM NaCl. Both IL 2 and BCGF activity eluted again in the same fraction ranging from 50-200mM NaCl concentration.

Active fractions from the DEAE-Sephacel column were pooled and applied to a Sephacryl S-200 column, followed by elution with 0.15M PBS containing 0.1% polyethyleneglycol. Considerable IL 2 and BCGF activity were again recovered in the same 20-25K M.W. range as a single peak (5) (Fig. 2). By these sequential procedures, IL 2 was purified up to 250-fold and BCGF up to 120-fold in terms of units/mg protein (Table 1). Finally, the active fractions from S-200 gel filtration was concentrated and chromatofocused using a polybuffer system (pH 8 to 5). Both IL 2 and BCGF were recovered together, at a pI of 7.8 and 6.9. IL 2 and BCGF were therefore inseparable by the above chromatography procedures.

ABSORPTION OF IL 2 WITH AN IL 2-DEPENDENT CELL LINE

When crude and partially purified IL 2 were absorbed twice with 1 x 10^7 CT-6 cells/ ml, IL2 activity was completely eliminated. Under the same condition, BCGF activity was not absorbed out by CT-6 cells from either the crude supernatant or partially purified fraction from Sephacryl S-200 gel filtration (Fig. 3). These results suggest that the two factors, even though biochemically very similar, presumably have different binding specificities.

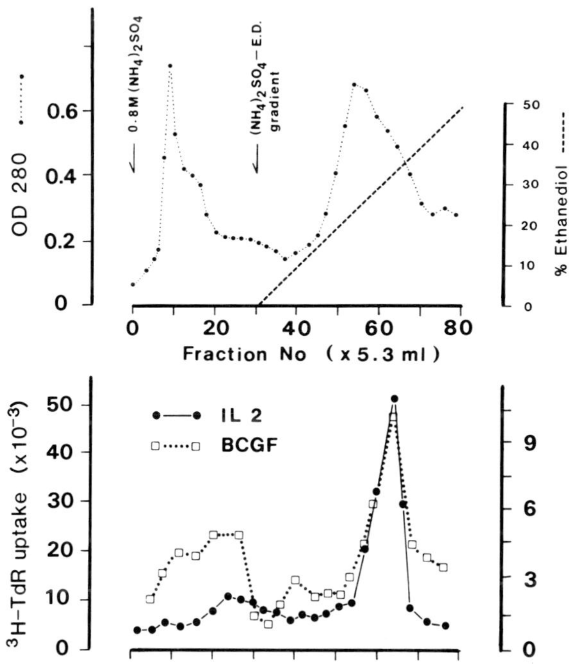

Figure 1. Phenyl-Sepharose chromotography of the 50 X-concentrated supernatant of PHA-stimulated peripheral blood lymphocyte culture. The OD280, IL 2, and BCGF activities in alternate fractions were determined.

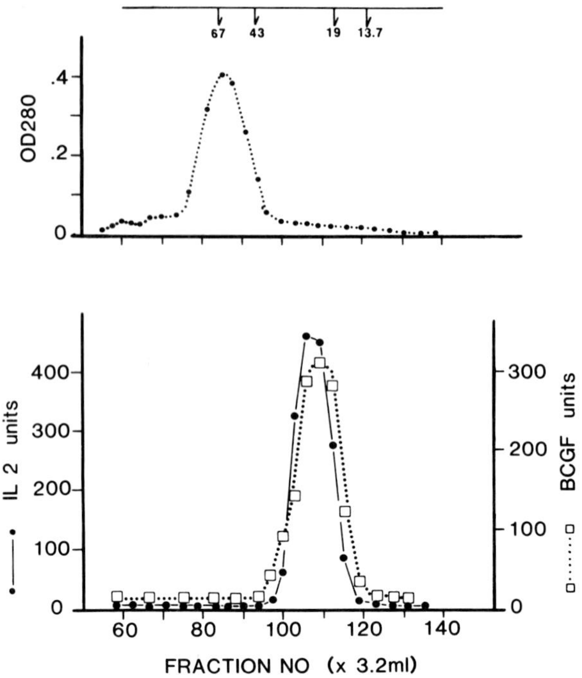

Figure 2. Sephacryl S-200 gel chromatography of pooled fractions containing Il 2 and BCGF activity after passage through Phenyl-Sepharose and DEAE-Sephacel.

COMMENTS

In contrast with our present findings and those of Maizel et al. (6), murine BCGF derived from the EL-4 cell line could be readily separated using Phenyl-Sepharose chromatography and has been shown by Farrar et al. (7) to be clearly distinct from murine IL 2.

TABLE I. Summary of Human BCGF Purification

Purification Step[a]	Volume ml	BCGF Activity u/ml	Specific Activity u/mg[b]	Total Activity u($x10^4$)	Yield %
1	2,200	25.6	109	5.60	100
2	50	744	123	3.75	67
3	62	468	1,061	2.90	52
4	22.5	780	5,735	1.76	31
5	24	512	13,128	1.20	20

[a]1) Crude PHA supernatant, 2) Amicon YM-10 concentration 3) Phenyl-Sepharose, 4) DEAE-Sephacel, 5) Sephacryl S-200.
[b]Specific activity expressed as units/mg protein.

CONCLUSION

Human IL 2 and BCGF activities from the PHA-stimulated culture supernatants could not be separated by hydrophobic Phenyl-Sepharose, DEAE-Sephacel, Sephacryl S-200 or chromatofocusing techniques, suggesting similarity in their biochemical characteristics. Nevertheless, since absorption of IL 2 activity with an IL 2-dependent cell line could selectively eliminate IL 2 activity alone, IL 2 and BCGF activities seem to be attributable to two different molecules.

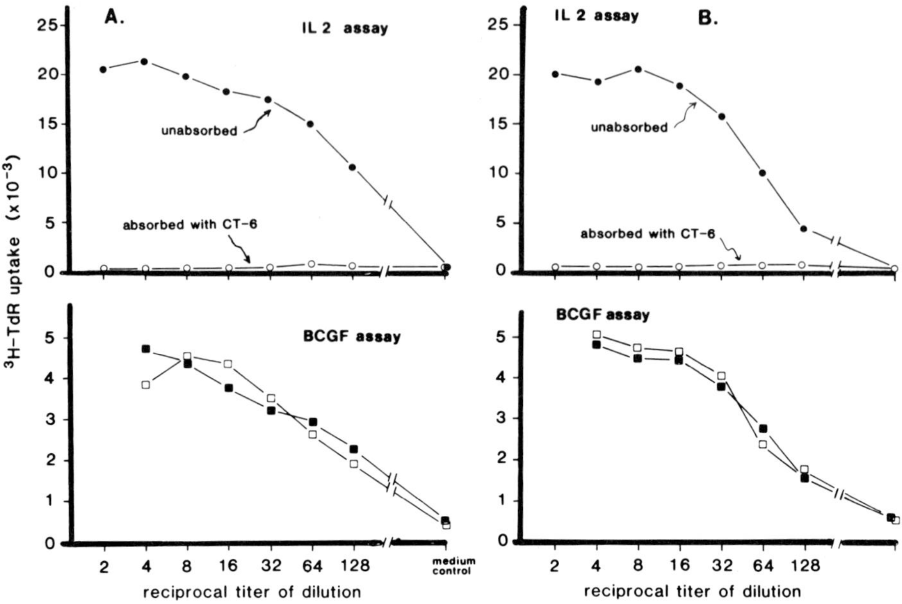

Figure 3. Effect of absorption with CT-6 cells on IL 2 and and BCGF activity in (A) crude PHA supernatant and (B) pooled partially purified eluate from Sephacryl S-200. A and B were absorbed twice with 1 x 10^7 CT-6 cells/ml at 37°C for 2 hr and the residual IL 2 and BCGF activities assessed.

REFERENCES

1. Sredni, B., Sieckmann, D. G., Kumagai, S., House, S. Green, I., and Paul, W. E., *J. Exp. Med.* 54, 1500 (1981).
2. Howard, M., Kessler, S., Chused, T., and Paul, W. E., *Proc. Natl. Acad. Sci. USA.* 78, 5788 (1981).
3. Ford, R. J., Mehta, S. R., Franzini, D., Montagna, R., Lachman, L., and Maizel, A. L., *Nature* 294, 261 (1981).
4. Muraguchi, A., and Fauci, A. S., *J. Immunol.* (in press).
5. Muraguchi, A., Kasahara, T., Oppenheim, J. J., and Fauci, A. S. *J. Immunol.* (in press)
6. Maizel, A., Sahasrabuddhe, C., Mehta, S., Morgan, J., Lachman, L., and Ford, R. (submitted).
7. Farrar, J. J., Benjamin, W. R., Hilfiker, M. L., Howard, M., Farrar, W. L., and Fuller-Farrar, J. *Immunol. Rev.* 63, 129 (1982).

DISCUSSION

FELDMANN: Has Kasahara tried the effect of BCGF and TCGF on antibody responses of human cells *in vitro*?

KASAHARA: As mentioned previously, our original BCGF preparation had but little TRF activity when assayed by the reverse plaque assay or by the determination of the immunoglobulin produced in the culture. Although we have not yet tested the partially purified BCGF and IL-2 preparations, we believe that factors other than BCGF, such as TRF, are also necessary for the differentiation of proliferating B cells to produce immunoglobulin.

Production of Monoclonal B Cell Growth Factor by a Human T Cell Hybridoma

Joseph L. Butler
Atsushi Muraguchi
Anthony S. Fauci

Laboratory of Immunoregulation
National Institute of Allergy and Infectious Diseases
National Institutes of Health
Bethesda, Maryland

I. INTRODUCTION

The technique of cell hybridization described by Köhler and Milstein (1) has been successfully applied to T lymphocytes of both murine and human origin (2-4). Recent studies have reported the development of human T cell hybridomas secreting a variety of immunoregulatory factors (5-7). The use of this technology to produce monoclonal T cell factors has proven to be of great value in extending our understanding of the mechanisms of immunoregulation of human immune responses.

Studies of human B cell activation and proliferation have demonstrated the need for exogenous growth factors to support proliferation of normal B cells placed in long-term culture (8). Supernatants from mitogen-stimulated mixed lymphocyte cultures (MLC) have been used as the source of B cell growth factor (BCGF). These conventionally generated supernatants contain other T-cell derived factors in addition to BCGF (9). Another disadvantage to the use of MLC supernatants is their variable ability to support B cell proliferation. To avoid these problems, we attempted to develop human T cell hybridomas that secrete monoclonal

Interleukins, Lymphokines, and Cytokines

ISBN 0-12-527540-4

immunoregulatory factors. The present study reports the successful development of a human T cell hybridoma secreting BCGF in the absence of other T cell factors.

II. HYBRIDOMA PRODUCTION

T cells were prepared from normal human peripheral blood by rosette formation with aminoethylisothiouronium bromide (AET)-treated sheep erythrocytes, followed by passage through a nylon wool column. The T-cell enriched population was stimulated with 10 μg/ml of concanavalin A (Con A) for 24 hours before fusion with the hypoxanthine-aminopterin-thymidine (HAT)-sensitive human T cell leukemic line, CEM-6. A cell fusion ratio of 2 normal T cells to 1 CEM-6 cell was used with polyethylene glycol as the fusing agent. Hybridoma cells were selected by incubation with HAT media in microtiter plates and cloned by limiting dilution.

III. ASSAY SYSTEMS EMPLOYED

The assay for the presence of BCGF in hybridoma supernatants was performed as previously described by Muraguchi *et al* (10). In brief, a B-cell enriched population was prepared from normal human peripheral blood by depletion of T cells and monocytes and stimulated with *Staphylococcus aureus* Cowan strain I (SAC). SAC binds to the Fab portion of the surface membrane immunoglobulin (Ig) of human B cells and consequently induces B cell proliferation in a T-cell independent fashion. Proliferation is maximal at day 3 and diminishes to undetectable levels by day 5. Hybridoma supernatants were added to SAC-stimulated B cell cultures on day 3, and cell proliferation was measured by tritiated thymidine incorporation on day 6.

T cell hybridoma supernatants were also tested for interleukin 2 (IL 2) and B cell differentiation factor (BCDF) production. Incorporation of tritiated thymidine by an IL 2-dependent T cell line (CT6) was used as the index of IL 2 activity in the hybridoma supernatants. A plaque-forming cell assay employing an Epstein-Barr virus-transformed B cell line (CESS) was utilized to test supernatants for BCDF activity (11).

IV. CLONING AND CHARACTERIZATION OF A BCGF-SECRETING HYBRIDOMA

Following hybridization of Con A-stimulated normal T cells with HAT-sensitive CEM-6 cells, the cell mixture was added to microtiter wells (Table I). Forty-one days after fusion, 184 wells contained viable hybrid cells. These wells were assayed for BCGF activity in the SAC assay system described above. Three wells were identified whose supernatants were capable of supporting B cell proliferation. These positive wells were cloned by limiting dilution, and 1 well with rapid growth characteristics ($2B_{11}$) was chosen for further cloning. Final cloning efficiency was 55% at a cell density of 0.5 cell/well.

TABLE I. Screening and Cloning of Human T Cell Hybridomas

Days after Fusion	Number of Wells Cultured	Number of Wells Producing BCGF	Cloning Efficiency[a]
			(%)
0	768	—	—
48	184 wells with hybrids screened	3	—
62	3 positive wells cloned, 10 cells/well (288 wells)	17	6
88	Well $2B_{11}$ recloned, 0.5 cells/well (120 wells)	26	55

[a]Calculated as number of positive wells/number of wells calculated to have $\geq$ 1 cell/well.

The functional activity of clone $2B_{11}$ is summarized in Table II. Supernatant from clone $2B_{11}$ significantly enhanced B cell proliferation in 6 day assays when compared to the proliferative response seen in mitogen-stimulated MLC supernatants. IL 2 activity in the hybridoma supernatant was negligible in comparison to the increased proliferation of CT6 cells induced by the addition of a PHA-stimulated MLC supernatant. Supernatant from clone $2B_{11}$ was also devoid of BCDF as measured in the CESS assay system.

The molecular weight of BCGF secreted by hybrid clone $2B_{11}$ was determined by application of concentrated culture supernatant to a G-100 Sephadex gel filtration column

TABLE II. Functional Activity of Hybridoma $2B_{11}$

Assay	Background	With Hybridoma Supernatant	With Standard Supernatant
		^{3}HTdR Incorporation	
BCGF[a]	1,466	27,613	5,762
IL 2	992	1,108	53,166
		Ig production (PFC per culture)	
BCDF[c]	3.4	5.2	44.8

[a]B cell proliferation was measured by incorporation of tritiated thymidine by purified human B cells triggered for 3 d with SAC with addition of test supernatant on day 3 and measurement of proliferation on day 6. Data are given as CPM per 10^5 B cells.

[b]IL 2 activity was assayed by measurement of tritiated thymidine incorporation by the IL 2 dependent T cell line, CT6. Data are given as CPM per 10^4 CT6 cells.

[c]BCDF activity was assayed on the CESS cell line which differentiates into Ig-secreting cells in the presence of a differentiation factor.

(90 x 2 cm). The column was eluted with phosphate-buffered saline (pH 7.4), and 2.5 ml fractions were collected. These fractions were assayed for the ability to maintain proliferation of SAC-activated B cells. Figure 1 shows a small peak of optical density in fractions 58-64. BCGF activity conicided with this protein peak which had a molecular weight of 18-20,000 daltons.

V. SUMMARY

The present study reports the development of a human T cell hybridoma secreting BCGF. Supernatant from the hybridoma was found to be capable of enhancing human B cell proliferation in the SAC assay system to a greater extent than conventionally derived BCGF. Of note is the fact that this hybridoma supernatant was devoid of IL 2 and BCDF activity. The molecular weight of the monoclonal BCGF as determined by gel filtration chromatography was 18-20,000 daltons.

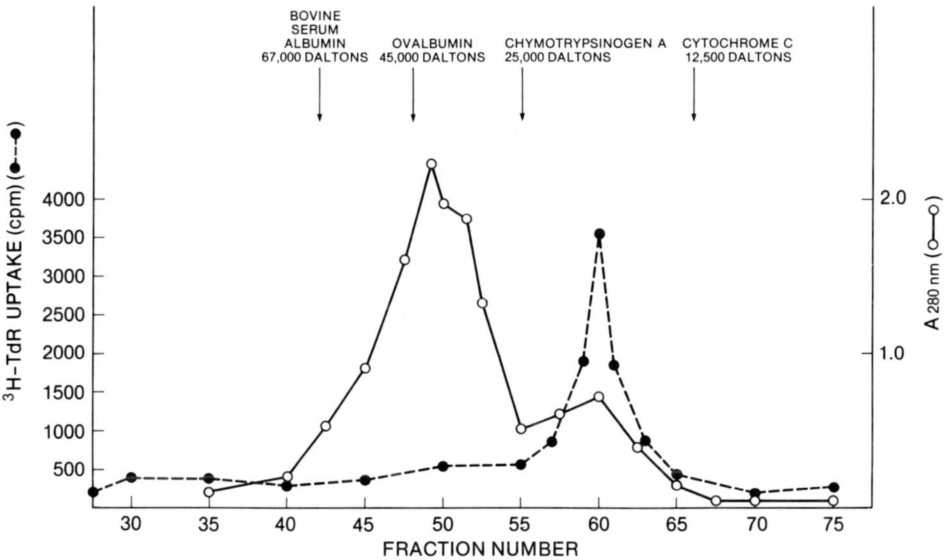

Figure 1. Sephadex G-100 gel chromatography of 20X-concentrated $2B_{11}$ hybridoma supernatant. Arrows indicate the elution profile of the proteins used for column calibration.

The availability of large quantities of purified BCGF, uncontaminated with other T cell factors, should prove to be of great importance in characterizing the biological and physical properties of this molecule. In addition, the development of other hybridomas secreting monoclonal T cell products will provide powerful tools with which to dissect the mechanisms of both normal and aberrant human immunoregulation.

VI. REFERENCES

1. Köhler, G., and Milstein, C., *Nature (Lond.) 256*, 495 (1975).
2. Taniguchi, M., and Miller, J.F.A.P., *J. Exp. Med. 148*, 373 (1978).
3. Kontiainen, S., Culbert, E.J., Ceck, M., Simpson, E., McKenzie, I.F.C., and Feldman, M., *Immunology, 43*, 747 (1981).
4. Grillot-Courvalin, C., and Brouet, J.C., *Nature (Lond.), 292*, 844 (1981).
5. Harwell, L., Skidmore, B., Marrack, P., and Kappler, J., *J. Exp. Med., 152*, 893 (1980).
6. Okada, M., Yoshimura, N., Kaieda, T., Yamamura, Y., and Kishimoto, T., *Proc. Natl. Acad. Sci. U.S.A., 78*, 7717 (1981).
7. Irigoyen, O., Rizzolo, P.V., Thomas, Y., Ragozinski, L., and Chess L., *J. Exp. Med., 154*, 1827 (1981).
8. Sredni, B., Sieckmann, D.G., Kumagai, S., House, S., Green, I., and Paul, W.E., *J. Exp. Med., 154*, 1500 (1981).
9. Mier, J.W., and Grallo, R.C., *Proc. Natl. Acad. Sci. U.S.A., 77*, 6134 (1980).
10. Muraguchi, A., and Fauci, A.S., *J. Immunol. 129*, 1104 (1982).
11. Muraguchi, A., Kishimoto, T., Miki, Y., Kuritani, T., Kaieda, T., Yoshizaki, K., and Yamamura, Y., *J. Immunol, 127*, 412 (1981).

DISCUSSION

SCHRADER: What is Butler's evidence that the proliferating cells were B cells?

BUTLER: Smears of the cells showed blastoid forms which were surface-Ig positive.

MAIZEL: Does IL-2 have any effect in Butler's BCGF assay?

BUTLER: Studies are now underway to assess the effect of IL-1 on the proliferation of B cells, as measured by our assay.

MAIZEL: Do the B cells used for the BCGF assay contain macrophages.

BUTLER: Yes. The cell population has <5% esterase-positive cells.

GREENE: What is Butler's evidence that the cell line producing BCGF is a hybridoma rather than a HAT-resistant variant of the parental CEM-fusing line?

BUTLER: One 2B11 was karyotyped and found to have 48-50 chromosomes as compared to 46 chromosomes in the CEM-6 line. Cell surface marker analysis is being done to ascertain whether the hybrid cells have shared markers from both parental cells.

HUMAN (T,G)-A--L SPECIFIC HELPER FACTORS

Edna Mozes
Dvora Katz
Daniel Suez
Ron N. Apte

Department of Chemical Immunology
The Weizmann Institute of Science
Rehovot, Israel

Zvi Bentwich

Department of Medicine C,
Kaplan Hospital,
Rehovot, Israel

Chaim Brautbar

Laboratory of Immunohematology and Immunology
Hadassah Medical School
Jerusalem, Israel

INTRODUCTION

Antigen specific T cell helper factors of murine origin have been useful in studies aimed at establishing the genetic regulation of immune responses as well as the nature of the recognition components of T lymphocytes. T cell replacing factors produced by murine activated T cells specific to the synthetic polypeptide poly(Tyr,Glu)-poly(DLAla)--poly(Lys) [(T,G)-A--L], the response to which is genetically controlled, have been extensively studied (1-5).

Information on immune response genes as well as on the function and specificity of T cells in man is very limited due to the ethical constrains of immunizing humans with antigens

ISBN 0-12-527540-4

and in view of the complexity of the antigenic systems investigated in relevance to disease occurrence. It appeared that an antigen specific functional factor produced by (T,G)-A--L activated T cells would be the desirable tool for approaching the mode of action as well as the specific recognition system of human helper T cells.

We have previously reported the successful production of (T,G)-A--L specific T cell factors secreted by human peripheral blood T cells of normal donors, which were activated *in vitro* on (T,G)-A--L pulsed autologous adherent cells. About 50% of the donors from a random population tested responded by the production of a T cell replacing helper factor suggesting differences in the immune response potential of the individuals tested (6).

In the present report we demonstrate the segregation in families of the ability to respond to (T,G)-A--L by the production of an antigen specific factor; we show that the (T,G)-A--L specific T helper factors of human origin possess HLA-DR and V region gene determinants, similarly to the murine T cell helper factors; and we describe the establishment and maintainance of a long term culture of (T,G)-A--L specific human helper T cells, which is also a constitutive secretor of an antigen-specific T cell replacing helper factor.

Antigenic Specificity of the Human T Cell Helper Factor

Peripheral blood leukocytes (PBL) of healthy donors previously found to respond to the synthetic polypeptide (T,G)-A--L by producing a helper factor, were activated *in vitro* for 5 days on antigen pulsed antologous adherent cells (6). The activated cells were thereafter incubated for additional 18 hours in presence of (T,G)-A--L and the supernatants collected were tested for their ability to replace (T,G)-A--L specific helper T cells in a hapten-carrier *in vitro* antibody production system. Since the (T,G)-A--L helper factor appeared to activate B cells across MHC and species barriers (1,7,8), the helper activity of the factors was tested using murine B cells. Thus, the factors were checked for their ability to cooperate with NIP-ovalbumin (OVA) primed spleen cells (used as a source of B cells) of C3H.SW, high responder mice to (T,G)-A--L, for eliciting NIP-specific antibodies when NIP-(T,G)-A--L was introduced into the cultures. NIP-human γ-globulin (HGG) was used as a control antigen in all experiments. To confirm the antigenic specificity of the factors of human origin, the active supernatants were passed through a column of (T,G)-A--L coupled to Sepharose. The effluents of the columns as well as the eluates obtained with 0.1 M ammonia were checked for their helper activity. Table I represents

TABLE I. The Antigenic Specificity of Human T Cell Helper Factors

NIP-OVA primed spleen cells cultured with:	NIP-specific PFC/10^6 viable cells[a]
-	113 ± 44[b]
NIP-OVA	1069 ± 125
NIP-HGG	117 ± 33
NIP-(T,G)-A--L	116 ± 12
(T,G)-A--L primed spleen cells + NIP-HGG	231 ± 35
(T,G)-A--L primed spleen cells + NIP-(T,G)-A--L	687 ± 134
Factor + NIP-HGG	311 ± 25
Factor + NIP-(T,G)-A--L	820 ± 147
Effluent from (T,G)-A--L-Sepharose + NIP-(T,G)-A--L	333 ± 56
Eluate from (T,G)-A--L-Sepharose[c] + NIP-(T,G)-A--L	862 ± 126

[a] Indirect plaque forming cells (PFC) were determined by using NIP-SRBC indicator cells.
[b] Results are expressed as the mean of quadriplicate tests ± S.D.
[c] Elution was performed with 0.1 M NH_4OH.

results of such experiments performed with a helper factor of one donor. As shown in the Table most of the helper activity of the factor was removed by the column and could be detected in the eluate of the immunoadsorbent.

Segregation of the Immune Response Potential to (T,G)-A--L in Family Members

The fact that donors of a random population differ in their ability to respond to (T,G)-A--L suggested a genetic regulation of the helper factor production. Studies on the response potential to (T,G)-A--L were performed with members of families. The results obtained till now suggested that the ability to respond to (T,G)-A--L is inherited as a dominant trait in association with the responder HLA haplotype, as demonstrated in Table II which represents the ability to produce (T,G)-A--L specific helper factors of members of 5 families.

TABLE II. (T,G)-A--L Specific Helper Factor Production by Activated T Cells of Family Members[a]

Family	*Father ab*	*Mother cd*	*Children 1*	*2*	*3*	*4*
			*a*c*	*a*c*	*a*d*	
1	+	-	+	+	+	
			*b*c*	*ac*	*b*c*	
2	+	-	+	-	+	
			ad	*b*d*		
3	+	-	-	+		
			*b*c*	*ac**	*bc**	*bd*
4	-	+	+	+	+	-
			*ac**	*ad*		
5	+	+	+	-		

[a]*The response potential of the (T,G)-A--L activated T cells tested by the production of helper factors is indicated by either + or -.*
[b]** indicates the responder haplotype.*

Human (T,G)-A--L Specific T Cell Helper Factors Possess V_H as well as HLA-DR Determinants

Antigen specific helper T cell factors of murine origin were found to possess V region as well as MHC determinants (3-5,9,10). It was therefore of interest to find out whether the human T cell factors specific to (T,G)-A--L possess variable heavy chain region determinants as well as DR determinants which are analogous to the Ia antigens of the mouse (11).

In order to determine whether the human antigen specific T cell helper factor possesses DR determinants, factors of donors previously typed for their HLA determinants were passed through Sepharose coupled to the relevant anti-DR antibodies. To determine whether the antigen specific factors carry V_H determinants, antisera against human immunoglobulin and the variable region of the heavy and light chain were coupled to Sepharose and the factors were passed through the latter columns. Antibodies specific to V_H and V_L fragments of human immunoglobulin were prepared and characterized (Y. Ben-Neriah, A. Cohen, N. Eshhar, D. Givol and Z. Bentwich, unpublished results) as previously described (12). Table III represents results of affinity chromatography experiments performed on

TABLE III. The (T,G)-A--L Specific Human Helper T Cell Factor Possesses V_H and DR Determinants

NIP-OVA primed spleen cells cultured with:	*NIP-specific PFC/10^6 viable cells*[a]
Factor	813 ± 117[b]
Effluent from anti V_H-Sepharose	344 ± 95
Eluate from anti V_H-Sepharose[c]	880 ± 75
Effluent from anti V_L-Sepharose	595 ± 134
Eluate from anti V_L-Sepharose	32 ± 26
Effluent from anti IgG-Sepharose	607 ± 128
Eluate from anti IgG-Sepharose	133 ± 35
Effluent from anti-DR5(BRA 677)-Sepharose	174 ± 64
Eluate from anti-DR5-Sepharose	781 ± 38
Effluent from anti-DR1(kZ 26182)[d]*-Sepharose*	300 ± 54
Eluate from anti-DR1-Sepharose	833 ± 167
Effluent from anti-DR3(MO 54784)[d]*-Sepharose*	680 ± 109
Eluate from anti-DR3-Sepharose	88 ± 31
Effluent from anti-Ia (L1.1.6)[e]*-Sepharose*	136 ± 67
Eluate from anti-Ia-Sepharose	691 ± 126
Effluent from anti-β_2M(M28)[e]*-Sepharose*	678 ± 156
Eluate from anti-β_2M-Sepharose	158 ± 51

[a]*Indirect PFC were determined using NIP-SRBC indicator cells.*
[b]*Results are expressed as the mean of quadriplicate tests ± S.D.*
[c]*Elutions were performed with 0.1 M NH_4OH.*
[d]*Kindly provided by Dr. J. van Rood.*
[e]*Kindly provided by Dr. J. Kalil.*

antigen specific T cell replacing factor of a donor which was typed to possess HLA-DR 1,5. The results shown in the Table indicate that the activity of the factor was removed by the anti-V_H immunoadsorbents and could be eluted from the column. However, the immunoadsorbents built of anti-V_L antibodies or of antibodies against the whole IgG molecule did not interact with the factor and the biological activity was detected in the effluents of the columns. The activity of the same factor was removed by columns of Sepharose coupled to anti-DR5 and anti-DR1 sera and could be eluted by 0.1 M ammonia, whereas no effect on the activity of the latter factor was observed when the irrelevant anti-DR3 was used as immunoadsorbent. Also is shown in the column that Sepharose coupled to a monoclonal antibody (L1.1.6) which recognizes a monomorphic determinant on HLA-DR antigens removed the activity of the factor whereas a monoclonal antibody (M28) shown to recognize β_2M did not affect the activity of the factor. These results suggest that the (T,G)-A--L specific factor of human origin possesses V_H as well as DR (Ia) determinants in its active moiety. Experiments performed with factors of different donors led to the same conclusions, however the molecular combination between the I region gene products and immunoglobulin variable region gene products is not yet understood. The minute amounts of T cell factors available from the (T,G)-A--L activated T cells are sufficient only for functional studies as well as for partial serological analysis. For the molecular characterization of the factors it is essential to have a source of large amounts of homogeneous factors.

Establishment and Maintenance of a (T,G)-A--L Specific Continuous T Cell Line

Activated T cells of murine and human origin can be cultivated for extended periods of time in presence of T cell growth factors (TCGF). We have recently reported the establishment of (T,G)-A--L specific helper T cell lines which constitutively secrete into the culture fluids antigen-specific T cell replacing factors (5,13). In attempts to establish a human T cell helper line, PBL of a donor previously shown to be a high responder to (T,G)-A--L were educated to (T,G)-A--L as previously described (6). The activated cells were recultured for an additional 24 hours period on (T,G)-A--L pulsed fresh adherent cells since this procedure was shown to enrich the helper activity of the stimulated cells (14). The cells were then transferred into medium supplemented with 30-50% TCGF. Every 4-5 days, the T cells of the newly established line were harvested, centrifuged and recultured. At times of slow growth, irradiated (5000 R, Co) autologous PBL were added to cultures, as filler cells. Supernatants were collected from the long term cultures 4 to 5 days after changing the TCGF-medium. The supernatants were found to be very potent in their biological activity and they could be diluted to 1:10000 before their antigen specific helper activity was reduced. The line's derived factors were shown to be similar in their characteristics to the factors obtained from educated T cells, since they are antigen specific and possess V_H as well as human Ia determinants. The established (T,G)-A--L specific continuous T cell line of human origin which constitutively secretes homogeneous T helper factors, provides the means for further studies on the mode of action as well as the nature of the antigen-specific receptors of human T cells.

REFERENCES

1. Taussig, M.J., Mozes, E. and Isac, R., *J. Exp. Med. 140,* 301 (1974).
2. Isac, R., and Mozes, E., *J. Immunol. 118,* 584 (1977).
3. Mozes, E., and Haimovich, J., *Nature 276,* 56 (1979).
4. Eshhar, Z., Apte, R.N., Löwy, I., Ben-Neriah, Y., Givol, D., and Mozes, E., *Nature 286,* 270 (1980).
5. Apte, R.N., Löwy, I., De-Baetselier, P., and Mozes, E., *J. Immunol. 127,* 25 (1981).
6. Katz, D., Bentwich, Z., Eshhar, N., Löwy, I., and Mozes, E., *Proc. Nat. Acad. Sci. USA 78,* 4505 (1981).
7. Kantor, F., and Feldmann, M., *Clin. Exp. Immunol. 36,* 71 (1979).
8. Taussig, M.J. and Finch, A.P., *Nature 270,* 151 (1977).
9. Taussig, M.J., Munro, A.J., Campbell, R., Davis, L.S., and Staines, N.A., *J. Exp. Med. 142,* 694 (1975).
10. Isac, R., Dorf, M.E., and Mozes, E., *Immunogenetics 5,* 467 (1977).
11. Winchester, R.J., and Kunkel, H.G., *Adv. Immunol. 28,* 221 (1979).
12. Ben-Neriah, Y., Wuilmart, C., Lonai, P., and Givol, D., *Eur. J. Immunol. 8,* 797 (1978).
13. Löwy, I., Apte, R.N., and Mozes, E., *J. Immunol. 128,* 190 (1982).
14. Apte, R.N., Dayan, M., and Mozes, E., *Cell. Immunol. 61,* 104 (1981).

DISCUSSION

ALTMAN: Since the TCGF preparation used to maintain the human T cell line contains mitogen, is it possible that helper factor production is not constitutive but rather mitogen-induced?

MOZES: We do not think so, because the mere addition of a mitogen will not itself elicit the production of the antigen-specific helper factor. Moreover, we collect routinely the supernatants on day 5 after changing the conditioned medium of the cultures. We have been able to demonstrate that by this time the TCGF has been consumed by the cells of the line. We have been unable to detect any TCGF in the supernatants, nor any mitogenic activity either.

SCHRADER: Why did Mozes express her results in terms of 10^6 viable cells present at the end of the culture? Don't the numbers of viable cells vary considerably with the addition of the different factors?

MOZES: There is little variability in the cell numbers at the end of the culture period, but we express the results as number of PFC/10^6 *viable* cells present at the end of culture because we think it more proper, as it expresses the *condition* of the cells at the time they are assayed.

OPPENHEIM: An issue that is of general concern is the relationship between the nonspecific and specific factors. The TCGF apparently induces a T cell line that in turn produces a specific factor. Does Mozes know whether the specific factor complex in turn includes a nonspecific helper lymphokine, or whether there are BCGF- or TRF-like factors acting in concert with the specific factor?

MOZES: We have not purified to homogeneity the human antigen-specific factor, therefore we are not certain that the supernatant does not contain still additional factors. However, it is not likely that the eluates of the antigen-specific columns or of the other specific columns contain additional non-specific factors. By assays for biological activity we could not detect any non-specific activity and from our work with the murine antigen-specific factor we know that a low molecular weight fraction (15-17K) possesses the specific activity and triggers the B cells in culture. The reasonable conclusion seems to be that the antigen-specific T cell helper factor can trigger the B cells in the absence of other factors in the culture.

FELDMANN: The remarks by Oppenheim prompt me to make a further comment on Mozes' work. There *is* a role for non-specific factors in the response induced by specific factors. We have found that certain human long term T helper cell lines make HLA-restricted helper factors, while others make specific, but unrestricted factors. Zanders has fractionated these supernatants, and found that the unrestricted factors consist of two components, a high MW (70k) antigen-specific and HLA-restricted factor, and a low MW (10 20k) non-specific factor. The synergy between these two results in the disappearance of the HLA restriction. We think it is possible that the non-restricted mouse-specific factors are explained by the same mechanism, i.e. B cells are already activated partly by other non-specific factors, perhaps from T cells, or in the fetal calf serum.

MOZES: I cannot argue against the proposition that FCS in the culture provides some non-specific activity which helps the triggering of B cells in the cultures by the antigen-specific factor, especially as we utilize FCS in *all* our *in vitro* antibody-producing systems. However, at least in the murine system, we know that an isolated low moelcular weight fraction (15-17k), which does not contain any non-specific activity, can trigger the B cells in culture for the production of *specific* antibodies.

ANALYSIS OF A CONTINUOUS T CELL LINE-DERIVED (T,G)-A--L SPECIFIC HELPER FACTORS

Ron N. Apte[1]
Ruth Lifshitz
Ilana Löwy[2]
Edna Mozes

Department of Chemical Immunology
The Weizmann Institute of Science
Rehovot, Israel

INTRODUCTION

Recent innovations in tissue culture techniques have indicated that it is possible to propagate *in vitro* for extended periods of time antigen or mitogen-activated T cells in medium containing a soluble lymphokine termed T cell growth factor (TCGF) or interleukin 2 (IL-2) (1,2). Using this mediator it was possible to establish normal antigen-specific continuous T cell lines with effector functions like cytotoxic, suppressor, helper or proliferating T cells (3-7).

We have attempted to generate antigen-specific factor-producing helper T cell continuous lines specific for the synthetic polypeptide (T,G)-A--L. Since antigen-specific T cell replacing factors are considered as the soluble form of the T cell recognition unit, we thought that by generating monoclonal antigen-specific helper T cells which secrete active products into the culture fluids, we can approach the T cell receptor issue.

One of the lines established by us is the E-9M(+) helper T cell line, which was derived from *in vitro* (T,G)-A--L activated

[1] *Present address: Department of Microbiology and Immunology, Faculty of Health Sciences, Ben-Gurion University, Beer-Sheva, Israel.*
[2] *Present address: Institute Pasteur, Paris, France.*

Interleukins, Lymphokines, and Cytokines

ISBN 0-12-527540-4

T cells, further enriched for helper activity by an additional 24 h pulse on antigen-bearing splenic adherent cells (8), and thereafter subsequently propagated in TCGF containing medium, without additional antigenic stimulations (9). The E-9M(+) line was cloned using the limiting dilution method in microwells containing TCGF and irradiated splenic cells, serving as fillers. Cells of the E-9M(+) line are in continuous culture by now for about 3 years without impairment in their helper activity or surface markers, and they can be easily frozen and thawed.

Cells and supernatants of the E-9M(+) line exerted a (T,G)-A--L specific helper effect when tested in an hapten-carrier *in vitro* antibody production system (9). As little as 10^2 cells of the E-9M(+) line or its active clones provided a potent helper signal to 10^6 NIP-OVA primed spleen cells challenged with NIP-(T,G)-A--L, while culture fluids could be diluted more than 1:1000 and still retained their antigen-specific effector function. Active factor preparations were constitutively generated by the proliferating cells, and they were collected from 5-6 dense old cultures, which had already absorbed or consumed the TCGF from the culture medium. Such preparations were indeed devoid of activities promoting T cell proliferation and nonspecific helper factors. Cells of the E-9M(+) line were shown to be Ly-$1^+2^-,3^-$, and they as well as factor derived from this line were found to activate B cells (anti-Thy-1.2 and complement treated hapten-primed spleen cells)(9). In the present report we describe the properties of cells and factors of the E-9M(+) line and its derived clones and attempts to purify the active moiety of the factor.

Properties of Cells and Factors of the E-9M(+) Line and its Derived Clones

Cells and factors of the E-9M(+) line were shown to bear V-region gene products. Cells of the lines and clones were selectively stained in immunofluorescence studies with antibodies directed against the variable region of the Ig heavy chain (V_H, 10) which are considered as general reagents with capacity to react with the T cell receptor. Such anti-V_H antibodies were previously shown to interact with functional T cells and with antigen-specific helper (9, 11) and suppressor factors (12). More specific probes of the V-region gene products, in regard to their preparation and antigenic-specificity are anti-idiotypic antibodies, which have been shown to be involved in the recognition system of antigen-specific effector T cells (13). Anti-idiotypic antibodies produced in C57BL/6 individual mice (H-2^b, Igh-1^b) against

TABLE I. Reaction with Anti-idiotypes of T Cell Lines and Clones and the Helper Activity of their Derived Factors

NIP-OVA primed spleen cells cultured with supernatants of:	NIP-specific PFC/10^6 cells[a]		FACS II staining with anti-idiotypes
	NIP-(T,G)-A--L	NIP-HGG	
None	101 ± 83[b]	133 ± 66	
Line E-9M(+)	790 ± 16	129 ± 109	+
Clone 4.3	987 ± 187	172 ± 187	+
Clone 4.8	396 ± 142	99 ± 128	+
Clone 4.23	1029 ± 308	302 ± 211	+
Clone 4.45	273 ± 38	114 ± 34	+
Clone 1.11[c]	113 ± 87	253 ± 91	-
Line E-14M(+)[d]	688 ± 199	113 ± 68	-
Line E-15M(+)[e]	126 ± 61	582 ± 268	-

[a] Indirect plaque forming cells (PFC) were determined by using NIP-SRBC indicator cells.
[b] Results are expressed as the mean of quadriplicates ± S.D.
[c] A non active clone derived from the E-9M(+) line.
[d] A (T,G)-A--L specific T cell line of CWB ($H\text{-}2^b$, $Igh\text{-}1^b$) origin.
[e] A C3H.SW ($H\text{-}2^b$, $Igh\text{-}1^j$) derived T cell line specific to HGG.

(T,G)-A--L specific antibodies of C3H.SW ($H\text{-}2^b$, $Igh\text{-}1^j$) mice (14) were used for staining of the E-9M(+) cells utilizing the fluorescence activated cell sorter (FACS II). The anti-idiotypic antibodies stained significantly cells of the E-9M(+) line but did not stain a helper T cell continuous line of C3H.SW origin specific for human gamma globulin (E-15M(+)), and they also did not stain a (T,G)-A--L specific helper T cell line of CWB (E-14M(+)) origin ($H\text{-}2^b$, $Igh\text{-}1^b$). Thus, as previously shown for the expression of idiotypic determinants on antibodies their expression on T cell lines is also antigen-specific and linked to the heavy chain allotypic marker (15, 16). Testing clones of the E-9M(+) line, a correlation was found between the expression of antigen-specific helper activity and the expression of cell surface idiotypic determinants. Table I represents the reaction with anti-idiotypes of the different T cell lines and clones and the helper activity of their derived factors. Similarly to the cells, factors derived from the E-9M(+) (T,G)-A--L specific helper T cell continuous line express V-region determinants and could be absorbed on and eluted from affinity columns built of (T,G)-A--L anti-V_H antibodies or anti-idiotypic sera (9).

TABLE II. Fine Antigenic Specificity of Cells and Factors of the E-9M(+) Line[a]

Antigen	*Cells of the E-9M(+) line*	*Factor of the E-9M(+) line*	*Factor of "educated" T cells*	*Antibody*
(T,G)-A--L	+[b]	+	+	+
(Phe,G)-A--L	+	+	+	+
(H,G)-A--L	±	±	*N.D.*	+
G-A--L	+	+	+	+
(T,G)-Pro--L	-	-	-	+
(T-T-G-G)-A--L	+	+	*N.D.*	+
(T-G-T-G)-A--L	-	-	*N.D.*	-

[a]*Cross-reactivity was tested as indicated in the text.*
[b]*(+) indicates cross-reactivity with the mentioned polypeptide.*

The fine antigenic specificity of the E-9M(+) and its secreted factors was compared using 2 experimental approaches: 1) NIP-OVA primed spleen cells were cultured with T cells of the line and its derived factor and challenged with NIP-conjugated to different polypeptide antigens related to (T,G)-A--L. 2) Supernatants of the E-9M(+) line were passed through immunoadsorbents built of the different synthetic polypeptide antigens and the (T,G)-A--L specific helper activity was tested in the effluents and eluates recovered from the columns.

Table II represents the pattern of cross reactivity with the different antigens in comparison to the specificity of factors obtained from "educated" T cells (17) and to that of (T,G)-A--L specific antibodies. The fine antigenic specificity of the T cells and their products was shown to be identical, however differing somewhat from that of antibodies. Using ordered synthetic polypeptides it was shown that the antigen specificity of the T cells and factors was mainly directed against the major antigenic determinant of the randomly polymerized polypeptide (T,G)-A--L, namely T-T-G-G, as shown for specific antibodies. Thus, (T,G)-A--L specific B cells and T cells, as well as their products, share V-region gene products.

In addition to V-region gene products, cells and factor of the E-9M(+) line were also shown to contain MHC markers (9). It was also evidenced that V-region genes and MHC products are linked together in the active moiety of the factor (9),however its exact molecular organization remains to be clarified.

Partial Purification and Characterization of the E-9M(+) Derived Factor

The antigen specific factors produced by the E-9M(+) continuous T cell line were partially purified by affinity chromatography, ammonium sulfate fractionation, and SDS-polyacrylamide gel electrophoresis. The active supernatants were passed through an irrelevant immunoadsorbent of goat anti mouse Fab-Sepharose. The active effluent was run twice on a (T,G)-A--L-Sepharose column. The iodinated active, and antigen specific 0.1 M ammonia eluate was subjected to SDS-polyacrylamide gel electrophoresis which revealed bands with molecular weight of 17 K and 15 K in addition to a difuse band of high molecular weight.

TABLE III. Antigen Specific Helper Activity of Electroeluates from Preparative SDS-Polyacrylamide Gel of the 65-75% Ammonium Sulfate Precipitate Derived from the Supernatant of the E-9M(+) T Cell Line

Fraction	*NIP specific IgG PFC/10^6 cells*[a]	
	NIP-(T,G)-A--L	*NIP-HGG*
65-75% ammonium sulfate precipitate	1000 ± 20[b]	260 ± 36
a	352 ± 32	240 ± 128
b	344 ± 104	208 ± 64
c	96 ± 16	0
d	776 ± 120	160 ± 32
e	96 ± 24	264 ± 16
f	512 ± 200	496 ± 168
g	360 ± 144	320 ± 224
h	224 ± 88	128 ± 24
i	1440 ± 96	112 ± 64

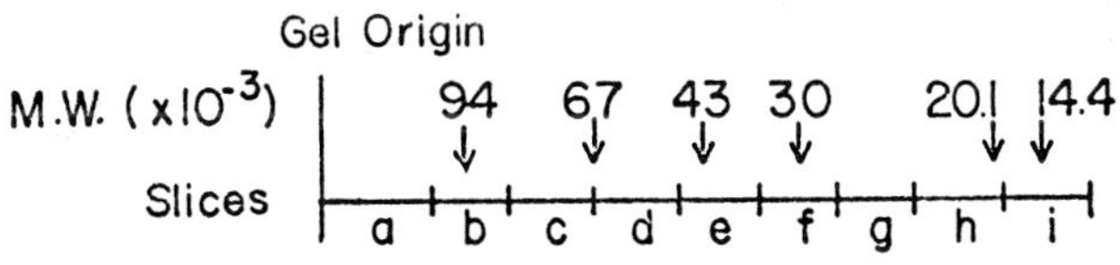

[a]Indirect plaque forming cells (PFC) were determined by using NIP-SRBC indicator cells.
[b]Results are expressed as the mean of quadriplicates ± S.D.

In another set of experiments the factor was subjected to sequential precipitation with ammonium sulfate. The (T,G)-A--L specific helper activity was associated with a 65-75% ammonium sulfate precipitate. Gel electrophoresis followed by electroelution of either the eluate of a (T,G)-A--L column or of the 65-75% salt precipitate indicated that in both preparations two fractions contained the biological activity of the factor, one of a high (less than 67 K) and the other of low (15-17 K) molecular weight. Table III represents the (T,G)-A--L specific helper activity of the fractions electroeluted from a gel of the 65-75% ammonium sulfate precipitate.

Culture supernatants of the ^{35}S-methionine internally labeled E-9M(+) line were subjected to a combined purification of sequential ammonium sulfate precipitation followed by affinity chromatography. Polyacrylamide gel patterns obtained of the eluates of the different salt precipitates demonstrated that the 65-75% ammonium sulfate precipitate contained two ^{35}S-labeled bands with apparent molecular weights of the range of 60K and 15K, similar to the activity pattern obtained by the gel electrophoresis fractionation experiments. Thus, the helper activity of the continuous T cell line E-9M(+) derived factors is associated with high (60-70K) and low (15-17K) molecular weight fractions. It is most striking that the antigen specific activity of the T cell line is fully preserved in a low molecular weight fraction. The molecular characteristics of this fraction is being currently investigated.

REFERENCES

1. Ruscetti, F.W., Morgan, D.R., and Gallo, R.C., *J. Immunol. 119,* 131 (1977).
2. Gillis, S., Ferm, M.E., Ou, W., and Smith, K.A., *J. Immunol. 120,* 2027 (1978).
3. Gillis, S., and Watson, J., *Immunol. Rev. 54,* 81 (1981).
4. Schreir, M.H., Iscove, N.N., Tees, R., Aarden, L., and Von Boehmer, H., *Immunol. Rev. 51,* 315 (1980).
5. Watson, J.D., *J. Exp. Med. 150,* 1510 (1979).
6. Frenso, M., Nabel, G., McVay-Boudreau, L., Furthmayer, H., and Cantor, H., *J. Exp. Med. 153,* 1246 (1981).
7. Sredni, B., and Schwartz, R.H., *Immunol. Rev. 54,* 187 (1981).
8. Apte, R.N., Dayan, M., and Mozes, E., *Cell. Immunol. 61,* 104 (1981).
9. Apte, R.N., Löwy, I., de Baetselier, P., and Mozes, E., *J. Immunol. 127,* 25 (1981).

10. Ben-Neriah, Y., Wuilmart, C., Lonai, P., and Givol, D., *Eur. J. Immunol. 8,* 791 (1978).
11. Apte, R.N., Eshhar, Z., Löwy, I., Zinger, H., and Mozes, E., *Eur. J. Immunol. 11,* 931 (1982).
12. Tada, T., Hayakawa, K., Okumura, K., and Taniguchi, *Mol. Immunol. 17,* 867 (1980).
13. Eichmann, K., *Adv. Immunol. 26,* 195 (1978).
14. Lifshitz, R., Parhami, B., and Mozes, E., *Eur. J. Immunol. 11,* 27 (1981).
15. Schwartz, M., Lifshitz, R., Givol, D., Mozes, E., and Haimovich, J., *J. Immunol. 121,* 421 (1978).
16. Lifshitz, R., Schwartz, M., and Mozes, E., *Immunogenetics 11,* 191 (1980).
17. Isac, R., and Mozes, E., *J. Immunol. 118,* 584 (1977).

DISCUSSION

SCHIMPL: Has Apte tried to precipitate the antigen-specific factors (internally labelled with ^{35}S) with anti-V_H, or anti-Gd, or anti-Ia, in order to show which bands carry which determinant?

APTE: We are presently raising Mc anti-Id antibodies for experiments to precipitate and isolate the specific bands.

MOZES: I would like to add a comment to Apte's response to Schimpl's query. We have recently used anti-idiotypic antibodies in immunoprecipitation experiments. Preliminary results with SDS-polyacrylamide gel electrophoresis indicate that we have precipiated from the internally labelled supernatant of the (T,G)-A--L specific murine T cell clone, the low (15-17k) molecular weight fraction.

ROSENWASSER: Does Apte's T cell factor express allotype markers and/or is it restricted in its activity with allotype markers?

APTE: The cells or factor of the E-9M(+) continuous helper T cell line are non-allotype restricted, since they (H-2^b, Igh-1^j) cooperate with cells of CWB mice (H-2^b, igh-1^b). However, cells and factors of the line bear cross-reactive idiotypic determinants with anti-(T,G)-A--L antibodies, and expression of such idiotypic determinants is linked to the heavy chain allotypic marker.

HAKIM: Earlier on Howard referred to the dichotomy of the Lyb5+ and Lyb^5- B cell responses to non-specific B cell growht factors. Has Apte considered the possibility that the antigen-specific, MHC-restricted factors he has found can act on the Lyb5- B cell, i.e., stimulating B cells by a mechanism other than by TRF or BCGF?

APTE: We have not yet tested the phenotype of the B cells which are triggered by the cells and factors of the E-9M(+) line.

VADAS: It seems to me that the most appropriate control for specificity in both the murine *and* human studies would be another factor induced to a different antigen. Were such controls performed in both species?

APTE: As we have shown, in our laboratory we include lines specific for HGG and they, as well as their products, serve as specificity controls for the (T,G)-A--L specific lines and factors.

ACTIVATION OF T HELPER CELLS BY A MONOCLONAL ANTIHELPER FACTOR ANTIBODY

Roger James, Geoff Sunshine, Susan Edwards, Eric Culbert, John Maudsley, Allison Kingsbury, Sirkka Kontiainen[1] and Marc Feldmann

ICRF Tumour Immunology Unit
Department of Zoology
University College London
London

[1] Dept of Bacteriology and Immunology
University of Helsinki
Finland

INTRODUCTION

There is still much confusion concerning the nature of antigen specific receptors on T cells and their relationship to factors found in T cell supernatants. The reasons for this are well known (reviewed 1), but modern techniques offer new approaches to resolving this problem. Thus T cell clones, generated in a variety of ways (reviewed 2), overcome the problems of T cell purity and heterogeneity. However, antibody probes which specifically detect T cell factors or receptors have not been widely available. We have previously generated rabbit anti helper or suppressor factor antisera (3) which may be rendered specific by extensive absorption (4), but to overcome the problems of purity and of restricted supply, we have attempted to generate monoclonal antibodies against T cell factors. We review here some of our studies with a rat monoclonal anti mouse helper factor, AF3.44.4, which binds to, and activates T helper cells.

METHODS AND RESULTS

1. Production of Monoclonal AF3.44.4

We have produced 2 types of antisera to factors. The heteroantisera, which recognize a constant region related to

Interleukins, Lymphokines, and Cytokines

ISBN 0-12-527540-4

the function of the factor, are more widely useful than syngeneic mouse antisera, which behave like anti-idiotypic antisera. These were therefore selected for conversion to monoclonals first. Rats were immunized with antigen (KLH) column eluates of the supernatants from spleen cells cultured in vitro with high doses of KLH. Fusion with NS-1 and selection procedures were as described by James et al (5) previously.

AF3.44.4 is a rat IgM, k antibody which binds to helper T cell clones and tumours but does not bind appreciably to suppressor T cell lines. This was somewhat unexpected because the high dose KLH cultures yield cells which act as suppressors and produce suppressor factor (SF) (6). However, there is ample evidence that such cell populations contain a mixture of both helper and suppressor T cells, with the net functional effects depending on the ratio of these cells (4). Thus the antigen column eluates used for immunisation

TABLE 1

Reaction of Monoclonal AF3.44.4 with Antigen Specific Helper Factor

STIMULUS		RESPONSE (IgM PFC/10^6 ± SE)	
BAB.14 HF_{KLH}	Immuno-adsorbent	Filtrate	Eluate
-	-	0	
+	-	136 ± 24	
+	Rab α SF	168 ± 27	44 ± 17
+	Rab α HF	19 ± 11	166 ± 29
+	Rat IgM	210 ± 14	20 ± 10
+	AF3.44.4	10 ± 10	177 ± 22

3 x 10^6 CBA spleen cells were cultured with TNP-KLH (2 ug/well) and BAB.14 HF_{KLH} (2.5% final concentration). The HF was absorbed and eluted [with acid glycine (pH2.8)] from Sepharose 4B coupled with rabbit anti-SF, rabbit anti-HF, control rat IgM or AF3.44.4. IgM DNP specific AFC/10^6 initially cultured spleen cells ± S.E. were assayed on day 4.

contained both helper and suppressor factors and the monoclonal antibody was produced against the minority component (HF).

2. Absorption of HF by AF3.44.4

AF3.44.4 was bound to Sepharose beads and used to absorb HF and SF. The results of absorptions of a BAB.14 HF_{KLH} are shown in Table 1. Rabbit anti HF and AF3.44.4 both absorbed HF activity (which could be recovered in full with acid elution). Rabbit anti SF or control rat IgM did not absorb any activity. In these sort of experiments AF3.44.4 does not absorb SF (data not shown).

3. Immunological Effects of AF3.44.4

A variety of studies have been performed to ascertain the effects of AF3.44.4 on in vitro immune responses. Although no effects were detected on either the induction or the effect of suppressor cells, primary or secondary antibody responses in vitro were augmented significantly, from 2-10 fold, depending on the amount of monoclonal antibody used and the kinetics of the response. In some circumstances there was also some suppression of the antibody response.

TABLE 2

Enhancement of in vitro 2° response to TNP-KLH by AF3.44.4

Antigen	IgG Response in Presence of Antibody ($AFC/10^6 \pm SE$)		
	NIL	YR 43	AF3.44.4
-	80 ± 42	60 ± 28	80 ± 56
TNP KLH	13,500 ± 955	11,300 ± 1,867	62,800 ± 2,867
DNP CGG	80 ± 56	120 ± 14	60 ± 28

CBA TNP-KLH primed spleen cells were cultured in vitro for 6 days at 5×10^5 cells/200 ul well in 96 well plates without antigen, with TNP-KLH (0.05 ug/ml) or with DNP-CGG (0.05 ug/ml). The TNP-KLH primed spleen cells were cultured in the absence of antibody (NIL), in the presence of control monoclonal rat IgM,k antibody (YR43) at 10 ug/ml or in the presence of AF3.44.4 at 10 ug/ml. Each triplicate group was assayed for the number of IgG DNP specific $AFC/10^6$ spleen cells ± S.E.

The effect on the secondary response to TNP-KLH is shown in Table 2. While striking enhancement was noted in the presence of the appropriate antigen, in the absence of antigen or in the presence of DNP-CGG no effect was observed, indicating that AF3.44.4 synergizes with antigen and does not activate primed T cells by itself.

As the antibody response was augmented in the absence of any effects on suppressor cells the augmentation of the response was not due to abrogation of suppression, and thus was presumably due to increased help. We investigated the primary induction of helper T cells, and the results indicate that helper cell induction is augmented, provided that antigen is present. We were also interested in the range of immunological effects which were influenced by AF3.44.4. Other responses which were found to be augented include the mixed lymphocyte reaction, T cell proliferation to protein antigens, and cytotoxic T cell responses (manuscript in preparation).

TABLE 3
Enhancement of Il-2 release by AF3.44.4

STIMULUS	RESPONSE			
	Proliferative (^{125}IUDR)		IL-2 Release	
	Syngeneic	Allogeneic	Syngeneic	Allogeneic
5 x 10^3 SAC	1025±45	6113±498	0.8	2.7
" + AF3.44.4	3591±1118	8952±2443	4.3	6.0
" + YM8	1184±284	1184±2320	0.7	2.2

IL-2 release was measured by the enhancement of mitogen dependent thymocyte proliferation. 10^6 thymocytes from 4-6 week old CBA mice were cultured in 200 ul wells in quadruplicate with phytohaemaglutinin (1/1000) for 3 days with test supernatants added at a dilution of 1/4. The enhancement of mitogen induced proliferation was assessed by adding ^{125}IUdR for 4 hours before washing and harvesting. Proliferation induced by supernatants from allogenic cultures produced in the absence of antibody in the presence of control antibody (YM8) (0.25 ug/ml) or in the presence of AF3.44.4 (0.25 ug/ml) were compared with proliferation induced by standard doses of IL-2. 1 unit of IL-2 was arbitrarily defined as an increment of 1000 cpm ^{125}IUdR over background proliferation. Stimulating spleen adherent cells (SAC) were from DBA/2 mice, syngeneic responding cells were from DBA/2 lymph nodes; allogeneic responding cells were from CBA lymph nodes.

4. Augmentation of Release of IL-2 by AF3.44.4

A common mechanism for the enhancing effects on immune responsiveness may be the production of more IL-2 which augments the growth of the appropriate T cells. Thus we tested whether in the presence of AF3.44.4 and antigen, more IL-2 would be released. As shown in Table 3, there was augmentation in IL-2 titers detected, but it is not dramatic, in the range of 2-4 fold, in parallel with the proliferative response. Thus we performed other experiments, adding IL-2 in optimal concentrations and determining whether AF3.44.4 could still augment the response. These experiments are not complete, but in preliminary experiments it is clear that AF3.44.4 can augment T cell cytotoxicity (James et al, in preparation) even in the presence of optimal concentrations of IL-2. Thus the mechanism of the immunostimulation is likely to be more complex than just enhanced IL-2 production, and a more detailed analysis is underway.

DISCUSSION

The production of monoclonal antibodies offers a new tool to analyse many of the unresolved questions concerning antigen specific factors. For example, it should be possible to establish the relationship of released factors to cell surface receptors for antigen, and biochemical analysis of these entities should be feasible with the appropriate antibody probe.

The results presented here reflect some of the unexpected properties of the monoclonal, which augments the activation of T cells (Tables 2,3). In view of fact that antibodies to receptors of the various tissue cells are able to stimulate them, e.g. antibody to TSH receptors stimulates thyroid cells in Graves' disease (7), it is not surprising that stimulation can also occur with lymphocytes as presumably happens here. However, the effect of AF3.44.4 also requires antigen, which suggests that reaction at the combining site of T cell receptors may also be required for T cell activation.

The synergy between AF3.44.4 and antigen suggests that antibodies of this type may be useful as a new method of augmenting responses of low or undectable amplitude, perhaps including those to tumour antigens.

ACKNOWLEDGMENTS

This work was supported by the ICRF, MRC, SERC and US PHS Grant Nos A1-13145-05, A1-15653-03 and A1-15636-03. We thank Ms Philippa Wells for preparing the manuscript.

REFERENCES

1. Feldmann, M. and Kontiainen, S. Lymphokines **2:** 87 (1981).
2. Feldmann, M. and Schreier, M. (eds) Lymphokines **5:** (1982).
3. Kontiainen, S. and Feldmann, M. Thymus **1:** 59 (1979).
4. Culbert, E.J., Kontiainen, S. and Feldmann, M. (1982) (Submitted for publication).
5. James, R.F.L., Culbert, E.J., Kontiainen, S, Maudsley, D.J. and Feldmann, M. (1982) (Submitted for publication).
6. Kontiainen, S. and Feldmann, M. Eur. J. Immunol. **6:** 296 (1976).
7. Doniach, D., Bottazzo, G.F. and Russell, R.C.G. Clinics Endocrinol. Metabl. **8:** 63 (1979).

DISCUSSION

SCHIMPL: Has Feldmann checked whether the monoclonal antibody, being IgM, could be made stimulatory by interacting it with a TCGF receptor rather than being inhibitory, as was the case with the IgG anti-receptor antibody described previously?

FELDMANN: We have certain lines of evidence that AF3.44.4 does not react with the TCGF receptors; it reacts with antigen-specific helper factor, not known to cross-react with the TCGF receptor; it reacts with the T helper hybridoma, not dependent on the TCGF receptor; it syndergizes with TCGF, and can augment cytotoxic responses even in the presence of optimal levels of TCGF.

OPPENHEIM: Is there any evidence that the postulated "shed" specific helper factors play a role *in vivo?*

FELDMANN: The evidence that specific helper factors play an *in vivo* role is scanty. Experiments have been done by Mozes and colleagues, and analogous ones for suppressor factor by Asherson, and by Benacerraf's group, by parenteral administration of the factor in mice. Determining the role of *in vitro* derived factor *in vivo* is a very difficult problem, and there is not much evidence that other mediators, such as TRF, or AEF are effective *in vivo*. The problem with experiments involving injection of the factor is that it is unlikely to get to the right microenvironment at the appropriate time or for the correct duration; this makes negative experiments difficult to interpret. It seems likely that a better approach to understand the role of these molecules is by *in vivo* use of monoclonal antibodies.

CHAIRMEN'S COMMENTS

The noteworthy feature and success of this session attributed to the considerable effort to generate monoclonal antibodies against lymphokines and factors. We consider this a significant advance, since the use of these monoclonal antibodies will permit more detailed analysis of the mechanism of lymphokine action, will aid biochemical analysis and will be a useful probe to examine the relationship of various lymphokines to each other. Perhaps the most important potential of these anti-lymphokine monoclonal antibodies will be their contribution towards developing simple, rapid and quantitative assays.

M. Feldmann
A. Schimpl

SECTION IV: LYMPHOKINE-INDUCED DIFFERENTIATION AND GROWTH OF TARGET CELLS

DIFFERENTIATION BETWEEN MACROPHAGE-ACTIVATION FACTOR AND MIGRATION-INHIBITION FACTOR IN LYMPHOKINE PREPARATIONS[1]

William E. Fogler
Jane S. Brandhorst
Isaiah J. Fidler

Cancer Metastasis and Treatment Laboratory
NCI-Frederick Cancer Research Facility
Frederick, Maryland

We determined whether the activities of macrophage-activation-factor (MAF) and migration-inhibition factor (MIF) in culture supernatants of Con-A-stimulated lymphocytes could be distinguished by physiocobiological criteria. Four days after the ip injection of Marcol 52, murine peritoneal macrophages (PEM) were harvested and incubated in suspension for 24 hr with lymphokines (LK) or control supernatants. The MIF activity was assayed by the agarose microdroplet assay, and MAF activity was determined by macrophage-mediated cytolysis of radiolabeled tumor cells. There was a two log difference in activity between titers of MIF (1:1000 dilution) and MAF (1:10 dilution). Treatment of LK at 56, 70, or 100°C for 30, 10, or 2 min, respectively, resulted in a linear decay to extinction of MIF activity. MAF activity was retained even after treatment at 100°C for 2 min. PEM cultures incubated for 24 hr with liposomes containing LK in the presence of 0.1 M α-L-fucose became highly tumoricidal. Presentation of LK in liposomes, however, did not generate MIF activity. We conclude that the activities of MIF and MAF can be distinguished on the basis of their dose-response characteristics, thermal stability, and mode of presentation to the macrophage.

[1]*Research sponsored by the National Cancer Institute, DHHS, under Contract No. NO1-CO-75380 with Litton Bionetics, Inc. The contents of this publication do not necessarily reflect the views or policies of the Department of Health and Human Services, nor does mention of trade names, commercial products, or organizations imply endorsement by the U. S. Government.*

ISBN 0-12-527540-4

I. INTRODUCTION

Lymphocytes incubated in vitro with a specific antigen (1-3) or nonspecific mitogen (2-6) release a variety of macromolecules classified as lymphokines (LK) into the culture medium. LK are also released spontaneously from lymphoid cell cultures (7) and from T-cell hybridomas (8). Many of the biological activities of LK are directed toward the cells of the monocyte-macrophage series and include among others, migration-inhibition factor (MIF) and macrophage-activation factor (MAF). To date, exact chemical definitions for LK acting in various manners on macrophages are lacking. The amount of MAF or MIF required to achieve macrophage activation for the desired phenotype cannot be measured directly inasmuch as the only available assay for their activities involves biological assay of their action on macrophages. From such assays it has been proposed that MIF and MAF activities are closely related and could well be induced by a single LK. In this report we present evidence to the contrary and demonstrate that MIF and MAF activities can be distinguished on the basis of dose-response activity, thermal stability, and mode of presentation to and interaction with macrophages.

II. MATERIALS AND METHODS

A. *Animals*

Specific-pathogen-free C57BL/6 mice and F344 rats were obtained from the NCI-Frederick Cancer Research Facility's Animal Production Area.

B. *Preparation of LK*

LK supernatants rich in MAF and MIF activity were harvested from concanavalin-A (Con-A)-stimulated cultures of normal rat splenocytes (NRSUP) as previously described (9). Control supernatant fluids were harvested from cultures of NRSUP that were not stimulated with Con-A but to which Con-A was added 2 hr prior to harvest. LK did not contain endotoxins as determined by the Limulus amebocyte lysate assay (sensitivity level 0.1 ng endotoxin/ml) (Associates of Cape Cod, Woods Hole, MA).

C. Lipids and Preparation of Liposomes

Chromatographically pure egg phosphatidylcholine (PC) and beef brain phosphatidylserine (PS) were purchased from Avanti Polar Lipids, Inc., Birmingham, AL. Multilamellar vesicles (MLV) were prepared as described previously (10). For encapsulation of rat NRSUP or LK within MLV, phospholipids (20 μmol total lipid, PC/PS 7:3 mol ratio) were first evaporated to dryness under nitrogen and then hydrated in the material to be encapsulated (5 ml). MLV were formed by mechanical agitation. Nonencapsulated material was separated from the MLV by 10 min centrifugation at 37,000 x g. The MLV were resuspended in medium and used immediately as described below. The internal volume of the MLV was determined to be 2.5 ± 0.4 μl/μmol phospholipid.

D. Preparation and Purification of PEM

Four days after the intraperitoneal injection of Marcol 52 (Litton Bionetics, Kensington, MD), murine peritoneal exudate cells (PEM) were harvested as previously described (9). The cells were plated in serum-free media onto plastic dishes. After 40 min all nonadherent cells were removed and the adherent PEM were gently scraped off the dish with a rubber policeman. Plating for 40 min in serum-free media preferentially selects for PEM since >97% of the adherent cells are phagocytic and exhibit a morphology typical of macrophages (9).

E. In Vitro Interaction of LK with PEM

PEM at a concentration of 2 X 10^6/ml (total volume of 10 ml) were suspended in supplemented Eagle's minimum essential medium (MEM), NRSUP, or LK. In the experiments employing liposome-encapsulated materials, the concentration of MLV was 0.5 μmol total phospholipid/2 x 10^6 PEM/ml (total volume of 10 ml). This amount of liposomes contained approximately 12.5 μl of encapsulated solution. Macrophages suspended in media with MLV-encapsulated MEM and 12.5 μl of LK in a total volume of 10 ml served as controls. The suspensions of PEM in polypropylene test tubes were kept on a rocking platform for 24 hr at 37°C. Following this period, the suspensions were centrifuged and resuspended in MEM. Nonphagocytosed liposomes were removed after centrifugation in 3 ml fetal bovine serum at 400 x g.[2] The upper phase containing nonphagocytosed

[2] *Kleinerman, E.S., Schroit, A.J., Fogler, W.E., and Fidler, I.J., J. Clin. Invest., submitted for publication.*

liposomes was aspirated and the pellet resuspended in MEM. Each culture group was divided into two aliquots. MIF activity was assayed by the agarose microdroplet technique (11); MAF activity was determined by a radiolabel-release assay from syngeneic tumor cells (9).

III. RESULTS

A. *Titration of MAF and MIF Activity in Rat Splenocyte LK*

The in vitro responses of murine PEM to MAF and MIF, following 24 hr incubation in dilutions of the LK, are shown in Table 1. PEM incubated in undiluted LK were rendered cytotoxic against syngeneic B16 melanoma cells (47%, $P < 0.01$). A tenfold dilution of the LK decreased the level of cytotoxicity to 18% ($P < 0.01$). No MAF activity was found with LK diluted further. In contrast, MIF activity was observed with LK diluted up to 1:1000.

TABLE 1. Dose Response of MIF and MAF Activity in Rat LK

Macrophage treatment group	*MIF activity (% inhibition of migration)*	*MAF activity (% cytotoxicity)*
MEM control	*0*	*0*
NRSUP	*0*	*0*
LK		
Undiluted	*61*	*47*
1:10	*56*	*18*
1:100	*46*	*0*
1:1000	*17*	*0*

The percentage of inhibition of migration and of cytotoxicity was calculated by comparing LK-treated PEM to those treated with NRSUP ($P < 0.01$).

TABLE 2. Thermal Stability of MIF and MAF Activity in Rat LK

Macrophage treatment group	*MIF activity (% inhibition of migration)*	*MAF activity (% cytotoxicity)*
MEM control	--	*0*
NRSUP	*0*	*0*
LK		
37°C	*49*	*42*
56°C, 30 min	*32*	*49*
70°C, 10 min	*21*	*48*
100°C, 2 min	*0*	*50*

The percentage of inhibition of migration and of cytotoxicity was calculated by comparing LK-treated PEM to those treated with NRSUP ($P < 0.01$).

B. Thermal Stability of MAF and MIF Activity

MIF and MAF activities demonstrate different thermal stabilities (Table 2). Both MIF and MAF activities were present in LK supernatants heated to 37°C. Treatment of the LK at 56°C for 30 min reduced MIF activity from 49% to 32%, treatment at 70°C for 10 min resulted in a further decrease in activity (21%), and MIF activity was completely abrogated when the LK supernatant was heated to 100°C for 2 min. In contrast, the level of macrophage-mediated cytotoxicity (a measurement of MAF activity) was unaffected by the heat treatments even when LK were subjected to 100°C for 2 min.

C. Response of PEM to LK Encapsulated in MLV

The endocytosis by macrophages of liposomes containing LK is very efficient in activation of tumoricidal properties. By the nature of the carrier vehicle for MAF, it is clear that the activation mechanism circumvents binding of the LK to a macrophage surface receptor (10,12,13). In the next set of experiments we investigated whether MLV-encapsulated LK that render macrophages tumoricidal will also produce MIF activity (Table 3). Incubation of macrophages with the nonencapsulated, free LK produced a 77% inhibition of migration (MIF) and 53% cytotoxicity (MAF). Incubation of macrophages with MLV-containing media or NRSUP resulted in no MIF or MAF activities. PEM incubated with liposome-encapsulated LK were highly cytotoxic (69%) but exhibited a low level (15%) of MIF activity. The total amount of LK contained within the liposome preparation was 12.5 μl, equivalent to a 800-fold dilution of the original material. Control incubation of PEM with liposome-encapsulated MEM and 12.5 μl of nonencapsulated LK did not activate macrophage tumoricidal properties but produced an 11% inhibition of migration ($P < 0.01$). As shown in Table 1, even a 1:1000 dilution of the original LK preparation produced significant MIF activity (17%, $P < 0.01$), suggesting that the level of migration inhibition observed with LK encapsulated in MLV may have been due to extracellular leakage of the entrapped material, which could interact with the macrophage surface.

TABLE 3. Modulation of MIF and MAF Activity by Encapsulation of Rat LK within MLV

Macrophage treatment group	*MIF activity (% inhibition of migration)*	*MAF activity (% cytotoxicity)*
MEM control	--	*0*
NRSUP	*0*	*0*
LK	*77*	*53*
MLV	*0*	*0*
MLV-NRSUP	*0*	*0*
MLV-MEM + LK	*11*	*0*
MLV-LK	*15*	*69*

The percentage of inhibition of migration and of cytotoxicity was calculated by comparing LK-treated PEM to those treated with NRSUP ($P < 0.01$).

TABLE 4. Effect of Fucose on the Modulation of MIF and MAF Activity by Encapsulation of Rat LK within MLV

Macrophage treatment group	*MIF activity (% inhibition of migration)*	*MAF activity (% cytotoxicity)*
MEM control	--	0
MEM + 0.1 M fucose	0	0
MLV-LK	14	59
MLV-LK + 0.1 M fucose	0	63

The percentage of inhibition of migration and of cytotoxicity was calculated by comparing LK-treated PEM to those treated with NRSUP ($P < 0.01$).

To investigate this possibility, PEM were incubated with liposome-encapsulated LK in the presence of α-L-fucose, which has been been shown to abolish the effect of free MIF and/or MAF (10,14,15). The addition of 0.1 M fucose to the culture incubated with liposome-encapsulated LK abrogated detectable MIF activity but had no effect on MAF activity (Table 4).

DISCUSSION

Our data demonstrate that the activities ascribed to MIF and MAF in rat LK supernatants can be distinguished by their dose-response characteristics, thermal stability, and requirement for the participation of the macrophage plasma membrane.

The relative thermal stability of MAF or MIF activity in lymphokine preparations has been controversial. MAF activity was reported to be stable to heating at 56°C for 30 min (3,8,16,17) or 100°C for 2 min (3,8). At the same time, a reduction in MAF activity was observed following heating to 56°C for 30 min (18) or 60°C for 60 min (17), and a complete loss of MAF activity was reported following treatment at 80°C for 5 min (16) or 70°C for 60 min (17). Similar discrepancies have also been reported for the thermal stability of MIF.

Heating at 56°C for 30 min resulted in either no effect or partial inactivation, while 80°C for 30 min destroyed all MIF activity (19). These discrepancies may be due to the fact that different preparations of LK were used in the different studies; it is entirely possible that LK produced as a constitutive product of a continuous cell line and those produced by lymphocytes in response to an antigen or a mitogen produce the same biological activities (MAF, MIF) but are not identical molecules.

Previous work from our laboratory demonstrated that liposomes containing LK can activate tumoricidal properties in macrophages that cannot respond to free LK (MAF) because their cell surface receptors to MAF have been destroyed or inactivated. The present data clearly demonstrate that when LK rich in MAF and MIF activity are encapsulated in liposomes and endocytosed by macrophages, only MAF activity is expressed; this suggests that for MIF activity to occur, interaction at the cell surface is mandatory. In contrast, these data and others (10,12,13) suggest that MAF activate macrophages subsequent to interaction with intracellular sites.

In conclusion, MAF and MIF activities in Con-A-stimulated splenocyte culture supernatants can be distinguished by dose-response studies, thermal stability, and mode of presentation to macrophages. These studies cannot determine whether MIF and MAF activities are produced by the same or different mediators. The exact nature of these LK await biochemical characterization.

REFERENCES

1. Piessens, W.F., Churchill, W.H., Jr., and David, J.R., *J. Immunol. 114,* 293 (1975).
2. Fidler, I.J., Darnell, J.H., and Budmen, M.B., *J. Immunol. 117,* 666 (1976).
3. Fidler, I.J., Darnell, J.H., and Budmen, M.B., *Cancer Res. 36,* 3608 (1976).
4. Pick, E., Brostoff, J., Krejci, J., and Turk, J.L., *Cell. Immunol. 1,* 92 (1970).
5. Remold, H.G., David, R.A., and David, J.R., *J. Immunol. 109,* 578 (1972).
6. Rocklin, R.E., *J. Immunol. 114,* 1161 (1975).
7. Papermaster, B.W., Holtermann, O.A., Klein, E., Parnett, S., Dobkin, D., Laudico, R., and Djerassi, I., *Clin. Immunol. Immunopathol. 5,* 48 (1976).
8. Ratliff, T.L., Thomasson, D.L., McCool, R.E., and Catalona, W.J., *Cell. Immunol. 68,* 311 (1982).

9. Raz, A., Fogler, W.E., and Fidler, I.J., *Cancer Immunol. Immunother.* *7,* 157 (1979).
10. Poste, G., Kirsh, R., Fogler, W.E., and Fidler, I.J., *Cancer Res.* *39,* 881 (1979).
11. McCoy, J.L., and Ames, R.S., *in "Manual of Macrophage Methodology"* (H.B. Herscowitz, H.T. Holden, J.A. Bellant, and A. Ghaffar, eds.), p. 417. Marcel Dekker, Inc., New York, (1981).
12. Poste, G., and Kirsh, R., *Cancer Res.* *39,* 1979.
13. Fidler, I.J., Raz, A., Fogler, W.E., Hoyer, L.C., and Poste, G., *Cancer Res.* *41,* 495 (1981).
14. David, J.R., and Remold, H., *in "Immunobiology of the Macrophage"* (D.S. Nelson, ed.), p. 401. Academic Press, New York, (1975).
15. Remold, H., *J. Exp. Med.* *138,* 1065 (1973).
16. Dy, M., Kamoun, P., Dimitriu, A., and Hamburger, J., *Transplantation* *21,* 273 (1976).
17. Kniep, E.M., Domzig, W., Lohmann-Matthes, M.L., and Kickhofen, B., *J. Immunol.* *127,* 417 (1981).
18. Leonard, E.J., Ruco, L.R., and Meltzer, M.S., *Cell. Immunol.* *41,* 347, 1978.
19. Pick, E., *in "Immunopharmacology"* (J.W. Hadden, R.G. Coffey, and F. Spreafico, eds.), p. 163. Plenum, New York, (1977).

DISCUSSION

REMOLD: What happens when Fogler increases the amount of MIF in the liposomes?

FOGLER: By increasing the amount of lymphokine encapsulated in the liposomes one also increases the amount of extracellular leakage which can then interact at the macrophage surface. Accordingly, we did see an increase in MIF activity. However, it should be emphasized that the activity was abrogated by fucose, which suggests that it was still primarily a surface interaction.

YOSHIDA: It is hard to conceive of Fogler detecting MIF activity at dilutions of the original supernatant as high as 1,000-fold. This is most unusual, compared with previous reports as well as our own experience. What were the circumstances? How did Fogler remove Con-A from his supernatants?

FOGLER: The dilution experiments stand by themselves. It may well be that our method of assay and responding macrophages are unusually sensitive to the activity of MIF. The Con-A was linked to Sepharose 4B, which we then removed by centrifugation.

SANDRU: Does Fogler have any evidence that anti-MIF antibodies (when bound to Sepharose 4B) could block or remove MAF activity?

FOGLER: No. I am not aware that any experiment of that nature has been done.

WOLSTENCROFT: I have always been fascinated by proteins that resist heating to 100°C - would Fogler tell us a bit more about this material?

FOGLER: The finding that MIF activity is retained following heating of the lymphokine preparation at 100°C for 2 minutes *is* intriguing. This was originally reported by Fidler some time ago and recently a group at Washington University have shown that MAF activity in supernatants obtained from a T cell hybridoma also resists exposure to 100°C for 2 min.

OPPENHEIM: Fogler has not convinced me that MAF or MIF are really different on the basis of his data. Differences in dose-response may merely be due to differences in the bioassays. The apparent difference in effect of fucose may be due to the different times at which these two activities are evaluated. Fucose may be able to inhibit only for the 24 hour MIF assay but not over the longer period of 72 hours required for the MAF assay. Finally, the response to heat treatment showed only partial retention and is compatible with overlapping effects.

FOGLER: Although it is true that dose-response differences may be due to the different bioassays in that our sensitivity level for the detection of MIF activity was higher than for MAF activity, this would not apply when the lymphokine is encapsulated in liposomes where we are working with an 800-fold dilution of the original material. Here we show that MIF activity is minimal, while MAF activity is actually increased. As to Oppenheim's second point, fucose was present only during the 24 hr incubation period of the macrophages with liposome-encapsulated lymphokine in order to abolish the effect of those MIF and MAF activities which may have resulted from leakage of the lymphokine from the liposomes. Following this incubation, fucose was removed and the resulting activities assayed. By that criterion fucose abolished MIF activity while MAF activity was unchanged; again suggesting that MIF activity is dependent on a cell surface interaction, while for MAF activity to be expressed this initial interaction is not essential. Finally, while there is only partial inactivation of MIF activity at 56° for 30 min., and 70° for 10 min, clearly treatment of the lymphokine at 100° for 2 min. totally abolished MIF activity while MAF activity remained unchanged.

PICK: My first comment refers to the dilution effect. I think that the high dilution of MIF still active, as found by Fogler, is dependent on the technique being used. The agarose microdroplet method is usually connected with high dilutions of active MIF. We are being offered by Fogler a new concept - i.e., that MIF acts via surface receptor in a classical receptor → second messenger way, while MAF (for cytotoxicity) requires uptake and might represent a post-phagocytic type of activation, essentially different from accepted forms of lymphokine action. This agrees with the premise that similar types of activation can be obtained with a variety of non-lymphokine materials incorporated into liposomes.

S. COHEN: In liposomes, MAF may bypass the receptor binding step. The effect of MIF may be a surface one. In other words, rather than activating an internal pathway associated with locomotion, MIF may simply make the cell surface "sticky". Thus, instead of distinguishing between two different mediators, Fogler may be distinguishing between two different loci of action.

FOGLER: I do not feel that our data distinguish whether MIF and MAF are similar or different; this awaits biochemical clarification. However, we do show that the activities due to these components can be distinguished, and on this basis I agree with Cohe's comment. In fact, this is the essence of my report.

GUINEA PIG MACROPHAGE-ACTIVATING FACTOR CAN BE DISTINGUISHED FROM MIGRATION INHIBITORY FACTOR BY ITS SENSITIVITY TO TRYPSIN[1]

W. Hallowell Churchill
Alma Mednis
Heinz Remold

Department of Medicine
Harvard Medical School
Division of Rheumatology and Immunology
Brigham and Women's Hospital
Boston, Massachusetts

Supernatants from Concanavalin A-stimulated guinea pig lymph node cells were fractionated by Sephadex G-100 gel filtration sucrose gradient electrophoresis and isoelectrofocussing. Macrophage activating factor (MAF) activity was found in two peaks having a pI of 3.0-4.5 (pH 3-MAF) and 5.0-5.5 (pH 5-MAF) coinciding with MIF activity. However, in contrast to pH 3-MIF, pH 3-MAF activity was decreased by trypsin treatment. Thus, pH 3-MAF can be distinguished from pH 3-MIF by its sensitivity to trypsin.

INTRODUCTION

Lymphocytes stimulated by antigens or mitogens secrete a number of soluble substances called lymphokines. One of these, a factor which enhances macrophage cytotoxicity for target cells, is called macrophage activating factor (MAF) (1). It was of interest to establish the relationship of MAF to migration inhibitory factor (MIF), another lymphokine which acts on macrophages. MIF activity is assayed by its ability to inhibit the migration of peritoneal cells out of capillary tubes (2). Guinea pig MIF can be separated into two distinct species. One, pH 3-MIF, has an isoelectric point of 3.0 to 4.5 and is resistant to treatment with trypsin. The other, pH 5-MIF, has an isoelectric point of 5.0 to 5.5 and is sensitive to trypsin (3, 4). We wanted to establish whether pH 3-MIF and pH 5-MIF also show MAF activity

[1]*This work was supported by U.S. Public Health Service Grants AI-12110 and AI-07685 and a Research Career Development Award, CA-00116-05.*

ISBN 0-12-527540-4

and furthermore, whether biochemical differences between MIF and MAF can be detected. Our results show that both pH 3 MIF- and pH 5-MIF-containing fractions can be distinguished from pH 3-MIF by its sensitivity to trypsin.

MATERIALS AND METHODS

Production of MIF-MAF Rich Supernatants

Mediator-rich supernatants were prepared as described by culturing 2.3 x 10^7 guinea pig lymph node cells/ml for 24 hr with Concanavalin A (10 μg/ml). Concanavalin A was added to the cell-free supernatants of control cultures after completion of the culture period. The mediator-rich and control supernatants were dialyzed against water, lyophilized and stored at 20°C (5).

Maintenance of Tumor

The ascites variant of the strain 2 diethylnitrosamine induced guinea pig hepatoma was maintained by serial passage of tumor in vivo (6,7) in inbred Sewall Wright stain 2 guinea pigs obtained from the colony maintained at Fort Dietrich, MD.

Harvest and In Vitro Activation of Peritoneal Exudate Cells

Mineral oil-induced peritoneal exudate cells (PEC) were obtained from random-bred Hartley guinea pigs as described (7). PEC were activated as described (8, 9), by culturing 5 X 10^6 cells/ml in 10 ml Spinner flasks (Bellco). The PEC cultures contained either MAF-rich fraction (activated cells) or the control fraction from unstimulated lymphocytes. After culture, the activated and control cells were collected, washed twice with medium, and tested for their cytotoxicity capacity.

Assay for Cytotoxic Macrophages and MIF

Using a described assay (8, 9), the target hepatoma cells prelabelled with ^{3}H-thymidine were cultured with activated or control PEC in microtiter plates for 18 hr. After discarding nonadherent material, the adherent cells were washed, the radioactivity measured and the percent cytoxicity assessed (8).

Purification and Fractionation of MIF

The purification of pH 3-MIF, pH 5-MIF and MAF were performed as previously described (3). Lyophilized MIF-rich control supernatants from 175 ml lymphocyte suspension were successively purified on Sephadex G-100 columns, sucrose density electrophoresis and isoelectrofocussing to separate pH 3-MIF from pH 5-MIF. The two species were collected into albumin-containing (0.125 mg/0.5 ml/tube) tubes and concentrated to 4 ml each. The amounts of pH 3-MIF and pH 3-MAF mentioned in this study are aliquots of these fractions. The MIF-containing fractions were diluted to 2 ml with minimum essential medium (Earle) containing 100 µg streptomycin and 100 U penicillin/ml (MEM-PS) and adjusted to 15% normal guinea pig serum. They were assayed on normal guinea pig PE cells with a capillary tube assay (5). A migration inhibition of greater than 20% is considered to be significant (5).

Statistical Analysis

Two way analysis of variance (10) was performed by Dr. Heinz Weiser, Center for Astrophysics, Cambridge, MA.

Treatment of pH 3-MAF and pH 3-MIF with Trypsin

0.5 ml fractions were diluted to 1.5 or 2.0 ml with Hanks balanced salt solution containing 16 units/ml trypsin (tosylphenylalanylchloromethylketone-trypsin, 202 units/mg, Sigma Chemical Co., St. Louis, MO). The supernatants were incubated with intermittent agitation at 37°C for 1 hr at a pH of 8.5. Diisopropylphosphofluoridate (Sigma Chemical Co., St. Louis, MO) was added to a final concentration of 10^{-4}M to inactivate trypsin. The mixtures were incubated 15 min at room temperature and then dialyzed against 300 volumes of MEM (4). A set of supernatants in which trypsin was omitted were treated in parallel.

RESULTS

MAF Activity is Found in the Fractions Containing pH 3-MIF and pH 5-MIF

In studies using isoelectrofocussing it has been shown that MIF activity is confined to a pI range of 3.0-4.5 (pH 3-MIF) and 5.0-5.5 (pH 5-MIF) (3). No activity was found in the intermediate fractions. In 14 experiments MAF activity was found in the same fractions (Fig. 1).

Trypsin Treatment Distinguished pH 3-MAF from pH 3-MIF

Trypsin sensitivity of pH 3-MAF was assessed next. Our results show that trypsin at a concentration of 16 U/ml consistently inactivated pH 3-MAF. Results of 6 experiments are

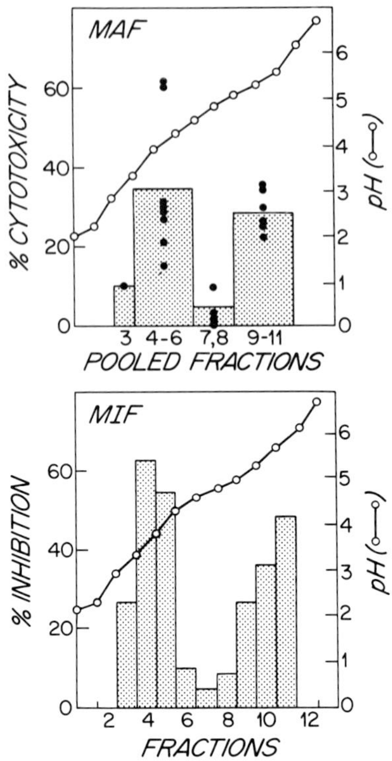

FIGURE 1. Isoelectrofocussing of MAF (upper panel and MIF lower panel) in a pH gradient from 2.0 to 6.2. MAF activity (14 experiments) as % cytotoxicity and MIF activity is demonstrated on the ordinate as indicated by the dots and the bars. The abscissa shows the fractions tested. There is significant statistical difference between the fractions 4-6 and 7-8 and 9-11 and 7-8 ($p<0.001$) indicating that MAF activity is found in two distinct peaks which have the same pH range as the MIF peaks.

shown in Table 1. In each case, trypsin treatment decreased MAF activity. The percent decrease ranged from 28% to 100% with a mean decrease of 58%. Pretreatment with trypsin had no effect on the activity of pH 3-MIF. The mean decrease in these experiments was 4% (experiments not shown). To exclude the possibility that differences in the dose response curves are the explanation for the apparent difference in sensitivity to trypsin, the dose response curves of three separate preparations of pH 3-MIF and one pool of pH 3-MAF were determined. Significant regressions were obtained for both MIF and MAF (Fig. 2). The slopes for the dose response for

TABLE I. Effect of Trypsin Treatment on pH 3-MAF[a]

	Percent Cytotoxicity		
Exp. #	*Buffer*	*Trypsin*	*% Decrease*
1	53	24	55
2	32	23	28
3	30	16	46
4	29	17	41
5	26	6	77
6	15	0	100

[a]*Fractions were incubated with 16 units trypsin for 1 hr after which the trypsin was neutralized with 10^{-3}M diisopropylphosphofluoridate and the fractions tested for MAF activity. There was statistically significant reduction of MAF activity after treatment with trypsin ($p<0.001$)*

MAF and MIF were, respectively, 22.4 and 36.0. Thus, the dose response for MIF is steeper than for MAF. Consequently, a small change in MIF concentration is easier to detect than a small change in MAF activity. It follows that a loss of MIF activity cannot be attributed to differences in the dose response curve.

DISCUSSION

Studies of MIF and MAF have demonstrated many similarities of these lymphokines. Both are produced by antigen or mitogen stimulated lymphocytes (7, 8). Surface fucose and sialic acid are essential for the action of both mediators on the macrophage (11, 12, 13). In addition, the response of the macrophage to both mediators is enhanced by pretreatment of macrophages with serine esterase inhibitors such as antithrombin III or soybean trypsin inhibitor (14, 15).

Our results show that both pH 3-MIF and pH 5-MIF also contain MAF activity. The dose response of these mediators was also similar. However, the sensitivity to trypsin of pH 3-MAF provided a clear criterion for its differentiation from pH 3-MIF. Because pH 5-MIF is trypsin-sensitive (3), trypsin sensitivity cannot be used as a criterion for the differentiation of pH 5-MIF and pH 5-MAF.

The similarity of the dose response curves for MIF and MAF exclude the possibility that assay differences can explain these results.

The amount of trypsin has been shown to be in excess since several doses exceeding the concentrations of trypsin used in our experiments had no effect on MIF activity (3).

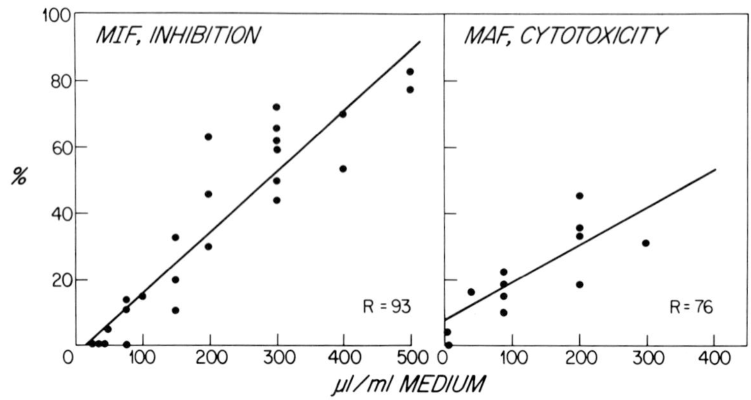

FIGURE 2. Dose response of pH 3-MIF (3 preparations) and pH 3-MAF (1 preparation). The ordinate indicates the activity, the abscissa the amount of mediator tested. R=correlation coefficient.

Thus, the lack of an effect of trypsin on MIF cannot be explained by the dose of trypsin used.

These results are compatible with a hypothesis which proposes that pH 3-MIF and pH 3-MAF are two closely associated molecules which are inseparable by various biochemical fractionation methods. The observation of Kniep et al. (16) support this hypothesis. These workers found that mouse MAF could only be separated from MIF by isoelectrofocussing in 8M urea.

REFERENCES

1. Piessens, W. F., Churchill, W. H., and Sharma, S. D., "Lymphokine Reports, Vol. 3" p. 293. Academic Press, New York (1981).
2. David J. R., and David R. A., *In* "Progress in Allergy" (P. Kallos, B. H. Waksman and A. deWeck, eds.) p. 300. S. Karger, Basel, (1972).
3. Remold, H. G., and Mednis, A. D., *J. Immunol. 93*, 264 (1977).
4. Remold, H. G., and Mednis, A. D., *J. Immunol. 122*, 1920 (1977).
5. Remold, H. G., Katz, A., Haber, E., and David, J. R., *Cell Immunol. 1*, 133 (1970).

6. Rapp, H. G., Churchill W. H. Jr., Kronman, B. S., Rolley, R. T., Hammond, W. G., and Borsos, T., *J. Natl. Cancer Inst. 41*, 1 (1968).
7. Piessens, W. F., Churchill, W. H. Jr., and David, J. R., *J. Immunol. 114*, 293 (1975).
8. Churchill, W. H., Piessens, W. F., Sulis, C. A., and David, J.R., *J. Immunol. 115*, 507 (1975).
9. Churchill, W. H., Piessens, W. F., and David, J. R., *In* "In Vitro Methods in Cell-Mediated and Tumor Immunity" (B. R. Bloom and J. R. David, eds.), p. 417. Academic Press, New York, (1976).
10. Snedecor, G. W., and Cochran, W. G., "Statistical Methods". The Iowa State University Press, Ames, Iowa, (1973).
11. Churchill, W. H., and Wong, C., *Cell. Immunol. 55*, 490 (1980).
12. Liu, D. Y., Petschek, K. D., Remold, H. G., and David, J. R., *J. Immunol. 124*, 2042 (1980).
13. Higgins, T. J., Liu, D. Y., Remold, H. G., and David, J. R., *Biochem. Biophys. Res. Commun. 93*, 1259 (1980).
14. Remold, H. G., and Rosenberg, R. D., *J. Biol. Chem. 250*, 6608 (1975).
15. Piessens, W. F., Remold, H. G., and David, J. R., *J. Immunol. 118*, 2078 (1977).
16. Kniep, E. M., Kickhofen, B. and Fischer, H., *In* "Biochemical Characterization of Lymphokines" (A. L. de Weck, F. Kristensen, M. Landy, eds.), p. 149, Academic Press, New York, (1980).

DISCUSSION

OLDHAM: Does Remold's Mo-MIF have MAF activity?

REMOLD: Yes, it does, but as we are presently working on this problem I do not have more definite information.

OLDHAM: What level of heat stability has Remold seen with the Mo-MIF and MAF activity, if the latter is present?

REMOLD: We have not yet compared the heat sensitivity of the Mo-MAF to Mo-MIF. Mo-MIF is stable at 56° for 1/2 hour.

STONE-WOLFE: Did Remold determine whether MAF binds to Con-A Sepharose?

REMOLD: We have not yet done the experiment to define whether MAF binds to Sepharose.

STONE-WOLFE: Since on the basis of trypsin-resistance Remold concluded that MIF must be glycosylated, he should also find out whether MAF, which was trypsin-sensitive, is also glycosylated--possibly by affinity-binding studies on different lectin columns.

REMOLD: We do not know whether human MAF is trypsin-sensitive, like guinea pig-MIF.

STONE-WOLFE: Trypsin-resistance is not an appropriate criterion for glycoslyation, since many glycoproteins are inactivated by trypsinization.

REMOLD: I agree that inactivation by trypsin is not a definitive criterion for glycosylation, but it most likely holds up in the case of the cell line MIF species.

PAULY: David Golde has reported that the unique human T cell line designated as Mo is capable of producing very high levels of human "immune" interferon ($HuIFN_{\gamma}$; ~40,000 U µl). By comparison with human blood lymphocytes, what are the levels of MIF and MAF produced by this macrophage line?

REMOLD: In our hands the Mo-cell supernatants contain about 2-10 times more MIF activity than suspensions of human Con-A-stimulated peripheral blood mononuclear cells--depending on the batch tested.

THE EFFECTS OF HUMAN INTERFERON ON THE MIGRATION OF GUINEA PIG PERITONEAL EXUDATE CELLS

Gary B. Thurman, Harold B. Stull, and Robert K. Oldham

Biological Response Modifiers Program
Division of Cancer Treatment
National Cancer Institute-FCRF
Frederick, Maryland 21701

I. INTRODUCTION

The evidence that IFN has significant effects on macrophage morphology and function has long been recognized (reviewed in ref. 1). IFN causes increased phagocytosis in vitro (2) and in vivo (3), accelerates spreading (4), increases lactate dehydrogenase levels (5), increases antibacterial activity (6), enhances antiviral action (7), and increases the capacity of macrophages to kill tumor cells in vitro (8). These actions indicate that IFN is intimately involved as a mediator of macrophage function and probably is one of the soluble mediators from lymphocytes (lymphokines) that is involved in directing and controlling the migration of the cellular components of the immune response.

Our interest in this area developed out of an effort to identify and quantitate the lymphokine activities in material being used in phase I clinical trials of lymphokine-containing material. Some of the preparations tested had both IFN and MIF activity and when a partially purified IFN was tested as a control, it also inhibited the migration of macrophages. This led to the initiation of a study on the effects of various IFN preparations in the MIF assay and what the possible relationships between MIF and IFN might be.

II. MATERIALS AND METHODS

A. *Lymphokine-Containing Materials*

Lymphokine-containing materials (9) prepared from the lymphoblastoid cell line, RPMI-1788, were supplied by

ISBN 0-12-527540-4

Dudley Dumonde, St. Thomas' Hospital, London, and by Dr. Frank den Hollander, Organon Scientific Development Group, Oss, The Netherlands. The materials were stored at 4°C in lyophilized form until reconstituted in RPMI-1640 medium for use in an assay. Once reconstituted, the preparations were aliquoted and stored at -20°C for use in subsequent assays.

B. *Interferons and Interferon Inducers*

IFN preparations used in this study were obtained from the sources indicated in Table II and their relative purity is shown. Poly I:poly C (Sigma Chemical Co.) was used as an IFN inducer.

Assays for IFN activity were performed by two different methods on the same material. One assay utilized an anti-human leukocyte IFN monoclonal antibody in a RIA assay (10) performed in the Fermentation Program at NCI-FCRF. The other assay measured viral neutralization in a plaque reduction assay (11) and was done by Biofluids, Inc., Rockville, MD.

C. *Measurement of Macrophage Migration Inhibition (MMI)*

The agarose droplet technique of Harrington and Stastny (12) was used. In brief, the macrophages were induced by injection i.p. of sterile light mineral oil into 3-9 month old female strain-2 guinea pigs, followed 3-5 days later by washing the peritoneum with Hanks' Balanced Salt Solution. In some experiments a human histiocytic lymphoma cell line (U937) was used (13). The cells were washed 3 times by centrifugation, counted and centrifuged in a conical bottom tube. The supernatant was completely removed and the cells were resuspended in 0.35% Sea Plaque Agarose (in RPMI-1640 medium) with 10% fetal calf serum. The cell concentration was 10^6 cells per microliter and one microliter droplets were centrally placed in the wells of microtiter plates. The plates were cooled to 4°C and then 0.1 ml of RPMI-1640, containing the appropriate concentration of test material, was added to each well. The microtiter plates were incubated for 14-24 hours in a humid, 5% CO2-in air atmosphere on a perfectly level shelf of a 37°C incubator. The areas of migration were computed using a TI-59 programmable calculator from the dimensions measured, using a Bausch and Lomb Tri-simplex Microprojector to project the image of the droplet and the cells onto a grid.

D. *Measurement of Factor Toxicity Using Indium-111*

Macrophages were labelled with In-111 using the technique of Wiltrout et al. (14). The cells were incubated for 24 hours at 37°C in microtiter plate wells at a cell concentration of 500,000 per 0.2 ml per well with various concentrations of the factors being tested for cytotoxicity. Toxicity was determined by measuring the percentage of In-111 released into 0.1 ml of the supernatant.

III. RESULTS

A. *MIF and IFN Activity of RPMI-1788 Lymphokines*

Five different RPMI-1788 lymphokine preparations were tested for IFN and MIF activities. The results, shown in Table I, indicate that 4 of 5 preparations had IFN activity.

TABLE I. Evaluation of MIF and IFN Activities in Various Preparations of Lymphokines Derived from Supernatants of the RPMI-1788 Cell Line

Preparations[a]	Endotoxin[b]	Interferon (U/mg)[c]		MIF[d]
	ng/mg	RIA	Plaque	μg/ml for 20% inhibition
10787A[e]	0.6	0	<5	>800
10787B	7.4	0	<5	>800
10787C	0.6	0	<5	>800
10786A	3.3	12,130	4,000	6.0
10786B	4.6	21,760	6,300	1.0
10786C	6.2	17,590	2,510	6.0
579.2	13.6	0	<25	0.2
488.2	N.A.[f]	3,375	1250	0.06

[a] *Lymphokine materials produced by Organon.*
[b] *Endotoxin levels measured by limulus lysate assay. Data supplied by Organon.*
[c] *See Materials and Methods for details.*
[d] *Determined as the concentration at which the % MMI dose response curve intersects the 20% migration inhibition level.*
[e] *Control preparations made by processing medium and serum in the same manner that the lymphokine material was processed.*
[f] *Data not available.*

One preparation, 579.2, had essentially no IFN activity detectable by either assay. All 5 lymphokine preparations had MIF activity, with 488.2 being the most potent (20% MMI at 0.06 μg per ml). None of the 3 control preparations had either IFN or MIF activity. There was some endotoxin present in the various preparations tested.

B. *Testing of Various Interferon Preparations on Guinea Pig Macrophages*

The results of testing partially purified and highly purified human IFN are shown in Table II. Each of the IFN's tested caused significant MMI in a dose dependent fashion with the % MMI dropping below the 20% cutoff for positive results between 100 and 10,000 units per ml. Only one of the IFN's gave any indication of being cytotoxic as measured by In-111 release and that was 7% at 100,000 units per ml of partially purified human βIFN. The recombinant DNA human αIFN seemed to stabilize the macrophages at higher concentrations, decreasing the level of In-111 release to 25.7% below that of spontaneous release control.

C. *Testing of Human α Interferon on the Migration of a Human Histiocytic Lymphoma Cell Line (U937)*

Two different preparations of human αIFN inhibited the migration of U937 significantly at concentrations of 500-1,000 units per ml (Figure 1). The more purified Burroughs Wellcome IFN appeared to be somewhat more active in the MMI assay per unit of antiviral activity. The dose response curves of MMI activity of alpha IFN on their U937 cell line paralleled those seen with guinea pig macrophages and the sensitivity of the U937 cells to inhibition of migration by IFN appeared to be approximately the same as for guinea pig macrophages.

DISCUSSION

The presence of IFN in various lymphokine preparations may mimic the action of MIF by inhibiting the migration of macrophages. This possibility may be one of the reasons it has been difficult to isolate and characterize MIF and may be why MIF has been described as having a variety of molecular sizes and having several different properties (15).

TABLE II. Comparison of Migration Inhibitory Capability and Toxicity of Various Human Interferon Preparations Using Guinea Pig Macrophages

Human interferon	Conc (U/ml)	% MMI[a]	% Toxicity[b]
Leukocyte (RX-x43618)	10^5	46.1	-3.2
(Warner-Lambert Co.)	10^4	40.9	-2.0
(4×10^5 U/mg protein)	10^3	25.4	-0.6
	10^2	-4.5	-0.6
Recombinant			
Leukocyte A	10^4	47.1	-25.7
(Hoffman-La Roche, Inc.)	10^3	34.0	-5.1
	10^2	18.1	-1.3
(2×10^8 U/mg protein)	10	5.5	-2.2
Namalwa Lymphoblastoid	10^3	38.5	-3.7
(Burroughs Wellcome)	10^2	25.7	-4.7
(10^8 U/mg protein)	10	10.3	-0.3
Fibroblast	10^5	73.0	7.0
(Rentschler)	10^4	55.7	-6.8
(10^5 U/mg Protein)	10^3	0.4	-0.3
Gamma	10^3	29.4	N.T.[c]
(Meloy Labs)	10^2	24.5	N.T.
(2.8×10^6 U/mg protein)	10	6.9	N.T.

$$^{a}\%\ \text{MMI} = \frac{\text{Area (medium)} - \text{Area (IFN)}}{\text{Area (medium)}} \times 100$$

$$^{b}\%\ \text{toxicity} = \frac{\text{CPM (IFN)} - \text{(Spon. release)}}{\text{CPM (Max. release)} - \text{CPM (Spon. release)}} \times 100$$

[c] Not tested

Studies that have evaluated the capability of various cell lines to make MIF (16) usually have not simultaneously measured IFN levels produced, thus making it difficult to determine if the activity being measured was due to MIF or IFN.

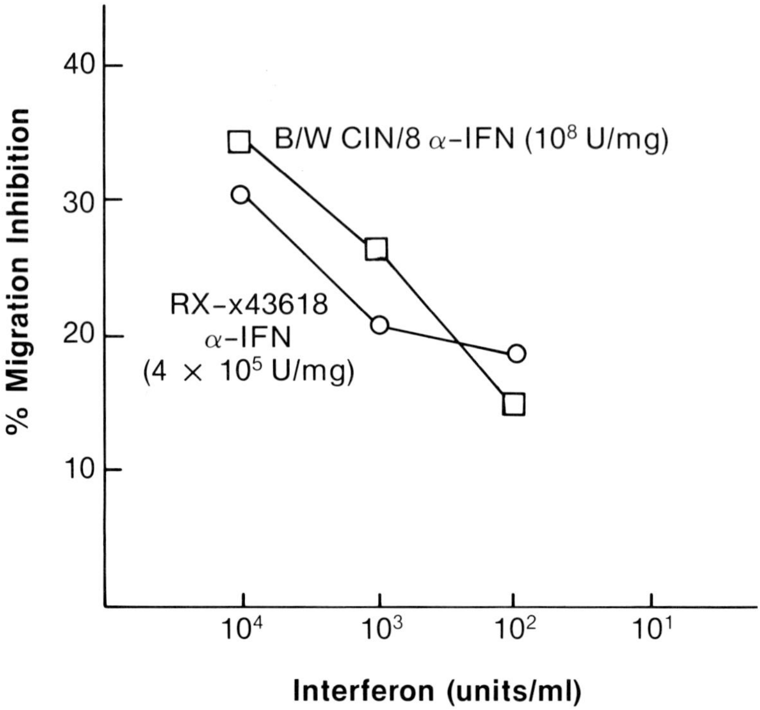

FIGURE 1. Effects of human leukocyte interferon (Warner-Lambert Co., RX-x43618) and Namalwa αIFN (Burroughs Wellcome, B/W CIN/8).

The cell line RPMI-1788 is an example of a cell line that evidently can make both IFN and MIF. Our results indicate that, depending on the processing steps used for purification, variable amounts of IFN can be present in RPMI-1788 lymphokine preparations and can make variable contributions to the amount of MMI measured. It is also clear from the data that the MIF made by the RPMI-1788 cells is a separate entity from IFN since one preparation has good MIF activity but no IFN activity. This does not rule out a possible close relationship between IFN and MIF.

The fact that all 3 types of IFN are effective in MMI suggests that this phenomenon is a part of a general mechanism whereby macrophages are induced to remain localized at the site where cells are making IFN. Any cell type capable of producing IFN would be capable of inhibiting the migration of macrophages. This would also provide one possible mechanism for the known effects of IFN on the development and expression of delayed-type hypersensitivity (DTH), cell-mediated immunity and B-cell responses to T-dependent antigens. This has been shown to be the case (17-19). Of particular interest is the study of De Maeyer-Guingard et al. (20) which showed that the induction or passive transfer of type I IFN to mice 1 day prior to injection of sheep red blood cells (sRBC) prevented sensitization as measured by the DTH footpad reaction. The degree of suppression correlated with the IFN levels, with strains of mice producing the highest levels of IFN showing the greatest suppression of DTH. This would suggest that IFN is able to preempt the DTH response by making the macrophages insensitive to other lymphokines such as MIF. Administration of MIF-containing material to sensitized guinea pigs also causes the transient loss of DTH response to both specific and nonspecific antigen (21). All of these studies would indicate that the effects of IFN and MIF on macrophages may be of prime importance in the development and expression of DTH.

The mechanism by which IFN induces the inhibition of macrophage migration was not investigated in this study. However, others (8) have found a dramatic increase in the degree of spreading of macrophages induced by IFN, even as early as 1 hour (22). This increased adherance might be the cause of the decrease in macrophage movement but other possible effects such as the blocking of microfilament or microtube production or function, or the changing of macrophage membrane potentials will also have to be considered.

These studies suggest that IFN has an important role in the development and expression of macrophage function. They indicate that there may be a parallel or sequential mode of action of IFN and MIF and support the conclusions of others (23) that they are not the same molecular species. Care will have to be taken by researchers in this area to ensure that the MMI activity being measured is indeed due to MIF and not due to the MIF-mimicing effects of IFN.

REFERENCES

1. Schultz, R.M., *Lymphokine Reports 1,* 63 (1980).
2. Huang, K.-Y., Donahue, R.M., Gordon, F.B., and Dressler, H.R., *Infect. Immun. 4,* 581 (1971).
3. Donahue, R.M., and Huang, K.-Y., *Infect. Immun. 13,* 1250 (1976).

4. Schultz, R.M., Chirigos, M.A., and Heine, U.I., *Cell. Immunol. 35,* 84 (1978).
5. Remold, H.G., and Mednis, A.F., *Inflammation 1,* 175 (1976).
6. Mizunoe, K., Hiraki, M., Nagano, Y., and Maehara, N., *Jpn. J. Microbiol. 19,* 235 (1975).
7. Morahan, P.S., Glasgow, L.A., Crane, J.L., and Kern, E.R., *Cell. Immunol. 28,* 404 (1977).
8. Schultz, R.M., Papamatheakis, J.D., and Chirigos, M.A., *Science 197,* 674 (1977).
9. Dumonde, D.C., Pulley, M.S., Hamblin, A.S., Singh, A.K., Southcott, B.M., O'Connell, D., Paradinas, F.J., Robinson, M.R.G., Rigby, C.C., den Hollander, F., Schuurs, A., Verheul, H., and van Vliet, E., in "Lymphokines and Thymic Hormones" (A.L. Goldstein and M.A. Chirigos, eds.), p. 301. Raven Press, New York, (1981).
10. Sccher, D.S., and Burke, D.C., *Nature 285,* 446 (1980).
11. Bridgen, P.J., Anfinsen, C.B., Corey, L., Bose, S., Zoon, K.C., and Riieggy, U.T., *J. Biol. Chem. 252,* 6585 (1977).
12. Harrington, Jr., J.T., and Stastny, P., *J. Immunol. 110,* 752 (1973).
13. Sundstrom, C., and Nilsson, K., *Int. J. Cancer 17,* 565 (1976).
14. Wiltrout, R.H., Taramelli, D., and Holden, H.T., *J. Immunol. Methods 43,* 319 (1981).
15. Remold, H.G., and Mednis, A.D., *J. Immunol. 122,* 1920 (1979).
16. Yoshida, T., Kuratsuji, T., Takada, A., Takada, Y., Minowada, J., and Cohen, S., *J. Immunol. 117,* 548 (1976).
17. DeMaeyer, E., DeMaeyer-Guignard, J., and Vandeputte, M., *Proc. Natl. Acad. Sci. USA 72,* 1753 (1975).
18. Hirsch, M.S., Ellis, D.A., Black, P.H., Monaco, A.P., and Wood, M.L., *Transplantation 17,* 234 (1974).
19. Sonnenfeld, G., Mandel, A.D., and Merigan, T.C., *Cell. Immunol. 34,* 193 (1977).
20. De Maeyer-Guignard, J., Cachard, A., and DeMaeyer, E., *Science 190,* 574 (1975).
21. Yoshida, T., and Cohen, S., *Fed. Proc. 41,* 2481 (1982).
22. Hovi, T., Saksela, O., Linnavuori, K., Vaheri, A., and Cantell, K., in "The Biology of the Interferon System" (E. DeMaeyer, G. Galasso, and H. Schellenkens, eds.), p. 217. Elsevier/North-Holland Biomedical Press, New York, (1981).
23. Block, L.H., Cantell, K., Bamberger, S., Ruhenstroth-Bauer, G., and Strander, H., *Arch. Virol.*

DISCUSSION

REMOLD: What evidence does Thurman offer that MIF and interferon are other than one and the same molecule?

THURMAN: We have found MIF activity in material that has no IFN activity which indicates that MIF is independent of interferon.

S. COHEN: Some time ago, we explored the ability of virus infection to induce nonlymphoid cells to make lymphokine-like factors. It was this work that brought the designation of cytokines into our vocabulatry. We found that one combination of cell and virus gave MIF but no IFN and another combination gave just the opposite. Thus assay sensitivity differences were seemingly ruled out and we concluded that the two activities were distinct. It is possible that in Thurman's system, IFN rather than directly blocking migration inhibition simply potentiates the effect of MIF itself on macrophages, or possibly induces macrophages to produce an MIF-like monokine.

THURMAN: Since migration inhibition was elicited by purified interferon, and by recombinant-DNA-produced interferon, the potentiation of MIF activity by these purified materials can be ruled out. In more crude preparations it certainly is a possibility that, to my knowledge, has not been tested. Whether the interferon acts directly to inhibit macrophage migration or acts through induction of another mediator remains to be determined.

OPPENHEIM: I am confused regarding Thurman's interpretation of his apparently simple, sound and straightforward observations. Since $_{\gamma}$IFN is a product of activated lymphocytes, and one which apparently has MIF activity, it *must* be considered an MIF. The fact that there are MIF activities in the absence of antiviral activities may only indicate that MIF, like many of the other lymphokine preparations, is heterogeneous. Thurman's findings should therefore lead one to conclude that IFN is indeed *one* of the group of MIF activities!

THURMAN: My observation suggests a possible relationship between MIF and IFN. If MIF's action were to induce IFN production, IFN could be the mediator that actually inhibits the migration of macrophages. Although there are many materials which will inhibit macrophage migration, the important point to investigate and understand is the mechanism and interactions of mediators which function *in vivo* to control macrophage mobility and function.

VILCEK: Thurman's results could be of conceptual importance in understanding the mechanism of interferon action. But he is observing a biological action of *human* interferons on guinea pig cells. Human fibroblast and γ-interferon would not be likely to have an antiviral effect on guinea pig cells. So if the inhibition of guinea pig macrophage migration is indeed due to these interferons then there must be a dissociation between the antiviral and non-antiviral activities. It would be interesting to know whether guinea pig macrophages have specific receptors for human fibroblast and gamma interferon.

SANDRU: Has Thurman sought to determine whether IFN could inhibit the migration of *other* target cells? Last year at a Symposium on human lymphokines, held in Dallas, a paper was presented concerning leukocyte migration inhibition by IFN. In that case, the inhibition induced by IFN of cell migration was non-specific and was apparently unrelated to the inhibition induced by MIF which has specificity for the target cell. Is the inhibition induced by IFN, of macrophage migration, reversible?

THURMAN: I have not examined cell types other than macrophages and macrophage-like cell lines. However, others have reported migration inhibitory effects of IFN on granulocytes. The macrophage migration inhibition by IFN is only partially overcome by longer incubation time, but this observation was only a preliminary one, made at higher IFN concentrations.

PICK: Thurman's is a potentially important observation. I would suggest that the following experiments need to be performed: 1) whether cells escape inhibition; 2) whether the IFN-induced inhibition is blocked by puromycin; 3) whether IFN-induced inhibition is obtainable in a pulsed situation; 4) conduct tests with IFN in the capillary tube system, to confirm agarose microdroplet assay data. I think that it is possible that IFN induces MIF production by lymphocytes in Thurman's peritoneal exudate cells. The only argument against it is the ability of IFN to block the migration of U937 cells.

THURMAN: We have similar data using purified human mononuclear cells that have but few contaminating lymphocytes and have made the same observations, suggesting that the IFN is acting directly on the macrophages and not being mediated by lymphoid cells.

REQUIREMENT OF γ-INTERFERON IN ACTIVATION OF MACROPHAGES FOR TUMOR CYTOTOXICITY

Daniela N. Männel
Werner Falk

Institut für Immunologie und Genetik
Deutsches Krebsforschungszentrum
Heidelberg, F.R.G.

The modulating effect of interferon (IFN) in the macrophage antitumor function has been extensively described (1,2). Nonspecific tumoricidal activity was induced in macrophages with semi-purified fibroblast IFN preparations and activation was inhibited by antibodies against IFN. The mechanism of activation seems to be different for fibroblast IFN and the lymphokine macrophage activating factor (MAF) (3). These two mediators can thus be distinguished by their way of action as well as by their physico-chemical properties. However, MAF and IFN-γ are very similar in regard to biochemical characteristics and coordinate production and have not yet been separated.

RESULTS

Antiviral titers of over 40 tested ConA-induced spleen cell supernatants correlated very well with their macrophage activating capacity. Monoclonal rat antibodies against murine IFN were utilized to test the requirement for IFN in macrophage activation in these lymphokine preparations. The antibodies neutralized the antiviral activity of all molecular weight species of IFN-α and IFN-β as well as the IFN-γ activity in ConA- and antigen-stimulated spleen cell supernatants (4). These monoclonal antibodies coupled to activated CH-Sepharose 4B absorbed IFN activity from ConA-lymphokines

Interleukins, Lymphokines, and Cytokines

ISBN 0-12-527540-4

TABLE I. Macrophage Activation with Lymphokine before and after Absorption on anti-IFN Antibodies

	Lymphokine present continuously dilution		
	1/128	1/64	1/32
ConA-LK	14±2[a]	51±3	68±4
ConA-LK absorbed	6±0	8±1	22±5
	Lymphokine present for 5 hours dilution		
	1/20	1/10	1/5
ConA-LK	17±3	31±2	33±1
ConA-LK absorbed	8±1	8±1	7±1

[a]Cytotoxicity (released [^{3}H] TdR) was estimated at 48 h and expressed as percentage of total incorporated counts.

(ConA-LK) when mixed for 2 hours at 37°C. Interleukin 2 (IL-2) activity of the ConA-LK was completely recovered demonstrating the specificity of the absorption.

Quantitative analysis of macrophage activation for tumor cytotoxicity was then performed with the ConA-LK before and after the absorption step. The absorbed ConA-LK had a markedly reduced activating capacity when the macrophages were cultured in the presence of the LK during 48 hours as seen in Table I. No activation was achieved with the absorbed ConA-LK when the macrophages were pulsed for 5 hours at the beginning of the culture.

TABLE II. Macrophage Activation with Lymphokine after Absorption on anti-IFN Antibodies and/or rat IgG

	Lymphokine present for 5 hours dilution		
	1/20	1/10	1/5
ConA-LK absorbed on rat IgG	11±1[a]	29±3	55±9
ConA-LK absorbed on rat IgG and anti-IFN	8±1	12±1	24±3

[a]Cytotoxicity (released [^{3}H] TdR) was estimated at 48 h and expressed as percentage of total incorporated counts.

An identical absorption protocol with commercial rat IgG coupled to activated CH-Sepharose 4B was followed to test for nonspecific absorption or destruction of IFN-γ. Table II shows the resulting macrophage tumoricidal activity when the cells were pulsed for 5 hours with the ConA-LK which was absorbed on rat IgG. This activating capacity was drastically reduced after a second absorption on anti-IFN antibodies.

These results show that IFN-γ is essential for activation of macrophages for tumor cell lysis. The question if IFN-γ and MAF are identical or if these two mediators act synergistically remains open as long as purified IFN-γ cannot be recovered from the immunosorbent.

REFERENCES

1. Schultz, R.M., in "Lymphokine Reports" (M. Landy, ed.), Vol.1, p.63, Academic Press, New York, (1980).
2. Boraschi, D., and Tagliabue, A., Eur. J. Immunol. 11, 110 (1981).
3. Boraschi, D., Soldateschi, D., and Tagliabue, A., Eur. J. Immunol. 12, 320 (1982).
4. Männel, D.N., DeMaeyer, E., Kemler, R., Cachard, A., and DeMaeyer-Guignuard, J., Nature, 296, 664 (1982).

DISCUSSION

OPPENHEIM: I would restate the analogous issue that I raised regarding the relationship of IFN and MIF, namely "Is that all there is?" In other words, is MAF activity ever seen in the *absence* of antiviral activity? If so, we must conclude that MAF activities are heterogeneous and that some of them are in fact due to interferon. In this regard, to reinforce Mannel's observations, it should be added that Herberman and coworkers are finding that clone leukocyte- and β-interferons also have MAF activities.

MANNEL: MAF activity can be detected in the absence of antiviral activity in supernatants of cloned cell lines that are induced with Con-A. Segregation of the two activities did not occur during purification procedures of the mediators from antigen- or mitogen-induced spleen cell supernatants.

KRAMMER: A comment in reply to the one by Oppenheim. By somatic cell genetics (constructing T-T cell hybridomas) we seem to be able to dissociate IFN_{γ} and MAF. An alternative but less likely explanation is the presence of separate inhibitors by these hybridoma clones.

McDANIEL: In the human system, I tested human MAF active fractions for the presence of interferon but found none.

MANNEL: Did McDaniel's MAF activate macrophages for tumor cell cytotoxicity?

McDANIEL: The MAF assay *was* a cytotoxic assay.

MANNEL: I cannot comment as I am not familiar with the human macrophage cytotoxicity assay used nor the nature of the lymphokine preparation tested.

BELLER: Mannel has used a monoclonal antibody that inhibits IFNα,β and γ. This broad reactivity against structurally distinct molecules makes it difficult to accept that its ability to block MAF activity indicates an identity of MAF and IFN.

MANNEL: The antibody has a broad spectrum of reactivity but *all* activities neutralized by the antibodies, tested so far, belong to the IFN family.

THE REGULATION OF MACROPHAGE Ia ANTIGEN EXPRESSION

David I. Beller

Department of Pathology
Harvard Medical School
Boston, Massachusetts

The ability of macrophages to synthesize and express Ia antigens is critical for their function as antigen-presenting cells. Results of studies both *in vivo* and *in vitro* have demonstrated that T cells, upon activation by antigen or mitogen, can release products that not only induce Ia expression on macrophages (1-3) but also attract increased numbers of readily inducible phagocytes to the site of an immune reaction (3). We have demonstrated that induction of primarily Ia(+) macrophage populations *in vivo* is a radiosensitive process requiring the attraction of developmentally younger monocytes from the circulation rather than the division of resident macrophages (4). The dynamic and transient nature of Ia expression was further supported by the finding that the *maintenance* of a previously induced Ia(+) population is similarly radiosensitive (4, 5). Because the target cells are acquired from peripheral sites, we refer to the active lymphokine as Macrophage (Ia+)-Recruiting Factor (MIRF).

Recent studies in our laboratory have been directed towards an evaluation of the cellular events leading to induction of Ia expression and the effects of Ia modulation on accessory cell function (6, 7). We have found that T cell lines and hybridomas produce substantial quantities of MIRF and have focused on T cell lines because their supernatants are active at dilutions of 1:1,000 or more, and because they can also be used to assess antigen presentation.

The ability of macrophages to be triggered to express Ia depends on their state of activation (Fig. 1). More activated cells, such as those elicited by bacterial infection, respond more readily (and to lower doses of MIRF) than inflammatory (peptone elicited) or resident cells. Once Ia expression is

ISBN 0-12-527540-4

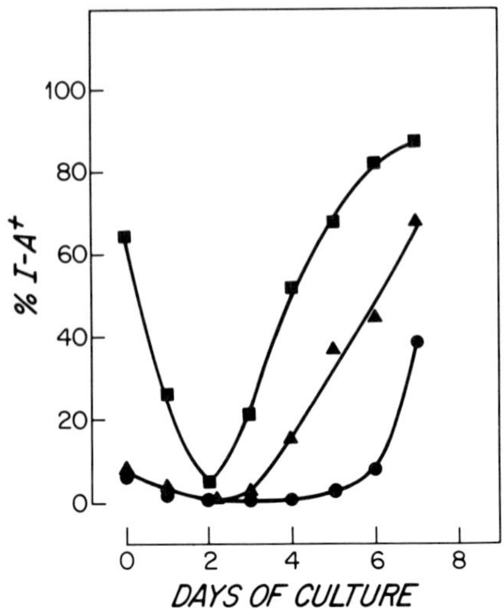

FIGURE 1. Kinetics of induction of Ia on different macrophage populations. PEC were elicited by Listeria followed one week later by peptone (■), peptone alone (▲), or no stimulus (resident; ●). Macrophages were cultured with or without MIRF (5% v/v), and I-A was evaluated daily by staining with fluorescent antibodies. In the absence of MIRF, all cultures showed less than 2% I-A-positive cells after Day 3.

initiated, however, the kinetics are quite similar in all groups. The distinguishing feature is thus the duration of the latent period before Ia is expressed.

Even in unstimulated mice, different phagocyte populations differ in their relative responsiveness, depending on their site of derivation: monocytes are more rapidly inducible than resident macrophages. In all cases, the more readily responsive cells are more activated in terms of expressing lower levels of 5′-nucleotidase (7). Moreover, the eventual onset of Ia expression in the resident macrophage correlates with a dramatic reduction in this enzyme.

Further studies on the nature of the latent period indicated that it is not due to an insensitivity of the macrophage to the inductive signal. The data in Fig. 2 reveal that all macrophage populations can be similarly triggered by a two-hour exposure to MIRF after overnight culture. However, the ability to become committed to a program of Ia expression is not inherent in any of these populations, since freshly adhered cells (or cells maintained non-adherent overnight) are very poorly responsive. Our feeling is that adherence may be an important physiologic event during Ia induction, perhaps mimicking the transient adherence of phagocytes at the site of immune reactions (8).

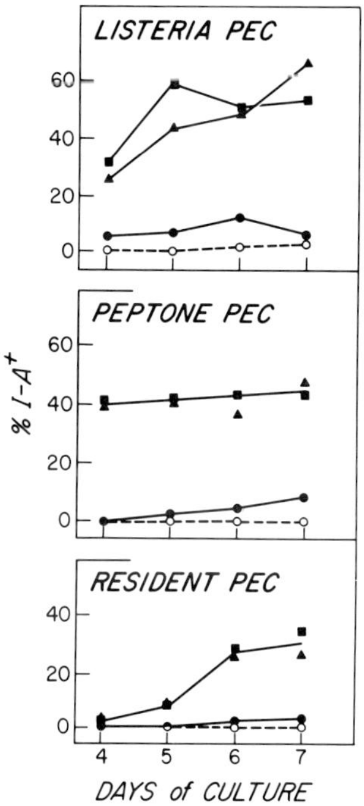

FIGURE 2. Induction of Ia expression by brief exposure to MIRF. Elicited (top two panels) or resident (bottom panel) macrophages were exposed to MIRF (5%) for the initial twenty-four hours of culture (▲) or only the first (●) or last (■) two hours of this twenty-four-hour period. Control cultures were kept in medium (○).

The functional importance of these events was evaluated in antigen presentation studies (Fig. 3). Macrophages cultured in medium alone lost both Ia and accessory cell activity, while the induction of Ia by MIRF was associated with antigen presentation function comparable to that of freshly isolated macrophages. Thus, the spontaneous transition from Ia(+) to Ia(-), and the induced expression of Ia, correlate extremely well with the functional competence of these macrophages in inducing antigen-dependent T cell proliferation.

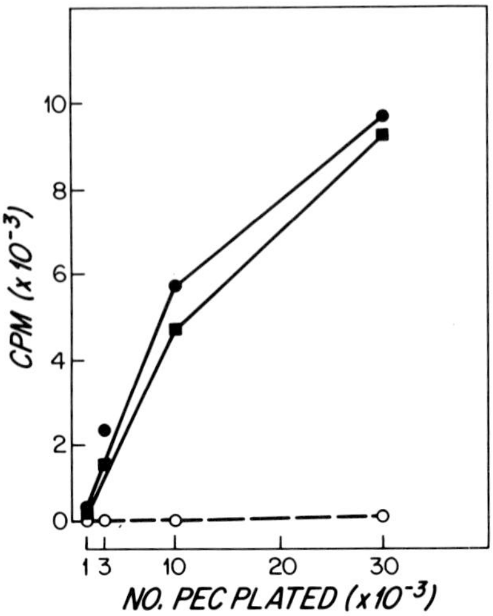

FIGURE 3. Antigen-presenting function of macrophages induced to express Ia determinants. Peptone-elicited macrophages were cultured in medium (○) or 5% MIRF (●) for six days and their ability to stimulate antigen-dependent proliferation of the Listeria-immune T cell line was compared to freshly isolated macrophages (■). After pulsing with antigen (heat-killed Listeria), macrophages were irradiated and then 5 x 10^4 immune T cells were added. ^{3}H-thymidine incorporation was assessed during the fourth day of culture.

ACKNOWLEDGMENT

I would like to thank Dr. Emil R. Unanue for his interest and suggestions during the course of this investigation.

REFERENCES

1. Steinman, R. N., Nogueira, N., Witmer, M., Tydings, J., and Mellman, I., *J. Exp. Med.* 152, 1248 (1980.
2. Steeg, P., Moore, R., and Oppenheim, J. J., *J. Exp. Med.* 152, 1734 (1980).
3. Scher, M. G., Beller, D. I., and Unanue, E. R., *J. Exp. Med.* 152, 1684 (1980).

4. Scher, M. G., Unanue, E. R., and Beller, D. I., *J. Immunol.* 126, 263 (1981).
5. Beller, D. I., and Unanue, E. R., *J. Immunol.* 126, 263 (1981).
6. Calamai, E. G., Beller, D. I., and Unanue, E. R., *J. Immunol.* 128, 1692 (1982).
7. Beller, D. I., and Ho, K., *J. Immunol.* in press (1982).
8. Nelson, D. S., and Boyden, S. V., *J. Immunol.* 6, 624 (1973).

DISCUSSION

SCHRADER: Does Beller find quantitative changes also in the expression of other antigens such as $H-2^k$?

BELLER: No, but Steinman has reported an increase in $H-2^d$ expression. We have looked at the effect of a T cell-derived IFN-γ-like molecule on P cells. We find that while the expression of Ia is *absolutely* dependent on the presence of the lymphokine, there is also a quantitative effect on $H-2^k$.

VADAS: What degree of cell proliferation takes place in Beller's cultures; could Ia expression by a new population of macrophages be responsible for the effects he has described?

BELLER: We have not attempted to quantitate possible cell division. However, lymphokine-induced Ia expression occurs with the same kinetics and to the same extent in normal and irradiated (1500R) macrophages.

YOSHIDA: Does Beller have any data on the molecular characteristics of his Ia-inducing lymphokine; is it similar to or different from e.g., MAF, etc.?

BELLER: The molecule has an Mr of about 40,000, is pH2-labile (loss of >90% of activity) and has *not* been successfully separated by biochemical means from MAF or IFN_γ. Its mode of action is also similar to that of MAF.

KRAMMER: Does the same macrophage increase its membrane Ia, which would also cause release of PGE?

BELLER: We do not know.

KRAMMER: I mention this since if it *were* the case we might have a situation in which the T cell releases lymphokines which induce in the same macrophage an expression of Ia as well as PGE release which then shuts off the Ia expression.

BELLER: Whether these two events occur in the same macrophage the point is that, as a population, macrophages can *autoregulate* their Ia expression. Presumably this affords them a means of reverting to their "normal", less activated state.

NIEDERHUBER: Is there any evidence for a differential re-expression of Ia determinants on the macrophage when stimulated by antigen-induced T cell lymphokine?

BELLER: No, there is not. Ia and Ie antigens seem to be regulated coordinately. Moreover, the use of A.TH anti-A.TL antibodies shows no quantitative difference compared to the monoclonal, sub-region-specific antibodies.

PICK: I have two questions for Beller: The first refers to the activity of TXB_2. This is a biologically inactive metabolite which should have no effect; it is therefore surprising that Beller gets a reversal of the PGE_2-induced inhibitor. My second query refers to the nucleotidase data. Is it not possible that the fall in 5'-nucleotidase is due to phospholipase released by Listeria? Phospholipase C is known to cause a fall in ectoenzymes such as 5'-nucleotidase.

BELLER: In answer to Pick's first question, the data suggest that we have defined a situation in which TXB_2 can be shown to function. It may be significant that it has no effect by itself, but only modulates the response to PGE_2. As far as invoking any complicated theory about 5-nucleotidase loss, we should remember that normal monocytes and thioglycollate-elicited macrophages likewise have virtually no 5-nucleotidase.

MODULATION OF HUMAN CORD BLOOD MONOCYTE DR ANTIGEN EXPRESSION *IN VITRO* BY LYMPHOKINES AND INTERFERON

Marcelo Sztein
Patricia Steeg
Joost J. Oppenheim

Laboratory of Microbiology and Immunology
National Institute of Dental Research
National Institutes of Health

Richard Stiehm

Department of Pediatrics
University of California at Los Angeles
Los Angeles, CA

Dean Mann
Michael Blaese

Division of Cancer Biology and Diagnosis
National Cancer Institute
National Institutes of Health

I. INTRODUCTION

The initiation of antigen specific, T-dependent human immune responses requires the participation of accessory cells bearing HLA-DR antigens (1-2). With regard to monocytes, a principal class of accessory cells, both DR antigen-expressing (DR^+) and DR antigen deficient (DR^-) subpopulations occur. This study was performed to examine conditions that determine the relative balance of DR^+ and DR^- human monocytes *in vitro*.

A novel source of human monocytes has been used to study the regulation of DR antigen expression, namely neonatal cord blood monocytes. These cells have reduced DR antigen expression as compared to peripheral blood monocytes (D. Mann, R. Stiehm and M. Blaese, personal communication). Cord blood monocytes are therefore suitable for detecting the immunoenhansive effects of elevating monocyte DR antigen expression.

Interleukins, Lymphokines, and Cytokines

ISBN 0-12-527540-4

The data presented in this paper indicate that lymphokine-containing culture supernatants, and IFN-α in particular, can elevate the percentage of DR antigen-expressing cord blood monocytes *in vitro*.

II. MATERIALS AND METHODS

Human neonatal cord blood mononuclear cells were isolated by dextran sedimentation and Ficoll-hypaque density separation. 5×10^5 cells in 0.1 ml of RPMI 1640 containing 10% fetal calf serum (FCS) were plated per well in microtitre plates, incubated for at least 1 hr at $37^\circ C$, and the nonadherent cells removed by washing four to six times. The remaining adherent cells were cultured in 0.2 ml RPMI 1640 containing 10% FCS and dilutions of either Concanavalin A stimulated PBMNC culture supernatant, a Concanavalin A supplemented (unstimulated) PBMNC culture supernatant or recombinant IFLrA (provided by Dr. S. Pestka, Hoffman La Roche) for 2-3d at $37^\circ C$ in a humidified atmosphere of 5% CO_2 in air.

To determine the percentage of DR antigen expressing cells the culture supernatants were removed, replaced with 50μl of RPMI 1640 containing 1% FCS and a 1:60 dilution of anti-DR (p23,30)(3-4), and were incubated at $4^\circ C$ for 1 hr. Subsequently, a 1:5 dilution of Low-Tox-H Rabbit Complement (Cedarlane Laboratories, Inc.) was added, and the cultures were incubated for an additional 1 hr at $37^\circ C$. Viability was determined by trypan blue exclusion, and the percentage of DR^+ cells was reported as a mean Cytotoxicity Index (CI), with 95% confidence limits, based on arcsine square transformation.

III. RESULTS

Freshly obtained adherent human neonatal mononuclear cells, in nine experiments, were approximately 50% nonspecific esterase-positive and 22.2% DR^+. The DR antigen expression of cord blood monocytes was significantly reduced as compared to that of human perpheral blood monocytes, which are approximately 75% DR^+. Table I indicates the effect of *in vitro* culture on cord blood monocyte DR antigen expression. In this experiment cord blood monocytes were incubated with dilutions of either Con A stimulated PBMNC culture supernatant or a control supernatant for 3 days, and the percentage of DR^+ cells determined. As shown, after 3 d of incubation with lymphokine containing culture supernatant, the percentage of DR^+ cord blood monocytes was enhanced in a dose-dependent manner, while the percentage of DR^+ cord blood monocytes in control cultures actually declined in comparison to day 0 DR antigen levels (CI=22.0 in this experiment). Thus, lymphokines enhance cord blood monocyte DR antigen expression *in vitro*, and in the absence of exogenous lymphokines DR$^+$ monocytes lose DR antigen expression.

TABLE I. In vitro Enhancement of Human Cord Blood Monocyte DR Antigen Expression By Lymphokines

	Mean Cytotoxicity Index (95% Confidence Interval)	
Dilution	*Con A stimulated PBMNC culture supernatant*	*Control PBMNC culture supernatant*
0	9.72 (4.3, 15.6)	
1:256	28.1 (21.2, 35.8)	8.4 (4.4, 12.5)
1:64	52.1 (34.2, 69.4)	10.1 (6.2, 14.1)
1:16	52.4 (43.8, 61.0)	6.2 (4.6, 7.4)

Table II indicates the kinetics of lymphokine enhancement of cord blood monocyte DR antigen expression. Cord blood monocytes were incubated for 1-9 d with a maximally active dilution of Con A stimulated PBMNC culture supernatant, and the percentage of DR$^+$ cells determined daily. As shown, significant enhancement of DR antigen expression by lymphokines was detected after one day of incubation, with optimal enhancement by day 5 of culture.

Subsequent investigation has focused on identification of the lymphokine(s) in the Con A stimulated PBMNC culture supernatant responsible for the enhancement of cord blood monocyte DR antigen expression. The availability of cloned human lymphokine gene products has enabled definitive determination of the DR antigen regulatory activity of certain human lymphokines. Based on previous experiments in the murine system, which indicated that an immune interferon (IFN-γ)-like lymphokine modulated macrophage Ia antigen expression (5), the effect of cloned human IFN (IFLrA) on cord blood monocyte DR antigen expression was tested. Table III shows a representative experiment, in which cord blood monocytes were incubated for 3d with dilutions of IFLrA, and the percentage of DR$^+$ monocytes determined. As shown, IFLrA enhanced cord blood monocyte DR antigen expression in a dose-dependent manner, although not to the same extent as crude lymphokine-containing supernatants. Maximal enhancement of cord blood monocyte DR antigen expression occured at approximately 1-2 U/ml IFLrA. The time course of IFLrA enhancement of cord blood monocyte

DR antigen expression paralleled that of crude culture supernatants (unpublished observation).

TABLE II. Time Course of In Vitro Enhancement of Cord Blood Monocyte DR Antigen Expression By Lymphokines

Day of culture	*Mean Cytotoxicity Index (95% Confidence Limits):* *Con A supernatant (1:4 Dilution)*	*Control supernatant (1:4 Dilution)*
0	*26.84 (23.9, 29.8)*	
1	*63.5 (57.2, 69.5)*	*25.4 (21.1, 30.0)*
3	*69.2 (63.2, 74.9)*	*24.1 (18.5, 30.1)*
5	*88.9 (85.6, 91.8)*	*24.0 (18.9, 29.5)*
7	*88.8 (84.6, 92.5)*	*25.7 (22.5, 29.0)*
9	*80.8 (74.8, 86.1)*	*12.4 (8.1, 17.6)*

IV. DISCUSSION

Human cord blood monocytes have significantly fewer DR^+ cells than peripheral blood monocytes. The reason for reduced DR antigen expression is presently unknown, although our data indicate that neonatal monocytes can be stimulated to express DR antigen at levels comparable to peripheral blood monocytes. Several possible explanations could therefore account for the observed reduced DR antigen expression of cord blood monocytes: (a) reduced levels of DR antigen modulatory lymphokines in cord blood, and/or; (b) the presence of inhibitors of DR antigen modulatory lymphokines in cord

TABLE III. Enhancement of Cord Blood Monocyte DR Antigen Expression In Vitro By Cloned Alpha Interferon

Units IFLrA	*Mean Cytotoxicity Index (95% Confidence Limits)*
0	*10.4 (2.5, 22.8)*
0.4	*29.3 (14.8, 35.9)*
1.6	*48.7 (38.1, 59.6)*
6.25	*43.3 (38.5, 48.7)*
25	*49.9 (39.1, 60.8)*
100	*50.4 (44.3, 56.3)*

blood. In this regard, prostaglandins have been demonstrated to inhibit lymphokine modulation of murine macrophage Ia antigen expression (6), and may therefore also inhibit human monocyte DR antigen expression. Alternatively, other inhibitors may be present.

Both crude lymphokine-containing culture supernatant and cloned IFLrA enhanced the percentage of DR^+ cord blood monocytes in vitro. These results indicate that IFN can modulate human cord blood monocyte DR antigen expression, but do not indicate that IFN is the only mediator in crude culture supernatants with DR antigen modulatory activity. Indeed, the observation that maximal enhancement of DR antigen expression by IFLrA is typically less than that by crude culture supernatants suggests that: (a) a different IFN type (IFN-γ for instance) may be a more potent DR antigen modulatory agent; (b) two types of IFN may act in concert to induce maximal DR expression, and/or; (c) a non-IFN lymphokine may act either independently (additively) or synergistically with IFN to elevate DR antigen expression.

REFERENCES

1. Bergholtz, B.O. and E. Thorsby, 1978. HLA-D restriction of the macrophage-dependent response of immune human T lymphocytes to PPD in vitro: Inhibition by anti-HLA DR antisera. *Scand. J. Immunol.* 8:63.

2. Rodey, G.E., L.K. Luehrman and D.W. Thomas, 1979. In vitro primary immunization of human peripheral blood lymphocytes to KLH: Evidence for HLA-D region restriction. *J. Immunol.* 103:2250.

3. Humphries, R.E., J.M. McCune, A.M. Chess, H.C. Herrman, A.J. Malenka, D.L. Mann, B. Parham, S.F. Schlossman, and J.L. Strominger, 1976. Isolation and immunological characterization of a human B lymphocyte specific cell surface antigen. *J. Exp. Med.* 144:99.

4. Lonney, J.K., D.L. Mann and D.H. Sachs, 1979. Sharing of Ia antigens between species. III. Ia shared between mice and human beings. *Scan. J. Immunol.* 10:403.

5. Steeg, P.S., R.N. Moore, H.M. Johnson and J.J. Oppenheim. Regulation of murine macrophage Ia antigen expression by a lymphokine with immune interferon activity. *J. Exp. Med.* In press.

6. Steeg, P.S., H.M. Johnson and J.J. Oppenheim. 1982. Regulation of murine macrophage Ia antigen expression by an immune interferon-like lymphokine: Inhibitory effect of endotoxin. *J. Immunol.* In press.

DISCUSSION

WAHL: Has Steeg monitored other monocyte functions in parallel with DR expression in cord blood or cell line monocytes?

STEEG: No, we have not.

PETER: Does Steeg have evidence that *all* monocytes can be modulated in their DR antigen expression, or is DR antigen modulation restricted to a monocyte subpopulation?

STEEG: Using a crude culture supernatant we have induced up to 95-98% of the monocytes to become DR^+.

VADAS: Does Steeg know whether DR expression is regulated during maturation; for example in semi-solid agar, under the influence of CSF of the monocyte series?

STEEG: We have not tested the effects of lymphokines or IFN on bone marrow cells. We have found that murine thioglycollate-induced, peritoneal-exudate macrophages, when cultured with MGF are Ia^-. If we then add IFN, the macrophages proliferate, and become Ia^+.

PETER: There are reports which claim a higher immunoregulatory activity of IFN_γ in comparison to IFNα or β. Can Steeg give us an idea of the relative potency of these IFNs and the IFN preparations employed to stimulate Ia-expression?

STEEG: Both IFNα and IFN_γ induce monocyte DR antigen expression. The dose-response characteristics of induction of monocyte DR antigen expression by IFNα and IFN_γ are similar--except that IFN_γ induces a greater percentage of monocytes to become DR^+.

KRAMMER: Do the macrophages, which on incubation with lymphokines become antigen-presenting cells, have similarities to dendritic cells?

STEEG: No. The monocytes become morphologically rounded when incubated with IFN or lymphokines. Further, two recent papers by Birmingham et al. and Walker et al. have established that macrophages have accessory cell capabilities.

CLONING OF THE CODING SEQUENCES OF A HUMAN LYMPHOKINE, OSTEOCLAST-ACTIVATING FACTOR: A PROGRESS REPORT

Glenn E. Nedwin
Marjorie A. Mohler
Richard A. Luben

Division of Biomedical Sciences
and Department of Biochemistry
University of California
Riverside, California

Supported by NIH Grants DE05467, AM26448, DE00057

I. INTRODUCTION

Osteoclast activating factor (OAF) is secreted by activated human peripheral blood leukocytes (1), human B-lymphoblast cell lines (2) and primary myeloma cells (3). This lymphokine stimulates resorption of bone by osteoclasts and inhibits synthesis of collagen by osteoblasts (4). There appear to be multiple species of bone-resorbing lymphokines secreted by human lymphoid cells; several of these molecules have been designated as OAF by separate laboratories (5,6), although their relationships of structure, source and mechanism are unclear at best. This laboratory's approach to investigating the nature of OAF has been to concentrate on a particular molecular species (M_r approx. 9000 daltons) which we have purified (5) and characterized as being correlated with pathologic bone resorption *in vivo* (6). Once the molecular details of this species of OAF are more fully known, it will be possible to determine more precisely its importance relative to the roles of other OAF species in disease processes.

An important means of characterizing biologically active peptides is to isolate and clone the nucleic acid sequences

ISBN 0-12-527540-4

which code for the primary structure of the peptide molecule. The current communication is a progress report on our program to clone the coding sequences of the 9000-dalton species of OAF. We have taken advantage of two OAF-producing cell lines and of sensitive biological and immunochemical assays for OAF, and have detected recombinant clones which hybridize with the OAF message in cellular mRNA isolates. These clones should serve as probes for future studies to isolate the complete coding sequences for OAF.

II. MATERIALS AND METHODS

Biologically active OAF was detected by a bone resorption assay (9) in which OAF stimulates release of previously incorporated Ca^{++} from cultured mouse cranial bones. This assay allows the detection of as little as 100 pg of OAF. Immunoreactive OAF was detected by immunoassay using monoclonal antibodies against OAF (10), or by immunoprecipitation of radiolabeled products as described below.

OAF and mRNA coding for OAF were isolated from either the human myeloma cell line RPMI 8226 (2) or a human-mouse hybridoma, CD6.20. The latter cell line was produced (11) by hybridization of PHA-activated human tonsil lymphocytes with the mouse plasmacytoma UCR-M3, a derivative of the MPC-11 45.6TG1.7 plasmacytoma. Clones of the human-mouse hybridomas were screened for their ability to produce bone-resorbing activity which could be neutralized by monoclonal antibodies against human OAF (12). The CD6.20 cell line retains human chromosomes 10 and 18 (12), secretes mouse IgG2b(k) but no human immunoglobulins, and can be propagated in Balb/C mice either as solid tumor or as ascites.

Cellular RNA was isolated from cell pellets of RPMI 8226 or CD6.20, or from tissue fragments of solid CD6.20 tumors, by the method of Chirgwin, et. al. (13). Cells were homogenized in 4 M guanidinium thiocyanate containing 0.5% sarcosyl, 0.1 M 2-mercaptoethanol, 0.1% Sigma antifoam A, and 25 mM sodium citrate, pH 7.0. The homogenates were layered over 5.7 M CsCl, 0.1 M EDTA, pH 7.0, and centrifuged at 15°C, 120,000 x g for 12 hours. RNA is the only cellular component dense enough to pellet through this CsCl layer, and the RNA pellets were thus easily collected, redissolved in 4M guanidinium thiocyanate and further purified by ethanol precipitation, followed by either oligo(dT) cellulose chromatography or glycerol gradient centrifugation. Translation of total or poly(A) RNA was carried out in nuclease-treated rabbit reticulocyte lysates (13) optimized for maximal incorporation of ^{35}S-methionine counts into translation products of CD6.20 RNA (11). Translation products were imunoprecipitated (when applicable) by overnight treatment at 4°C with anti-OAF monoclonal antibody, followed by anti-mouse IgG for two hours. The pellets were washed twice and then either counted directly or redissolved for SDS-PAGE. Gels were visualized using En^{3}Hance fluorography, with exposure times of 7 to 28 days.

Cloning procedures were carried out using minor modifications of the procedures described by Goodman, et. al. (15) and Dugaiczyk, et. al. (16). Poly(A) RNA isolated from RPMI 8226 or from CD6.20 were transcribed into double-stranded cDNA using AMV reverse transcriptase. The cDNA was tailed with poly(dC) using terminal transferase and then inserted into the Pst I site of plasmid pBR322, which had been tailed with poly(dG) after linearization. Ligated plasmids were transfected into E. coli LE392, and recombinants selected by differential tetracycline and ampicillin sensitivity testing. Recombinant colonies were pooled into approx. 100 separate cultures containing 5 colonies per culture for further assay and plasmid purification.

Recombinant plasmids were isolated from transformed E. coli after amplification with chloramphenicol (15). Identification of cultures containing OAF cDNA was performed using hybrid-arrested translation (17). Briefly, appropriate amounts of plasmid DNA and CD6.20 mRNA were mixed, heat denatured and then allowed to hybridize at 48°C for 2 hr. Each sample was then divided into two fractions, one of which (hybridized sample) was translated in the reticulocyte lysate system, the other of which (heat melted) was denatured at 100°C and quick-chilled prior to translation. Translation products were analyzed as above; cDNA-dependent disappearance of immunoreactive OAF counts in the hybridized sample, with recovery of those counts in the heat-melted sample, was considered evidence for the presence of cDNA which could hybridize with OAF mRNA and thus block its translation.

III. RESULTS AND DISCUSSION

The two cell lines chosen were excellent sources of OAF, as shown in Table 1.

TABLE 1. OAF Activity from Various Sources

	Bone Resorbing Activity, Treated/Control[b]			
	Without Anti-hOAF[c]		With Anti-hOAF(1:1K)	
Tonsil Cell OAF (1 ng/ml)	1.96*	(.09)	1.01	(.03)
CD6.20 supernatant (1:10)[a]	1.86*	(.11)	0.98	(.11)
RPMI 8226 supernatant (1:10)[a]	2.01*	(.13)	1.05	(.07)
Parathyroid hormone (10 ng/ml)	1.98*	(.12)	1.89*	(.10)

[a]Culture medium from confluent cell cultures, diluted 1:10 with fresh bone culture medium for resorption assay.
[b]Mouse cranial bones were cultured for 72 hr in the presence or absence of the bone resorbing agent indicated. Values given are means (S.E.M.) of 3-4 pairs of treated vs. control bones, for the ratio of percentage $^{45}Ca^{++}$ release. * denotes significant stimulation of bone resorption ($p<0.05$).
[c]Anti-hOAF denotes monoclonal antibody against human tonsil cell OAF, added as a 1:1000 dilution (final concentration) of clarified mouse ascites fluid.

These cells were used to isolate mRNA which was translated in the rabbit reticulocyte lysate system for analysis of the content of OAF mRNA as analyzed by immunoprecipitation with anti-hOAF monoclonal antibody. These experiments showed (Table 2) that the OAF message comprised approximately 0.1-0.3% of the total translatable message in these cells. Moreover, interestingly, the translation products contained biologically active OAF as assessed by bone resorption in culture (Table 2). These results, though surprising, are consistent with the yields of immunoprecipitable OAF: less than 100 pg or 10^{-14} moles of OAF are required to stimulate bone resorption, a value consistent with the immunoprecipitable counts obtained. The results thus strongly suggest that synthesis of biologically active OAF does not require post-translational modification. In all cases, the bone resorbing activity of OAF translation products were blocked by anti-hOAF monoclonal antibody (data not shown).

TABLE 2. Translation of OAF mRNA In Vitro[a]

Source of RNA	Total cpm (TCA)	Immunoprecipitate cpm (% of total)[b]	Bone 45-Ca % release[c]
Lysate only	--	--	10.2
Globin mRNA	2.4×10^6	180 (.007%)	9.1
CD6.20	8.4×10^5	840 (.10%)	26.2*
RPMI 8226	6.2×10^4	240 (0.37%)	23.5*

[a]100-750 ug of mRNA from each source was translated in the nuclease treated rabbit reticulocyte lysate system in the presence of carrier free ^{3}H-leucine. Parallel samples were translated without label for assay of bone resorption.
[b]Immunoprecipitated using anti-hOAF monoclonal antibody, anti-mouse IgG second antibody and protein A.
[c]Lysate was sterile filtered and added to bone resorption assay after translation incubation. * denotes significant difference from control (no lysate), $p<0.05$.

Glycerol gradient fractionation of CD6.20 and RPMI mRNAs, with translation and immunoprecipitation of each fraction (not shown) demonstrated that the most active regions of the gradient were at 3S-4S, consistent with a low molecular weight for the primary translation product of OAF. SDS-PAGE revealed three major bands of immunoprecipitable OAF with molecular weights of <10,000, 18-20,000 and 37-40,000 daltons, as well as other minor bands of intermediate weights. The mRNA fractions isolated from OAF-producing cell lines were transcribed, inserted into plasmids, cloned and screened as described in the Methods. At least two of some 500 plasmid clones analyzed showed some ability to block translation of OAF by hybridization with the OAF message, with recovery of the product by denaturation of the mRNA-cDNA hybrids (Table 3). Agarose gels of the recombinant plasmids indicated that they contained 300-500 base pair inserts. Unfortunately, attempts to release the recombinant inserts from these plasmids using Pst I released only very small (about 100 bp) fragments and totally destroyed the ability of the samples to arrest translation of OAF. The latter data suggest that the OAF message contains an internal Pst I site which makes it very difficult to isolate cDNA fragments suitable for sequencing or other critical work. We are currently engaged in further cloning of OAF, using alternative restriction enzymes and linkers; this work should be facilitated by the availability of the previously isolated plasmids, which can still be used as screening probes.

TABLE 3. Hybrid-Arrested Translation Screening of OAF Clones

	Immunoprecipitable Label (Arbitrary Units)[a]		
Clone No.	Hybridized Sample	Melted Sample	After Pst I
Control[b]	1.00	0.83	--
pBR322	1.12	0.88	0.91
p02-12	0.09	0.82	0.88
p02-48	0.22	0.80	0.84
p01-30	0.98	0.86	0.90

[a]Samples were treated as described in the methods section, with translation products separated by SDS-PAGE after immunoprecipitation with anti-hOAF monoclonal antibody. Immunoprecipitable label was determined by densitometric scanning of the fluorograms, with the entire density of each lane (5000-80000 daltons) included.
[b]CD6.20 mRNA translated after incubation without plasmid or recombinant DNA, but with or without heat melting.

The principles of isolation and assay of low-abundance messages which are central to this project should be applicable not only to lymphokines such as OAF, but to other low-abundance messages as well. The major keys to cloning a low-abundance message such as a lymphokine, hormone or growth factor are good sources of mRNA (e.g., active cell lines), a sensitive assay for the product (e.g., monoclonal antibody immunoprecipitation) and the ability to detect positive recombinant clones with a reliable and sensitive technique (e.g., hybrid-arrested translation). If these prerequisites can be acquired, it should be possible to clone almost any sequence, regardless of its scarceness, and to ultimately determine its character and relevance to regulatory processes.

REFERENCES

1. Horton, J.E., Raisz, L.G., Simmons, H.A., Science 177:793 (1972).
2. Mundy, G.R., Luben, R.A., Raisz, L.G., Oppenheim, J.J., Buell, D.N., N. Engl. J. Med. 290:867 (1974).
3. Mundy, G.R., Raisz, L.G., Cooper, R.A., Schecter, G.P., Salmon, S.E., N. Engl. J. Med. 291:1041 (1974).
4. Raisz, L.G., Luben, R.A., Mundy, G.R., Dietrich, J.W., Horton, J.E., Trummel, C.L., J. Clin. Invest. 56:408 (1975).
5. Mundy, G.R., Raisz, L.G., J. Clin. Invest. 60:122 (1977).
6. Nimberg, R.B., Humphries, D.E., Lloyd, W.S., et. al., Cancer Res. 38:1983 (1978).
7. Luben, R.A., Biochem. Biophys. Res. Comm. 84:15 (1978).
8. Luben, R.A., in "Hormonal Control of Calcium Metabolism" (D.V. Cohn, et. al., eds.), p. 315. Academic Press, New York, (1981).
9. Luben, R.A., Chen, M.C.-Y., Rosen, D.M., Mohler, M.A., Calcif. Tiss. Int. 28:23 (1979).
10. Luben, R.A., Mohler, M.A., in "Biochemical Characterization of Lymphokines" (A.L. deWeck, et. al., eds.), p. 55. Academic Press, New York, (1980).
11. Nedwin, G.E., Ph.D. Dissertation, University of California, Riverside (1982).
12. Nedwin, G.E., Luben, R.A., Miami Winter Symposia 18:88 (1981).
13. Chirgwin, J.M., Przybyla, A.E., MacDonald, R.J., Rutler, W.J., Biochemistry 18:5294 (1979).
14. Pelham, H.R.B., Jackson, R.J., Eur. J. Biochem. 67:247 (1976).
15. Goodman, H.M., MacDonald, R.J., Meth. Enzymol. 68:75 (1979).
16. Dugaiczyk, A., Yu, L.-Y., Woo, S.L.C., in "Manual for Hormone Action and Molecular Endocrinology" (W.T. Schrader and B.W. O'Malley, eds.), p.5-1. Baylor University, Houston, Texas (1980).
17. Paterson, B.M., Roberts, B.E., Kuff, E.L., Proc. Nat. Acad. Sci. 74:4370 (1977).

DISCUSSION

WEBB: The level of cloning efficiency (500 clones) seems rather low compared to that observed by other workers; does Luben think this contributed to his fragmented cDNA sequences?

LUBEN: The number of clones is appropriate for the relative frequency of the message, i.e., 0.1-0.4% of the translation products appear to be OAF and thus 1/1000 to 1/250 of the clones should correspond to OAF.

WEBB: In addition to the use of hybrid arrest of translation, one can also use hybrid selection-positive hybridization and re-isolation of the relevant mRNA, followed by cell-free translation.

LUBEN: We have not been successful with hybrid-selected translation in this particular system. The results appear to vary from message to message and from laboratory to laboratory.

GARLAND: In our laboratory, work by Testu on osteoclasts indicates that Ca release may be due largely to macrophages, and the osteoclasts are difficult to activate. In Luben's own system, what is the likelihood that OAF is activating macrophages rather than osteoclasts?

LUBEN: Ca release in our system is due to osteoclasts; macrophages do not contribute much to the release. OAF may activate macrophages to become phagocytic, but in our experience this is not highly reproducible.

OPPENHEIM: Now that Luben has in hand this purified 9000 MW lymphokine that activates osteoclasts, has he had the opportunity to ascertain whether OAF has any *other* biological activities, exerted particularly on other target cells?

LUBEN: We have looked for other activities and have found none, as yet, except that OAF may facilitate Fc-mediated phagocytosis by macrophages. However, we are continuing exploratory experiments seeking resemblances between OAF and other lymphokines.

PAETKAU: It is curious that the Pst I-cleaved recombinant DNA (carrying OAF sequences) does not inhibit RNA translation. The same sequences in intact plasmid do register by HART. What is the size of the insert? Can Luben comment on its lack of inhibition after PstI excision?

LUBEN: The inserts appear to be 300-500 base pairs in the intact plasmid, but we can find only very small (<=100 bp) fragments after PstI treatment. One interpretation of these data would be that PstI has a recognition site within the OAF sequence and, consequently, treatment with the enzyme destroys the message, at least in terms of blocking translation activity.

MITOGENIC AND CHEMOKINETIC EFFECTS OF SECONDARY MLC SUPERNATANTS ON CULTURED ENDOTHELIAL CELLS

Sandra L. Watt
DeAnna Kirchmayer
Robert Auerbach

Department of Zoology
University of Wisconsin
Madison, Wisconsin

Both activated lymphocytes and certain tumors induce angiogenesis (new vessel growth) in vivo. Mouse lymphocytes were previously shown to induce angiogenesis when injected intradermally into allogeneic recipients, and this response was shown to correlate with a graft-versus-host (GVH) type response. Tumor cells and extracts have been shown by several investigators to induce angiogenesis in vivo using such systems as the rabbit, mouse, or rat cornea or the chick chorioallantoic membrane implant.

In this study, three in vitro systems were used to study the nature of the cell product(s) produced by lymphocytes which are responsible for the neovascular response. Fetal bovine aortic endothelial cells (FBAE) or adult bovine adrenal capillary endothelial cells (BCE) were used to test materials for their mitogenic and chemokinetic activities. It has been shown that cultured endothelial cells proliferate and migrate in response to tumor-derived angiogenesis factors, and hence these two systems can be used to screen for angiogenesis factors. Also, mixed lymphocyte culture (MLC) supernatants were collected and assayed to determine their mitogenic activity towards endothelial cells in vitro. The MLC system has been shown to be an in vitro correlate of the GVH reaction.

The results from this study show that secondary MLC supernatants, collected under conditions optimal for IL-2 production, contain both mitogenic and chemokinetic activity for bovine endothelial cells in vitro. Preliminary results indicate that the material is non-dialyzable (12,000 MW cutoff), and that it may not be the same molecule as IL-2.

Interleukins, Lymphokines, and Cytokines

ISBN 0-12-527540-4

I. INTRODUCTION

When immunocompetent lymphocytes are injected intradermally into irradiated allogeneic mice, a complex vascular response results which includes changes in the number of visible blood vessels, alterations in capillary histology, and the generation of new vessels (1-3). In addition, chemical compounds which elicit a delayed-type hypersensitivity (DTH) reaction have been shown to have similar effects on the local vasculature (4,5).

That this lymphocyte-induced angiogenesis (LIA) is a manifestation of a GVH reaction was shown by Sidky and Auerbach (1). Further, they have shown that the major antigenic stimulus leading to LIA is associated with I-region alloantigens (6), as had previously been shown for the GVH reaction (7).

The identification of the LIA effector cells as Lyt 1^+, 2^- (8) is consistent with similar observations for the GVH reaction (9), and also suggests that the LIA response is mediated by a soluble factor(s) (lymphokine(s)) (10,11). Soluble mediators have been implicated in the in vitro GVH reaction (12,13). In addition, work by Pliskin et al. (14) and by Nishioka and Katayama (15) has shown that activated spleen cells and their supernatants induce neovascularization in vivo.

While these studies strongly suggest that LIA is elicited via the release of a lymphokine(s) from activated T cells, the identity of the factor(s) has not been established. The aim of this investigation is to further define the factor(s) responsible for LIA.

II. MATERIALS AND METHODS

A. Secondary MLC Supernatants

Secondary MLC supernatants using the allogeneic combination BALB/c anti CBA-T_6 splenocytes were obtained as described by Ryser et al. (16): supernatants were collected at 24 hr. All mice used were obtained from our own inbred colony. In some experiments, these supernatants were dialyzed overnight against an equal volume of medium at 4°C using Spectrapore 2 membranes (Spectrum Medical Ind.) (MW cutoff 12,000): both the dialysate and the dialyzed materials were sterile filtered and tested in the in vitro assays. The media used for secondary MLC supernatants was RPMI 1640 with 2% FCS.

B. Endothelial Cell Mitogen Activity Assay

Fetal bovine aortic endothelial cells (FBAE) and adult bovine adrenal capillary endothelial cells (BCE) were obtained by standard methods (17,18) and grown in Dulbecco's MEM (GIBCO) supplemented with penicillin, streptomycin, and 10% FBS (GIBCO). Cells were endothelial by morphology and were shown to express angiotensin converting enzyme activity.

The assay for endothelial cell mitogenic activity was performed as previously described (19). Briefly, endothelial cells were plated at low density in 24-well dishes and allowed to attach. The medium was replaced with fresh medium to which test additions were made. Two days later, the cells were pulse-labeled with tritiated thymidine (New England Nuclear), and the amount incorporated per well determined.

C. Endothelial Cell Migration Assay

Endothelial migration was assayed by the methods of Zetter (20) and Albrecht-Buehler (21) with the following modifications. The glass coverslips were coated with a solution of a 10 μg/ml poly-L-lysine (Sigma) in water for 5 min and then dried by carefully removing the solution. The endothelial cells were allowed to attach to the coverslips for 2-3 hr in DMEM with 10% FBS, then transfered to test media and incubated for 18 hr, after which the migration was assessed using a computer-assisted calculation for area and pattern of migration as described by Zetter (20).

III. RESULTS

A. Proliferative Response of Endothelial Cells to Secondary MLC Supernatants

It has been previously shown (17) that under the assay conditions used, tritiated thymidine incorporation was proportional to actual endothelial cell numbers, and hence the assay can be used to assess mitogenic activity towards endothelial cells. When 24 hr secondary MLC supernatants were added to microwell cultures of FBAE cells, these cells were stimulated to divide. This response was dose-dependent (Fig. 1), and at the highest concentration the response was 4-fold that of controls. The same results were obtained with BCE cells (data not shown).

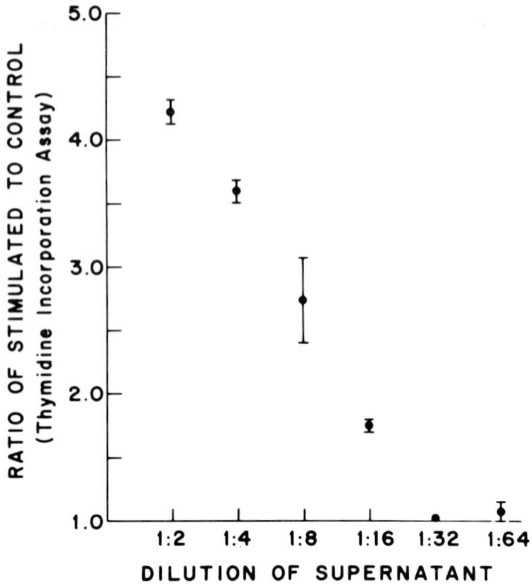

FIGURE 1. Mitogenic Response of FBAE Cells to Secondary MLC Supernatants. Data are expressed as mean ± Standard Deviation.

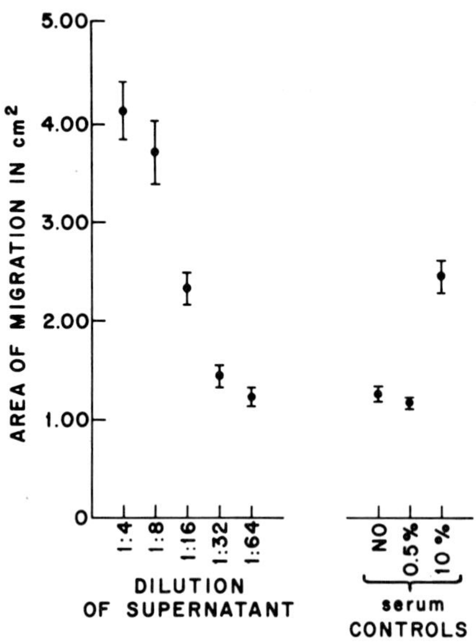

FIGURE 2. Chemokinetic Response of FBAE Cells to Secondary MLC Supernatants. Data are mean ± S.E.. Area of migration is calculated from tracings at 160 X magnification.

When secondary MLC supernatants were dialyzed against medium and tested, the activity was found to be non-dialyzable (MW cutoff 12,000). At a 1:2 dilution (equivalent to a 1:4 dilution of the original supernatant), the proliferative response of FBAE cells was 2.52 ± 0.25, whereas the dialysate gave a response of 0.81 ± 0.01 (relative to medium controls). These data suggest that the active factor(s) is larger than 12,000 MW.

B. Chemokinetic Response of Endothelial Cells to Secondary MLC Supernatants

When secondary MLC supernatants were tested in the migration assay, these supernatants were found to stimulate the migration (chemokinesis) of FBAE cells. Migration in this assay is measured as the area of migratory "tracks" made on gold-coated coverslips by individual cells.

The response of the endothelial cells to the secondary MLC supernatants was even greater than their response to 10% FBS (Fig. 2). Additionally, one can see that the chemokinetic response to the MLC supernatants is dose-dependent, and that the dose-response curve is very similar to the dose-response curve obtained using the proliferation assay.

IV. DISCUSSION

Fenselau et al. (19) have previously shown that tumor-derived materials which stimulated the proliferation of FBAE cells in vitro also stimulated angiogenesis in vivo. Additionally, Zetter (20) has reported that angiogenic materials stimulate endothelial cell migration in vitro. It was therefore decided to use these two in vitro systems to investigate the nature of the factor(s) leading to lymphocyte-induced angiogenesis (1,6,8). We have shown here that secondary MLC supernatants (the in vitro correlate of the GVH (and LIA) response) stimulate both the proliferation and chemokinesis of cultured endothelial cells. Both of these responses are dose-dependent with respect to the amount of secondary MLC supernatant added. In fact, the dose response curves for the proliferative and migratory responses of FBAE cells to secondary MLC supernatants are almost superimposable, suggesting that perhaps the same factor(s) is responsible for both activities. We have begun to investigate the biochemical nature of the factor(s) inducing these responses, and have shown that the mitogenic factor(s) have a molecular weight greater than 12,000 Daltons.

While the lymphokine(s) we have characterized by in vitro tests for mitogenesis and chemokinesis have not yet been tested in in vivo assays, further studies using in vitro tests appear to be critical before in vivo results can be interpreted. Secondary MLC supernatants are complex mixtures of lymphokines including, for example IL-1, IL-2, MIF, etc. (22). Standard fractionation procedures can be used to delineate somewhat more precisely the particular cell product(s) that can induce either proliferation or cell movement of endothelial cells. Later experiments can then be carried out to determine the in vivo effects of active material, to generate serological reagents that can be used to interfere with the action of the material, and to assess the relevance of the material to in vivo processes of angiogenesis. It should be emphasized here that angiogenesis has been shown to be a component of many disease states including psoriatic lesions, arthritic synovial fluid, multiple ocular diseases and wound lesions (10). In addition the importance of angiogenic processes to the growth of solid tumors has been stressed repeatedly (10,18,19).

In summary, the mechanisms involved in in vivo angiogenic responses are still not well understood. We have described active preparations obtained following secondary stimulation of lymphocytes in mixed lymphocyte culture. By further characterization of the specific factor(s) in these lymphocyte-derived supernatants we hope eventually to be able to gain greater understanding of the complex and important question of induced neovascular responses accompanying both immunological reactions and the growth of tumors.

REFERENCES

1. Sidky, Y.A., and Auerbach, R., *J. Exp. Med. 141*, 1084 (1975).
2. Kaminiski, M., Majewski, S., Kaminska, G., Bem, W., and Szmurio, A., *Arch. Immunol. Ther. Exp. (Warsz) 26*, 1075 (1978).
3. Fromer, C.H., and Klintworth, G.K., *Am. J. Path. 82*, 157 (1976).
4. Dvorak, A.M., Mihm, M.C., and Dvorak, H.F., *Lab Invest. 34*, 179 (1976).
5. Polverini, P.J., Cotran, P.S., Gimbrone, M.A., and Unanue, E.R., *Nature 269*, 804 (1977).
6. Auerbach, R., and Sidky, Y.A., *J. Immunol. 123*, 751 (1979).
7. Klein, J., and Park, J.M., *J. Exp. Med. 137*, 1213 (1973).
8. Roehm, N.W., Sidky, Y.A., and Auerbach, R., *Cell. Immunol. 63*, 272 (1981).
9. Kisielow, P., Hirst, J.A., Shiky, H., Beverly, P.C.L., Hoffman, M.K., Boyse, E.A., and Oettgen, H.F., *Nature (London) 253*, 219 (1975).
10. Auerbach, R., *in* "Lymphokines", Vol. 4, p. 69. Academic Press, New York (1981).
11. Adelman, N.E., Ksiazek, J., Yoshida, T., and Cohen, S., *J. Immunol. 124*, 825 (1980).
12. Auerbach, R., and Globerson, A., *Exp. Cell Res. 42*, 31 (1966).
13. Roehm, N., *Transplantation 23*, 49 (1977).
14. Pliskin, M.E., Ginsberg, S.M., and Carp, N, *Transplantation 29*, 255 (1980).
15. Nishioka, K., and Katayama, I., *J. Pathol. 126*, 63 (1978).
16. Ryser, J.E., Cerottini, J.C., and Brunner, K., *J. Immunol. 120*, 370 (1978).
17. Fenselau, A., and Mello, R.J., *Cancer Res. 36*, 3269 (1976).
18. Folkman, J., Haudenschild, C.C., and Zetter, B.R., *Proc. Natl. Acad. Sci. (USA) 76*, 5217 (1979).
19. Fenselau, A., Watt, S., and Mello, R.J., *J. Biol. Chem. 256*, 9605 (1981).
20. Zetter, B.R., *Nature 285*, 41 (1980).
21. Albrecht-Buehler, G., *Cell 11*, 395 (1977).
22. Cerottini, J-C., "Biochemical Characterization of the Lymphokines", p. 513, Academic Press, New York (1980).

DISCUSSION

LANDY: The biologic system devised by Watt and Auerbach for exploring chemokinetic effects of cell products is admirable. However, as regards the "tentative" conclusion they have arrived at (as given in their abstract) viz. "that the mitogenic activity in their secondary MLC supernatants may *not* be the same molecule as IL-2" is, in my view, arriving at a questionable conclusion. I gather that this rests largely on their filtration separations and its indication of molecular size of the activity. I would, however, point out that Kendall Smith has marshalled impressive evidence in support of the view that the activity we have previously designated "blastogenic factor" is in fact attributable to IL-2. It seems to me that Watt's initial indication of molecular size of the activity she measures would be a most inadequate basis for coming to even a tentative judgment.

WATT: Let me emphasize that in our abstract we said it *may* not be IL-2. Certainly, we must do fractionation experiments before IL-2 can be ruled out as being angiogenic. In support of that statement I can tell Landy that we have been screening hybridoma supernatants provided by Kappler. There was one supernatant which had IL-2 activity but it did *not* have endothelial cell proliferation activity.

YOSHIDA: Does Watt's lymphokine preparation have chemotactic activity for endothelial cells? To understand angiogenesis *in vivo,* it seems to me that unidirectional movement may be more relevant than the increase in random movement.

WATT: This chemokinesis assay has already been used to assess angiogenic factors from tumor cells. However, the point raised by Yoshida is a good one. Sometime soon we do intend to extend our studies to parallel assays for chemotaxis using the Boyden chamber-type system.

R. T. SMITH: Are the gold bugs (endothelial cells) being assessed for migration in Watt's assay, consuming, or simply moving the gold particles about? If the former, why not simply measure their uptake a la Aschoff et al.?

WATT: They *do* ingest the gold, and after a time they release the gold, leaving a little trail. With regards to assessing this assay by measuring uptake, a graduate student in our laboratory plans to do just that; he has already observed that these cells migrate on coverslips coated with fluorescent-labelled microspheres in a manner similar to the way they migrate on gold. It is intended to assess the uptake of these beads using FACS analysis.

LYMPHOKINES AND ENDOTHELIAL CELL PRODUCTS IN THE REGULATION OF LYMPHOCYTE-ENDOTHELIAL INTERACTIONS

Dudley C Dumonde, Rene Anderson, Josephine Cox, Anne Hamblin, Eva Kasp-Grochowska, Melanie Pulley and Robin A Wolstencroft

Department of Immunology, St Thomas' Hospital and Medical School, London SE1, United Kingdom

In this paper we draw attention to the importance of lymphocyte-endothelial interactions in regulation of the immunological response and to a number of experimental approaches which lead us to infer that metabolic products of both lymphocytes and endothelial cells may modulate interactions between these two cell types. We propose that the microenvironment of lymphocyte-endothelial interaction is a critical site at which the induction and regulation of T-cell immunity may be effected. We explore the view that lymphokines act locally as both cell cooperators and traffic regulators within lymphoendothelial microenvironments and that lymphokines may also trigger extrinsic (neuroendocrine) mechanisms by which lymphocyte compartmentation and activation may be governed homeostatically at sites of lymphocyte-endothelial interaction. We point to the need to devise simplified systems by which the complex interaction between lymphocytes and endothelial cells may be examined.

I. VASCULAR ENDOTHELIUM AND LYMPHOCYTE FUNCTION

Vascular endothelium forms the natural barrier between blood and tissues; and it is upon vascular endothelium that much immunological action takes place. There is renewed interest in mechanisms which govern the compartmentation of lymphocytes between blood and tissues (Butcher and Weissman, 1980) and in how lymphocyte activation and positioning might be

Interleukins, Lymphokines, and Cytokines

ISBN 0-12-527540-4

Table I. Lymphocyte-endothelial interactions in vivo implicating local or indirect actions of lymphokines (LK)

Recirculatory lymphocytes preferentially migrate into delayed hypersensitivity granulomas where endothelial hypertrophy/hyperplasia accompanies LK production (1)
Intralymphatic injection of LK increases the entry of recirculatory lymphocytes into lymph node paracortex (2)
Intradermal injection of LK induces sustained perivascular lymphocyte accumulation and endothelial hypertrophy (3)
Injection of activated lymphocytes induces angiogenesis (4)
Intravenous injection of LK into rats (5) *or humans* (6) *induces elevation of plasma cortisol and temporary lymphopenia; glucorticoids arrest recirculatory lymphocytes outside postcapillary venular endothelium* (7)
Broad range of MHC-linked immunoinflammatory diseases characterized histopathologically by widespread lympho-endothelial inflammation (8)

(1) *Issekutz et al (1982);* (2) *Kelly et al (1972);*
(3) *Dumonde et al (1982a);* (4) *Auerbach (1981);*
(5) *Besedovsky et al (1981);* (6) *Pulley et al (1982);*
(7) *Cox and Ford (1982);* (8) *Dumonde (1982).*

related to each other (de Sousa, 1981). Certain features of lymphocyte-endothelial interaction have been conserved phylogenetically (Butcher and Weissman, 1980) yet little is known of the factors which regulate the compartmentation and activation of recirculatory lymphocytes during their physiological diapedesis.

Tables I and II summarize a number of experimental approaches which, when taken together, point to the endothelial cell as an important participant in the induction and regulation of T-cell immunity. By virtue of the microanatomical positioning of endothelial cells, in the 'front line' of lymphocyte diapedesis, it seems plausible that endothelium should present antigen to T cells *in vivo*. It also seems plausible that metabolic products of endothelial cells should influence the fact, nature and kinetics of specific T cell

Table II. Lymphocyte-endothelial interactions in vitro of potential relevance to immunoregulation

---

Cultured endothelial cells are competent to present antigen for T-lymphocyte sensitization or specific recall with genetic restriction at Ia (DR) loci (1,2)

Cultured endothelial cells generate high concentrations of E type prostaglandins (3) *which suppress lymphocyte activation* (4); *and also other factors which can potentiate lymphocyte transformation* (5)

Activated lymphocytes generate lymphokines which are mitogenic for cultured endothelial cells (6) *and which will inhibit their migration* (7)

Activated lymphocytes preferentially penetrate confluent cultures of vascular endothelium (8)

Some lymphokine preparations induce transendothelial chemotaxis of lymphoblasts and lymphocytes in vitro (9)

---

(1) *Hirschberg et al (1980);* (2) *Burger et al (1981);* (3) *Jose et al (1981);* (4) *Goodwin et al (1978);* (5) *Kasp-Grochowska et al (1982a);* (6) *Watt and Auerbach (1982);* (7) *Cohen et al (1982);* (8) *Dumonde et al (1982b);* (9) *Dumonde et al (1982c)*

activation during the physiological events of lymphocyte-endothelial interaction that are continually taking place.

On the other hand, it is also clear that endothelial cell (and hence microvascular) function is modulated by interaction with activated lymphocytes and by lymphokines themselves (Auerbach, 1981). As both endothelial cells and lymphocytes can generate autopharmacological products (eg prostaglandins and lymphokines respectively) which can affect the reciprocal cell type, it seems probable that feedback homeostatic mechanisms exist by which lymphocyte: endothelial interaction is normally regulated. The fact that endothelial cells are competent to present antigen for T-cell activation implies that they may fulfil the triple requirements of antigen processing, MHC-dependent signal recognition and cytokine ('endokine') production.

On this basis, a central role for the endothelial cell in the regulation of T-cell immunity would have implications for the physiology, pathology, developmental biology, genetics and autopharmacology of the immune response as a whole. The microenvironment of lymphocyte-endothelial interaction offers opportunity for locally-produced mediators to exert immunoregulatory effects. It also offers opportunity for extrinsic factors, such as hormones and acute phase proteins (Pulley *et al*, 1982) to exert effects on one or both interacting cell types. A 'lymphoendothelial' hypothesis of immunoregulation would be compatible with an interaction-modulation theory of lymphocyte positioning (Curtis, 1978), in which a degree of specificity of receptor-ligand interaction would be modulated by locally produced and extrinsically generated mediators. Recent evidence suggests that lymphokines may activate neuroendocrine mechanisms which may be involved in homeostasis of the immunological response (Besedovsky *et al*, 1981). This discussion therefore carries the corollary that extrinsic factors modulating immune function may also be found to exert their effects at these critical microenvironmental sites of lymphocyte-endothelial interaction.

II. IN VITRO APPROACHES TO THE INVESTIGATION OF LYMPHOCYTE-ENDOTHELIAL INTERACTIONS

Our interest in this area is to develop simplified approches to the study of lymphocyte-endothelial interaction that may indicate how some of the aforementioned mechanisms may be operating *in vivo*. A major problem has been to devise suitable methods of co-culture, involving both lymphocytes and endothelial cells, by which effects of one cell type upon the other may be studied. A characteristic function of vascular endothelium *in vivo* is the selective compartmentation of large and small molecules and of formed elements of the blood; whilst characteristic features of lymphocyte populations involve their selective traffic across vascular endothelium as well as their functional differentiation when metabolically or immunologically stimulated. Accordingly we have devised three approaches to the study of lymphocyte-endothelial interactions in which these functional attributes of the cell populations can be examined.

A. *Modulation of lymphocyte transformation by products of cultured endothelial cells*

The first model consits of studying the ability of pig aortic vascular endothelial culture supernatants to affect the transformation of porcine peripheral blood lymphocytes. We found that endothelial culture supernatants inhibited PHA- and periodate- induced lymphocyte transformation in partial concordance with their content of prostaglandin E2; and that these suppressive effects were preferentially exerted on weakly transforming lymphocytes. When endothelial cell prostaglandin production was abolished by indomethacin, some endothelial supernatants actually stimulated lymphocyte transformation. We concluded that endothelial cells have the capacity to produce diffusible products which might well exert controlling effects upon lymphocyte function within the microenvironment of lymphocyte-endothelial interaction (Kasp-Grochowska *et al*, 1982a).

B. Formation of lymphoendothelial aggregates in vitro

The second model consists of co-culturing resting or periodate-stimulated pig peripheral blood lymphocytes with confluent monolayers of pig aortic endothelium and of assessing the degree of cellular interaction by phase contrast microscopy and by the use of ^{111}In-labelled lymphocytes. We found that after co-culture for up to 24 hours, 2 populations of associated lymphocytes could be identified: 'adherent' and 'integrated'. Adherent lymphocytes lay on top of the endothelial cells whilst integrated lymphocytes lay between the endothelial cells in the same phase contrast plane. The adherent cell population could be separated from the integrated population by means to treating the cultures briefly at pH4 with a glycine buffer. Prior activation of lymphocytes (with periodate) led to a greatly increased adherence during the early hours of co-culture which was accompanied by a more rapid integration of the lymphocytes in between the endothelial cells. In this situation the co-cultures formed interwoven lymphoendothelial layers in which both lymphocytes and endothelial cells appeared in a state of cytoplasmic enlargement; and in some experiments, integration of activated lymphocytes also led to increased PGE production by the endothelial cells (Kasp-Grochowska *et al*, 1982b). Experiments are now in progress to determine whether lymphocyte populations which 'adhere' to endothelium are functionally different from lymphocyte populations which 'integrate' into the cultured endothelial monolayers. It seems likely that both lymphocytes and endothelial cells will exert modulatory effects upon each other; and these await characterization.

C. Transendothelial lymphocyte chemotaxis

This third model has been devised in order to simulate conditions under which lymphocyte-endothelial interaction may take place during diapedesis at sites of inflammation.

Vascular endothelial cells (from pig aorta) are grown on thin (20 μ) microporous PTFE membranes (pore size 12-15 μ) until confluence is reached as judged by restriction of fluid flow across the membrane (McCall *et al,* 1981a). A plastic holder bearing the endothelialized membrane forms the upper compartment of a chemotaxis assembly, the lower compartment simply being a Linbro well containing tissue culture medium (McCall *et al,* 1981b). Experiments consist of adding lymphocyte or leucocyte suspensions to the upper compartment, putative chemotaxin to the lower compartment, and of measuring cell migration across the endothelial monolayer in relation to its degree of confluence.

Several interesting characteristics of this model are emerging. Unless the PTFE membrane is endothelialized, it fails to support leucocyte chemotaxis; but even partial endothelialization permits the chemotaxis of granulocytes, monocytes and lymphoblasts across the membrane in response to zymosan-activated serum present in the lower compartment. Experiments with a lymphokine preparation especially rich in TCGF (a gift of Dr A Boyleston, St Mary's Hospital, London) have shown that both small and large lymphocytes will undergo transendothelial chemotaxis across a confluent monolayer in response to the lymphokine, without impairing the state of confluency. This co-culture system will enable the study of how metabolic peturbation of either endothelial cells or lymphocytes affects transendothelial lymphocyte compartmentation *in vitro*.

III. CONCLUSIONS

Microenvironmental interactions between recirculatory lymphocytes and vascular endothelium are viewed as being modulated by metabolic products of both cell types as well as by endocrine and other humoral substances generated at a distance from the microenvironment. These interactions are clearly of central importance in regulation of the immunological response and may well involve lymphokines as both 'traffic regulators' and 'cell cooperators'. There is a need to employ simplified systems as models by which lymphocyte compartmentation and activation can be studied alongside alterations in vascular endothelial function. Such models will be expected to accommodate antigen-presenting functions of vascular endothelium as well as its ability to modulate lymphocyte activation and traffic. The experimental approaches summarized in this paper are being pursued with these objectives in mind.

REFERENCES

Auerbach, R. *Lymphokines 4*, 69 (1981).

Besedovsky, H. O., del Rey, A. and Sorkin, E. *J. Immunol.*, *126*, 385 (1981).

Burger, D. R., Ford, D., Vetto, R. M., Hamblin, A., Goldstein, A., Hubbard, M. and Dumonde, D. C. *Human Immunol.*, *3*, 209 (1981).

Butcher, E. C. and Weissman, I. L. *Ciba Found. Symp. 71*, 265 (1980).

Cohen, M. C., Picciane, P. T., Douglas, W. J., Yoshida, T., Kreutzer, D. L. and Cohen, S. *Science 215*, 301 (1982)

Cox, J. H. and Ford, W. L. *Cell. Immunol.*, *66*, 407 (1982)

Curtis, A. S. G. *Symp. Soc. Exp. Biol.*, *32*, 51 (1978)

De Sousa, M. *Lymphocyte circulation*, New York: Wiley (1981).

Dumonde, D. C., *Scand. J. Rheumatol.*, *42 (suppl.)* 14 (1982).

Dumonde, D. C., Hamblin, A. S., Kasp-Grochowska, E., Pulley, M. S. and Wolstencroft, R. A. (1982b) In *Immuno-regulation* (ed. N. Fabris). Plenum Press, New York, in press.

Dumonde, D. C., Pulley, M. S., Paradinas, F., Southcott, B. M., O'Connell, D., Robinson, M. R. G. and den Hollander, F. In *Human Lymphokines* (eds. A. Khan and N. O. Hill) New York: Academic Press, p. 665 (1982a).

Dumonde, D. C., Cox, J. H., Darby, H. and Anderson, R. L. (1982c) MS in preparation.

Goodwin, J. S., Bankhurst, A. D. and Messner, R. P. *J. exp. Med.*, *146*, 1719 (1978).

Hirschberg, H., Bergh, O. J. and Thorsby, E. *J. exp. Med.*, *152*, 249 (1980).

Issekutz, T. B., Chin, W. and Hay, J. B. *Immunology*, *46*, 59 (1982).

Jose, P. J., Page, D. A., Wolstenholme, B. E., Williams, T. J. and Dumonde, D. C., *Inflammation*, *5*, 363 (1981).

Kasp-Grochowska, E., Page, D. A., Povey, J., Dumonde, D. C., Jose, P. and Williams, T. J. (1982a) MS in preparation.

Kasp-Grochowska, E., Anderson, R. L. and Dumonde, D. C. (1982b) MS in preparation.

Kelly, R. H., Wolstencroft, R. A., Dumonde, D. C. and Balfour, B. M. *Clin. exp. Immunol.*, *10*, 49 (1972)

McCall, E., Povey, J. and Dumonde, D. C., *Thrombosis Res.*, *24*, 417 (1981a).

McCall, E., Povey, J. and Dumonde, D. C., *Agents and Actions*, *11*, 654 (1981b).

Watt, S. and Auerbach, R. (1982): this symposium.

DISCUSSION

PICK: Did Dumonde try to use capillary endothelial cells instead of those derived from the aorta?

DUMONDE: No, we have not done this yet - but I have discussed it with Auerbach and Watt and, hopefully, we will collaborate on this.

PICK: Has Dumonde considered the possibility that the observed suppressive effect of endothelial cell supernatants could be due to PGI_2, not only to PGE_2?

DUMONDE: Yes; there is a substantial amount of $PGF_1\alpha$ in our endothelial supernatants and we have shown that this metabolite can suppress pig lymphocyte transformation. PGI_2 is unstable, as you know, so we did not use this separately. We have tested mixtures of PGE_2, $PGF_2\alpha$ and $PGF_1\alpha$ in appropriate concentrations and we can increase the suppressive effect of PGF_2 by this means.

OPPENHEIM: Since IL-1-like activities appear to be produced by a number of cell types, it becomes relevant to point out that not all cells have this capacity. For example, we were unable to induce fibroblasts to produce such an activity. Furthermore, since endothelial cells are able to present antigens, we also tested their capacity to produce a thymocyte-activating factor. We were unable to detect such an activity in crude supernatants nor in the endothelial cell-derived factor which is mitogenic for fibroblasts. Has Dumonde tested his endothelial cell-immunohancing activity for IL-1 function?

DUMONDE: In view of the high content of prostaglandins in endothelial cell culture supernatants, one could well miss IL-1 activity in crude or unmodified culture fluids. We do not have a standardized IL-1 assay in our laboratory at present. I would be most grateful if Oppenheim were to test some of our products for IL-1 activity.

WATT: Do Dumonde's endothelial cells put down a basement membrane in these chambers?

DUMONDE: We cannot see a basement membrane by electron microscopy (McCall et al., 1981), but endothelial supernatants do contain a factor which assists the attachment of endothelial cells to PTFE membranes. This might be extracellular matrix material, but we have not set out to isolate it.

ROSENWASSER: In one of his protocols, Dumonde showed that TCGF induced lymphocyte chemotaxis across non-confluent pig endothelial cell monolayers. Would he tell us more about the purity of this TCGF preparation - and what other biological activities or lymphokines are present in it.

DUMONDE: The TCGF was a gift from Boyleston at the St. Mary's Hospital, London. It is a human lymphokine made in MLC with additional lectin stimulation, and from which the lectin has been removed. The TCGF is not purified, but it is certainly very active in maintaining blast cell growth *in vitro*. It probably contains a variety of lymphokine activities, which is why we used it.

FIBROBLAST GROWTH FACTOR PRODUCTION BY HUMAN PERIPHERAL BLOOD AND LEUKEMIC CELL LINE LYMPHOCYTES

Sharon M. Wahl
Celia L. Gately
William E. Helsel

Laboratory of Microbiology and Immunology
National Institute of Dental Research
National Institutes of Health
Bethesda, Maryland

I. INTRODUCTION

Lymphokines, which are polypeptide hormones produced by lymphocytes, regulate proliferative and functional processes in inflammatory cells and also in noninflammatory target cells. Recent evidence suggests that lymphokines may modulate both the formation (1) and destruction (2) of connective tissue. Activated lymphocytes, as well as monocytes, produce a variety of mediators with important regulatory effects in intercellular communication leading to a fibrotic response. These inflammatory cells can recruit fibroblasts through the release of chemoattractants (3) and mediate expansion of the fibroblast population through the production of growth factors which trigger fibroblast proliferation (1). The consequence of this increase in fibroblast numbers is an increase in matrix and collagen synthesis necessary for tissue repair. Collagen synthesis has also been shown to be influenced by signals from inflammatory cells (1). Thus, the characterization of these lymphocyte-derived regulatory molecules will further the understanding of the mechanisms whereby fibroplasia and fibrogenesis are regulated by immunologic systems. Furthermore, the identification of permanent lymphoid cell lines which produce such mediators will facilitate studies of the biochemical and molecular properties of this lymphokine. In these studies, a human T cell line (HSB2) has been found to produce a fibroblast growth factor with properties similar to those of the human peripheral blood lymphocyte product.

ISBN 0-12-527540-4

II. PRODUCTION OF FIBROBLAST ACTIVATING FACTOR (FAF) BY HUMAN PERIPHERAL BLOOD LYMPHOCYTES

Peripheral blood mononuclear cells isolated from buffy coat preparations were stimulated with concanavalin A (Con A) to generate fibroblast activating factor (FAF) containing supernatants (4). When these supernatants were assayed for their ability to stimulate a population of quiescent dermal fibroblasts to actively proliferate, as determined by the incorporation of tritiated thymidine into fibroblast DNA, it was clear that the supernatants contained significant amounts of activity (Table I). Depletion of adherent cells from the mononuclear cell preparation did not result in loss of FAF production and purified T lymphocytes ($OKT3^+$) obtained by counterflow centrifugation elutriation also produced the factor, indicating that the T cell is the source of this mediator. Furthermore, the FAF containing supernatants could be generated by culturing sensitized lymphocytes with specific antigen. Whether stimulated by antigens or nonspecific mitogens, the production of this factor was maximal at 72 hr.

Table 1

Production of FAF by Human Peripheral Blood Lymphocytes

Supernatant Dilution	*Fibroblast TdR^3H Incorporation (cpm)*
1:5[a]	*48592+288*[b]
1:10	*36255+2099*
1:20	*20471+1809*
1:40	*8973+1343*
1:80	*3453+ 653*
1:160	*3152+ 204*
Control	*850+ 152*

[a] *$4X10^6$ human peripheral blood mononuclear cells were stimulated with 5 μ g/ml Con A for 48 hr and the cell free supernatants collected.*

[b] *$5X10^4$ nonconfluent, serum-free fibroblasts were incubated with supernatants or medium only (control) for 48 hr, pulsed 4 hr with 1μCi/ml TdR^3H and processed for determination of TdR^3H incorporation.*

III. PRODUCTION OF FIBROBLAST ACTIVATING FACTOR BY A CONTINUOUS T LYMPHOCYTE CELL LINE

HSB2, a human T cell line (5), can be induced to produce several lymphokines by mitogenic agents such as concanavalin A (6,7). When supernatants from this cell line were assayed for their ability to induce fibroblast proliferation, some activity was detected in supernatants from nonstimulated cultures. The titer of FAF was enhanced, however, by activating the HSB2 cells with Con A (Table 2). Maximal production of this factor occurred 48 hr after mitogen stimulation.

Table 2

Stimulation of Fibroblast Proliferation by HSB2 Supernatants

HSB2 Supernatant	*Fibroblast TdR^3H Incorporation (cpm)*
Con A Stimulated[a]	*12797 ± 1098*[b]
Nonstimulated	*5729 ± 919*
Nonstimulated + Con A	*1930 ± 182*
Medium Only	*2137 ± 165*

[a] *HSB2 cells ($1X10^6$) were cultured with or without Con A for 48 hr and the supernatants diluted 1:5 and assayed for their effect on fibroblast TdR^3H incorporation.*

[b] *$5x10^4$ nonconfluent, serum-free fibroblasts were incubated with HSB2 supernatants or medium only for 48 hr, pulsed 4 hr with 1μCi/ml TdR^3H and processed for determination of TdR^3H incorporation.*

IV. CHARACTERIZATION OF FAF PRODUCED BY PERIPHERAL BLOOD LYMPHOCYTES AND THE HSB2 CELL LINE

Because of the ability of both human peripheral blood T lymphocytes and the HSB2 cells which are a T cell line to generate soluble mediators of fibroblast proliferation, experiments were carried out to establish whether the two populations of cells were producing a similar mediator of fibroblast growth. Supernatants from both populations of cells were concentrated and analyzed by molecular sieve chromatography. When the peripheral blood lymphocyte product was chromatographed on Sephacryl S200 columns, a single peak of fibroblast mitogenic activity near 40,000 was observed. A similar peak of activity with an apparent molecular weight of 40,000 was found after gel filtration of the HSB2 supernatants (Table 3). In subsequent studies, both blood lymphocyte and HSB2 cell products were analyzed by isoelectric focusing in a 5-50% sucrose gradient containing pH 3.5-10 ampholines. A primary peak of activity which stimulated fibroblast proliferation was found between pH 5.0 and 5.5 for both cell populations. Furthermore, both lymphocyte-derived FAF and HSB2 FAF were heat labile and trypsin sensitive, indicating that the mediators were protein-like with an apparent molecular weight of 40,000. Thus, by all criteria evaluated, it was apparent that the HSB2 derived FAF was indistinguishable from the factor produced by normal human peripheral blood lymphocytes.

Table 3

Characterization of Human Peripheral Blood Lymphocyte and HSB2 Cell Line Supernatant Factors Which Stimulate Fibroblast Growth

	Blood Lymphocytes	*HSB2 Cell Line*
Molecular Weight	*40,000*	*40,000*
pI	*5.0-5.5*	*5.0-5.5*
Heat Labile	+	+
Trypsin Sensitive	+	+
Con A Induced	+	+

V. DISCUSSION

Evidence is accumulating that inflammatory cells can modulate fibroblast growth and matrix synthesis. A mediator generated by human peripheral blood lymphocytes, FAF, can stimulate mitosis in a quiescent population of fibroblasts. This factor is a heat labile protein with an apparent molecular weight of 40,000 and pI of 5-5.5. In addition, it has been found that this T cell-derived factor which stimulates fibroplasia appears to be indistinguishable from a product of the HSB2 continuous cell line which also induces fibroblast proliferation. Thus, the HSB2 cell line provides a readily available and homogeneous source for obtaining sufficient quantities of FAF to facilitate a more detailed analysis of its structure and function. Further characterization of FAF and its interactions with the target cells may contribute to a better understanding of the molecular link between inflammatory cells and the fibrotic response associated with inflammation.

REFERENCES:

1. Wahl, S. M and Wahl, L. M., *Lymphokines 2*, 279 (1981)

2. Wahl, L. M., Wahl, S.M., Mergenhagen, S. E. and Martin, G. R., *Science 187*, 261 (1975).

3. Postlethwaite, A. E., Snyderman, R. and Kang, A. H., *J. Exp. Med. 144*, 1188 (1976).

4. Wahl, S. M. and Gately, C. L., In Press (1982).

5. Adams, R. A., Potheir, A., Flowers, A., Lazarus, H., Farber, S. and Foley, G. E., *Exp. Cell Res. 62*, 5 (1970).

6. Stadler, B. M., Dougherty, S. F., Farrar, J. J. and Oppenheim, J. J., *J. Immunol. 127*, 1936 (1981).

7. Gately, C. L. and Oppenheim, J. J., *Fed. Proc. 41*, 768 (1982).

DISCUSSION

DAVID: Has Wahl explored the characteristics of the fibroblast growth factor made by the monocyte-rich fractions? How does it compare with the T cell product?

WAHL: The human monocyte-derived mediator which induces fibroblast proliferation has an apparent molecular weight of 15,000, whereas the T cell product with similar activity chromatographs in the 40,000 MW region. A single isoelectric point is represented by the T cell activity, whereas the monocyte activity occurs at 3 isoelectric points.

WOOD: Has Wahl looked for synergy between IL-1 and the lymphocyte-derived, fibroblast-activating factor?

WAHL: Unfractionated mononuclear cell supernatants contain both T cell and monocyte-derived mediators, but we have thus far not taken the two purified mediators and sought synergistic effects on fibroblast proliferation.

PICK: Could Wahl summarize for us the essential differences between platelet-derived growth factor (PGF) and her T cell-derived fibroblast growth factor? Could she also tell us whether the production of PFG can be stimulated?

WAHL: The platelet-derived growth factor has a molecular size similar to the T cell product, which stimulates fibroblast growth. However, PGF is stable to boiling, which destroys the T cell product; it also has a very alkaline pI (~9) whereas the pI for the T cell product is near 5. PDGF is generally obtained from boiled platelets, by extraction rather than by stimulation of platelet populations.

CATHCART: How do Wahl's observations on the T cell factor compare with the macrophage factor reported previously? Do T cells activated in her system produce fibronectin?

WAHL: The macrophage-derived product Cathcart refers to, which is chemotactic for fibroblasts, has been characterized as the glycoprotein, fibronectin. It is distinct from the macrophage product which stimulates fibroblast proliferation and which appears to be inseparable from IL-1, as shown by Schmidt and coworkers. There is no evidence that T lymphocytes produce fibronectin, but they do produce a chemoattractant for fibroblasts as shown by Postlethwaite.

ACTIVATION OF NUCLEAR DNA SYNTHESIS[1] BY CYTOPLASMIC PROTEIN

Janice K. Gutowski[2]
Stanley Cohen

Department of Pathology
University of Connecticut Health Center
Farmington, Connecticut

It is well known that resting lymphocytes can be stimulated to proliferate by a wide variety of extracellular agents including antigens, mitogens, lymphokines, and growth factors. Little is known, however, about the intracellular systems that relate the triggering event (membrane receptor binding) with the induction of nuclear DNA synthesis. A number of mechanisms have been implicated as putative "second signals" for lymphocyte activation, including changes in cyclic nucleotides, receptor-ligand internalization, ion fluxes, and cytoskeletal changes. However, the specific involvement of these events in mechanisms of mitogenesis remains unclear. Another approach is suggested by studies of cytoplasmic control of nuclear activity in other cellular systems. Experimental evidence has accumulated in a variety of biological settings which suggests that the entry into the S phase of the cell cycle is controlled by factor(s) in the cytoplasm (1). Cytoplasmic extracts prepared from embryonic tissue (2), continuously proliferating cells (3), and mitogenic hormone-treated cells (4) have been shown to be capable of inducing DNA synthesis in isolated nuclei. For example, Das has shown (4) that the activation of 3T3 fibroblasts by epidermal growth factor (EGF) leads to the appearance of a cytoplasmic protein, distinct from EGF or receptor,

[1] *Supported in part by NIH Grants AI-16706, AI-12477 and CA 09205*
[2] *Supported by NIH Grant CA 09205*

ISBN 0-12-527540-4

which is capable of initiating DNA synthesis in isolated quiescent nuclei.

Our studies were undertaken to determine whether such cytoplasmic intermediates play a role in lymphocyte activation for proliferation. As the manuscripts describing these studies in detail are in press, the results will be briefly reviewed in this communication. Cytoplasmic extracts were prepared from spontaneously proliferating lymphoblastoid cell lines and from mitogen-activated normal lymphocytes, and tested for the ability to induce DNA synthesis in quiescent nuclei isolated from adult frog splenocytes. Extracts were prepared as described by Das (4). Briefly, lymphoid cells were incubated in a hypotonic buffer and disrupted in a Dounce homogenizer. The homogenate was subjected to a low-speed centrifugation, and the supernatant was then recentrifuged at 140,000 x g for 1 hour. The resulting high-speed supernatant was used as cytoplasmic extract in the assay.

Isolated nuclei were prepared by detergent lysis as described by Benz and Strominger (5). Briefly, frog spleen cells were incubated in a buffer containing Brij-58 (Sigma) and disrupted by vigorous agitation on ice. The nuclei were then washed free of the detergent, resuspended in buffer and used immediately. Frog spleen cell nuclei were used in these studies because they are easier to prepare and more stable in culture than mammalian nuclei, and because in other systems, cytoplasmic stimulatory factors have been shown to lack species specificity.

Nuclei (2×10^5) and cytoplasmic extract (50 µl) were cultured in triplicate wells of a microtiter plate. Because this is a cell-free assay using isolated nuclei, the cultures also contained an "incorporation mixture", which contains ATP, phosphoenolpyruvate, pyruvate kinase, KCl, sucrose, $MgCl_2$, dithiothreitol, and all of the deoxyribonucleoside triphosphates. Three µCi tritiated thymidine triphosphate (^{3}H-TTP) was also added to each well, and DNA synthesis was measured by the ATP-dependent incorporation of ^{3}H-TTP into TCA-precipitable material.

Cytoplasmic extracts were prepared from three different proliferating cell lines: P3 (a murine plasmacytoma line), 8392 (a human B-cell lymphoma line), and MOLT 4 (a human T-cell line). Extracts prepared from all three lines induced high levels of DNA synthesis in the isolated nuclei, as shown in Table I. The incorporation of ^{3}H-TTP was rapid, reaching maximal levels within 90 minutes. In contrast, incubation of nuclei without extracts, or extracts without nuclei, resulted in only minimal amounts of ^{3}H-TTP incorporation. We also studied extracts prepared from resting cells (unstimulated

TABLE I. Induction of Nuclear DNA Synthesis by Cytoplasmic Extracts Prepared from Proliferating Lymphoblastoid Cell Lines[a]

Source of Extract	*Cultures*	^{3}*H-TTP Incorporation* Δ*cpm/culture* $x\ 10^{-3}$ $(\bar{x} \pm SEM)$
P3	*nuclei alone*	2.7 ± 0.6
	extract alone	1.5 ± 0.6
	nuclei plus extract	15.8 ± 1.6
8392	*nuclei alone*	1.2 ± 0.2
	extract alone	0.2 ± 0.1
	nuclei plus extract	12.3 ± 0.6
MOLT 4	*nuclei alone*	1.3 ± 0.1
	extract alone	3.5 ± 1.0
	nuclei plus extract	22.3 ± 2.3

[a] *$2x10^{5}$ frog spleen cell nuclei were cultured for 90 minutes with 50 µl cytoplasmic extract prepared from proliferating P3, 8392, or MOLT 4 cells (10^{8}/ml). Background ^{3}H-TTP incorporated for time zero incubations have been subtracted.*

murine spleen cells and splenic T-cells). These extracts failed to induce nuclear DNA synthesis.

Dose-response analysis revealed that a given number of nuclei can be saturated with respect to the cytoplasmic stimulatory activity. When 200,000 nuclei are cultured with increasing doses of extract (5µl-160µl), the result is a sigmoidal DRC which plateaus out to a dose which is about 13 times the ED_{50}.

Having demonstrated that continuously proliferating cells contain a cytoplasmic activator of DNA replication (ADR), we next asked whether the mitogenic activation of normal lymphocytes would lead to the appearance of this activity. Human peripheral blood lymphocytes were cultured for 66 hours in the presence and absence of a mitogenic dose of phytohemagglutinin (PHA). Extract prepared from the PHA-stimulated cells induced significant levels of nuclear DNA

TABLE II. Induction of Nuclear DNA Synthesis by Cytoplasmic Extracts Prepared from Mitogen-Activated Lymphocytes[a]

Culture	^{3}*H-TTP Incorporation Δcpm/culture* $x\ 10^{-3}$ $(\bar{x} \pm SEM)$
Nuclei alone	1.3 ± 0.2
Extract (unstimulated) alone	0.4 ± 0.1
Extract (PHA-stimulated) alone	0.6 ± 0.2
Nuclei plus extract (unstimulated)	2.6 ± 0.9
Nuclei plus extract (PHA-stimulated)	10.1 ± 0.6

[a]*Human peripheral blood lymphocytes were cultured for 66 hours in the presence and absence of a mitogenic dose of PHA. Extracts were prepared from the unstimulated and PHA-stimulated cells at* 5×10^{7} *cells/ml. 200,000 nuclei were incubated for 90 minutes with 50 µl extract. Background* ^{3}*H-TTP incorporation for time-zero incubation have been subtracted.*

synthesis, as shown in Table II. In contrast, extract prepared from the unstimulated cells contained little ADR activity. The stimulatory effect of the activated cell extracts was not due to carryover of residual mitogen from the PHA-stimulated cultures, as PHA alone is incapable of activating isolated nuclei. Similar results have also been obtained using concanavalin A (Con A)-stimulated murine thymocytes.

ADR activity was totally abrogated by trypsin treatment (50µg/ml). This effect was shown not to be due to the soybean trypsin inhibitor used to stop the digestion, nor to any residual tryptic effects on the nuclei. Other characterization studies revealed that the active factor is heat labile (60°C for 20 minutes), non-dialyzable, stable at 4°C for 24 hours, and precipitable with ammonium sulfate at 30-50% saturation. When subjected to Amicon ultrafiltration, all of the ADR activity present in the whole extract was retained on an XM100A filter, suggesting that the active component has a molecular weight $\geq$ 100,000 daltons. Taken together, these results strongly suggest that the active component is a protein.

We were also interested in determining whether ADR is released from the cell, as is the case for conventional growth factors and growth-stimulatory lymphokines. To this end, proliferating 8392 cells were cultured in serum-free medium for 24 hours. The supernatants were collected, dialyzed against extract buffer, and tested for the ability to activate DNA replication in the isolated nuclei. Under these conditions, we were unable to recover any stimulatory activity for nuclei in the supernatants of cultured cells whose cytoplasm was known to contain high levels of such activity. We tested a wide range of dilutions and concentrations, but could never detect any ADR activity in the culture supernatants.

Finally, we asked whether ADR was capable of stimulating DNA synthesis when added to intact cells. For these studies, intact frog spleen cells were cultured for 1-4 days with medium alone, cytoplasmic extract, or with a mitogenic dose of Con A. Cytoplasmic extracts containing ADR activity for isolated nuclei failed to induce DNA synthesis in the intact cells, even when concentrated 5X. The cells were sensitive to mitogenic signals, as evidenced by their positive response to Con A. Although the ADR lacks species specificity with respect to its effect on isolated nuclei, it was possible that when whole cells are concerned, a binding step may show such specificity. Thus, these experiments were repeated using a murine source of ADR and murine splenocytes as targets. In no case could we detect a stimulatory effect on intact cells.

In conclusion, we have shown that mitogen-activated normal lymphocytes, as well as continuously proliferating lymphoid cells, contain an extractable cytoplasmic factor which is capable of inducing DNA synthesis in isolated, quiescent nuclei. This factor, which we have called ADR, is protein in nature and is not detectable in the cytoplasm of resting cells. ADR has no stimulatory effect on intact cells, and cells that contain high levels of ADR in the cytoplasm do not release it into the surrounding milieu. Thus, the factor appears to be distinct from conventional growth factors and lymphokines, and may represent an important new class of intracellular growth-regulatory molecules.

REFERENCES

1. Ord, M.J., *In* "International Review of Cytology, Supplement 9" (G.H. Bourne and J.F. Danielli, eds.), p. 221. Academic Press, N.Y., (1979)

2. Benbow, R.M. and Ford, C. C., *Proc. Natl. Acad. Sci. USA 72*, 2437 (1975).
3. Jazwinski, S.M., Wang, J.L., and Edelman, G.M., *Proc. Natl. Acad. Sci. USA 73*, 2231 (1976).
4. Das, M., *Proc. Natl. Acad. Sci. USA 77*, 112 (1980).
5. Benz, W.C., and Strominger, J.L., *Proc. Natl. Acad. Sci. USA 72*, 2413 (1975).

DISCUSSION

STADLER: Does Gutowski know whether these intracytoplasmic DNA-stimulating proteins might be identical in EGF-stimulated cells, mitogen-stimulated lymphoid cells and tumor cell lines?

GUTOWSKI: We cannot say for sure, yet. However, the cytoplasmic factors from lymphoid cells that I have described share many physicochemical properties with factors that have been found by ourselves and others in non-lymphoid and tumor cells. I think it is intriguing to speculate that neoplastic cells may have "learned" to synthesize constitutively these growth regulatory factors, which are present only transiently in normal cells when they are activated.

WOOD: I wonder if your nuclear factor is a marker of early events? Has Gutowski analyzed the kinetics of its induction?

GUTOWSKI: Yes, we have. The appearance of the cytoplasmic stimulatory factor in PHA-stimulated human PBL correlates very closely with the mitogenic response of these cells. That is, the greatest amount of activity is found when the cells are at peak proliferative response to the mitogen. This suggests that the factor is rather late-acting.

PAETKAU: Has Gutowski checked to see that the new DNA synthesis involved replication, as distinct from repair?

GUTOWSKI: Our evidence that we are measuring bona fide DNA synthesis is that ^{3}H-TTP uptake is stringently dependent on the presence of ATP, all of the deoxyribonucleoside triphosphates, and divalent cations in the incorporation mixture.

GARLAND: Most of the cells Gutowski used were lymphoblastoid. Is there any possibility that the activity is terminal transferase?

GUTOWSKI: Some activity is also present in *non*-lymphoid cells, e.g. hepatoma, Ehrlich's ascites, and macrophage cell lines.

GREENE: Do the partially purified cytoplasmic extracts from activated cells contain detectable protein kinase activity?

GUTOWSKI: We have not yet explored this issue. However, Greene's question raises an important possibility that we plan to investigate when we have more purified material.

AMNIOTIC FLUID CELL GROWTH FACTOR (ACGF): A NEW MEDIATOR RELEASED BY STIMULATED HUMAN MONONUCLEAR LEUKOCYTES (MNL).

Hinnak Northoff*, W.Schmidt**, H.D.Hager*** and G.Hänsch*
*Institut für Immunologie, ***Institut für Humangenetik, Universität Heidelberg, and **Universitätsfrauenklinik, 6900 Heidelberg, FRG.

INTRODUCTION

Human amniotic fluid usually contains a few viable cells of fetal origin. These amniotic fluid cells (AFC) can be obtained by amniocentesis and cultured in medium, containing fetal calf serum (FCS) and processed to enable the prenatal diagnosis of chromosome aberrations and enzyme defects. By now, this procedure is routinely performed in many laboratories, however, since it is often tedious, timeconsuming and occasionally unsuccessful, means to accelerate growth of AFC and to improve the rate of success are needed. Here we show that addition to routine cultures of 5-10 % lymphokine-rich conditioned medium significantly increased the number of clones, the speed of growth and the rate of success of cultures. This could be demonstrated by counting clones and mitoses directly (in primary AFC-cultures) and by determining incorporation of 3H-TDR (in secondary cultures). The activity which we preliminarily term Amniotic Fluid Cell Growth Factor, (ACGF) is not due to contaminating mitogen and is very likely different from IL-2.

METHODS

The lymphokine-rich supernatant used (CM) was prepared as described in (1) (sup. of human MNL, stimulated for 24hrs by 10 μg Con A/ml following 3d. of preincubation. The medium throughout was RPMI 1640 c̄ 10 % FCS). Primary AFC cultures were performed in leighton tubes following routine culture techniques in Ham's FIO or RPMI 1640 c̄ 20 %

ISBN 0-12-527540-4

FCS, each individual culture starting from 2.5 ml of amniotic fluid (16-20 wks of gestation). Parallel cultures of individual samples were set up c̄ 10 % CM, or 10 % CM which was absorbed twice to human RBC to remove Con A (CMa) or control media as indicated. The time of culture is indicated in figures or text. For harvest colchicine was added to block cells in mitotic configurations, and the cells were washed and stained and mitoses counted. Secondary AFC cultures were obtained by trypsinisation and reculturing from primary cultures. Cells within 2-10 passages were used for the 3H-TDR incorporation assay: AFC were incubated in quadruplicates at $2x10^5$/ml in flat bottom microtiter plates in RPMI c̄ 10 % FCS ± CM or control fluid for 48 hrs with a 3H-TDR pulse for the final 24 hrs. For harvest, the cells were washed by means of a multiple automated harvester and eventual nonadherent cells collected on glassfibre paper. Then trypsin was added, incubated, and the cells washed again. Filter dots from both washes were collected and counted together by liquid scintillation counting.
IL-2 was measured on an IL-2 dependent T-cell line (M-2) as described in (2).

RESULTS

Growth of primary AFC Cultures.

Table 1 shows the effect of CM, of Con A absorbed CM (CMa; c̄ less than 1 µg Con A after absorption, as determined in an agglutination assay with Anti-Con A), of a Con A reconstituted control medium and of control medium containing only 10 % FCS, on rate of mitoses and number of clones. All parallel cultures were harvested at the same day, between d.9 and d.13 of culture. Cultures grown in CM-containing medium showed a significantly higher number of mitoses than control cultures. Removal of most of the Con A from CM (CMa) seems to have a positive effect, although this has to be claryfied further. In any case these results make unlikely that the observed enhancement was due to the contaminating mitogen. The number of clones/culture was increased as well by CM. We then investigated the kinetik of growth of primary AFC in presence or absence of CM. (Fig. 1). At d.9, seven of the CM-cultures showed more than 15 mitoses (and could be

Tab.1. EFFECT OF LYMPHOKINE-RICH SUPERNATANT ON RATE OF MITOSIS AND CLONES IN PRIMARY AMNIOTIC FLUID CELL CULTURES

	CM	CMa	CMtot	Con A	Ctr
N	12	9	22	13	22
Mitoses ($\bar{x} \pm$ SEM)	264 ±72	418 ±235	332 ±93	95 ±43	107 ±38
Clones ($\bar{x} \pm$ SEM)	6,2 ±0,9	9,9 ±3,0	7,3 ±1,1	4,6 ±1,3	4,5 ±0,9

All cultures originated from 2,5 ml amniotic fluid/culture. Highest and lowest value of each group rejected before calculation. CM = cultures $\bar{c}$ 10 % crude conditioned medium. CMa = cultures containing 10 % Con A-absorbed CM. CMtot = pooled results of CM and CMa.

successfully harvested for analysis) while in the controls only two had reached a comparable state. Seen the fact that, in standard technique, it is advisable to have 16-30 mitotic configurations for safe chromosome analysis, this means that with use of CM, 7 of 9 cultures could be harvested at d.9, while without it was only 2. We have now additional data (not presented) showing that the enhancement can be shown in fact considerably earlier than d.9.

Using conventional techniques it occasionally happens, that cultures of AFC show no growth at all or stagnate after some growth before a sufficient number of mitoses at a given time can be obtained. We have now 4 cases in which growth of stagnating cultures could be restimulated by later addition of CM, leading to successful harvest within 2-4 days after addition of CM.

So far we found no hint that the ACGF activity itself or any component of CM might influence chromosome configuration (e.g. induce aberrations). This remains however to be definitely proven.

Effect of Con A

The results obtained on primary cultures had suggested that the enhancement was not due to contaminating Con A. We further investigated this question using secondary cultures. AFC were set up with serial dilutions of crude

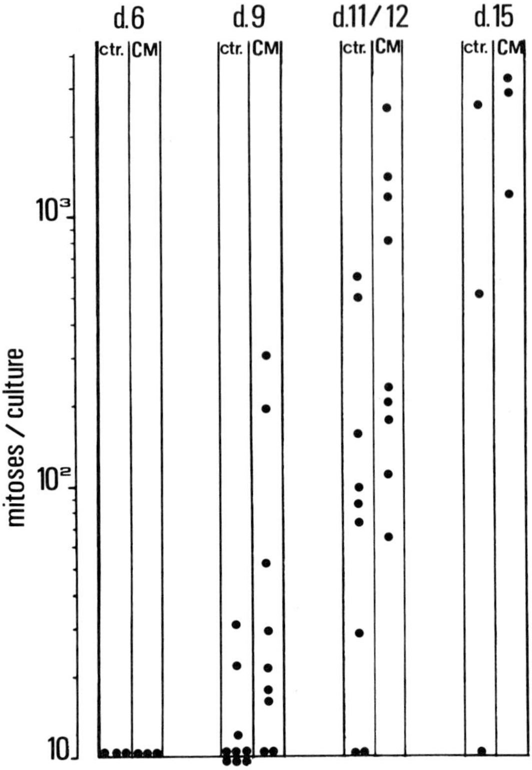

Fig. 1

supernatant (CM), Con A-reconstituted control supernatant or control medium ± α-methyl-D-mannoside (MAM) (Fig.2). Addition of MAM had a slight depressing effect; this was, however, observed in the controls as well. The data show that Con A itself was not responsible for the enhancement, and that the presence of Con A is very likely not necessary for the observed effects. Here it should be mentioned that ACGF-active sups could be obtained by stimulation with PHA as well.

The relation between ACGF and IL-2

As a first step to get some basic information about how ACGF relates to other lymphokines we measured units of IL-2 and titers of ACGF in 12 different crude lymphokine rich supernatants. (Tab. 2). Obviously supernatants containing high titers of IL-2 tend to contain high titers of ACGF, as well. We found, however, exceptions to this rule (sup. No 4). So far we do not know if this was due to the presence of an inhibitor or lack of activity. ACGF is, however, after all we can say, not identical with IL-2: a crude CM, containing high titers of both activities was purified by Sephadex G 100 column fractionation. The eluted peak of IL-2 showed no ACGF activity. So far we did not succeed in positively purifying ACGF. Further work is, however, in progress.

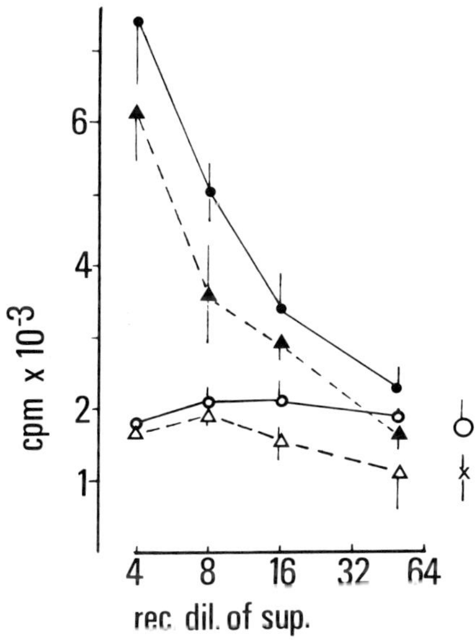

Fig.2.Effect of MAM 0,05 M.
—●— crude sup.(CM)
—○— ctr.sup.(Con A reconst.)
○ ctr.medium
-▲- crude sup. + MAM
-△- ctr.sup. + "
x ctr.medium + "

Stability of ACGF
When kept at 4 °C, ACGF is stable for at least 3 months, at 37 °C for at least 16 hrs. It is destroyed by heat (1 hr 56 °C) and partially stable (40 % loss) when exposed to PH2 for 16 hrs. The latter may indicate that ACGF is different from -interferon, which was also present in CM preparations used.

Species specificity
We also tested if lymphokine-rich CM from other species have ACGF activity. So far, Con A stimulated, IL-2 containing supernatants of mouse and rat spleen cells were tested and found to be completely negative when assayed on human primary or secondary AFC-cultures.

SUMMARY AND CONCLUSIONS

Lymphokine-rich supernatants of human origin contain an activity (ACGF) enhancing the growth of amniotic fluid cells. This could be shown with primary cultures by counting mitoses and with secondary cultures by measuring 3H-TDR incorporation. The activity found in crude CM is not due to contaminating mitogen and not identical with IL-2. Since ACGF is quite stable, it is a useful tool to accelerate and improve the diagnosis of chromosomal aberrations. The relationship to other lymphokines remains to be established. Apart from the practical use of ACGF, our results open the question which role lymphokines may play in normal and pathological fetogenesis. Interesting, to this respect, may be the old observation

Tab. 2. ACGF AND IL-2 IN SUPERNATANTS OF STIMULATED HUMAN MNL

SUP.NO	IL-2*	ACGF**
1	220	64
2	180	128
3	125	128
4	95	ø !
5	90	64
6	90	16
7	75	32
8	50	16
9	35	32
10	25	8
11	10	ø
12	ø	4
13	ø	ø
14***	ø	ø

* tested on IL-2 dependent mouse T-cell line M 2; activity expressed in units.
** tested on secondary amniotic fluid cells; activity expressed as reciprocal titer.
*** negative control (Con A reconstituted medium).

that offsprings of inbred mice are generally weaker and smaller at birth when compared to outbred newborn. It is thinkable that the outbred fetus profits from maternal lymphokines generated in the surrounding immunological reaction zone.

References:

1. H. Northoff, C. Carter, and J.J. Oppenheim, J. Immunol. 125; 125 (1980)
2. H. Northoff, J.J. Hooks, G. Jordan, and J.J. Oppenheim Behring Institut Mitteil. 67, 90-94 (1980)

This work was in part supported by the Deutsche Forschungsgemeinschaft, grant No 109/2-1.

Present address of first author: H. Northoff, DRK-Blutspendezentrale, Oberer Eselsberg 10, 7900 Ulm, FRG

CHAIRMEN'S COMMENTS

This session centered around the analysis of molecular identity of different lymphokine activities such as migration inhibitory factor (MIF), macrophage activating factor (MAF), and γ-interferon (IFN-γ). Further studies will require homogenous lymphokines in order to answer this question satisfactorily. Until this is accomplished, it will not be possible to determine whether different activities can be attributed to one or to different molecules.

Heinz Remold
S. Wahl

SECTION V:
REGULATION OF SUPPRESSION BY LYMPHOKINES

ASSOCIATIVE RECOGNITION IN THE GENERATION OF LYMPHOKINES WHICH INHIBIT THE TRANSFER OF CONTACT SENSITIVITY AND THE PRODUCTION OF IL-2: REQUIREMENT FOR CROSSLINKING OF ANTIGEN RECOGNITION SITES AND THE PRESENCE OF I-J SUBREGION DETERMINANTS

Geoffrey Asherson
Vittorio Colizzi[1]
Miroslav Malkovsky[1]
Marek Zembala[1]

Division of Immunological Medicine
Clinical Research Centre
Harrow, U.K.

There are three elements in the T suppressor circuit which blocks the expression stage of the contact sensitivity reaction to picryl chloride. There is the T suppressor cell which acts at the efferent stage, Ts-eff, which is probably analogous to Ts_3 and produces antigen-specific T suppressor factor (TsF). This arms the T acceptor cell (Tacc). Immunization is required to generate the Tacc but its specificity is entirely due to the TsF which arms it and the antigen used to produce the Tacc is immaterial. The Tacc armed with TsF liberates nonspecific inhibitor(s) when exposed to haptenized cells whose I-J genotype corresponds to the strain of mouse used to generate the TsF. The requirements for triggering the armed Tacc are 1) exposure to antigen which crosslinks separate molecules of TsF 2) exposure to an I-J subregion product which corresponds in genotype to the source of the TsF. The nonspecific inhibitors are detected by their ability to inhibit the transfer of contact sensitivity and the production of interleukin-2 (IL-2) by spleen cells stimulated with Concanavalin A (Con A).

[1]*This work was supported by grants from the Wellcome Trust and European Molecular Biology Organization (EMBO)*

ISBN 0-12-527540-4

The T Suppressor Circuit which Affects the Expression Stage of the Contact Sensitivity Reaction

Two distinct suppressor cell circuits influence the contact sensitivity skin reaction. There is the T suppressor cell (afferent), Ts-aff, which only acts when given at the time of immunization and prevents the production or action of antigen-specific T helper factor (1). It is distinct from the T suppressor cell which acts at the expression or effector stage (Ts-eff) and does not act through the Ts-eff (2).

The Ts-eff is induced by the injection of water-soluble, chemically reactive haptene intravenously (3,4), the injection of haptenized cells (5) and by painting the skin with high or conventional doses of contact sensitizer (6,7). The Ts-eff produces antigen-specific TsF. This factor does not block contact sensitivity directly but acts through an acceptor cell.

The T Acceptor Cell Produces an Inhibitor of IL-2 Production

An acceptor cell is defined as a cell which acquires its immunological specificity by virtue of a factor produced by another cell. The classical example is the mast cell. This lacks intrinsic immunological specificity but, after arming with IgE, liberates nonspecific factors such as histamine when exposed to antigen. Similarly the antigen-specific TsF arms the T acceptor cell. This then releases nonspecific inhibitor when exposed to antigen corresponding to the TsF (8). The unexpected finding is that exposure to an I-J subregion product of the major histocompatibility complex (MHC) corresponding to the source of the TsF is also required (9).

The nonspecific inhibitor produced by the Tacc can be assayed by its ability to block the production of IL-2 by lymphocytes stimulated with Con A. In one experiment T acceptor cells from BIO.A(3R) mice were armed with syngeneic TsF and triggered by picrylated cells of various genotypes. Table I shows that picrylated 3R cells, which are syngeneic with the source of the TsF (and the Tacc), and 4R cells which only share $I\text{-}B^b$ and $I\text{-}J^b$ trigger the release of inhibitor, while 5R cells which are syngeneic with 3R except at the I-J subregion are ineffective. It was concluded that the triggering of the release of the inhibitor of IL-2 production, like the triggering of the inhibitor of contact sensitivity, required matching in the I-J subregion between the picrylated cell used for triggering and the source of the TsF. The

TABLE 1. Inhibitor of Interleukin-2 Production Generated by T Acceptor Cells Armed with Picryl T Suppressor Factor when Exposed to Antigen and the Appropriate I-J Subregion Products

Production of nonspecific inhibitor				Testing of inhibitor by blocking of production of IL-2
Oxazolone Tacc	Pic TsF	Picrylated cell	MHC homology[a]	IL-2 assay
No IL-2	(negative control)			6982 ± 1238
3R	3R	3R	total	12890 ± 2894
3R	3R	--		55878 ± 4818
3R	--	3R	total	55503 ± 5479
3R	--	--		56213 ± 3223
--	--	3R		56836 ± 5165
3R	3R	4R	I-B I-J	14859 ± 2146
--	--	4R		56211 ± 4847
3R	3R	5R	total except I-J	54070 ± 5508
--	--	5R		55753 ± 3929

The BIO.A(3R) T acceptor cells were nylon wool purified lymph node and spleen cells taken 4 days after immunization with oxazolone. They were armed (2.5×10^7/ml) with picryl TsF. The release of nonspecific inhibitor was triggered by picrylated cells of various genotypes. The supernatant taken at 2 hours (1ml) was tested for its ability to inhibit the generation of IL-2.

Normal spleen cells were pulsed with the inhibitory supernatant (50%) and Con A (5μg/ml). After 4 hours the supernatant was removed and replaced with serum free medium. The supernatant after a further 20 hours was tested for its ability to support thymidine incorporation by 4 day Con A blast cells at a final concentration of 20% (15). The negative control was supernatant of cells incubated without Con A. The results are expressed as the mean and standard deviation of thymidine incorporation (cpm) in quadruplicate.

[a]MHC homology between the picrylated cell used as a source of antigen and the source of the TsF and the Tacc.

possibility that the essential matching was between the Tacc and the picrylated cell was excluded in other experiments (10). These results imply that the TsF has recognition sites for both antigen and I-J products and there is formal evidence that these recognition sites occur on one molecule or one molecular complex (9).

The nonspecific inhibitor can also be assessed by its ability to block Con A induced DNA synthesis when given at the beginning of culture. Using this assay it has a molecular weight on gel filtration around 50kD and a pI of 6.4-6.9.

Molecular Requirements for Triggering the Tacc

In the IgE mast cell system the essential requirement for triggering the release of histamine is crosslinking of two separate IgE molecules (11). Crosslinking of both sites of one molecule is not sufficient. In the present system the requirement for both antigen and MHC can be met by a picrylated cell or by antigen and MHC given separately. The antigen can be provided by haptenized albumin or bis-haptene-lysine in the picryl and oxazolone (4-ethoxymethylene-2-phenyloxazolone) TsF systems and in the NP (4-hydroxy-3-nitrophenylacetyl) TsF_3 system of Dorf (12). The requirement for the MHC can be met by an intact unhaptenized cell, cell membrane or KCl extract.

To investigate the requirements for crosslinking, bifunctional monovalent haptenes were prepared by substituting one amino group of lysine, a C_6 dibasic aminoacid, with "oxazolone" and the other amino group with NP. Table II shows a typical experiment in which Tacc armed with TsF_3 directed against the NP determinant were triggered by bis-NP-lysine together with a KCl extract of CBA cells as the source of the relevant MHC. Monovalent (NP-oxazolone-lysine) and the irrelevant haptene (bis-oxazolone-lysine) failed to trigger. It was concluded that bivalent but not monovalent haptene triggered the release of the nonspecific inhibitor which blocked the transfer of contact sensitivity.

Bis-NP-lysine may trigger the armed Tacc either by crosslinking recognition sites on one molecule of TsF or sites on two separate molecules. To distinguish between these possibilities, Tacc were armed with both NP TsF and oxazolone TsF. Table II shows that these cells were triggered by NP-oxazolone-lysine. This agent, being univalent for each of its two haptenic groups, cannot cross-link the two sites on one molecule of TsF but may crosslink a molecule of NP TsF with a molecule of oxazolone TsF. Other experiments indicated that the antigen may be given

TABLE II. Triggering of the release of nonspecific inhibitor from the armed T acceptor cell with simple chemical haptenes

Tacc	TsF	antigen	MHC	Contact sensitivity
Ox BALB	NP	NP,NP	CBA	4.2 ± 0.57
Ox BALB	NP	NP,Ox	CBA	6.5 ± 1.76
Ox BALB	NP	Ox,Ox	CBA	6.6 ± 1.34
Ox BALB	NP + Ox	NP,Ox	CBA	3.4 ± 0.96
Ox BALB	NP + Ox	NP,Ox	-	6.0 ± 1.55
Ox BALB	NP + Ox	--	CBA	6.0 ± 0.35
Ox BALB	NP + Pic	NP,Ox	CBA	6.5 ± 0.71
No cell transferred (negative control)				3.3 ± 0.76

The T acceptor cells were nylon wool T lymph node and spleen cells taken 4 days after immunization with oxazolone. They were armed (1.5×10^8) with 3 ml of monoclonal NP TsF_3 diluted in 3 ml medium or in 3 ml Ox TsF made in CBA mice. The cells were then exposed to bis-NP-lysine or α-NP-ε-oxazolone-lysine or bis-oxazolone-lysine (0.5mg/ml) together with KCl extract (5 ml) of 1.5×10^8 cells as a source of MHC. After 2 hours the supernatant was harvested and tested for its ability to inhibit the passive transfer of contact sensitivity. In practice 2.5×10^8 cells were incubated in 4 ml inhibitory supernatant for one hour and injected into group of 5 mice. They were tested immediately afterwards by painting the ears with 1% oxazolone and the increment of ear thickness at 24 hours measured in units of 10^{-3}cm ± standard deviation.

sequentially instead of simultaneously. It was inferred that there are two requirements for triggering
1) occupation of both the antigen and the MHC recognition sites of a given molecule of TsF
2) crosslinking of the antigen binding sites of two distinct molecules of TsF. It is not known whether there is also a requirement for the crosslinking of the MHC recognition sites.

These results are relevant to "associative recognition" in which the occurrence of an immune event requires recognition of both antigen and MHC. In the present system associative recognition is based on a single molecule or molecular complex which recognizes both antigen and MHC. However these two elements can be presented separately. This shows that linked recognition of antigen and MHC is not essential and suggests that these two elements are recognized by two distinct sites. In fact this is an example of a particular version of the two receptor hypothesis in which the receptors for antigen and MHC are part of the same molecule or molecular complex. The best evidence for this model in other systems comes from the findings of Lonai et al (13) and it may be applicable in some of the situations in which a one acceptor hypothesis has been evoked (14).

ACKNOWLEDGMENTS

We wish to thank Mrs. B. James and Mrs. M.C. Watkins who helped in the present work and Dr. M.E. Dorf for allowing us to report the results with TsF_3.

REFERENCES

1. Colizzi, V., (unpublished data).
2. Thomas, W. R., Watkins, M. C., and Asherson, G. L., *Immunology 127,* 217 (1981).
3. Zimber, C., Ben-Efraim, S., and Weiss, D. W., *Immunology 45,* 449 (1982).
4. Asherson, G. L., and Zembala, M., *Immunology 41,* 1005 (1980).
5. Miller, S. D., Sy, M. S., and Claman, H. N., *J. Immunol. 121,* 265 (1978).
6. Sy, M. S., Miller, S. D., and Claman, H. N., *J. Immunol. 119,* 240 (1977).
7. Zembala, M., Asherson, G. L., Colizzi, V., and Watkins, M. C., *Immunology 47,* In press (1982).
8. Asherson, G. L., and Zembala, M., *Ann. New York Acad. Sci. 392,* 71 (1982).
9. Zembala, M., Asherson, G. L., and Colizzi, V., *Nature 297,* 411 (1982).
10. Malkovsky, M., Asherson, G. L., Stockinger, B., and Watkins, M. C., Submitted for publication (1983).
11. Siraganian, R. P., Hook, W. A. and Levine, B. B., *Immunochemistry 12,* 149 (1975).
12. Okuda, K., Minami, M., Furusawa, S., and Dorf, M. E., *J. exp. Med. 154,* 1938 (1981).
13. Lonai, P., Bitton, S., Savelkoul, H. F., Puri, J., and Hammerling, G. J., *J. exp. Med. 154,* 1910 (1981).
14. Kappler, J. W., Skidmore, B., White, J., and Marrack, P., *J. exp. Med. 153,* 1198 (1981).
15. Granelli-Piperno, A., Vassalli, J. D., and Reich, E., *J. exp. Med. 154,* 422 (1981).

DISCUSSION

ALTMAN: Since one of Asherson's readout systems for the antigen-nonspecific suppressor material is the inhibition of a Con-A-induced proliferative response, can this inhibition be reversed by adding an exogenous TCGF source? In other words, does nsINH block TCGF production, or the ability to respond to TCGF?

ASHERSON: Yes, the non-specific inhibitor acts by inhibiting the production of IL-2. In keeping with this, it inhibits only DNA synthesis when given on the first day of culture; its effect is prevented by the addition of Con-A supernatant. It does not act by inducing an inhibitor which blocks the action of IL-2. This is shown by two observations: a) IL-2 can be isolated from Con-A supernatants by isoelectric focusing and then assayed. It is not found in supernatants of spleen cells pulsed with Con-A and non-specific inhibitor. b) This supernatant, lacking IL-2, behaves just like tissue culture medium.

SCHRADER: What type of cell presents antigen and IJ-coded products to the armed acceptor cell? Will any cloned lymphomas, etc. work in this system, thus demonstrating the effective suppression of IJ products on these cells? The important cells appear to be IJ^+ T cells.

ASHERSON: Formal experiments show that picrylated, purified T cells, which are I J^+, "present" antigen while the same number of I J^- cells are inactive. In fact, picryl albumin can also be used; after two washes by a KCl extract of cells of appropriate I J genotype. Under these circumstances it is likely that it is not the presenting cell that is critical, but rather that it is the antigen and the I J gene product interacting directly with the T-suppressor factor on the surface of the T-acceptor cell.

STADLER: Can Asherson abrogate the release of non-specific inhibitor from the T-acceptor cell armed with TSF, using monovalent haptens in order to prevent crosslinking?

ASHERSON: It is certainly the case that monovalent haptens will *not* release non-specific inhibitor, while the bivalent hapten, $\alpha\varepsilon$-*bis*-picryl-N, N-lysine will do so.

PURIFICATION AND CHARACTERIZATION OF A LYMPHOKINE PRODUCED BY FcRγ-STIMULATED B CELLS

Tohru Masuda
Muneo Miyama-Inaba[1]
Takayo Suzuki[1,2]

Department of Immunobiology[1]
Institute for Immunology
and
The First Department of Internal Medicine[2]
Faculty of Medicine
Kyoto University

I. INTRODUCTION

The biological functions of Fc receptor (FcR) on T and B lymphocytes are not yet completely understood, although many observations are in agreement that FcR may play a role in the regulation of immune responses. Enhancing and/or inhibiting effects of antibody molecules on immune responses depend in part on the Fc portion of the molecules; suggesting the involvement of FcR in these effects (1).

We have reported that B cells bearing FcR for IgG (FcRγ) release an antigen-nonspecific suppressor factor, termed SBF (suppressive B cell factor) when stimulated by an immune-complex (2, 3). In addition, we have also demonstrated that expression of FcR on T cells is regulated by the corresponding isotype and that FcR bearing T cells isotype-specifically regulate immunoglobulin production by releasing factor(s) which specifically bind to the corresponding isotype (4, 5). These findings suggest that there may be FcR-dependent regulatory mechanism in immune responses.

[1,2] *Present adress: Kyoto University, Sakyo, Kyoto, Japan.*

Interleukins, Lymphokines, and Cytokines

ISBN 0-12-527540-4

The point of the experiments presented here was focussed on the regulatory mechanism of immune responses obtained by B cells. Biological and biochemical properties of SBF, prepared from splenic FcRγ(+) B cells and hybridoma cells producing SBF, are described.

II. MATERIALS AND METHODS

The details of these experiments have been published elsewhere (2, 3, 5). Only a brief description of techniques follows.

A. Preparation of SBF.

Ten million FcRγ(+) B cells separated from normal spleen cells of either BALB/c or C3H/He mice by the Ficoll/Isopaque centrifugation method after rosetting with sheep red blood cells (SRBC) sensitized with IgG class of rabbit or mouse anti-SRBC antibody (EAγ), were suspended in RPMI 1640 medium and cultured at 37°C in 5% CO_2. Twenty four hours later, the supernatant were collected and used as SBF. Cells to be tested were mixed with an equal volume of SBF at the concentration of $5X10^7$/ml at 4°C for 1 hr.

B. SBF-producing hybridoma.

FcRγ(+) B cells from normal nu/nu spleens (BALB/c) were fused with HAT-sensitive 3T3-4E cells (TK^-) using PEG-1000, immediately after the preparation of EAγ-rosetting cells. Before the fusion, EAγ or free erythrocytes were removed by the treatment of cells with hemolytic Gey's solution. A clone which has the highest activity as SBF, TS-4.44, was selected by the limiting dilution methods, out of 18 clones (6).

III. RESULTS AND DISCUSSION

SBF, obtained from FcRγ(+) B cells of normal BALB/c or C3H/He mice, suppresses in the antigen-nonspecific manner, the production of antibody especially belonging to IgM and/or IgG class (2, 3). As can be seen in TABLE I, the treatment of DNP-Asc primed FcRγ(-) splenic B cells with SBF at 4°C for 1 hr, results in 66% suppression of IgG anti-DNP PFC response in adoptive secondary responses, in comparison with the response

TABLE I. Target Cells for SBF in anti-DNP response

Exp.	DNP-primed	SBF	KLH-primed	SBF	% Suppression of Ind. PFC
1.	FcRγ(-) B 1x10^7	-	spl. T 3x10^6	-	0[1]
		+		-	66
		-		+	14 (n.s)
	DNP-primed	SBF	absorbed with[3]		
2.	spleen cells 1x10^7	-	-		0[2]
		+	-		54
		+	Thymus		54
		+	Bone Marrow		60
		+	Splenic T		58
		+	Splenic B		0

1) 350,000 IgG PFC. 2) 76,000 IgG PFC.
3) 1 ml of SBF, absorbed with 10^8 cells at 4°C for 1 hr.

of SBF-untreated FcRγ(-) B cells reconstituted with KLH primed helper T cells (350,000 IgG PFC). In contrast, no significant suppression is attained by the treatment of KLH primed T cells with SBF; indicating that SBF somehow impaires the function of precursor B, but not helper T cells. Absorption tests of SBF with various kind of cells further confirm this possibility that the target of SBF is B cells; only splenic B, but not thymocytes, bone marrow cells, nor splenic T cells, completely eliminate the suppressive activity of SBF. In mitogenic response of murine spleen cells, SBF also dose-dependently suppresses only the response against B cell mitogens, such as LPS, but not Con A nor PHA. Thus, 50% suppression of LPS-response by $5x10^5$ normal spleen cells in 96 well microculture plates is used for the evaluation of SBF.

Since SBF suppresses B cell responses to both antigen and mitogen, the effect on the growth of tumor cells, including B and non-B cells was examined. As shown in TABLE II, the proliferation of tumor cells of the B cell lineage is drastically suppressed by coculturing with 50% of 1:10 diluted SBF, obtained from H-2 matched mice. In contrast, T-origin thymoma or lymphoma cells and fibroblasts (3T3-A31) are not affected by SBF at all. Furthermore, SBF inhibits the growth of L-1210 cells at the G_1-S junction of the cell cycle; implying the cell-cycle dependent appearance of receptor for SBF in L-1210 cells. These findings suggest that the suppression of either immune or mitogen responses by SBF is perhaps due to the inhibition of the proliferation of one of B cell subsets.

TABLE II. Growth Inhibition of Various Tumor Cell Line Cells In Vitro by SBF.

Cell	*Type*	*Strain*	*H-2(SBF)*	*% Inhibition (Day 3)*
L-1210	*B*	*DBA/2*	*d*	*95 %*
DL-4	*B*	*DBA/2*	*d*	*60*
DL-5	*B*	*DBA/2*	*d*	*80*
MLA	*B*	*DBA/2*	*d*	*85*
X-5563	*B**	*C3H/He*	*k*	*90*
MOPC-31c	*B**	*BALB/c*	*d*	*100*
MOPC-315	*B**	*BALB/c*	*d*	*90*
NS-1	*B**	*BALB/c*	*d*	*100*
X63.653	*B**	*BALB/c*	*d*	*95*
L-1210	*B*	*DBA/2*	*k*	*0*
X-5563	*B**	*C3H/He*	*d*	*0*
WC-2	*T**	*AKR*	*k*	*0*
DL-1	*T*	*DBA/2*	*d*	*0*
DL-3	*T*	*DBA/2*	*d*	*0*
DL-8	*T*	*DBA/2*	*d*	*20*
3T3-A31	*non-T,B*	*BALB/c*	*d*	*0*

B: myeloma, T*: thymoma.*

SBF obtained from hybridoma cells, TS-4.44, provides almost the same biological properties as these conventional SBF. Biochemical properties of SBF derived from the hibridoma and conventional FcRγ(+) B cells are summerized in TABLE III. By gel chromatography using Sephacryl S-200, the molecular weight of SBF was evaluated as 43,000 daltons, because the active component suppressing LPS response was eluted in the fraction coresponding to ovalbumin used as one of the markers. After concentration, this fraction was further fractionated by DEAE chromatography. The activity was found in the fraction eluted between 0.1 and 0.2 mol. NaCl (pH 7.2) as one peak, but SDS-PAGE demonstrated one main band with MW of 43K and two faint bands with MW less than 43K.

SBF, thus fractionated, is resistant to RNase and DNase, while sensitive to both protease and neuraminidase, suggesting the possible involvement of sialic acid in the active site of SBF which is glycoprotein. However, only a portion of SBF molecules binds to Con A, implying that SBF may not be high-mannose type of molecules. Furthermore, it is noted that SBF

TABLE III. Biochemical Properties of SBF.

1. *MW: 43,000 (determined by Sephacryl S-200)*
2. *Sensitivity to enzymes:*
   Sensitive to Protease *and* neuraminidase
   Resistant to RNase and DNase.
3. *Able to bind partially to Con A.*
4. *Ia antigen coded by I-C subregion gene.*
   (H-2 restriction between SBF and targets)
5. *Unabsorbable with immune complex, heat-aggregated, rabbit anti-mouse IgG, and mouse IgG.*

binds to neither mouse IgG and rabbit anti-mouse IgG, nor heat-aggregated HGG; indicating that SBF is quite different from either FcR or IgG itself and also from IBF, reported by Fridman et al. (7). SDS-PAGE of immune-precipitate between SBF from internally labeled TS-4.44 with ^{3}H-leucine and relevant antiserum to H-2 haplotype (anti-H-2^d) demonstrated that MW of main peak corresponds to 43K and that no such a peak is observed in the precipitate between the same SBF and H-2 irrelevant antiserum (anti-H-2^k). In the case of the latter, the peak of 43K appeared in the supernatant, but not in the precipitate. Thus, this result is well accordant with that obtained by gel chromatography.

Another characteristic of SBF is the presence of H-2 antigen on the molecules. Therefore, there is a H-2 restriction between the target and SBF. By the absorption test using various kind of anti-H-2 alloantisera raised between congeneic mice, it has been proven that SBF bears an antigen coded by I-C subregion gene.

From these observations we would like to conclude that there is a immunoregulatory cirkit operated by FcR-network and that immunological homeostasis, such as the constant level of immunoglobulins in blood, would be maintained by receiving isotype-specific regulation for immunoglobulin production from FcR bearing lymphocytes or lymphokines synthesized by them. SBF seems to be one of these immunoregulatory lymphokines. In the case of FcR on T cells, it has been shown by Dr. Fridman that FcRγ(+) T cells release a factor which suppresses antigen-nonspecifically IgM/IgG antibody production (7). This factor is named IBF, because of specific binding to IgG. Isotype-related immunoregulatory factor involved in IgE responses, has been also reported to be produced by FcRε T cells in rat by

Ishizaka and his colleague (8). Moreover, we have recently found that IgA responses is also class-specifically regulated by a factor released from FcRα T cells (5).

In the case of B cells, we have not yet direct evidences proving isotype-specific regulation by SBF, except that SBF obtained from FcRγ(+) B cells does not inhibit IgE response, and that FcRγ(+) B cells still possess precursor activity for IgE forming cells even if coupled with EAγ, while they lose the precursor activity for IgG/IgM forming cells by binding EAγ (9). These results suggest that the isotype specific regulation by FcR bearing B cells is likely. Each class of monoclonal antibodies against SRBC, DNP etc seems to be a useful tool to prove the pertinence of this hypothesis. The works are in progress.

ACKNOWLEDGMENTS

This work was supported by Grant in Aid of the Ministry of Education of Japan and by the technical efforts of Mayumi Kyoizumi, Mihoko Koike and Junko Hatazawa.

REFERENCES

1. Theofilopoulos, A. N., and Dixon, F. J., *in* "Advances in Immunology, Vol. 28" (F. J. Dixon, and H. G. Kunkel, ed.), p. 89. Academic Press, New York, (1979).
2. Masuda, T., Miyama, M., Kuribayashi, K., Yodoi, J., Takabayashi, A., and Kyoizumi, S., Cell. Immunol. 39, 238 (1978).
3. Miyama, M., Yamada, J., and Masuda, T., Cell. Immunol. 44, 51 (1979).
4. Yodoi, J., Adachi, M., and Masuda, T., J. Immunol. 123, 888 (1982).
5. Yodoi, J., Adachi, M., Teshigawara, K., Masuda, T., and Fridman, W. H., submitted.
6. Suzuki, T., and Masuda, T., submitted.
7. Fridman, W. H., Rabourdin-Combe, C., Neauport-Sautes, C., and Gisler, R. H., Immunol. Review 56, 51 (1981).
8. Yodoi, J., and Ishizaka, K., J. Immunol. 124, 1322 (1980).
9. Miyama-Inaba, M., Masuda, T., Inada, T., Takatsu, K., and Hamaoka, T., Int. Arch. Allergy Appl. Immunology 66, 282 (1981).

DISCUSSION

YOSHIDA: How pure is Masuda's B cell population in producing SBF? Is there any evidence of contamination by macrophages?

MASUDA: By using the Ficoll/Isopaque technique after EA-rosetting, we usually obtain the $FcR_\gamma^{(+)}$ B cell fractions with 60% purity; repeating the same procedure for this fraction, increases purity to ~80%. However, sedimentation at 1XG yields almost 100% pure FcR_γ B cells. In any case, all fractions thus prepared, produce a factor possessing the same function. The likelihood that contamination with non-B cells, especially macrophages, might lead to production of SBF, is ruled out by the finding that Sephadex G-10-passed nu/nu spleen cells effectively yields SBF, whereas peritoneal macrophages fail to produce SBF, after stimulating FcR_γ with EA.

IFN-γ MEDIATED SUPPRESSION OF EPSTEIN-BARR VIRUS INDUCED ACTIVATION OF B LYMPHOCYTES BY OKT4$^+$ T LYMPHOCYTES

RONALD PALACIOS
OTONIEL MARTINEZ-MAZA

Department of Immunobiology
Wallenberg Laboratory
Karolinska Institute
Stockholm - Sweden

INTRODUCTION

The immune responses to Epstein-Barr virus (EBV) include the following: a. antiviral neutralizing antibodies that prevent the spread of free virus. The neutralizing antibodies persist lifelong, as does a small fraction of EBV-infected B cells (1); b. generation of cytotoxic T cells which do not persist beyond the acute phase of the disease (2); and c. development of suppressor T cells that limit the outgrowth of EBV-infected cells (3,4). EBV is known to activate B cells to both DNA synthesis and Ig production (5). Such activation can be suppressed by T cells from either adults or neonates. Nevertheless, cord blood B cells are not susceptible to this T cell mediated regulation (6,7). The characteristics of the cell and the mechanism of escape of B cells from neonates have not been defined.

This study was undertaken to define the phenotype of the T cell that suppresses EBV-induced activation of B cells and the mechanism by which this function is exerted.

MATERIALS AND METHODS

A. Separation of Cells

Human mononuclear cells (MNC), T cells and B cells were isolated from either the peripheral blood of healthy adults or from the cord blood of full term neonates as described previously(8,9). Purified OKT4$^+$ (>92% OKT4$^+$ and < 1% OKT8$^+$) and OKT8$^+$ (>90% OKT8 and <3% OKT4) T lymphocytes were obtained by treatment of total T cells with either OKT8 or OKT4 monoclonal antibodies plus complement, as described previously(8,9).

Interleukins, Lymphokines, and Cytokines

ISBN 0-12-527540-4

B. Culture Medium

Cells were washed three times in BSS and resuspended in RPMI-1640 medium supplemented with glutamine,gentamicin, and sodium bicarbonate buffer solution (culture medium,CM).Culture medium supplemented with 10% heat inactivated human AB serum will be referred to as complete culture medium (CCM).

C. IFN-α, IFN-γ, and anti-IFN-γ antibody

IFN-α and IFN-γ preparations were the kind gift of Dr. Kari Cantell and were obtained as previously described (10) Purified rabbit (IgG) anti-human IFN-γ antibody (kind gift of Dr. Marc De Ley) was used in this study. The production,specificity and characteristics of this antiserum have been described (11).This antibody does not crossreact with IFN-α or IFN-β (11).

D. Cell Culture Experimental Assays

B lymphocytes ($5x10^5$) from adults or neonates, suspended in 1 ml of CM, were placed in 24 well Limbro tissue culture plates. Two different concentrations ($2x10^6$ and $5x10^6$) of the following types of autologous or allogeneic T cells from adults or neonates were added to the cultures: $OKT4^+$ cells,$OKT8^+$ cells or total T cells. The cultures were exposed to cell-free supernatants (B95-8 cultures) containing EBV and incubated at 37°C for 12 h. After this, 1 ml of CCM was added to each well.In some experiments purified IgG anti-IFN-γ antibody (neutralizing titer for 16 IU,32IU and 48 IU of IFN-γ) or rabbit anti-human IgA serum (final dilution 1:100)(control) were added at the start of the cultures established as described above.The cultures were incubated at 37°C for 6 days and the number of Ig secreting cells was determined by the protein-A reverse hemo lytic plaque assay (12).

E. Detection of IFN-γ secreting cells by a protein-A hemolytic plaque assay

This was performed essentially as described earlier (13). Peripheral blood MNC ($3x10^6$ cells) were incubated in CM containing 50 ng/ml of the OKT3 antibody in 12x75 mm round bottom plastic tubes at 37°C for 24 h. The cells were washed three times and resuspended in 1 ml of BSS. Protein-A coupled SRBC were obtained by exposing one part washed and packed SRBC to one part staphylococcal protein-A (Pharmacia,Uppsala,Sweden) (0.5 mg/ml in 10 parts of CrCl3). Rabbit anti-IFN-γ antibody (final dilution 1:130) was used as developing antiserum.Guinea pig serum absorbed with SRBC was the source of cpmplement. IFN-γ plaque forming cells (IFN-γ PFC) were counted with a four fold magnification lens; the results were expressed as IFN-γ PFC per $2x10^6$ initially cultured cells. In some experiments, MNC were treated with OKT3,OKT4,OKT8 or OKIa monoclonal antibodies (kind gift Dr. Gideon Goldstein,Ortho Pharmaceutical Co.,Raritan New Jersey) and rabbit complement as described (13). The cells were washed and the number of IFN-γ PFC was determined by the protein-A plaque assay as described (13, see also above).

RESULTS

A. Suppression of EBV-stimulated Ig production is mediated by OKT4$^+$ T lymphocytes

The results depicted in Table I show that OKT4$^+$ T cells from either adults or neonates strongly suppressed EBV-induced Ig production of B cells from adults but not from neonates.In contrast, OKT8$^+$ T cells from adults or neonates did not inhibit Ig secretion by EBV-activated B cells from either adults or neonates.Lack of suppressor activity of OKT8$^+$ T cells were observed at the two distinct cell concentrations studied.We conclude that OKT4$^+$ but not OKT8$^+$ T cells exerted suppression on EBV-induced Ig production of adult B lymphocytes and that EBV-infected cord blood B cells were not susceptible to the suppressor activity of OKT4$^+$ T cells.

TABLE I. OKT4$^+$ T lymphocytes suppress EBV-induced Ig production

effector cells		target cells for suppressor activity adult B	 neonatal B
		PFC/10^6 cells	
ADULT			
medium		3740	2150
total T	$2x10^6$	785	2000
	$5x10^6$	720	1905
OKT4$^+$ T	$2x10^6$	820	1850
	$5x10^6$	740	1910
OKT8$^+$ T	$2x10^6$	3690	1900
	$5x10^6$	3700	1815
NEONATE			
medium		1320	1510
total T	$2x10^6$	550	1470
	$5x10^6$	430	1445
OKT4$^+$ T	$2x10^6$	640	1520
	$5x10^6$	510	1400
OKT8$^+$ T	$2x10^6$	1500	1450
	$5x10^6$	1420	1495

B. Anti-IFN-γ antibody abrogates suppression on EBV-infected B lymphocytes by OKT4$^+$ T lymphocytes

Table II shows that the addition of anti-IFN-γ antibody to EBV-stimulated adult B cells at the start of the cultures (antiserum with capacity to neutralize 16 IU of IFN-γ or higher concentrations) eliminated the suppressor activity on Ig production exerted by OKT4$^+$ T cells from either adults or neonates. In contrast, treatment of similar cultures with control rabbit antiserum (anti-human IgA) did not affect the suppressor activity of OKT4$^+$ T cells (Table II). Nevertheless, the addition of anti-IFN-γ antibody to cultures of EBV-stimulated B cells cocultured with OKT8$^+$ T cells had no effect on the production of Ig (Table II), eliminating the possibility that the anti-IFN-γ antibody abrogated suppressor activity of OKT4$^+$ lymphocytes on EBV-induced Ig production by acting directly on B cells. In addition, these findings together with previous studies showing that the anti-IFN-γ antibody is specific for IFN-γ (11,13,see below), ruled out the possibility that this antiserum acted nonspecifically. Thus, it can be concluded that the anti-IFN-γ antibody interfered with the mechanism by which OKT4$^+$ lymphocytes exert suppression on EBV-induced activation of B lymphocytes.

TABLE II. Anti-IFN-γ antibody abrogates suppressor activity on EBV-stimulated B cells by OKT4$^+$ T lymphocytes

effector cells		anti-IFN-γ antibody neutralizing effect	Ig synthesis adult B PFC/10^6	Cord B cells
ADULT	medium	-	560	350
	OKT4$^+$T	-	70	420
		16 IU	118	-
		32 IU	480	-
		48 IU	730	400
		anti-human IgA(control)	50	320
	OKT8$^+$T	-	540	375
		48 IU	571	358
NEONATE	medium	-	1300	350
	OKT4$^+$T	-	200	300
		16 IU	664	-
		32 IU	985	-
		48 IU	1250	330
	OKT8$^+$T	-	1285	410
		48 IU	1360	380

C. Characterization of the cell secreting IFN-γ as studied at the single cell level

Recently, we have described an indirect hemolytic plaque assay for the quantitation of IFN-γ secreting cells (13). In this study we have attempted to define the phenotype pf the circulating cells which produce IFN-γ after stimulation with the OKT3 monoclonal antibody using the protein-A plaque assay as indicator system. TableIIIoutlines the results of a representative experiment. As can be seen, treatment of MNC with OKT3 or OKT4 antibodies plus complement significantly decreased the number of IFN-γ plaque forming cells (IFN-γ PFC). In contrast, treatment of MNC with OKT8 or anti-HLA-DR antibodies and C did not decrease the number of IFN-γ PFC as compared to those observed in control cultures. Furthermore, the results depicted in Table 3 show that the addition of IFN-γ to the plaque assay effectively blocked the development of IFN-γ PFC, whereas the addition of IFN-δ to this assay did not. These results point out that the plaque assay was specifically detecting IFN-γ secreting cells, and further document that the anti-IFN-γ antibody is highly specific for IFN-γ (11,13). Thus, the circulating cell secreting IFN-γ after mitogen stimulation is an OKT3$^+$,4$^+$,8$^-$,DR$^-$ T lymphocyte, as determined at the single cell level. These findings are in keeping with the results presented before showing that OKT4$^+$ lymphocytes suppress EBV-induced activation of B cells and that this suppressor activity was abrogated by the anti-IFN-γantibody. In addition, they agree with the recent report of Wang et al (14), who found that OKT4$^+$ T cells and monocytes were necessary for optimal production of IFN-γ(assessed by conventional procedures) after OKT3 antibody stimulation. Thus, we conclude that OKT4$^+$ but not OKT8$^+$ T cells secrete IFN-γ.

TABLE III.Detection of IFN-γ secreting cells at the single cell level

cell assay	treatment	IFN-γ PFC/2x10^6 cells
MNC + OKT3	medium	6560
	C	6520
	OKT3 +C	315
	OKT4 +C	300
	OKT8 +C	6450
	anti-HLA-DR + C	6510
	IFN-& 500 IU	6550
	1000 IU	6475
	IFN-γ 150 IU	815
	300 IU	405

CONCLUSIONS

The results of the present study have demonstarted that OKT4$^+$ T lymphocytes from either adults or neonates are the cells which exert suppression of the EBV-induced activation of adult B lymphocytes.This suppressive activity seems to be mediated via IFN-γ since: a. a highly specific anti-IFN-γ antibody abrogated the inhibitory activity of OKT4$^+$ T cells on EBV-infected cells;b. OKT4$^+$ but not OKT8$^+$ T cells produce IFN-γ as determined at the single cell level; and c. we previosuly showed that IFN-γ prevented B cells from adults from being activated by EBV (10).

EBV-infected cord blood B cells were not susceptible to the suppressor activity of OKT4$^+$ T cells from either adults or neonates.These findings, together with our previous observations (10) and those of Thorley-Lawson (7) that neither IFN-γ nor IFN-& prevent cord blood B cells from being activated by EBV, suggest that EBV-induced activation of cord blood B cells is not suppressed by OKT4$^+$ T lymphocytes due to the insensitivity of B cells from neonates to IFN-γ.

REFERENCES

1. Hewetson JF, Roche G, Henle W, and Henle G. J.Infect.Dis. 128:283,1973.
2. Sedmyr E, and Jondal M. Proc.Natl.Acad.Sci.USA.72:1622,1975
3. Thorley-Lawson DA,Chess L, and Strominger J. J.Exp.Med. 146:495,1977.
4. Haynes BF, et al. J.Immunol.123:2095,1979.
5. Bird G, and Brittom S. Immunol.Rev.45:41,1979.
6. Andersson U,Bird G, Brittom S, and Palacios R. Immunol.Rev. 57:5,1981.
7. Thorley-Lawson DA. J.Immunol. 126:829,1981.
8. Palacios R, et al. Immunology 42:127,1981.
9. Palacios R. Immunogenetics 14:309,1981.
10. Palacios R, Andersson U, and Brittom S.J.Clin.Invest(submitted).
11. De Ley M. et al. J. Virol.Methods.3:149,1981.
12. Gronowicz E, Coutinho A, and Melchers F.Eur.J.Immunol.6:588 1976.
13. Palacios R, Martinez-Máza O, and De Ley M. Submitted for publication.
14. Wang TW. et al. J.Immunol.128:585,1982.

DISCUSSION

ORSON: What is Palacios explanation for the difference between his observations detecting suppression at 5-7 days, and those of others -- who only found suppression much later, at 14 days?

PALACIOS: The query involves two different systems. The system I have described acts very early and it is related to the effect of IFNγ on the mechanism by which EBV gets into the genome of the cell. In this sense, IFNγ will prevent EBV from activating B cells. The effects in the system described by Tosato et al. occur only after 8-14 days of culture. It seems to me that both the mechanism and the cell that exerts suppression in their system, are likely to be distinct from those operative in the system I described. Moreover, in the delayed type of suppression it is not excluded that it could be due to killing of EBV-infected cells.

SCHRADER: The argument used to establish the specificity of the plaque assay seems circular, in the sense that the rabbit anti-IFNγ (although undoubtedly specific for IFNγ versus IFNα or β) could also contain antibodies to contaminants that could well be present in the IFNγ preparation utilized to block plaque formation.

PALACIOS: There *is* a problem unless one has pure IFNγ. However, there is documented evidence that this particular anti-IFNγ antibody is highly specific for IFNγ (see references in our report in this volume). Furthermore, the blocking experiments were performed with different preparations of IFNγ than those used in the induction of anti-IFNγ antibody. Therefore, the possibility raised by Schrader is unlikely to be a factor.

REEM: Has Palacios conducted similar experiments with B cells not infected with EB virus?

PALACIOS: No, we have not. We have however stimulated IFNγ synthesis in human T lymphocytes and in lectin-activated thymocytes with B lymphoblastoid cells. This IFNγ synthesis was abrogated by cyclosporin A.

PURIFICATION OF RADIOLABELED SOLUBLE IMMUNE RESPONSE SUPPRESSOR[1]

Thomas M. Aune[2]
Carl W. Pierce

Department of Pathology and Laboratory Medicine
The Jewish Hospital of St. Louis
Department of Pathology
Washington University School of Medicine
St. Louis, MO

David R. Webb

Roche Institute of Molecular Biology
Nutley, NJ

Soluble immune response suppressor (SIRS) is produced by concanavalin A-activated murine T lymphocytes. After oxidation by macrophages in a H_2O_2-dependent process, SIRS suppresses immune responses and inhibits cell proliferation by normal and neoplastic cells. A T cell hybridoma producing SIRS was constructed to obtain sufficient material for purification and characterization. Hybridoma-derived SIRS, internally labeled with ^3H-leucine ($\sim 10^8$ cpm of non-dialyzable radioactivity and $\sim 10^6$ units of activity), was purified to radiochemical homogeneity by a combination of molecular sieve chromatography and high performance liquid chromatography on RP-18 and diphenyl columns. SIRS was eluted from an RP-18 column with 20% propanol and from a diphenyl column with 30%

[1]*Supported in part by grants RR-05491 and AI-15353 from the NIH and grant PCM-8119606 from the NSF.*
[2]*Recipient of a Junior Faculty Research Award from the American Cancer Society.*

ISBN 0-12-527540-4

propanol. This final purification step yielded one peak of radioactivity which co-migrated with SIRS-bioactivity. SDS-PAGE revealed a single radioactive band with an Mr of 15,000 daltons. This purification procedure resulted in a 5000-fold enrichment of SIRS which contained about 50 units SIRS/cpm ^{3}H-leucine.

INTRODUCTION

Soluble immune response suppressor (SIRS), a product of interferon (IFNβ) or concanavalin A-activated Ly 2^+ T cells, suppresses immune responses and the division of both normal and neoplastic cells (1-7). Considerable evidence suggests that SIRS must be activated by H_2O_2 produced by macrophages before it can inhibit cell division and immune responses (8). Further, activated SIRS ($SIRS_{ox}$) is inactivated by sulfhydryl compounds, reducing reagents and a variety of peroxidase substrates (5-8). Since these compounds can serve as electron donors, it is possible that $SIRS_{ox}$ acts as an oxidizing agent and that $SIRS_{ox}$ inhibits cell division by oxidizing certain cellular components necessary for division. This notion is further supported by the observation that $SIRS_{ox}$ mediated inhibition can be reversed by high concentrations of 2-mercaptoethanol or dithiothreitol under appropriate conditions (6).

To obtain sufficient quantities of SIRS for purification and biochemical analyses, con A-activated spleen cells were fused with the AKR thymoma BW5147 and a hybridoma producing SIRS was cloned in soft agar. Supernatant fluids from this cell line 393D2.6 contain 10-50 fold greater SIRS than conventional sources and produces SIRS without stimulation and in the absence of serum. By all criteria available, hybridoma and conventional SIRS appear identical (9,10).

Hybridoma-derived SIRS has been radiolabeled with ^{3}H-leucine; in this report we describe the purification of SIRS to apparent homogeneity by a combination of gel filtration and high performance liquid chromatography.

MATERIALS AND METHODS

Mice

C57BL10 mice were bred in the Animal Resources Facility at Jewish Hospital and used when 2-4 mo. of age.

SIRS Assay

Single cell suspensions of spleen were incubated under modified Mishell-Dutton conditions and stimulated with sheep erythrocytes for 4-5 days. IgM plaque-forming cell (PFC) responses using SRBC as indicator cells were assayed with the slide modification of the Jerne hemolytic plaque assay. One unit of SIRS activity is defined as that amount necessary to suppress antibody responses by 50% when added at culture initiation (11).

Preparation and Purification of Radiolabeled SIRS

The SIRS-producing T cell hybridoma, 393D2.6, was grown in RPMI 1640 media supplemented with antibiotics and 10% horse serum to a concentration of 1-2 X 10^6 cells/ml, centrifuged and resuspended in MEM without leucine or serum, but with 100 μCi/ml ^{3}H-leucine (137 Ci/mmole, Amersham) at a density of 2 X 10^6 cells/ml in a final volume of 50 ml. Cells were cultured for 3 days and supernatant fluids were harvested by centrifugation, dialyzed against water and lyophilized.

Lyophilized SIRS (10^8 cpm ^{3}H-leucine) was suspended in 0.4 M pyridine, 0.4 M acetic acid, pH 5.0 (0.4/0.4 M Pyr-HAc buffer) and applied to a Sephadex G-50 column (3 X 45 cm) equilibrated in Pyr-HAc buffer. Fractions of 4 ml were collected and tested for bioactivity after appropriate dilution in medium. Active fractions were pooled and lyophilized.

SIRS fractions were suspended in 1 M pyridine, 0.5 M acetic acid, pH 5.5 (1/0.5 M Pyr-HAc buffer) and applied to either a Lichrosorb RP-18 (10μ or 5μ) column or a Si-100 diphenyl (10μ) column equilibrated in 1 M Pyr, 0.5 M HAc buffer; protein was eluted with an increasing stepwise n-propanol gradient. Fractions from HPLC were lyophilized and tested individually for SIRS bioactivity. SIRS containing fractions were pooled and re-chromatographed on HPLC as described in the text.

RESULTS

^{3}H-leucine labeled SIRS (50 ml) from the T cell hybridoma was prepared as described above. SIRS was dialyzed against water, lyophilized and resuspended in 5 ml of (0.4/0.4 M) Pyr-HAc buffer and fractionated by gel filtration on Sephadex G-50 (Fig. 1). In this high ionic strength buffer, SIRS migrated with an apparent Mr of 10,000-20,000. In low ionic strength buffer (150 m0), hybridoma-derived SIRS migrates with Mr > 150,000 and between 45,000-55,000. Fractions with SIRS

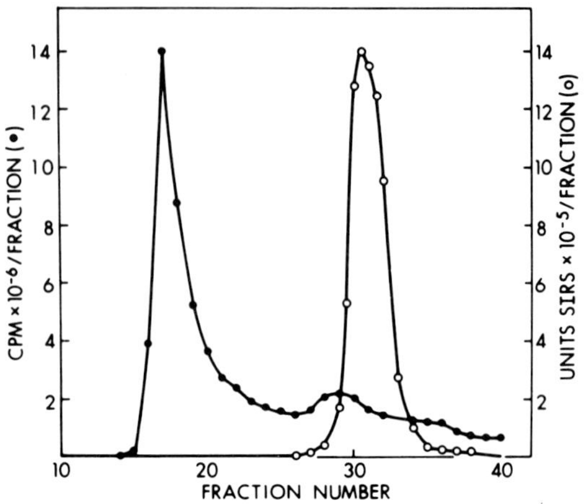

FIGURE 1. Fractionation of SIRS by gel filtration. Fractions of 4 ml were collected and assayed for radioactivity (●) or SIRS activity (o).

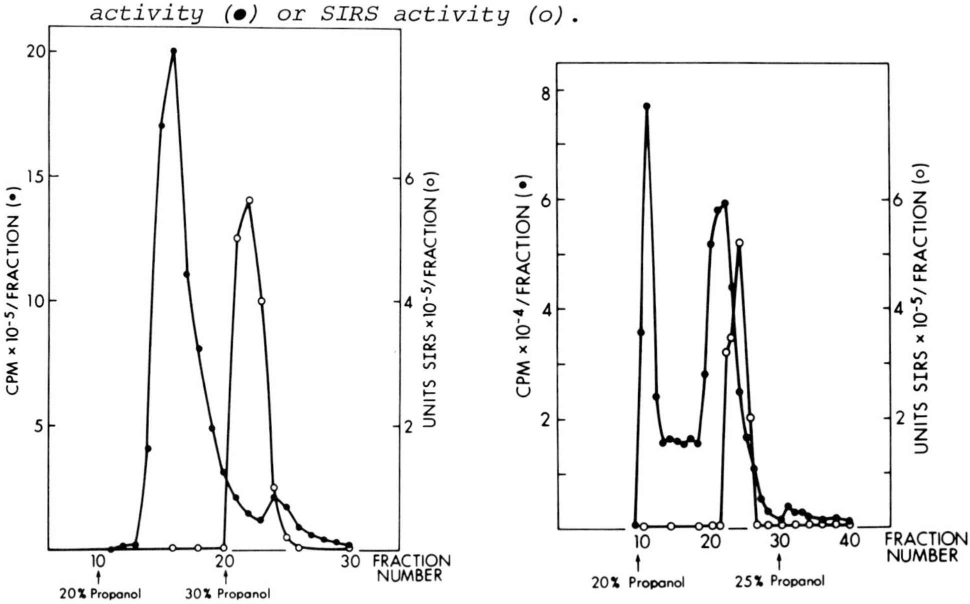

FIGURE 2a. Fractionation of SIRS by reverse phase HPLC. SIRS-rich fractions from figure 1 were applied to a Lichrosorb RP-18 (10μ) column in 1.0 M pyridine, 0.5 M acetic acid, pH 5.5 and eluted with a stepwise propanol gradient (flow rate: 1 ml/min). Fractions of 2 ml were collected and assayed for radioactivity (●) or SIRS activity (o). b. SIRS containing fractions from 2a were pooled and chromatographed on a RP-18 column eftsoon with a flow rate of 1 ml/min.

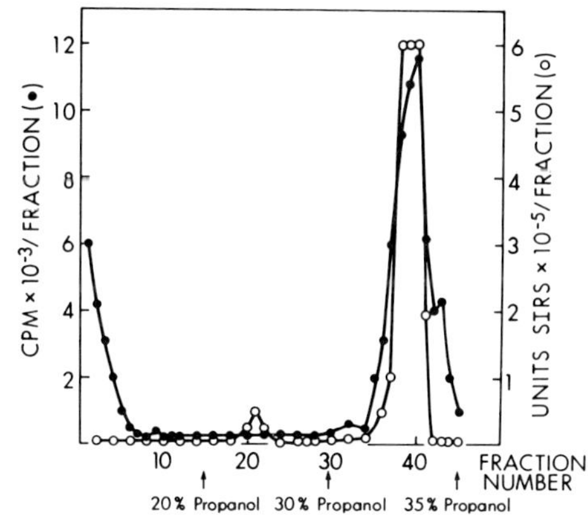

FIGURE 3. Fractionation of SIRS on a Si-100 diphenyl HPLC column. SIRS fractions from figure 2a were applied to a Si-100 diphenyl column and eluted with a stepwise propanol gradient. Individual fractions of 1 ml were tested for radioactivity (●) or SIRS activity (o).

activity were pooled, lyophilized and resuspended in (1/0.5 M) Pyr-HAc buffer and further purified by HPLC on a RP-18 column (4.6 X 250mm). SIRS activity was eluted with a step-wise gradient of propanol at a flow rate of 30 ml/hr. Figure 2a shows that SIRS activity eluted at 20% propanol between the two major peaks of radioactivity. Fractions 21-23 were pooled and re-fractionated on an RP-18 column at a flow rate of 13 ml/hr; SIRS activity was detected in fractions 22-25 between 25-30% propanol (Fig. 2b). Fractions with SIRS activity were pooled and applied to a diphenyl column (4 X 250 mm) in (1/0.5 M) Pyr-HAc buffer and eluted with a stepwise propanol gradient. Only one major peak of radioactivity was detected and this peak also contained all of the SIRS activity (Fig. 3). Analysis of these fractions by SDS-PAGE showed a single peak of radioactivity with Mr of approximately 15,000 daltons.

Table I summarizes purification of radiolabeled SIRS. T cell hybridoma supernatant fluids contained 1 X 10^8 cpm of acid precipitable radioactivity and 1 X 10^6 units of SIRS activity. Following the gel filtration step and three purification steps by HPLC total cpm were reduced to 4 X 10^4 while SIRS activity increased to 2 X 10^6 units. This represents a purification of 5000-fold and a SIRS specific activity of 50 units/cpm. Radiochemically purified SIRS has properties identical to either partially purified hybridoma-derived SIRS or SIRS in culture supernatant fluids. Fractions 38-40 from

Table I. Summary of SIRS Purification

	cpm	SIRS[a] (units)	Purification
Culture supernatant fluids (50 ml)	1×10^8	1×10^6	1
Sephadex G-50	1×10^7	1×10^6	10
RP-18 I	1.5×10^5	1.7×10^6	1133
RP-18 II	7×10^4	1.4×10^6	2000
Diphenyl column	4×10^4	2×10^6	5000

[a] *SIRS activity was assayed by suppression of PFC responses by spleen cells to sheep erythrocytes. One unit of activity is defined as the reciprocal of the dilution which yields 50% suppression when added on day 0.*

the diphenyl HPLC purification step could be converted to $SIRS_{ox}$ by H_2O_2 and suppression of PFC responses by fractions 38-40 was blocked by 2-mercaptoethanol or catalase (data not shown).

REFERENCES

1. Pierce, C.W. and Kapp, J.A., *Contemp. Topics Immunobiol. 5,* 91 (1976).
2. Rich, R.R. and Pierce, C.W., *J. Exp. Med. 137,* 649 (1973).
3. Tadakuma, T. and Pierce, C.W., *J. Immunol. 117,* 967 (1976).
4. Tadakuma, T., Kuhner, A.L., Rich, R.R., David, J.R., and Pierce, C.W., *J. Immunol. 117,* 323 (1976).
5. Aune, T.M. and Pierce, C.W., *J. Immunol. 127,* 1828 (1981).
6. Aune, T.M. and Pierce, C.W., *J. Immunol. 127,* 368 (1981).
7. Aune, T.M. and Pierce, C.W., *Proc. Natl. Acad. Sci. U.S.A. 79,* 3808 (1982).
8. Aune, T.M. and Pierce, C.W., *Proc. Natl. Acad. Sci. U.S.A. 78,* 5099 (1981)
9. Aune, T.M. and Pierce, C.W., *in* "Monoclonal Antibodies and T-Cell Hybridomas" (G.J. Hammerling, U. Hammerling and J.F. Kearney,eds.), p. 516. Elsevier,North Holland, (1981).
10. Aune, T.M. and Pierce, C.W., *in* "Isolation, Characterization and Utilization of T Lymphocyte Clone" (G. Fathman and F. Fitch. eds.) in press, (1982).
11. Aune, T.M. and Pierce, C.W., *J. Immunol. Methods.,* in press (1982).

DISCUSSION

ASHERSON: After activation by macrophages, is Aune's SIRS preparation bactericidal? Does he know the nature of the modification to SIRS produced by H_2O_2? Is an epoxide or peroxide produced?

AUNE: The bactericidal activity of SIRS-treated macrophages has not been tested. Certain circumstantial evidence would suggest that SIRS contains metal ions which are necessary for SIRS activation by H_2O_2 and subsequent SIRS activity. Presumably H_2O_2 would bring such metals to a higher oxidative state, thereby activating SIRS.

PICK: Did Aune try to purify $SIRS_{OX}$ (the SIRS processed by macrophages) by passing the macrophage-derived suppressor factor through the same purification procedure which he had used for the purification of the hybridoma-derived SIRS?

AUNE: Macrophage-processed SIRS loses all activity on HPLC; we cannot use this purification schema for $SIRS_{OX}$.

PAETKAU: Can Aune estimate the molar concentration of SIRS required to evoke an effect? If he can detect 1 unit (=50% inhibition), and the specific activity is 5×10^{13} units/mg, the quantity of SIRS involved must be incredibly low. My own calculations indicate that 1 unit/ml would correspond to 10^{-18}M SIRS. This comes out to 600 molecules/ml, or *less* than one SIRS molecule per cell! Accordingly, there must be something strange about his calculation of specific activity.

AUNE: Suppression by SIRS does not dilute out in a linear fashion: it stays at ~50% over a range of several logs. This suggests that SIRS may act catalytically (other data also support this notion). Consequently, calculations based on SIRS molecules/cell may not be interpreted in the same manner as typical hormone-receptor type binding studies.

ANTIGEN-SPECIFIC AND NON-SPECIFIC REGULATION OF LYMPHOKINE PRODUCTION[1]

Takeshi Yoshida
Matsunobu Suko
Stanley Cohen

Department of Pathology
University of Connecticut Health Center
Farmington, Connecticut

Regulatory mechanisms that control both the production of lymphokines and the expression of lymphokine activity have not yet been elucidated, although a feed-back type regulatory pathway has been found to represent at least one of them (1,2). We will discuss the results indicating that there are both antigen non-specific and antigen-specific down regulation of lymphokine production in vivo.

Two in vivo systems have been used to study such regulatory mechanisms: desensitization of delayed hypersensitivity reactions, and granuloma formation. In a desensitization model, groups of doubly immunized animals were prepared by injecting dinitrophenylated bovine serum albumin (DNP-BSA) and dinitrophenylated egg albumin (DNP-EA) with complete Freund's adjuvant (CFA). Seven days later, these animals were desensitized with egg albumin (EA) and skin tested with DNP-EA, DNP-BSA, PPD, and a lymphokine preparation, skin reactive factor, (SRF). For the first 24 hours of desensitization, skin reactivities to all antigens as well as SRF were suppressed. While the reactivity to SRF recovered to normal after 24 hours of desensitization, skin reactions to specific and non-specific antigens (DNP-EA, DNP-BSA and PPD) remained suppressed up to 4 days, when unresponsiveness to non-specific antigens (PPD and DNP-BSA) was over. Reactivity to specific antigen (DNP-EA) remained suppressed as late as 7 days after desensitization.

[1] *This work is supported by NIH Grant AI-16706*

Interleukins, Lymphokines, and Cytokines

ISBN 0-12-527540-4

If we confine our discussion to antigen non-specific suppression of delayed hypersensitivity, the suppression was found to be mediated by circulating lymphokines (1). The suppressive effects of the circulating lymphokines on expression of cutaneous delayed hypersensitivity reaction are caused by many factors: loss of mediator gradient between blood and local tissues, pre-emption of effector inflammatory cells, impaired function of such effector cells, and the induction of suppressive factor(s) which can inhibit the production of lymphokines. Thus, we could find not only the decrease in number of circulating monocytes and the suppressed chemotactic function of monocytes/macrophages but also the serum factor which could inhibit the production of lymphokines (MIF and chemotactic factor(s)) by immune lymphocytes. Such a suppressive factor(s) was present in the circulation for 1 ∿ 4 days after desensitization.

The temporal relationship, the appearance of the serum suppressive factor subsequent to the appearance of circulating lymphokines, has prompted us to speculate that the former may be generated from the target cells affected by the latter. Various studies have indicated that macrophages treated _in vitro_ with lymphokines could produce suppressive factors including PGE_2 to inhibit the production of lymphokines by immune lymphocytes (2). The present studies have revealed that macrophages from the desensitized animals could release a factor to inhibit MIF production. Thus, peritoneal macrophages obtained from the animals desensitized 48 hours previously were cultured in RPMI-1640 medium supplemented with normal guinea pig serum for 24 hours. The supernatant separated from these cultures was shown to contain a factor which could suppress the MIF production when it was added into the culture of immune lymphocytes stimulated with specific antigen. Macrophages from either immune or normal animals did not have significant effect on the MIF production by immune lymphocytes. More directly, macrophages were treated _in vivo_ by intraperitoneal injection of lymphokine supernatants. Such macrophages and/or the culture supernatant from these cells were found suppressive of _in vitro_ lymphokine (MIF) production. Vigorous _in vitro_ cell separation studies confirmed that the cells responsible for production of this suppressor factor are an adherent, nonspecific esterase staining macrophage population.

Since _in vitro_ studies have indicated that metabolites of arachidonic acid, particularly PGE_2, may be the major factor responsible for suppression of lymphokine production, we have examined whether or not the treatment by indomethacin _in vivo_ can abrogate the appearance of the serum suppressive factor. This was found to be the case. Serum from the desensitized guinea pigs, if treated with daily injections of 5 mg of indomethacin for 5 days, could not inhibit the production of MIF by immune lymphocytes whereas the sera from the desensitized, but not indometahcin-treated animals contained the suppressive factor as discussed above.

The evidence supporting the importance of lymphokine-activated macrophages in the generation of the serum suppressive factor was also obtained from our recent studies on another *in vivo* model, namely, granuloma formation induced nonimmunologically by muramyl dipeptide (MDP). The serum obtained from guinea pigs bearing pulmonary granulomas induced by the injection of MDP-coated agarose beads was shown to contain a suppressive factor to lymphokine production by immune lymphocytes. These animals were also shown to be partially suppressed of their skin reactivity to phytohemagglutinin (PHA), suggesting that the circulating suppressive factor may be responsible for such an anergic state in granulomatous inflammation. In analogy to the desnsitization model, we have speculated that the activated macrophages in the granulomatous lesion may be the source of such a serum inhibitory factor. It was found that macrophages incubated with MDP *in vitro* were found to release a suppressive factor to lymphokine production. However, a more direct study to eamine macrophages from granulomaous lesions is necessary to confirm our hypothesis.

Finally, again in the desensitization model, it is important to clarify the mechanism by which the suppression of antigen specific delayed hypersensitivity skin reaction lasts longer than that of antigen-nonspecific reaction. Although we could not find any serum factor which is nonspecifically suppressive of *in vitro* MIF production, we have obtained the preliminary result which indicates the presence of antigen-specific suppresoor lymphocytes in the circulation. When the T-cell rich population of guinea pig peripheral blood lymphocytes, collected 5 days after desensitization, was added to immune lymphocytes, the production of MIF upon stimulation with the specific antigen was reduced significantly. Further studies have revealed that these suppressor lymphocytes could release an antigen-specific factor which is able to suppress the MIF production . These findings seem to be consistent with the fact that the third stage of desnsitization (suppression of cutaneous DTH reaction to specific antigen) continues to be present for at least one week.

The present results together with our previous observations, therefore, strongly suggest that there exist at least two separate mechanisms to suppress the lymphokine production by lymphocytes stimulated with antigens, namely, antigen-non specific suppression by the factors released from lymphokine-activated macrophages and antigen-specific suppression by suppressor lymphocytes (probably T cells) or by their soluble products.

REFERENCES

1. Yoshida, T. and Cohen, S. *Fed. Proc.*, *41*, 2480 (1982).
2. Yoshida, T., Suko, M., Baba, T. and Cohen, S. in " Proceedings of The Second Naito International Symposium on Self-Defense Mechanisms: Role of Macrophages", (D.Mizuno and N.Ishida, ed.), Tokyo Univ. Press, in press.

DISCUSSION

ASHERSON: Does Yoshida suggest that the entire phenomenon of unresponsiveness in desensitization is mediated by circulating lymphokines?

YOSHIDA: We believe that at least the antigen-nonspecific suppression of skin reactivity is mediated directly or indirectly by circulating lymphokines. This is brough about by their direct action on effector cells (e.g. macrophages) and also by a suppressor factor generated from macrophages stimulated by circulating lymphokines.

WOLSTENCROFT: Normally, when guinea pigs are immunized with protein antigens in Freund's complete adjuvant and subsequently challenged by the intravenous route, the animals die of anaphylaxis, which is, I suppose, one form of non-specific anergy! How did Yoshida avoid this situation in his studies?

YOSHIDA: In our work, anaphylactic reactions or antibody-mediated reactions, have been avoided by injecting intravenously only a carrier-protein, (e.g., BSA) into guinea pigs immunized with hapten-carrier conjugates (e.g., DNP-BSA) in Freund's complete adjuvant. Intravenous challenge was made as early as 7 days after immunization, at a time when the antibody level is still rather low. Furthermore, in some experiments, antigen-antibody complexes (e.g., diphtheria toxoid and antitoxin) were used to immunize guinea pigs in order to minimize (or essentially nullify) the production of antibody during the first week. In this system, antigen alone (e.g., diphtheria toxoid) was injected intravenously. Even more convincingly, however, "passive" desensitization was achieved by injecting lymphokines into immune animals as discussed in my report. Since the lymphokine preparations do not contain the antigen utilized for immunization of recipient guinea pigs, there is no threat of anaphylactic or other antibody-mediated reactions being involved. Thus, at least the first two stages of desensitization can be analyzed by the use of this system (injection of exogenous lymphokines).

ASHERSON: Is the desensitization that Yoshida observes affected by adult thymectomy or cyclophosphamide? My reason for asking is that in the mouse, desensitization cannot be demonstrated *(in vitro)* after adult thymectomy or cyclophosphamide treatment. The basic system is to incubate cells which transfer contact sensitivity to picryl chloride by means of picrylated cells *in vitro* for one hour, and then to inject these cells intravenously. The recipient's ears are then immediately painted with picryl chloride in order to assess contact sensitivity. The desensitization is brought about by a cyclophosphamide-resistant, adult thymectomy-sensitive, antigen-specific, T suppressor cell and an antigen-nonspecific T acceptor cell which is sensitive to both adult thymectomy and cyclophosphamide (Zembala et al., Immunology, 1982 in press). It is likely but not proved that macrophages may also act as acceptor cells in this system).

YOSHIDA: We have not yet tried adult thymectomy nor cyclophosphamide treatment. However, it is likely that the third stage of desensitization, namely the antigen-specific phase of the anergic state, may be affected by the treatment Asherson suggests, since this stage of desensitization is mediated by antigen-specific suppressor T cells.

CHAIRMEN'S COMMENTS

The control of the immune response proceeds from specific factors, usually made by T cells, to the secondary production of nonspecific factors which may be stimulatory or inhibitory. It is well known that nonspecific factors may be made by T cell, macrophages and mast cells. The session provided evidence that B cells also make nonspecific suppressor factors.

Speakers discussing antigen specific T cell factors drew attention to the paradox that these factors carry Ia determinants and may be genetically restricted by similar or identical determinants. The properties of an antigen specific T suppressor factor was used to provide a model of associative recognition in which the antigen recognition site and the major histocompatibility product (I-J) recognition site were distinct but part of the same molecule. This finding favoured the double receptor model of associative recognition. In this system, as in the IgE mast cell system, an antigen-specific factor armed an intermediate or *acceptor* cell which then produced nonspecific inhibitors of contact sensitivity and interleukin 2 when exposed to antigen. This acceptor cell system provided biological amplification and it is an interesting question whether analogous acceptor cell systems are involved in the production of other nonspecific regulatory molecules.

The question of the modification of nonspecific regulatory molecules after secretion is usually discussed in terms of the production of the active factor from a pro-factor by enzymatic attack. However an interesting situation in which soluble immune response suppressor (SIRS) effectively acts as a carrier of "activated oxygen" was described which involved modification of a T cell product by a macrophage.

The participants also discussed phenomena in the whole animal. Desensitization was originally described by Koch in patients given large doses of old tuberculin. It refers to the loss of delayed hypersensitivity which occurs when immune animals are injected with antigen. It is due in part to the production of a factor by macrophages which inhibits delayed hypersensitivity. This is an example of the specific turning on of a nonspecific suppression and may be analogous to the T suppressor circuits described by Dorf and others. Desensitization *in vivo* draws attention to the general point that failure of the production of a particular lymphokine may be due to a suppressor circuit and not to direct loss of the cell which produces the lymphokine. An example of such a suppressor circuit is the nonspecific inhibitor produced by the T acceptor cell armed with antigen-specific T suppressor factor and then exposed to antigen. This factor then suppresses IL-2 production. Similarly the macrophage, when armed with antigen specific T suppressor factor, and then exposed to antigen liberates a nonspecific inhibitor of the transfer of contact sensitivity.

Some scientists study lymphokines because they believe that with increasing knowledge the systems will become simple and the effects will be attributed to a few key lymphokines. Other scientists are motivated by the though that there are many lymphokines to be discovered. The evidence of this session suggests that Occam's razor does not apply in the immune system and that in a defense system a more complex mechanisms is *a priori* more likely to be correct than a simple mechanism. This may be a requirement of any sophisticated defense system which needs to incorporate a high degree of redundancy.

Geoffrey Asherson
Takeshi Yoshida

SECTION VI:
SOURCES, ACTIVITIES, AND CHARACTERISTICS OF INTERLEUKIN-1 AND ENDOGENOUS PYROGEN

INTERLEUKIN-1 PRODUCTION BY A HUMAN MONOCYTIC LEUKEMIA CELL LINE, THP-1[1]

Steven B. Mizel
Barbara J. Andersen

Microbiology Program
Pennsylvania State University
University Park, Pennsylvania

I. INTRODUCTION

Within the last few years, substantial progress has been made in the chemical and biologic characterization of the macrophage-derived mediator, interleukin 1 (IL 1) (1). Recently, murine IL 1 was purified to apparent homogeneity and a partial amino acid analysis was obtained for one of the charge species of IL 1, termed IL 1α (2). From a biologic point of view IL 1 appears to be an essential participant in the initial activation of T lymphocytes as well as an important regulatory mediator of inflammatory responses (1,3). The production and chemical characterization of murine IL 1 have been facilitated by the availability of murine macrophage cell lines, such as $P388D_1$ that produce relatively high levels of IL 1. Interleukin 1-producing human macrophage cell lines have not previously been described. In this report, we present the results of our studies on the IL 1 produced by a human monocytic leukemia cell line, termed THP-1, which was originally described by Tsuchiya and co-workers (4). The THP-1 cell line was derived from the blood of a 1 year old male with acute monocytic leukemia. These cells possess F_c and C3b receptors, produce lysozyme and non-specific esterase, phagocytize latex beads as well as sensitized sheep red blood

[1]*Supported by USPHS Grant AI 17559.*

ISBN 0-12-527540-4

cells, and are able to restore the concanavalin A responsiveness of human T lymphocytes (4). This latter activity was the obvious stimulus for our interest in this cell line as a possible source of human IL 1.

II. METHODS

THP-1 cells were kindly provided by Dr. Shigeru Tsuchiya (Department of Pediatrics, Tohoku University School of Medicine, Sendai 980, Japan). The cells were maintained in stationary suspension culture in RPMI 1640 containing 5% fetal calf serum (FCS), $5x10^{-5}$M 2-mercaptoethanol, and 50 μg/ml gentamicin sulfate. For IL 1 induction or morphologic examination, the cells were harvested, centrifuged, and resuspended in medium with or without FCS at a concentration of $1x10^{6}$ cells/ml (IL 1 induction) or $2x10^{5}$/ml (morphologic examination). Stimulants [mezerein or lipopolysaccharide (LPS)] were added at the indicated times. All chromatographic and electrophoretic procedures were done at 4-7°C.

III. RESULTS

The murine macrophage cell line, $P388D_1$ can be superinduced to produce relatively high levels of IL 1 (2). We therefore examined the possibility that THP-1 cells could also be superinduced to synthesize IL 1. The results from several experiments inducted that THP-1 cells cannot be superinduced to make IL 1. Our attempts to induce the cells with LPS also met with failure. However, we did notice that the normally round, non-adherent THP-1 cells would rapidly attach and spread out on the tissue culture plates following the addition of mezerein, but not LPS. Within 2 hrs after addition of 10^{-8}M mezerein, the THP-1 cells began to adhere and spread out on the plastic surface. By 24 hrs the cells were tightly adherent and extremely flat. We then asked if these morphologically transformed cells would be capable of producing IL 1. Our results indicated that THP-1 cells could be induced to release IL 1 if they were incubated with 10^{-8}M mezerein for 24 hrs, washed to remove unbound mezerein, and then incubated with 10 μg/ml LPS for 6-48 hrs. However, the IL 1 activity in the culture fluid was relatively low and it appeared as if an inhibitor of IL 1 action might also be present. The conclusion that an inhibitor might be presented was based on the observation that the IL 1 activity of the THP-1 culture fluids exhibited a bell-shaped curve and did not reach the level of thymocyte proliferation obtained with the optimal concentration of the IL 1 standard.

Nonetheless, we prepared large-scale cultures of the THP-1 cells, stimulated the cells with mezerein followed by LPS, and then chromatographed the concentrated culture fluids obtained after different times on a 2.5 x 90 cm AcA54 column. As shown in Figure 1, the major species of IL 1 present in the

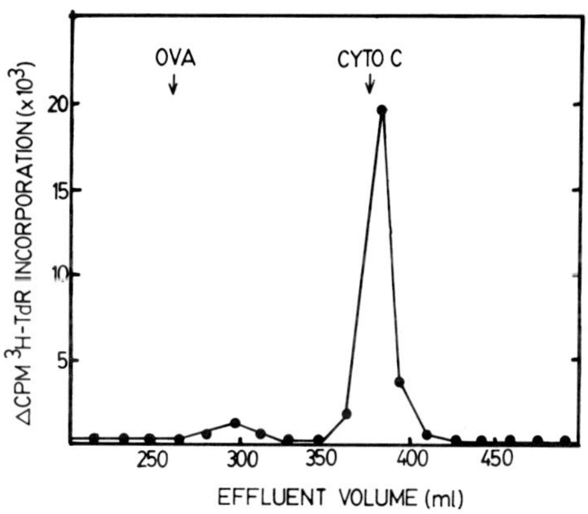

FIGURE 1. Ultrogel AcA54 chromatography of THP-1 derived IL 1.

culture fluid 6 hr post-LPS addition exhibited a molecular weight of 13,200, a value that is quite similar to the molecular weight of normal human macrophage-derived IL 1 (5). A small peak of IL 1 activity was also observed to elute with an apparent molecular weight of 32,000. In contrast to these findings, the elution profile of IL 1 activity in 24 hr post-LPS culture fluids exhibited a number of different peaks with apparent molecular weight of approximately 52,000, 28,000, 18,000, 13,000 and 8,500 (data not shown). The basis for this molecular weight heterogeneity is not known. Of interest is our observation that a very potent inhibitor of thymocyte proliferation eluted with an apparent molecular weight of approximately 22,000 (data not shown). The presence of such an inhibitor is consistent with the results obtained with unfractionated culture fluid.

When the 13,200 molecular weight IL 1 was examined by isoelectrofocusing in a granulated gel, we found that this IL 1 exhibited considerable charge heterogeneity (Fig. 2).

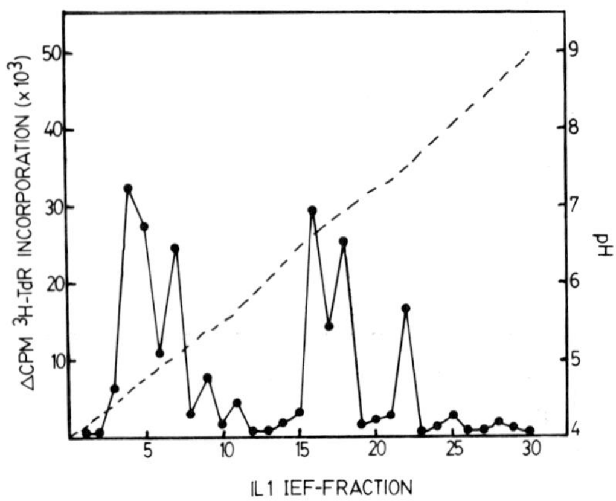

FIGURE 2. Isoelectrofocusing of THP-1 derived IL 1.

Peaks of IL 1 activity were associated with pI values of 4.7-5.1, 6.7-7.0, and 7.5. Titration analysis of each peak revealed that the major charge species of IL 1 was in the pI 6.7-7.0 fraction. Interestingly, the maximal activity of the pI 4.7-5.1 material was significantly lower than that obtained with the pI 6.7-7.0 IL 1. It should be mentioned that normal human macrophage-derived IL 1 shows a very similar pattern of pI species (5).

IV. DISCUSSION

The data presented in this communication demonstrate that a human monocytic leukemia cell line, THP-1, can be converted from a non-adherent IL-1-non-producing cell to an adherent IL 1 producing cell using mezerein and LPS. From the standpoint of studying human macrophage differentiation, the THP-1 cell should prove to be a very useful cellular system. In a recent publication that appeared during the course of our studies, Tsuchiya et al (6) reported that 12-0 tetradecanocylphorbol-13-acetate (TPA) also induced a morphological change in THP-1 cells as well as an increase in phagocytic activity. In terms of IL 1 production the THP-1 line has proven to be somewhat disappointing; the levels of IL 1 being rather low--even after removal of the inhibitory activity. Nonetheless, this cell line may prove of value in the study of IL 1 production during human macrophage differentiation.

REFERENCES

1. Mizel, S.B. Immunol. Rev. 63, 51 (1982).
2. Mizel, S.B., and Mizel, D. J. Immunol. 126, 834 (1981).
3. Mizel, S.B., Dayer, J., Krane, S.M., and Mergenhagen, S.E., Proc. Natl. Acad. Sci. (USA) 78, 2474 (1981).
4. Tsuchiya, S., Yamabe, M., Yamaguchi, Y., Kobayashi, Y., Konno, T., and Tada, K. Int. J. Cancer 26, 171 (1980).
5. Schmidt, J.A., Mizel, S.B., Cohen, D., and Green, I., J. Immunol. 128, 2177 (1982).
6. Tsuchiya, S., Kobayashi, Y., Goto, Y., Okumura, M., Nakae, S., Konno, T., and Tada, K. Cancer Res. 42, 1530 (1982).

DISCUSSION

PAULY: Has Mizel tested phorbol esters other than Mezerin for IL-1 production? Is there any change in IL-1 production when the THP-1 cells are raised in spinner cultures rather than in static cultures in which the cells are allowed to adhere to the polystyrene surface?

MIZEL: TPA (phorbol-myristate actetate) exerts the same effects as Mezerin. The inactive molecular species of these tumor promoters do not induce IL-1 production. Our experience with the human line as well as with the murine macrophage line, $P388D_1$, is that when one puts these cells into suspension culture they cease to make much IL-1. I do not know why this is so, but it may well be that a number of interactions with the surface are also sending signals to the cell interior.

SALVIN: Do the IL-1 fractions of different MW have different degrees of biological activity? I ask this as the MW Mizel reports varies from 8,000 to 52,000, with the main peak at 13,000.

MIZEL: All these species, because we see them via the assay, have that activity. My own feeling relates back to some experiments done a number of years ago with Rosenstreich, in which we found a number of MW species of IL-1 *inside* the cells, i.e. not released. These species dissociate at 37°. My prediction is that this probably dissociates and it is really the smallest MW species that is biologically active. One sees the higher ones because one takes it from that range of the column, but I believe that it is the dissociated species that is really active.

KURNICK: Has Mizel looked at the effect of T lymphocyte products on THP-1? With Amento and Krane I have utilized a T cell product (from lectin-stimulated human T cell clones) to stimulate a monocyte cell line (U937) to become adherent and produce a monokine (MCF) which expresses activity on rheumatoid synovial cells. This MCF activity is also expressed on thymocytes in an IL-1 assay, provided it is first separated, by column chromatography, from a higher moelcular weight-inhibitory component.

MIZEL: We have been very interested in the situation described by Kurnick. We have not, to this point, tried this experimental approach, but I think that is where his line might prove quite interesting; looking at the product of T cells in terms of changes that might occur in these cells.

BARTOGLIO: What assurances can Mizel offer that the THP-1 cell line is mycoplasma free?

MIZEL: These cells are routinely tested and have proved negative in every instance where they were examined.

PLACENTAL MACROPHAGES AS A SOURCE OF INTERLEUKIN-1[1]

Arthur Flynn
James H. Finke
Mary L. Hilfiker

Department of Molecular and Cellular Biology
Cleveland Clinic Foundation
Cleveland, Ohio

Activated macrophages produce a number of immunologically important factors (1). Interleukin-1 (IL-1), a 12,000 to 18,000 dalton glycoprotein is produced by macrophages and is responsible for a number of T and B lymphocyte actions (2). Further study of human IL-1, both immunologically and biochemically, requires a source of cells which produces sufficient quantities of macrophage culture supernatants.

Human placenta contains approximately 65 percent mononuclear phagocytes (3). The majority of cells released from tissue rich in chorionic villi had C3 and Fc receptors, morphologically resembled mononuclear phagocytes and phagocytized opsonized erthrocytes. Wood (4) subsequently reported that large amounts of IgG are bound to Fc receptors on mononuclear phagocytes in the placenta, suggesting a mechanism for the removal of incompatible antibodies. Since in pregnancy many events mimic the effects of IL-1 (5), we questioned whether the macrophages found in placenta were producing or could produce IL-1.

Placenta from C57Bl/6 mice, which were taken two days prior to term, and human placenta collected after normal delivery were tested to determine macrophage IL-1 activity. Whole mouse placenta and tissue from human placenta separated from connective tissues and membranes were minced and digested

[1]*This work was supported in part by grants from U.S.P.H.S. NIH CA 24474 and CA 34107.*

ISBN 0-12-527540-4

with trypsin (0.125 percent) and collagenase (50 mg/dl) in calcium-free Hank's balanced salt solution for 30 minutes at room temperature (3). The cell suspension was washed twice with iced serumless media, plated on tissue culture plates at 1.0×10^6 cells/ml and cultured for two hours at 37° C and 5% CO_2 and 95% air. The plates were then washed with media to remove non-adherent cells and the percent adherent cells estimated by counting the cells collected from the washing. Macrophage characteristics were determined by counting the number of adherent cells that phagocytized latex beads (6) and positively stained for non-specific esterase (7). The adherent cells (in human placenta it was possible to obtain up to 10^{11} adherent cells) were cultured for 48 hours, supernatants collected, dialyzed (40x) and filtered. Supernatant fluids were tested for IL-1 activity using a thymocyte proliferation (TP) assay (8) and Interleukin-2 (IL-2) using a long-term IL-2 dependent T cell line (9). Phagocytosis has been described as a method to increase the production/release of IL-1 from macrophages (10) and we tested this concept by using latex bead (0.22 μ diameter) ingestion in a four hour assay (5).

TABLE 1: Thymocyte Proliferation and IL-2 Assays of Supernatant Fluids from Mouse Placental Mononuclear Phagocytes.

Assays/Groups	*H^3-thymidine incorporation CPM ± S.D.*	
Thymocyte proliferation		
Control supernatants	*305 ± 33*	
Placental cell sup.	*2895 ± 884*	
Stimulation index[a]		*9.5*
IL-2		
Control supernatants	*1733 ± 195*	
Placental cell sup.	*2046 ± 804*	
Stimulation index		*1.2*

N = 18, all means are the result of triplicate analysis. All assays reported at 1:4 dilutions of placental cell supernatants, which was the optimal dilution of four dilutions tested.

[a]Stimulation index = CPM experimental cultures/CPM control cultures

Placental macrophages from C57Bl/6 mice contained a mean of 67 percent adherent cells after the two hour culture. At

48 hours a mean of 86 percent of the adherent cells were capable of phagocytizing latex bonds. Culture supernatant fluids from the placental macrophages contained a significant amount of IL-1, but did not contain measurable amounts of IL-2, as indicated in Table 1.

Human placental cells cultured for a two hours adherence phase contained 77 percent adherent cells. Eighty-five percent of the adherent cells phagocytized latex beads and a mean of 88 percent of the adherent cells were positive for the non-specific esterase stain. IL-1 activity was significantly elevated in the culture supernatants from the placental macrophages and little or no IL-2 was detectable in the supernatant fluids, as reported in Table 2.

TABLE 2: Thymocyte Proliferation and IL-1 Assays of Supernatant Fluids from Human Placental Mononuclear Phagocytes.

Assays/Groups	*H^3-thymidine incorporation CPM ± S.D.*	
Thymocyte proliferation		
Control supernatants	*168 ± 42*	
Placental cell sup.	*1949 ± 138*	
Stimulation index[a]		*11.6*
IL-2		
Control supernatants	*1540 ± 282*	
Placental cell sup.	*1935 ± 408*	
Stimulation index		*1.3*

N = 18, all means are the result of triplicate analysis. All assays reported at 1:4 dilutions of placental cell supernatants, which was the optimal dilution of four dilutions tested.

[a]Stimulation index = CPM experimental cultures/CPM control cultures.

Latex bead phagocytosis by the human placental macrophages increased the production/release of IL-1, Table 3.

Cultured supernatant fluids from placental macrophages were chromatographed on an Ultragel AcA 54 gel filtration column, 1.0cm X 60cm, and compared against a mixture of four proteins (albumin, ovalbumin, chymotrypsinogen A and ribonuclease) to determine an approximate molecular weight for the IL-1 active protein. One liter of culture supernatants from 48 hour cultures was concentrated on an Amicon ultrafiltration membrane system 500 fold and applied to the calibrated gel

TABLE 3: Thymocyte Proliferation and IL-1 Assays of Supernatant Fluids from Human Placental Mononuclear Phagocytes.

Thymocyte proliferation	*H^3-thymidine incorporation CPM ± S.D.*	
Control supernatants	*168 ± 42*	
Placental cells treated with latex beads - S.I.[a]	*2638 ± 210*	*15.7*
Placental cells not treated with latex beads - S.I.	*1462 ± 183*	*8.7*

N = 3, all means are the result of triplicate analysis. All assays reported at 1:4 dilutions of placental cell supernatants, which was the optimal dilution of four dilutions tested.

[a]S.I. = Stimulation index.

filtration column. The protein was eluted with 0.1M phosphate/glycine buffer, pH 7.4, and collected in 2.0 ml fractions. The fractions were assayed for IL-1 activity using the TP assay and approximate molecular weight determined. Chromatographic separation of proteins in the culture supernatants indicated that IL-1 activity from placental macrophages was evident in the column fractions with a major peak at 17,000 daltons and a minor peak at 48,000 daltons, which is consistent with the molecular weight of human IL-1 (2).

In addition, to determine if IL-1 is pre-formed in the placental macrophages, cells collected after enzyme digestion (1×10^9/ml) were lyzed by sonication using a Heat Systems Ultrasonics, Inc. cell disrupter. Solutions were filtered and the filtrate tested for IL-1 activity. IL-1 activity in the lysate of placental macrophages was also significantly elevated, as indicated in Table 4, which indicates the protein is pre-existing in these cells.

Supernatant fluids from placental macrophages of both mice and humans contain the factor IL-1, but not IL-2. Gel filtration of the crude human placental macorphage culture supernatants confirmed that the IL-1 activity measured was related to a protein(s) with a major peak of 17,000 dalton and a minor peak of 48,000 dalton. Phagocytosis initiated by stimulating the cells with latex beads resulted in greater production/release of IL-1 into the culture supernatants. Lysing of the human placental macrophages gave indication that IL-1 is constituatively produced by placental macrophages. These data support the concept that placental derived macrophages are

TABLE 4: Thymocyte Proliferation of Placental Macrophage Lysate.

Dilution of lysate[a]	*H^3-thymidine incorporation CPM ± S.D.*	*S.I.*[b]
Control	*162.7 ± 14.9*	-
1:4	*215.3 ± 84.7*	*1.32*
1:8	*619.3 ± 137.2*	*3.81*
1:16	*1838.7 ± 209.1*	*11.30*
1:32	*700.0 ± 90.4*	*4.30*
1:64	*434.0 ± 111.1*	*2.67*
1:128	*406.0 ± 169.2*	*2.50*

[a]*Lysate of 10^9 cells/ml of placental cells.*
[b] *S.I. = Stimulation index.*

producing IL-1 without stimulation and can be stimulated to release greater quantities thus making them a useful source of human IL-1 and perhaps other macrophage derived factors.

ACKNOWLEDGEMENTS

We are grateful to M. Loftus for excellent technical assistance in conducting this research and to C. Reinhard for secretarial assistance in the preparation of this manuscript.

REFERENCES

1. Rosenstreich, D.L., in "*Cellular Functions in Immunity and Inflammation*" (J.J. Oppenheim, D.L. Rosenstreich and M. Potter, Eds.) p.127 Elsevier/North Holland, New York, (1981).
2. Mizel, S.B. and Rosenstreich, D.L. *J. Immunol.* 122,2173 (1979); Togawa, A., Oppenheim, J.J. and Mizel, S.B. *J. Immunol.* 122,2112 (1979).
3. Wood, G.W., *Placenta* 1,113 (1980).
4. Wood, G.W., *Placenta* 1,309 (1980).
5. Flynn, A., in *Inflammatory Diseases and Copper*, J.R.J. Sorenson, Ed. (Humana, Clifton) p.17 (1982).
6. Finke, J.H., Sharma, S.D. and Scott, J.W., *J. Immunol.* 127,2354 (1981).
7. Yam, L.J., Lin, C.Y. and Crosby, W.H. *Amer. J. Clin. Pathol.* 55,283 (1971).
8. Farrar, J.J., Simon, P.L., Koopman, W.J. and Fuller-Bonar, J. *J. Immunol.* 121,1353 (1978).

9. Watson, J., Gillis, S., Marbrook, J., Mochizuki, D. and Smith, K.A. *J. Exptl. Med.* 150,849 (1979).
10. Gery, I., Davies, P., Derr, J., Krett, N. and Barranger, J.A. *Cell Immunol.* 64,293 (1981).

THYMOCYTE ACTIVATING FACTOR PRODUCED BY GLOMERULAR MESANGIAL CELLS

David H. Lovett
John L. Ryan
R. Bernd Sterzel

Department Internal Medicine
West Haven VAMC and
Yale Univ. School of Medicine
New Haven, CT.

INTRODUCTION

The glomerular mesangium is centrally involved in the pathogenesis of immune-mediated glomerular injury. Consisting of a cellular component embedded within a loose extracellular matrix, the mesangium represents a specialized interstitial space localized between the glomerular capillary loops (1). The mesangium is frequently exposed to circulating immune deposits and phlogogenic materials because of the high capillary luminal pressures and the absence of a basement membrane at the endothelial-mesangial interface. The mesangial cell (MC) is a contractile, endocytic vascular smooth muscle-like cell, which contains myosin filaments, and secretes fibronectin and types III and IV collagen (2).

Tissue macrophages have been considered to be significant effector cells in the pathogenesis of certain forms of immune-mediated glomerulonephritis (3,4). The mesangial cell has been viewed primarily as a responder cell, particularly in terms of the induction of cellular proliferation and the enhancement of matrix production (2). In contrast to this generally accepted viewpoint, we considered the MC as a potential effector cell under certain circumstances. These multipotential mesenchyme-derived cells possess many macrophage-like properties. They have a high lysosomal enzyme content (5),

ISBN 0-12-527540-4

secrete extracellularly active neutral proteinase (6), and release substantial levels of PGE_2 (7). We therefore investigated the ability of mesangial cells to produce non-specific immunoregulatory factors which might affect lymphocyte activation in a fashion analogous to the cytokine Interleukin 1 (IL-1).

MATERIALS AND METHODS

Glomerular mesangial cells were cultured using a modification of the method of Striker, et al. (2,6). Whole isolated glomeruli from saline perfused blood-free rat kidneys were separated from renal cortex homogenates using serial metal sieves and differential centrifugation. The epithelial cell component was removed by incubation with bacterial collagenase and centrifugation, after which the glomerular remnants were placed into tissue culture. The medium consisted of minimal essential medium-Eagle supplemented with 100 U/ml penicillin, 100 μgm/ml streptomycin, 300 μgm/ml fresh glutamine, 2×10^{-5}M mercaptoethanol (complete-MEM), and 20% heat-inactivated fetal calf serum (FCS). Cell outgrowths appeared by the second day of culture, and formed confluent monolayers by day 8-10. These cells were homogenous morphologically, with a stellate to fusiform shape. Growth occurred in tightly interwoven bundles. Using immunofluorescence (IF) techniques, intracellular staining of coarse myosin fibrils was noted (6). The extracellular matrix components, fibronectin and collagen types III and IV, were detected using monospecific antisera. Fibroblast contamination was excluded by the ability of MC to grow in D-valine-containing selective medium (8). Endothelial cell contamination was excluded morphologically and by the absence of IF staining for factor VIII. Macrophages were excluded by morphology and the absence of latex bead phagocytosis or neutral red ingestion. IF studies were negative for cells bearing surface Ia antigen at the time of study (6).

The determination of the kinetics of MC growth was assessed by serially measuring the uptake of ^{3}H-thymidine (^{3}H-TdR) into cellular DNA. Following a lag phase of 1-2 days, the MC entered log-phase growth which lasted until confluency was reached by day 8-10 of culture. At given time points the growth medium was removed, the MC washed extensively, and fresh serum-free medium added. The conditioned medium thus produced was harvested at intervals, spun at 400gx10 minutes, and stored at -80 degrees C until bioassay.

The effects of MC-conditioned medium on lymphocyte activation were assessed using a rat thymocyte proliferation assay (9). In this assay, thymocytes obtained under sterile conditions from 150-200 gram Sprague-Dawley rats were suspended in complete MEM +5% FCS and incubated in plastic dishes for 60 minutes to deplete adherent cells. Washed nonadherent thymocytes were plated at a density of 5×10^5 cells per microtiter well in serum-free complete MEM along with serial dilutions of test supernates and 0.5 µgm/ml PHA. After 48 hours of culture at 37°C the wells were pulsed with 1 µCi ^{3}H-TdR, (sp. act. 2 miCi/mmol), harvested after a further 24 hours, and quantitated by liquid scintillation. All samples were assayed in triplicate, and compared to control wells consisting of cells plus mitogen alone.

RESULTS

In our initial studies, we observed that culture supernates of MC enhanced rat thymocyte proliferation in the presence of sub-optimal doses of PHA. Subsequently, using conditioned media from 5 independently derived lines of cultured MC, a 595±62% stimulation of proliferation was noted (9). As seen in Table 1, the degree of thymocyte proliferation was dependent on the concentration of the MC-conditioned medium.

TABLE 1. Concentration Dependence of Thymocyte Stimulation[a]

Supernate dilution	Δcpm±SD ^{3}H-TdR
1:4	2605 ± 443
1:16	1757 ± 323
1:32	914 ± 324

[a]Thymocytes incubated with mitogen alone incorporated 386±118 cpm ^{3}H-TdR.

The amount of mesangial cell derived thymocyte activating factor (MC-TAF) detected in the culture supernates was dependent on the duration of culture in serum-free medium. The percent increase in thymocyte ^{3}H-TdR corporation from supernates harvested at 24, 48, and 72 hours were 285±35%, 495±45%, and 675±30%, respectively. We observed that the release of

MC-TAF into the medium was maximal during the period of log-phase growth, while little activity was found in supernates from confluent, slowly proliferating cultures (9).

CONCLUSIONS

These data indicate that growing rat glomerular mesangial cells produce a cytokine, MC-TAF, that enhances the proliferation of PHA-primed rat thymocytes. The degree of stimulation was dependent on MC-TAF concentration and the duration of serum-free culture. Analysis of the physicochemical properties and target cell specificities of MC-TAF reveal close similarities, but not identity, with macrophage IL-1 (9).

Macrophage IL-1 has been shown to have both immunologic and nonimmunologic properties, including the enhancement of IL-2 production, stimulation of the hypothalamic fever center, and stimulation of fibroblast proliferation (10). The possibility that MC-TAF may share some of these properties is presently under investigation. These studies may be expected to provide new insights into the pathogenesis of immune-mediated glomerular diseases.

REFERENCES

1. Sterzel, R.B., Lovett, D.H., Stein, H., and Kashgarian, M., Klin. Woch. (in press).
2. Striker, G.E., Killen, P.D., Farin, F.M., Trans. Proceed. 12, Suppl. 1, 88 (1980).
3. Hunsicker, L.G., Shearer, T.P., Plattner, S.B., Weisenberger, D., J. Exp. Med. 150, 413 (1979).
4. Holdworth, S.R., Neale, T.J., Wilson, C.B., J. Clin. Invest. 68, 686 (1981).
5. Lovett, D.H., Ryan, J.L., Kashgarian, M., Sterzel, R.B., Amer. J. Pathol. 107, 161 (1982).
6. Lovett, D.H., Sterzel, R.B., Kashgarian, M., Ryan, J.L., Kidney Int. (in press).
7. Sraer, J., Foidart, J., Chonsel, D., Mahieu, P., Kouznetzova, G., Ardaillou, R., FEBS Lett. 104, 420 (1979).
8. Gilbert, S.F., Migeon, B.R., Cell 5, 11 (1975).
9. Lovett, D.H., Ryan, J.L., Sterzel, R.B., (submitted).
10. Oppenheim, J.J., Stadler, B.M., Siraganian, R.P., Mage, M., Mathieson, B., Fed. Proceed. 41, 257 (1982).

DISCUSSION

LUGER: In terms of the similarity between MC-TAF and E-TAF, did Lovett's crude MC-TAF preparation contain serum? Was it possible to induce increased factor production by stimulating the cells with "injurious" agents such as LPS, silica, mechanical disruption or hydrolysis?

LOVETT: The cells produce the factor during exponential growth, but they were not otherwise stimulated. Studies are under way at present applying some of the more classic probes (used with macrophages) to these cells. Our work with purified endotoxin suggests that it causes enhancement of this mesangial cell factor.

LUGER: Did Lovett observe the presence of an additional inhibitory activity in his supernatants?

LOVETT: That is conceivable, particularly in view of the plateau we reported. We have not investigated this point, but is should be looked into. I would point out that mesangial cells make large quantities of PG2. We determine the PG2 levels in our preparations but were unable to change their curves by dializing our preparations so it did not appear to be the PG2, or thymidine, that was inhibiting our assay.

SALVIN: Has Lovett had an opportunity to test the effect of a monoclonal anti-macrophage-IL-1 on the macrophage factor and on the mesangial cell factor?

LOVETT: Not at this time.

ASHERSON: Is Lovett's factor, with IL-1-like activity, a growth factor for mesangial cells? This question arises because confluent mesangial cells produce little or no IL-1-like material.

LOVETT: Those studies are actually in progress and we will be reporting on them later on.

WILLOUHGBY: One of the cardinal changes associated with immune glomerular injury is swelling of the endothelial walls. Has Lovett tested his mesangial cell factor for its effects on endothelial cells in culture?

LOVETT: No, we have not.

OPPENHEIM: How does Lovett distinguish the mesangial cells from eipthelial cells? They look similar to me.

LOVETT: Glomerular epithelial cells grow in very large polygonal monolayers. Our cells, however, have a different shape and grow in interwoven bundles. Also, the presence of intracellular fibrils is limited to smooth muscle-like cells, or mesangial cells, and can be detected using the myosin stain in conjunction with certain techniques and antisera. We have also demonstrated, by electron microscopy of these glomerular remnants after enzyme treatment, that they are free of the epithelial cell component.

OPPENHEIM: What is Lovett's evidence that MC-TAF is biologically and biochemically distinct from macrophage-derived IL-1?

LOVETT: We attempted to exclude any potential macrophage contribution. We used perfused blood-free kidneys; we know that esterase stains of normal rat glomeruli contain fewer than .01% esterase-positive cells. In addition, I have been able to reproduce these results using serial-passaged mesangial cells derived from clones. Accordingly, I feel we have excluded macrophage contamination.

Whether the biological activities are the same or separate is going to require more work. The biochemical activities are similar, but we do not have a pI on this yet, so we will have to hold off on deciding their difference or similarity.

CULTURED ASTROCYTES AND C_6 GLIOMA CELLS SECRETE INTERLEUKIN-1 LIKE FACTORS

Adriano Fontana
Rolf Dubs
Markus Glantschnig
Elisabeth Weber
Peter J. Grob

Section of Clinical Immunology,
University Hospital
Zurich/Switzerland

Flemming Kristensen
Alain L. de Weck

Institute for Clinical Immunology
Inselspital
Bern, Switzerland

I. INTRODUCTION

The capacity of the central nervous system (CNS) to mount cellular and humoral immune responses has been studied by several groups with conflicting results. Recently, it has been reported that the injection of T-dependent antigens into the brain parenchyma or into the subarachnoidal space resulted in antibody levels similar or even higher to those achieved by subcutaneous immunization (1). In addition there is recent evidence that the CNS is capable of rejecting histoincompatible grafts or tumor cells (2). Thus, intracerebral amplification systems might be important to overcome the priviledged site of the brain which lacks a lymphatic drainage and is shielded from the immune system by the blood

Interleukins, Lymphokines, and Cytokines

ISBN 0-12-527540-4

brain barrier. We here report the secretion of interleukin-1 (IL-1) like factors by cultured murine astrocytes and C_6 glioma cells. Thus, by producing helper factors astrocytes may be important in the generation of intracerebral immune reactions.

II. RESULTS AND DISCUSSION

First passages of primary cultures of newborn mouse brain (BALB/c) develope into monolayer cultures composed mainly of astrocytes as defined by immunofluorescence using an antibody against the astrocyte specific glia fibrillary acidic protein (3). When treating these cultures with lipopolysaccharide (LPS) or lipid A an increase of the RNA synthesis was observed (4). An evaluation of the supernatants (SN) of glia cultures indicate that at least two separate immuneregulatory activities become detectable during LPS stimulation (5). The glia cells were found to produce prostaglandin E, - potent inhibitors of the lectin and IL-1 induced thymocyte proliferation. In the CNS prostaglandin E, released by astrocytes, may act as a negative signal on the ongoing immune reaction. Impressive was the finding of an IL-1 like factor elaborated by LPS treated astrocyte cultures (Table 1). 600'000 thymocytes [of C3H/HeJ mice] in RPMI 1640 supplemented with 1% L-glutamin, $1x10^{-5}$ M 5 -mercaptoethanol and 5% fetal calf serum were incubated in microtiter plates for 72 hr in the presence of phytohaemagglutinin and glia cell SN (diluted 1:4), which had been collected from first passage glia cell cultures treated for 24 hr with LPS in the presence of indomethacin (IM, 10µg/ml). As a control medium of untreated glia cells was harvested and supplemented with LPS and IM. In contrast to medium control, the SN of LPS-treated glia cells induced an augmentation of the PHA-response of thymocytes. The effect was dependent on the dose of LPS used to stimulate the glia cells and on the dilution of the SN added to thymocytes, - still 50% of the augmentatory effect of SN (diluted 1:4) was observed when

TABLE I. RNA-synthesis and production of IL-1 by glia cells

Stimulants of glia cells	Concentration (μg/ml)	RNA-synthesis in glia cells [a] [SI]	IL-1 activity in glia cell supernatant [b] [SI]
none	-	1,0	1,0
LPS	0,001	2,7	1,6
LPS	0,1	8,3	5,2
LPS	10	11,2	10,9
PHA	5	7,6	7,4
Protein A	100	12,0	12,5
PPD	8	8,4	10,2
ConA	5	1,1	2,2
Streptokinase-Streptodornase	1:1,000 dil.	1,4	1,1
PWM	1:100 dil.	2,5	1,8

[a] SI= stimulation ratio (mean value of cpm of ^{3}H-uridin incorporated in treated glia cultures divided with the mean value of cpm in control cultures)

[b] glia cell supernatants were asseyed on thymocyte in the presence of PHA

$$\left[SI = \frac{\text{cpm } (^3\text{H-Tdr}) \text{ in the presence of glia SN}}{\text{cpm } (^3\text{H-Tdr}) \text{ in the presence of medium control}}\right]$$

adding a 1:256 dilution. When testing other agents that induce the release of the helper factor, a general parallelism was observed between increased RNA synthesis in stimulated glia cell cultures and the level of helper factor activity in the conditioned medium (Table 1).

Although 95% to 98% of first passage cells of primary brain cell cultures are glia fibrillary acidic protein positive, contaminating macrophages could still contribute to the effects observed. The finding of helper factor secretion by C_6 cells support the statement of the helper factor being an astrocyte product. The C_6 glioma cell line, originally cloned from a rat astrocytoma, represent undifferentiated astrocytic cells which synthesize S-100 protein and glial fibrillary acidic protein. A helper factor activity for the PHA-response of thymocytes was detected in the supernatant of C_6 cells cultured for 3 to 4 days in Ham's F-12 medium containing 5% fetal calf serum,IM (10µg/ml) and 0,5% L-glutamin. Unlike primary astrocyte cultures, the C_6 cells released the factor spontaneously during exponential cell growth (Table II).

TABLE II IL-1 activity in C_6 glioma cell supernatant

C_6 cells per culture	PHA-response of thymocytes (mean cpm ± SD) treated with	
	3 day SN	4 day SN
0	5,085 ± 356	5,615 ± 383
10^3	6,981 ± 165	6,652 ± 582
10^4	13,101 ± 1,205	17,881 ± 286
10^5	43,077 ± 3,134	64,366 ± 7,706
10^6	44,507 ± 4,455	57,660 ± 2,053

In further experiments we investigated the possibility that the glia cell derived helper factors were related to IL-1 secreted by macrophages or to interleukin - 2 (IL-2) released by T helper cells. Our results indicate that 1) both, SN of primary astrocyte cultures and C_6 cells do not support the growth of an IL-2 dependent cytotoxic T-cell line (CTLL cells) but 2) they both augment the lectin-induced IL-2 secretion by peanut agglutinin negative thymocytes and 3) the helper factor generated by C_6 cells in serum-free medium has a m.w. between 13,500 and 18,000 d. Thus we have concluded that the astrocyte derived helper factors are IL-1 like mediators (5).

These observations suggest that in addition to cells of the immune system other cell types may be involved in the release of immunoregulatory factors. There is evidence that epidermal cells (keratinocytes) do also release an IL-1 like factor (6,7). By producing IL-1 both astrocytes and keratinocytes may play a role in the modulation of the local immune response in the CNS or the skin.

REFERENCES

1. Quirico Santos,T., and Valdimarsson,H. Neuroimmunol. 2, 215 (1982).
2. Hasek,M., Chutna,J., and Sladecek,M., Nature 268, 68 (1977).
3. Fontana,A., Dubs,R., Merchant,R., Balsiger,S., and Grob,P.J., J. Neuroimmunol. 2, 55 (1982).
4. Fontana,A., Bosshard,R., Dahinden,C., Grob,P.J., and Grieder,A., J. Neuroimmunol. 1, 343 (1981).
5. Fontana,A., Kristensen,F., Dubs,R., Gemsa,D., and Weber,E. (1982). J. Immunol. submitted (1982).
6. Luger,T.A., Stadler,B.M., Katz,S.I., and Oppenheim,J.J. J. Immunol. 127, 1493 (1981).
7. Luger,T.A., Stadler,B.M., Luger,B.M., Mathieson,B.J., Mage , M., Schmidt,J.A., and Oppenheim, J.J., J. Immunol. 128, 2147 (1982).

DISCUSSION

OPPENHEIM: Could Fontana tell us the difference between astrocytes and microglial cells and whether the microglial cells also produce an IL-1-like factor?

FONTANA: Astrocytes are of neuro-ectodermal origin; are Thy-1 positive and do not have Fc- and C3b-receptors. The microglial cells are thought to be of the monocyte-macrophage lineage, express Fc/C3b receptors and have phagocytic properties. Until now, microglial cells have not been tested for ability to produce IL-1 mainly because of difficulties in achieving enrichment of microglial cells.

DAVID: Has anyone examined cerebrospinal fluid to ascertain whether there is any IL-1 activity in individuals with neurologic immune inflammation or infection?

FONTANA: In collaboration with Weiner and Hauser, Harvard Medical School, I have investigated the cerebrospinal fluid of patients in various stages of disease for the presence of helper factor activities. All we found was a suppression of the normal PHA response of thymocytes due to the presence of a (>MW10,000) component in the cerebrospinal fluid.

SALVIN: How would Fontana define his C-6 glial cells?

FONTANA: C-6 glioma cells were originally cloned from a rat astrocytoma. The cells are rich in S-100 protein and glial fibrillary acidic protein (GFA). In certain tissue culture systems, they form GFA-positive processes. Accordingly they represent undifferentiated cells with the potential to develop into astrocytes. The presence of some enzyme typical for oligodendrocytes speaks in favor of their being stem cells.

KURNICK: Because Fontana showed IL-1-like activity from astrocytes, does it necessarily follow that he should postulate a T cell target for this product? Has he considered the possibility that non-T cells may be target cells for this astrocyte product?

FONTANA: Both primary astrocyte cultures and C-6 glioma cells were found to elaborate factors supporting the growth or differentiation of neuronal cells *in vitro*. However, the biochemical characteristics of the factors reported are not worked out yet. Therefore their relationship to the IL-1-like factor still remains to be established.

CHARACTERISTICS OF PMN-DERIVED HELPER T CELL POTENTIATION FACTOR IN AN INFLAMMATORY SITE*

Masaru Yoshinaga
Susumu Ohkawara
Shigeki Nakayama
Fumimasa Goto

Department of Immunopathology, Kumamoto University Medical School, Kumamoto, Japan

Inflammatory and immune responses are respectively the prime examples of non-specific and specific mechanisms involved in the recognition and elimination of foreign substances. Usually, if a foreign antigenic substance is introduced into a tissue, first inflammation is induced and then immune responses come into play. Cooperation between the two mechanisms may be expected. Although the immune mechanism is known to support the function of the inflammatory response in various ways, the reciprocal situation in which inflammation may modulate the immune response has hardly been investigated.

I. EFFECT OF PERITONEAL INFLAMMATION ON SPLENIC PLAQUE FORMING CELL (PFC) RESPONSE

Inflammation was induced by an intraperitoneal injection of 0.2% sodium caseinate or 0.05% aluminum hydroxide gel (alum). In the control animals, 2 ml of phosphate buffered saline (PBS) was injected. Then, antigen (3 millions sheep red blood cells, SRBC) was introduced into the peritoneal cavity at various intervals after the phlogistic stimuli.

This work was supported in part by a Grant-in-Aid for Scientific Research from Japanese Ministry of Education, Science and Culture.

ISBN 0-12-527540-4

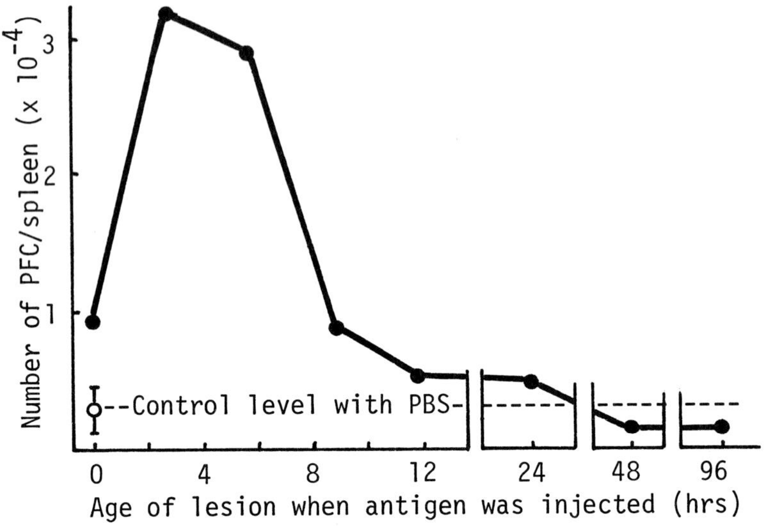

FIGURE 1. Effect of an ongoing inflammation on splenic PFC response.

At indicated intervals after the intraperitoneal injection of 2 ml 0.2% casein, the antigen (3 millions SRBC) was injected and splenic direct PFC was determined 4 days after the injection of antigen.

As shown in Fig. 1, the injection of SRBC into PBS-treated control mice resulted in the production of 3,720 ± 652 PFC, irrespective of the interval between PBS and SRBC administrations. The splenic PFC response was markedly enhanced in mice which had received the antigen at the same time as, or within 9 hr of the casein administration. Upon the successive injection of casein and SRBC, the enhancement of the antibody response may be due to an activation of the nonspecific helper T cells by casein as an antigen. Therefore, we tested the effect of an intraperitoneal injection of alum as a non-antigenic phlogistic stimulus. When SRBC was injected 3 hr after the injection of alum, the resulting PFC ranged from 38,072 to 86,935; in the controls, it was 246 to 4,365. The mean enhancing index was 17.

This enhancement of PFC response by casein or alum pretreatment was observed only with the antigen amounts between

1 to 10 millions SRBC and there was no difference between the casein-pretreated and control animals when one hundred millions SRBC was adminstered or when SRBC was injected intravenously (1).

II. REPRODUCTION OF THE ENHANCED PFC RESPONSE BY THE ADOPTIVE TRANSFER OF EARLY INFLAMMATORY EXUDATE CELLS

3 hr after the intraperitoneal injection of casein, peritoneal exudate cells (PEC) were harvested. The 3-hr PEC (2 to 32 millions), consisting of 93% PMN, 4% lymphocytes, 2% macrophages and 1% mast cell-other cells, were adoptively transferred into the peritoneal cavity of sex-matched syngeneic mice together with 3 millions SRBC. The resulting enhanced splenic PFC on the 4th day depended on the number of PEC transferred. The highest enhancement was approximately 8.2 ± 3 fold at doses of 16 and 32 millions PEC. On the other hand, the transfer of 96-hr PEC, consisting of 25.7% lymphocytes and 74.3% macrophages, produced almost no enhancement of the PFC response. This enhancing capability of 3-hr PEC **did not** decrease after treatment with anti-Thy 1 plus complement (1).

III. EFFECT OF INFLAMMATORY CELL-FREE EXUDATE ON THE *IN VITRO* PFC RESPONSE

After an intraperitoneal injection of 2 ml 0.05% alum, the cell-free exudate (PEF) was harvested at various intervals by washing the peritoneal cavity with 2 ml PBS. The PEF was dialysed against PBS for 24 hr. The PEF preparations (25 μl) were added to the culture (200 μl) of one million spleen cells and one million SRBC in Click's medium. The resulting direct PFC were estimated 4 days later.

As shown in Fig. 2, the PEF taken from early inflammatory lesions (1 to 6 hr) produced a marked enhancement of the PFC response, while PEF taken from later lesions (48 to 96 hr) produced a slightly suppressed PFC response. In the control, 12.5 μg alum, which might be carried over to the spleen cell culture, were added to the culture without any measurable effect on the PFC response. The PFC enhancement was dependent on the amount of 3-hr PEF used. The same PEF preparations were also tested for the thymocyte DNA synthesis. The early PEF also produced a marked enhancement of the thymocyte response, but late PEF did not. The activity profile for the DNA synthesis potentiation in the course of alum-induced inflammation, closely resembled to that for PFC enhancement.

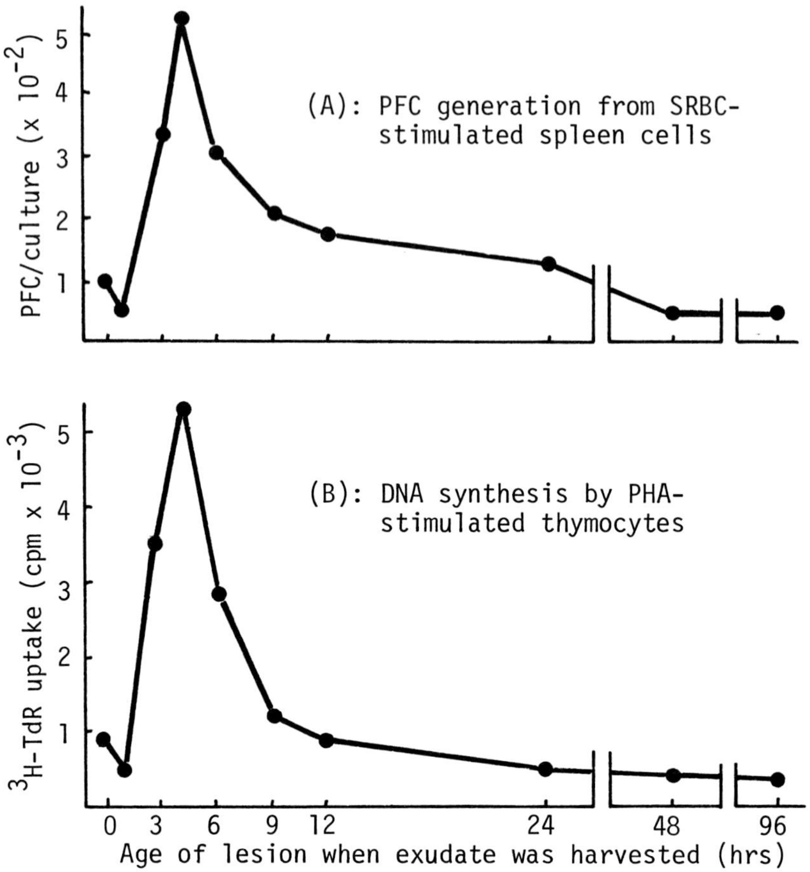

FIGURE 2. Effect of alum-induced cell-free peritoneal exudate on in vitro primary PFC response and thymocyte DNA synthesis.

(A): C3H spleen cells (one million) and SRBC (one million) were cultured in 200 μl of Click's medium in the presence of 25 μl inflammatory exudate for 4 days.
(B): C3H thymocytes (3 millions) were cultured for 3 days in one ml of RPMI medium containing one μl of PHA and 100 μl of inflammatory exudate.

IV. PRODUCTION OF THE PFC ENHANCING FACTOR BY THE PMN PREPARATION *IN VITRO*

3 hr after the intraperitoneal injection of casein, the PEC were harvested and washed. To exclude the possibility that the contaminated macrophages may produce the active factor, the 3-hr PEC was purified on a discontinuous gradient of Percoll. The PMN was recovered in a fraction heavier than 65% Percoll. The purified PMN preparation consisted of 99% PMN and 1% lymphocytes-mast cells. Contamination with macrophages was less than 0.02%. The PMN (3.2 millions/ml) were stimulated *in vitro* with 0.3 mg/ml alum and the culture supernatant was harvested 3 hr later. When the supernatant was added to the spleen cell culture, which was stimulated with one million SRBC, the resulting PFC at 4 days were enhanced in a dose-dependent manner.

V. ENHANCEMENT OF THE GENERATION OF HELPER CELLS BY PMN FACTOR

Thymocytes (2 millions) were preincubated with one million SRBC for 3 to 7 days in the presence or absence of the PMN factor. The cells were washed and transferred to the

TABLE I. Generation of Helper Activity from Thymocytes in the Presence of PMN Factor

Education culture*			Assay culture**		
thymocytes (million)	*SRBC (million)*	*PMN factor (µl)*	*B cells*** (million)*	*SRBC (million)*	*PFC/culture*
2	1	25	0.2	0.5	250 ± 34
none	1	25	0.2	0.5	5 >
2	none	25	0.2	0.5	22 ± 8
2	1	none	0.2	0.5	15 ± 8
2	1	25	none	none	0
none	none	none	0.2	0.5	5 >

**Thymocytes were preincubated for 7 days and transferred to the assay culture after washing with HBSS.*
***B cells were cultured for 4 days with the educated cells and SRBC.*
****Spleen cells which were passed through Sephadex G-10 and killed with anti-Thy 1 plus complement.*

second culture, which contained 200 thousands purified B cells and further cultured for 3 to 4 days. The representative data is presented in Table I. These thymocytes manifested helper activity only when they were educated with SRBC and the PMN factor. No helper activity was generated in the presence of SRBC without the PMN factor.

Similar experiments were carried out by using isolated spleen T cells which were non-adherent to nylon column and negative with surface Ig. The T cells were activated to generate helper T cell activity in the presence of SRBC only when they were supplemented with the PMN factor and 8 thousands X-irradiated spleen adherent cells as antigen presenting cells.

VI. CHARACTERIZATION OF THE HELPER CELL-ENHANCING PMN FACTOR

The alum-stimulated PMN culture supernatant was tested for its molecular weight on a Sephadex G-75 column and for its isoelectric point on an LKB multiphor apparatus. The helper-cell enhancing activity was eluted in m.w. ranging from 24,000 to 12,000. On the isoelectrofocusing flat bed, the activity was recovered at a pI 9.0 to 10.0 and 6.0 to 4.5. The second peak, pI 5 showed a rather broad distribution. Both the active fractions at pI 10 and 5 contained also the DNA synthesis-enhancing activity for the PHA-stimulated thymocytes (2). But, these fractions failed to support the DNA synthesis of interleukin 2-dependent C3H cytotoxic cell line against the EL-4 cells, which were established in our laboratory and have been maintained for the past 11 months. This suggests that the PMN factor is similar to the interleukin 1 of macrophages in its biologic nature, though one of the PMN factor showed very high pI value.

Thus, the inflammation-induced enhancement of the primary PFC response to SRBC is explicable in terms of a function of PMN in the early inflammatory exudate cells which were shown to produce *in vivo* and *in vitro* a T cell-stimulating factor.

REFERENCES

1. Nakayama, S., Rodriguez-Pinzon, J., Nakamura, S., and Yoshinaga, M., *Immunology*, 45, 669 (1982).
2. Nakamura, S., Goto, F., Goto, K., and Yoshinaga, M., *J. Immunol.* 128, 2614 (1982).

DISCUSSION

WOOD: Does the PMN factor exhibit other IL-1-like activities besides the mitogenic ones?

YOSHINAGA: As far as we have determined, both of these PMN factors enhance the generation of alloantigen-specific cytotoxic cells in the MLC, in addition to the activities reported here. But we have not yet determined their ability to induce fever or serum amyloid protein A.

WOOD: Do the 24-hr PMN fail to make the active factor, or do they inhibit the activity of the factor?

YOSHINAGA: 24-hr PMN failed to produce the lymphocyte-stimulating activity when treated with alum or kaolin. Furthermore, we have failed to show any suppressive action against the lymphocyte-stimulating activity in the culture supernatant from 24-hr PMN.

RODRICK: My comment relates to Wood's query on whether PMN in the 24 hr cell population may not only have lost stimulatory ability, but actually be suppressive. Our own work has shown that PMN supernatants can contain suppressive as well as enhancing activities; moreover, these two factors can be separated immunochemically.

DAVID: Can Yoshinaga obtain the potentiation factor by stimulating human peripheral blood PMN with kaolin?

YOSHINAGA: We performed this type of experiment with human PMN. Curiosly enough, human circulatory PMN do not respond to kaolin stimulation with production of lymphocyte-stimulating activity. However, inflammatory exudate human PMN, which were collected from the inflamed periotneal cavity, *do* respond by producing the activity. Therefore, we assume that PMN *can* be triggered by some unknown mechanism to respond to kaolin stimulation during the emigration process from blood vessels, or just after the emigration into an inflamed site.

PHYSICOCHEMICAL CHARACTERISTICS OF PMN-DERIVED T-LYMPHOCYTE STIMULATING MOLECULES*

Shuji Nakamura
Fumimasa Goto
Kumiko Goto
Masaru Yoshinaga

Department of Immunopathology, Kumamoto University Medical School, Kumamoto, Japan

We found that a lymphocyte stimulating factor in an inflammatory site could be detected only in the early stage of inflammation where the exudation of polymorphonuclear leukocytes (PMN) are predominant (1). Therefore, we assumed that PMN were the source of the lymphocyte stimulating factor. In fact, the 3-hr peritoneal exudate cells (PEC), consisting of 93% PMN, 4% lymphocytes, 2% macrophages and 1% other cells, released a lymphocyte stimulating factor *in vitro* and this release was enhanced by an *in vitro* stimulation with alum particles (2). This factor was shown to enhance the DNA synthesis of PHA-stimulated thymocytes and antigen-stimulated T cells, but not the DNA synthesis of LPS-stimulated B cells (3).

Recently, it was reported that rabbit endogenous pyrogen (EP) and rabbit lymphocyte stimulating factor (interleukin 1, IL 1) could not be separated chromatographically and that goat anti-rabbit EP blocked the IL 1 activity (4). It was also demonstrated that rabbit EP was produced by macrophages but not by purified PMN which were separated from the blood leukocytes or from rabbit PEC which were from 24-hr inflammatory lesion (5).

Therefore, we investigated the question whether PMN produce a lymphocyte stimulating factor or not, and what kind of lymphocyte stimulating factor PMN produce, if any.

* *This work was supported in part by a Grant-in-Aid for Scientific Research from the Ministry of Education, Science and Culture and by a Grant for intractable diseases from the Ministry of Health and Welfare, Japan.*

ISBN 0-12-527540-4

In this report, we tested PMN for the production of the lymphocyte stimulating factor, in comparison with macrophages. Then, we reached the conclusion that PMN, as well as macropahges, produce the lymphocyte stimulating factors. The evidence to support this is described in this report. Furthermore, we found that this PMN factor was mainly composed of isoelectrically distinct two factors and determined their physicochemical characteristics.

I. EVIDENCE FOR SUPPORTING THE PRODUCTION OF A LYMPHOCYTE STIMULATING FACTOR BY PMN

A. *Lymphocyte Stimulating Activity in the Cell-Free Inflammatory Exudate*

The cell-free exudate were harvested at various intervals after intraperitoneal injection of 2 ml 0.05% alum or 0.2% casein. When this exudate was assessed for their potentiation of the thymocyte DNA synthesis in the presence of 1 μl PHA, the exudate which was collected from 1 to 9 hr lesion after the casein stimulation, showed the DNA synthesis potentiation activity. The exudate from 12 to 24 hr lesion showed neither potentiation or inhibitory activity on the thymocyte DNA synthesis. The exudate from 48 to 96 hr lesion showed slight suppressive effect.

This kinetics of the production of the DNA synthesis potentiation activity in the inflammatory exudate well coincided with that of the emigration of PMN into the peritoneal cavity.

B. *Release Kinetics of the Lymphocyte Stimulating Factor from 3-hr PEC*

When casein-induced 3-hr PEC was stimulated by 0.3 mg/ml alum *in vitro*, most of the lymphocyte stimulating activity was released into the culture supernatant within 3 hr after the stimulation. On the other hand, the release of the lymphocyte stimulating activity (IL 1) from LPS-stimulated macrophages continued more than 24 hr.

C. *Protein Synthesis-Independent Production of the Factor by 3-hr PEC*

In the presence of 20 μg/ml puromycin or 20 μg/ml cycloheximide, the production of the lymphocyte stimulating factor by alum-stimulated 3-hr PEC was not inhibited at all. While, the production of IL 1 by LPS-stimulated macrophages reduced greatly in the presence of the above inhibitors of protein synthesis. This data suggests that lymphocyte stimulating factor production by 3-hr PEC did not depend on their protein synthesis. In fact, an extract from sonicated 3-hr PEC, which were not stimulated with alum, showed to contain the lymphocyte stimulating factor.

D. A Macrophage Cytotoxic Agent, 2-chloroadenosine Failed to Inhibit the Production of the Factor by 3-hr PEC

To test a possible participation of the macrophages, which were contaminated by 2% in the 3-hr PEC, in the production of the activity, we used 2-chloroadenosine to kill the macrophages. This drug is known to be cytotoxic for macrophages but not for PMN or lymphocytes.

A striking reduction of the activity in the supernatant was observed when the macrophages were stimulated with LPS in the presence of 10^{-4} M 2-chloroadenosine. On the contrary, the alum-stimulated 3-hr PEC produced the activity almost to the same degree, irrespective of the presence of 2-chloroadenosine. A trypan blue dye exclusion test revealed more than 90% of the macrophage preparation and 3% of the 3-hr PEC were judged as dead cells after 3-hr incubation with 2-chloroadenosine.

E. Stimulation Spectrum for Production of the Lymphocyte Stimulation Factor by 3-hr PEC and Macrophages

Various stimulations for 3-hr PEC and macrophages were compared in the production of the factor. As stimulants, 2 mg/ml kaolin, 0.3 mg/ml alum, 2.5 μg/ml LPS and 0.2 μg/ml muramyl dipeptide (MDP) was chosen after several preliminary experiments. In the production of the factor, kaolin was the best stimulation for 3-hr PEC and alum was the second. MDP and LPS enhanced the production of the factor by 3-hr PEC in a lesser degree. Kaolin stimulation of 3-hr PEC resulted in about 16-fold increase in the production in comparison with LPS-stimulated 3-hr PEC.

On the contrary, MDP was the best stimulation for the macrophages and LPS was the second. Alum stimulated the macrophages in a lesser degree. Kaolin inhibited the production rather than enhanced.

F. Production of the Factor by Purified PMN Preparation

From the casein-induced 3-hr PEC of mouse or rabbit, the PMN were purified by centrifugation at 400 g for 60 min on a discontinuous gradient of Percoll, consisting of 5 layers, 30%, 50%, 60%, 65% and 70%. The resulting mouse PMN in the most heavy Percoll, showed 99% in the purity. The contamination of the PMN preparation by macrophages was less than 0.02%. The rabbit PMN preparation showed a purity more than 99.99%. Practically, no contamination by other cells was observed. Both the highly purified mouse and rabbit PMN produced well the lymphocyte stimulating factor by stimulation with kaolin or alum.

All of the above evidence suggest that PMN produce the lymphocyte stimulating factor.

II. PHYSICOCHEMICAL CHARACTERISTICS OF THE PMN FACTOR

A. Isoelectric Point

The culture supernatant from alum-stimulated mouse 3-hr PEC was eluted through a column of Sephadex G-75 and the active fractions between the m.w. of 24,000 and 12,000 were collected. The active fraction was subjected for isoelectrofocusing by using an 110 ml column of 1% carrier ampholite spanning a pH range from 3 to 10. The lymphocyte stimulating activity appeared in two positions with distinct pH values; pH 10.5 and 5.4. Therefore, we called them as cationic- and neutral PMN factor, respectively. The exact pI value for the cationic PMN factor was determined as pI 9.8 by using electrofocusing at pH 9 to 11.

B. Stokes Radius

For determination of molecular characteristics of the two PMN factors, we prepared the purified ^{125}I-labeled PMN factor by using repeated gel-filtration and isoelectrofocusing. All the determination were done in the parallel study on the profiles of ^{125}I-activity and biologic activity.

Both cationic and neutral PMN factor was eluted through a calibrated column of Sephadex G-75. The Kav of cationic and neutral PMN factor was plotted against the Stokes radius. By interpolation from the plot, the Stokes radius for cationic and neutral PMN factor was calculated to be 20.2 Å and 22.2 Å, respectively.

C. Sedimentation Coefficient, S_{20W}

This was determined by ultracentrifugation in a linear sucrose density gradient. Interpolation of the relative sedimentation of the protein markers gave values of 2.3 S both for the cationic and neutral PMN factors. The biologic activity in a sucrose gradient, roughly coincided with the ^{125}I-profile, although they showed rather vague distribution.

D. Cesium Chloride Isopycnic Ultracentrifugation

The buoyant density of the two PMN factors was determined by measuring their positions in a CsCl gradient generated by ultracentrifugation. The valeus were 1.326 g/cm^3 for the cationic and 1.296 g/cm^3 for the neutral PMN factor.

Calculation of the molecular weight of the two PMN factors were done by the Svedberg equation, with the use of physical constants obtained in the above experiment. Diffusion coefficient, D_{20W} was calculated by using the value of Stokes radius. The partial specific volume of the factors were calculated from their buoyant density, assuming the molecules of the PMN factors had a net hydration value of 0.20 g H_2O/g. From these data, the molecular weight of the cationic PMN factor was calculated to be 19,000 and that of neutral PMN factor, 21,000.

The frictional ratio was estimated by comparing the experimentally determined frictional coefficient with the

TABLE I. Physicochemical Properties of PMN Factors

	Cationic PMN factor	*Neutral PMN factor*
Isoelectric point (pH)	*9.8*	*5.4*
Stokes radius (A)	*20.2*	*22.2*
D_{20W} *(cm²/sec)*	10.62×10^{-7}	9.66×10^{-7}
S_{20W} *(10^{-13} sec)*	*2.3*	*2.3*
Buoyant density (g/cm³)	*1.326*	*1.296*
Partial specific volume	*0.705*	*0.726*
Molecular weight	*19,000*	*21,000*
Frictional ratio	*1.16*	*1.22*

From Ref. (2) Nakamura et al

theoretical coefficient. The ratio was 1.16 and 1.22 for the cationic and neutral PMN factor, respectively. These values were within the range of the frictional ratio for usual proteins. The physicochemical properties of the cationic and neutral PMN factors are listed in Table I.

The biological activity of the PMN factors are closely resembled to that of macrophagic IL 1, because both of the PMN factors enhanced the DNA synthesis of PHA-stimulated thymocytes. But, the PMN factors failed to support the DNA synthesis of interleukin 2-dependent cytotoxic cell line. Both of the PMN factors were similar to IL 1 in the physical characteristics related to molecular size, however, cationic PMN factor was markedly different from IL 1 with respect to the pI value. At present, except for mouse cationic PMN factor, no IL 1-like factor with such a basic nature are known. When purified mouse PMN was used as a starting material, almost the same amount of the neutral active factor was obtained in comparison with neutral factor from 3-hr PEC. Therefore, most of the neutral factor seemed to originate from PMN.

REFERENCES

1. Yoshinaga, M., Nakamura, S., and Hayashi, H., *J. Immunol.*, 115, 533 (1975)
2. Nakamura, S., Goto, F., Goto, K., and Yoshinaga, M., *J. Immunol.*, 128, 2614 (1982)
3. Yoshinaga, M., Nishime, K., Nakamura, S. and Goto, F., *J. Immunol.*, 124, 94 (1980)
4. Murphy, P. A., Simon, P. L., and Willoughby, W. F., *J. Immunol.*, 124, 2498 (1980)
5. Hanson, D. F., Murphy, P. A., and Windle, B. E., *J. Exp. Med.*, 151, 1360 (1980)

IN VIVO PROPERTIES OF HUMAN INTERLEUKIN-1

Lawrence B. Lachman

Immunex Corporation
51 University Bldg.
Seattle, Washington

Elisha Atkins

Department of Medicine
Yale University School of Medicine
New Haven, Connecticut

Ralph F. Kampschmidt

The Samuel Roberts Noble Foundation
Ardmore, Oklahoma

I. PRODUCTION OF HUMAN INTERLEUKIN 1

Interleukin 1 (IL 1) was prepared by culturing human acute monocytic leukemia (AMoL) cells in minimal essential medium (MEM) containing 5% human serum (NHS) and 10μg/ml of *E. coli* LPS for 48 hrs. Previous studies have demonstrated that AMoL cells release large quantities of IL 1 (1) and that similar to peripheral blood monocytes, AMoL cells require stimulation with LPS to release IL 1 (2). All studies described in this report were conducted using cells obtained by theraputic leukophoresis of a single AMoL patient. The cells were stored frozen at -56° in MEM containing 5% NHS with 10% DMSO.

ISBN 0-12-527540-4

II. PURIFICATION OF HUMAN IL 1

IL 1 was purified from AMoL cell conditioned medium exactly as previously described (3). Briefly, IL 1 was purified by diafiltration and ultrafiltration using a hollow fiber device of 50,000 MW cut-off, ultrafiltration with a 10,000 MW cut-off membrane and sucrose gradient isoelectric focusing (IEF). The IEF-purified IL 1 had been purified about ~ 50,000-fold, contained ~ 4% of the original biological activity and was about 10% pure (2). The IEF-purified IL 1 could be further purified to homogeneity by SDS-polyacrylamide gel electrophoresis, but with very large loss of biological activity (2).

III. IEF-PURIFIED IL 1 IS ENDOTOXIN-FREE

Although no special precautions were taken to purify the IL 1 using endotoxin-free conditions, the IEF-purified IL 1 was found to be endotoxin-free. A 0.2 ml sample was consistantly negative in the Limulus amebocyte lysate assay for endotoxin. The explanation for this finding was that IEF effectively separates IL 1, which has an isoelectric point of pH 7, from endotoxin which has an isoelectric point of pH 3.0 to 5.0. The absence of endotoxin is of critical importance for *in vivo* testing since endotoxin contamination could result in spurious effects.

IV. THE PYROGENIC PROPERTY OF IEF-PURIFIED IL 1

The ability to elevate the body temperature of rabits was tested for the IEF-purified IL 1 from two purification procedures. A 1.0 ml sample of IEF-purified IL 1 was mixed with an equal volume of endotoxin-free isotonic saline and the sample was injected in equal portions intraveneously into two rabbits. A monophasic fever of 1°, typical of that described for EP (4), was experienced by both rabbits. The fever was not characteristic of an endotoxin fever (4). In a subsequent experiment, rabits made tolerant to endotoxin (4) were injected with IEF-purified IL 1. These animals also experienced a fever, which further demonstrates that the pyrogenic property of the IEF-purified IL 1 was not due to endotoxin.

V. THE EFFECT OF IEF-PURIFIED IL 1 ON THE LEVEL OF CIRCULATING NEUTROPHILS

Increasing amounts of IEF-purified IL 1 from 0.01 to 0.10 ml were injected intraveneously into normal rats. One hour following injection the level of circulating neutrophils in a heart blood sample was determined. As shown in Table I, IEF-purified IL 1 resulted in a dose-dependent increase in the number of circulating neutrophils. It should be noted that endotoxin cannot elevate the level of circulating neutrophils at one hour after injection (5), and thus this property of IL 1 clearly distinguishes the factor from residual endotoxin which might have been undetectable in the Limulus lysate assay.

Table I

The Effect of Intravenously Injected IEF-Purified IL 1 on the Level of Circulating Neutrophils[a]

Number of Trials	Dose of IL 1 Injected	Number of Blood Neutrophils	Difference from Normal
	(ml/animal)	(cells/mm^3±SE)	
7	----	1334 ± 154	0
7	0.010	1335 ± 184	0
5	0.020	3405 ± 481	2070
4	0.050	5557 ± 422	4222
3	0.075	6881 ± 1384	5476
4	0.100	8727 ± 1439	7392

[a]200 gm female Holtzman rats were injected with IEF-purified IL 1, or saline, in the amount indicated. One hour following injection the number of circulating neutrophils was determined on a sample of heart blood by counting the number of leukocytes and by differential counting of a Wright's stained smear.

VI. THE EFFECT OF IEF-PURIFIED IL 1 ON FIBRINOGEN LEVELS IN PLASMA

The plasma levels of fibrinogen were measured 24 hours following the intravenous injection of 0.25 ml of IEF-purified IL 1, or saline, into rats (6). The plasma levels of fibrinogen were significantly increased over controls, showing an average increase of 44 mg/100 ml of plasma.

VII. ESTIMATION OF THE BIOLOGICAL ACTIVITY OF IL 1 IN THE THYMOCYTE ASSAY AS COMPARED TO THE *IN VIVO* ASSAYS

Table II summarizes the effect of IL 1 in four assay systems. The IEF-purified IL 1 used in these experiments was from two separate purifications. The thymocyte stimulating activity of the IEF-purified IL 1 was nearly identical for both purifications and equaled 63.0 units/ml. The results demonstrate that the mouse thymocyte assay is the most sensitive assay for IL 1 since it can readily detect 0.01 units of IL 1 activity. This is demonstrated by noting that the IEF-purified IL 1 was 63.0 units/ml and the thymocyte stimulating activity was measurable when 0.1 ul of this material was added to 0.1 ml of thymocytes (0/1% v/v). In order of increasing sensitivity the thymocyte assay>neutrophil increase>fibrinogen increase>fever. The thymocyte assay is of course a microassay requiring only microliter amounts of IL 1.

VIII. CONCLUSION

The data presented, and other published reports (7), offer strong evidence that IL 1 released by activated monocytes or tissue macrophages may initiate a spectrum of responses which include fever, granulocytosis, and the release into the plasma of acute phase proteins. Thus, IL 1 may be implicated as a mediator of inflammation in chronic inflammatory diseases such as arthritis and glomerulonephritis. Agents which either neutralize IL 1 or inhibit the release of IL 1 from macrophages could be future anti-inflammatory drugs.

Table II

Comparison of the Units of IL 1 Injected and The Biological Properties Observed

Units of IL 1 Injected[a]	Response
31.0	1° Fever
6.3	Increase of 8,000 Neutrophils/ mm^3 of blood
15.8	Fibrinogen increase of 44mg/100ml of plasma

[a]1.0 ml of IEF-purified IL 1 contained 63.0 units of thymocyte stimulating activity. An aliquot of 0.01 units can stimulate (^{3}H)TdR incorporation by mouse thymocytes equal to five times the level of unstimulated cells.

REFERENCES

1. Lachman, L.B., Moore, J.O., and Metzgar, R.S., Cell. Immunol. 41, 199, (1979).
2. Lachman, L.B., Fed. Proc., in press.
3. Lachman, L.B., Page, S.O., and Metzgar, R.S., J. Supramol. Struct. 13, 457, (1980).
4. Bernheim, H.A., Block, L.H., Francis, L., and Atkins, E. J. Exp. Med., 152, 1811, (1980).
5. Kampschmidt, R.F. and Upchurch, H.F., J. Reticuloendothelial Soc., 28, 191, (1980).
6. Kampschmidt, R.F. Upchurch, H.F., and Pulliam, L.A., Ann. New York Acad. Sci., 389, 338, (1982).
7. Sztein, M.B., Vogel, S.N., Sipe, J.D., Murphey, P.A., Mizel, S.B., Oppenheim, J.J., and Rosenstreich, D.L., Cell. Immunol., 63,164, (1981).

DISCUSSION

GERARD: Did Lachman try to mix his six different monoclonal antibodies in order to effect even greater inhibition of IL-1 activity?

LACHMAN: No, we did not.

SCALA: Are all AMML IL-1 producers?

LACHMAN: All AMML and AMoL are positive for IL-1 production. AML, i.e., acute myelocytic leukemia is negative for IL-1 production.

SCALA: What were Lachman's controls for ELISA and for immuno-affinity columns?

LACHMAN: I used clones which were not inhibitory in the assay.

PALACIOS: We have tested several clones from U937 cells for IL-1 production. We found no clone of U937 cells which produced high levels of IL-1 activity both spontaneously and after PMA stimulation (Europ. J. Immunol., in press).

LACHMAN: Some U937 cells are mycoplasma-contaminated and will not make IL-1. I am pleased to know that Palacio's is mycoplasma-free and does produce IL-1.

FARRAR: In seeking to make antibody to IL-1, has anyone tried to create a "toxoid" out of IL-1 by some type of denaturation so as to maintain antigenicity but destroy the biological activity?

LACHMAN: Not to my knowledge.

OPPENHEIM: In response to Farrar's query, I would say that the problem with "denaturing" the IL-1 would be that it would likely also destroy or greatly reduce its immunogenicity.

EPIDERMAL CELL THYMOCYTE ACTIVATING FACTOR (ETAF) EFFECTS DIFFERENT INFLAMMATORY CELLS

Thomas A. Luger[1]
Marcelo B. Sztein
Jacques A. Charon
Joost J. Oppenheim

Laboratory of Microbiology and Immunology
National Institute of Dental Research
National Institute of Health

I. INTRODUCTION

The predominant cell type within the epidermis, the keratinocyte has recently been shown to perform nonspecific immunoregulatory functions through the secretion of factors similar to macrophage-derived Interleukin 1 (IL 1) (1,2). Since this activity is detected by the same thymocyte costimulator assay which is used for the detection of IL 1 activity, this factor was named epidermal cell derived thymocyte activating factor (ETAF) (1,2). ETAF is produced by freshly isolated epidermal cells devoid of the macrophage-like Langerhans cells (2) and by human squamous cell carcinoma cell lines (SCC) (3) as well as murine transformed keratinocyte cell lines (Pam 212) (1). ETAF is released spontaneously _in vitro_ by normal and malignantly transformed keratinocytes. However, mechanical disruption of a confluent monolayer, plating keratinocytes under serum free conditions or addition of endotoxin (LPS), silica, phorbol myristate acetate (PMA) or Concanavalin A (Con A) to keratinocyte cultures all significantly increased supernatant ETAF activity (4). Although the exact mechanism of action of these stimulants is not clear, increased ETAF production always was associated with diminished cell viability and either with

[1]On leave from the II Department of Dermatology, Univerity of Vienna, Austria. Recipient of a Max Kade Foundation Fellowship.

Interleukins, Lymphokines, and Cytokines

ISBN 0-12-527540-4

a prolongation of the G_1 phase of the cell cycle or a shift of an increased number of keratinocytes into the G_1 phase, suggesting that ETAF release by keratinocytes is cell cycle dependent.

II. BIOCHEMICAL CHARACTERIZATION OF ETAF

The biochemical properties of human as well as murine ETAF have recently been reviewed (5,6). Studies of ETAF derived from either human (SCC) or murine (Pam 212) keratinocyte cell lines revealed that this factor is a low molecular weight polypeptide which is relatively resistant to attack by proteolytic enzymes and sodium dodecyl sulfate (SDS) (6). Both human and murine ETAF elute from DEAE sephacel at a sodium chloride concentration of approximately 150 mM (1,6) and interact with hydrophobic phenyl sepharose (6). The only biochemical difference between murine and human ETAF is that murine ETAF, like murine IL 1, has only one peak on isoelectrofocusing with a pI of 5.2 (4), whereas human ETAF, like human IL 1, has 3 isoelectric points at pI 7.2, 6.0 and 4.8 (3). Since the biochemical properties ot ETAF are identical to those of IL 1 (7) keratinocytes appear to release the same immunopotentiating signal as macrophages.

III. BIOLOGICAL EFFECTS OF ETAF

ETAF affects a wide variety of lymphocytic as well as non-lymphoid cells (5,6). According to these studies ETAF promotes the activation of the subset of immunocompetent peanut lectin non-agglutinating (PNA^-), Lyt 2^- thymocytes (4). ETAF also enhances lectin stimulated peripheral blood lyphocytes to produce significantly increased levels of Interleukin 2 (IL 2)(1,2). In addition ETAF has been shown to directly affect nonlymphocytic inflammatory cells. ETAF is chemotactic for polymorphonuclear leukocytes (PMN) and mononuclear leukocytes (MNL) (3,8). ETAF also is directly mitogenic for fibroblasts in culture (4). In vivo administration of ETAF induces acute phase protein production such as serum Amyloid A (SAA) by murine hepatocytes (9) and causes fever in rabbits (6). These biological properties of keratinocyte derived ETAF are identical with those of macrophage derived IL 1, suggesting that both ETAF and IL 1 are identical (10).

It is not yet clear whether this diversity of biological effects is due to one molecule or due to several species of ETAF or IL 1. Therefore in the present study we investigated the effect of biochemically different fractions of ETAF on different target cells.

IV. DISSOCIABLE EFFECTS OF ETAF ON INFLAMMATORY CELLS

IL 1 has been demonstrated to be directly mitogenic for human dermal fibroblasts (CRL 1445) (11). Similarly human as well as murine ETAF were mitogenic for fibroblasts. The fibroblast activating effect of human ETAF coeluted with the thymocyte growth enhancing activity upon AcA 54 gel filtration chromatography. Subsequently to gel fitration the three moieties of ETAF activity detected by chromatofocusing (3,4) in the thymocyte costimulator assay were also active in the fibroblast growth stimulating assay (table I). In addition, ETAF stimulated fibroblasts *in vitro* to produce more fibronectin (Charon *et al*. unpublished observation). Thus ETAF released by keratinocytes may participate in normal cutaneous wound healing and scar formation during chronic inflammation.

TABLE I. Effect of human ETAF on human fibroblasts

	ETAF (U/ml)	^{3}H-TdR uptake by fibroblasts (cpm ± SE)[a]
RPMI 1640	0	1,060 ± 111
ETAF (S-200)[b]	25	5,287 ± 300
ETAF[c] α (pI 7.2)	8	4,612 ± 379
β (pI 6.0)	12	7,573 ± 410
γ (pI 4.8)	7	3,305 ± 21

a) $1x10^4$ CRL 1445 fibroblasts were incubated at 37°C in 5% CO_2 containing atmosphere for 48 hours and then pulsed with ^{3}H-TdR for the final 18 hours of incubation.
b) Pooled activity in fractions eluted off Sephacryl S-200 gel filtration.
c) Pooled activities of ETAF obtained by chromatofocusing.

Partially purified human ETAF as well as IL 1 were chemotactic for polymorphonuclear leukocytes (PMN) and mononuclear leukocytes (MNL) (3,8) as determined by a modified Boyden chamber assay. However when separated by isoelectrofocusing or chromatofocusing into 3 peaks which all contained similar levels of thymocyte growth enhancing activity, in two experiments the chemotactic activity for PMN differed from that for MNL. Human ETAF at pI 4.8 was most chemotactic for MNL, whereas ETAF at pI 7.2 was more active on PMN (table II). This

TABLE II. Chemotactic effect of human ETAF on human peripheral blood polymorphonuclear leukocytes (PMN) and mononuclear leukocytes (MNL).

	ETAF (U/ml)	Chemotactic response (cells/field ± SE)	
		PMN	MNL
Control[a]	–	22 ± 5	20 ± 3
ETAF (S-200)[b]	15	96 ± 10	44 ± 6
ETAF[b] α (pI 7.2)	12	220 ± 21	20 ± 5
β (pI 6.0)	12	85 ± 15	25 ± 7
γ (pI 4.8)	12	30 ± 7	72 ± 8

a) Phophate buffered saline (PBS)
b) Pooled activity in fractions eluted from Sephacryl S-200 gel filtration and dialysed against PBS.
c) Pooled activities of ETAF obtained by chromatofocusing and dialysed against PBS.

data suggests that distinct species of ETAF may exhibit different biological activities.

In vivo admninistration of partially purified ETAF, like IL 1 (pooled activity off gel filtration) resulted in increased serum Amyloid A production by hepatocytes of C3H/HeJ mice (9). ETAF at pI 7.2 and pI 6.0 were most active in inducing SAA production. In contrast ETAF at pI 4.8, although active in the thymocyte costimulator assay induced only low levels of serum amyloid A (table III). Therefore ETAF may consist of different molecules with varied in vivo capacity to induce hepatocytes to produce acute phase proteins.

In addition to the data presented herein, indicating distinct biological effects of different isoelectric forms of ETAF, variations of different moieties of endogenous pyrogen (EP) have been reported. EP at pI 7.2 was most active in releasing bone marow neutrophils, whereas EP at pI 4.8 enhanced thymocyte proliferation but did not increase peripheral blood PMN (12). Recently it has been demonstrated that EP and IL 1 may be identical (13,14). Similarly ETAF containing preparations were able to induce fever when injected into rabbits (6) suggesting that ETAF or some species of ETAF may be involved in inducing fever.

TABLE III. Induction of murine serum amyloid A (SAA) production by human ETAF.

	ETAF (U/ml)	SAA[a] (μg/ml)
LPS (2.5 μg/ml)	-	1.8 ± 0.8
SAA Inducer[b]	10.0	45.0 ± 8.3
ETAF (S-200)[c]	6.0	42.1 ± 4.2
ETAF[d] α (pI 7.2)	1.5	37.7 ± 5.1
β (pI 6.0)	3.0	47.5 ± 8.8
γ (pI 4.8)	6.0	11.0 ± 3.3

a) SAA activity was measured by a radioimmunoassay using an affinity chromatography purified anti-AA antibody. The results are expresed as mean ± SE of 2 different experiments each consisting of 5 mice.
b) Serum Amyloid A inducer, supernatant from LPS stimulated thioglycolate induced C3H/HeN peritoneal adherent cell cultures.
c) Pooled activity in fractions off Sephacryl S-200 gel filtration.
d) Pooled activities of ETAF obtained by chromatofocusing.

In a more recent study the chemotactic effect of the 3 different pools of ETAF was also tested by an agarose micro droplet assay (8). These data also indicated that ETAF, partially purified by gel filtration exhibited significant chemotactic effects on both PMN and MNL. In contrast, assays of the different pools obtained by chromatofocusing in two experiments revealed only ETAF at pI 6.0 to be chemotactic for PMN, whereas all 3 pools of ETAF stimulated MNL to migrate. The reasons for these differences from the previous results are unclear. Perhaps they can be attributed to differences in the bioassays. For example, the exposure time in Boyden chamber is only 1 - 2 hours, whereas migration in the micro droplet assay is evaluated after 18 hours. Doses of ETAF yielding optimal responses in the agarose assay may also differ. In addition the charge heterogeneity of ETAF may represent an artifact that is attributable to in vitro biochemical manipulations. Although ETAF clearly is chemotactic for PMN as well as MNL, further experimentations are needed to establish whether the electrophoretically different ETAF activities consistently have different biological activities.

ACKNOWLEDGMENTS

We would like to thank Drs. T. Kasahara, T. Krakauer and P. Steeg for the critical discussion of the manuscript.

REFERENCES

1. Luger, T.A., Stadler, B.M., Katz, S.I. and Oppenheim, J.J., J. Immunol. 127, 1493 (1981).
2. Sauder, D.N., Carter, C., Katz, S.I. and Oppenheim, J.J., J. Invest. Derm. 79, 34 (1982).
3. Luger, T.A., Stadler, B.M., Luger, B.M., Sztein, M.B., Schmidt, J.A., Hawley-Nelson, P., Grabner, G. and Oppenheim J.J., J. Invest. Derm. (submitted) (1982).
4. Luger, T.A., Stadler,B.M, Luger, B.M., Mathieson, B.J., Mage, M., Schmidt, J.A. and Oppenheim J.J., J. Immunol. 128, 2147 (1982).
5. Luger, T.A. and Oppenheim, J.J., in "Advances in inflammation research" Vol. 6, (G. Weissmann, ed.) (in press) Raven Press, New York (1982).
6. Luger, T.A., Sztein, M.B., Schmidt, J.A., Murphy, P., Grabner, G. and Oppenheim, J.J.,Fed. Proc. (in press)(1982).
7. Mizel,S.B. in "the Lymphokines, Biochemistry an biological activity" (J.W. Hadden, W.E. Steward II, ed.), p. 347 Humana Press, New Jersey, (1981).
8. Luger, T.A., Charon, J.A. and Oppenheim, J.J., J. Immunol. (submitted) (1982).
9. Sztein, M.B., Luger, T.A. and Oppenheim, J.J., J. Immunol. 129, 87 (1982).
10. Oppenheim, J.J. and Gery, I., Immunol. Today 3, 113 (1982).
11. Schmidt, J.A., Mizel, S.B., Cohen, D. and Green, I., J. Immunol. 128, 2177 (1982).
12. Kampschmidt, R.F. in "the physiologic and metabolic role of the host" (M.C. Powanda, P.G. Canonico, ed.), p. 55. Elsevier, North Holland, Biomedical Press (1981).
13. Murphy, P.A., Simon, P.L. and Willoughby, W.F., J. Immunol. 126, 2498 (1980).
14. Rosenwasser, L.J., Dinarello, L.A. and Rosenthal, A.S., J. Exp. Med. 150, 709 (1979).

DISCUSSION

DAVID: Do the α, β, and γ species of IL-1 exert the same effect on PMN and monocyte chemotaxis and SAA production?

LUGER: Partially purified IL-1, pooled activity off gel filtration, was also chemotactic for PMN and mononuclear cells. The different immunofocusing or isoelectric focusing pools of IL-1 have not yet been tested.

ROSENWASSER: Dinarello, McAdam and I have demonstrated that human LP/IL-1 purified by chromatography and isoelectric focusing, have manifested pI's of 6.8-7.0, 5.6 and 4.8-5.0. All three pI's produced fever, IL-1(LAF) activity and the production of SAA; moreover these activities were equivalent for three molecular species of LP/LAF/IL-1/SAASF.

LUGER: Our experience is generally similar to the results of Rosenwasser and associates. We too find that all 3 CF pools of ETAF were able to induce SAA. Moreover, the finding that ETAF at pI.4.8 was less active, may be due to varied forms of ETAF inducing different species of SAA. The effect of the 3 CF pools on fever induction has not yet been examined.

VADAS: Could Luger enlarge on his studies of chemotactic responses to ETAF? Did he measure random migration and response of ETAF-incubated cells to a known chemoattractant? Without such controls one could interpret his studies as showing that all moieties are equally chemoattractant. However, some do "activate" neutrophils to become more sticky and thus prevent all forms of its movement.

LUGER: We did not measure migration of ETAF-activated cells to a chemoattractant. In our experience, PMN were stimulated to release enzymes and became more sticky only when much higher doses of BTAF (>60u/ml) were used. Therefore it is not very likely that BTAF, at the concentrations used in these experiments, would prevent migration.

OPPENHEIM: I really do not understand what Vadas is driving at, since as I comprehend Luger's work, the same cells can migrate in response to equivalent units/ml of one but not another of the species of ETAF.

VADAS: I do not deny the heterogeneity of the various ETAF's. What I am disputing is the *interpretation* of the "chemotactic" data. It is of some significance whether these substances *cause* or *prevent* cell movement.

MURPHY: What amount of ETAF is required to produce a thymocyte response? Is there a difference between ETAF and monocyte-derived IL-1 in specific activity?

LUGER: ETAF, like IL-1, is active at very low, not readily measureable protein concentrations.

GERARD: Did Luger try to reduce ETAF heterogeity by neuraminidase treatment?

LUGER: No. We have not yet done neuraminidase treatment of ETAF.

McADAM: Luger's data and the title of his report suggest that he has demonstrated that ETAF stimulates hepatocytes to synthesize SAA. Did he demonstrate the activity of ETAF on isolated hepatocytes in culture, or by injecting ETAF into test animals? Since there is evidence for SAA heterogeneity and different sites of SAA synthesis, it may be premature to conclude that ETAF necessarily stimulates hepatocytes or that it *only* stimulates hepatocytes to produce SAA.

LUGER: We induced SAA production by injecting different preparations of ETAF into C_3H/He mice. Although in this system it is possible that SAA production is stimulated at different sites, it remains most likely that acute phase protein, such as SAA, is due to stimulation of hepatocytes.

EFFECTS OF RABBIT ENDOGENOUS PYROGENS WITH DIFFERENT ISOELECTRIC POINTS UPON PLASMA IRON, NEUTROPHILS, FIBRINOGEN, AND IL1 ACTIVITY

Ralph F. Kampschmidt
Herbert F. Upchurch
Larry Pulliam
Roger Mitchell

Biomedical Division
The Samuel Roberts Noble Foundation, Inc.
Ardmore, Oklahoma

Most of the studies on endogenous pyrogen (EP) have been done with either cells from human peripheral blood monocytes or from rabbit acute peritoneal exudates induced with shellfish glycogen (1)(2). Both of these sources yield a pyrogen with a molecular weight of 12,000 to 16,000 Daltons and an isoelectric point at approximately 7.3 (3)(4). In addition to its pyrogenic activity purified preparations of these EP's will also promote increased acute phase protein synthesis, alterations in plasma metals and neutrophilia (5-8). The mediator for these activities has been called leukocytic endogenous mediator (LEM), but it now seems likely that EP and LEM may be the same protein (5-8). The LEM activities of these pyrogens can be shown in rabbits, rats and mice (9-11).

Recently it has been found that rabbit peritoneal cells removed four days after the injection of mineral oil produced four different endogenous pyrogens (12). One of these seems to be the same pyrogen as the one obtained from rabbit peritoneal exudates elicited with shellfish glycogen (12). The others have isoelectric points (pI) near pH 5 and this report presents some of their biological effects when injected into rabbits or rats.

ISBN 0-12-527540-4

I. COMPARISON OF THE ALTERATIONS PRODUCED IN RABBITS AND RATS AFTER INJECTING pI 7 OR pI 5 RABBIT ENDOGENOUS PYROGEN

Procedures and special precautions for producing pI 7.3 EP/LEM have been previously described (13)(14). Crude material containing predominately pI 5 endogenous pyrogens was obtained by injection of 50 ml sterile mineral oil into rabbits. After four days the cells were washed from the peritoneal cavity with 0.15 M saline and the oil removed in a separatory funnel. The cells (approximately 90% macrophages) were filtered through sterile gauze, centrifuged and washed with Eagles minimal essential medium (MEM). They were activated by incubating 5 x 10^6 macrophages/ml in MEM containing 5% rabbit serum and 0.2 μg/ml of *Salmonella abortus* endotoxin. The cells were activated for two hrs. at 37°C, centrifuged and washed with serum-free MEM. The cells were incubated at 37° for 16 hrs. and the supernatant was the crude pI 5 EP.

Both of the crude preparations of EP were further purified by Sephadex G-50 chromatography (11), and isoelectric focussing with 3-10 pH range ampholytes on a bed of G-75 superfine Sephadex gel with an LKB 2117 multiphor apparatus. (LKB Instruments Inc., Rockville, MD.)

Methods for determining temperature (15), plasma iron and zinc (13), blood neutrophils (15) and fibrinogen (11) have been previously described. A comparison of the effects of PI 5 and PI 7 pyrogens in rabbits is presented in Table I.

TABLE I. Comparison of the Effects of pl 5 and pl 7 Rabbit Endogenous Pyrogen in Rabbits

Biological manifestation	*No. trials*	*Saline controls*	*EP at pl 7*	*5*
Pyrogenic unit[a]	*17*	*0*	*1.0*	*1.0*
Neutrophils/mm^{3}x10^{-2}	*23*	*39±4*	*94±5*[b]	*106±13*[b]
Plasma Iron μg/100ml	*13*	*198±17*	*88±8*[b]	*102±19*[b]
Fibriongen mg/100ml	*23*	*319±23*	*441±32*[b]	*302±17*

[a]*Amount needed to produce 1°fever (3).*
[b]*Significantly different from control p<0.01.*

TABLE II. Comparison of the Effects of pI 5 and pI 7 Rabbit Endogenous Pyrogen in Rats

Biological manifestation	*No. trials*	*Saline controls*	*EP at pI 7*	*EP at pI 5*
Pyrogenic unit[a]	36	0	1.0	1.0
Plasma Iron µg/100ml	31	235±10	127±6[b]	231±9
Plasma Zinc µg/100ml	24	100±2	43±2[b]	107±3
Plasma fibrinogen mg/100ml	36	225±7	345±14[b]	231±7

[a]*One unit equals 1°fever/rabbit (3).*
[b]*Significantly different from control $p<0.001$.*

Both pyrogens were injected iv at a dose of one pyrogenic unit (3). These pyrogens produced approximately equal effects upon plasma iron and peripheral blood neutrophils. The pI 5 EP apparently had little or no effect upon fibrinogen. When one pyrogenic unit of these materials was injected into rats it produced the effects shown in Table II. Rats showed consistent and predictable increases in plasma fibrinogen and decreases in plasma iron and zinc when pI 7 EP prepared from rabbit acute peritoneal exudates was injected at a dose of one pyrogenic unit. The pI 5 EP at a dose of one pyrogenic unit had no effect on any of these biological parameters. The effects of varying pyrogenic units of both of these EP's upon rat peripheral blood neutrophils at 1 hr. after an iv injection is shown in Figure 1. The pI 7 pyrogen caused a significant increase in neutrophils when a dose of 0.07 pyrogenic units was used. Doses of 0.1 pyrogenic unit or more resulted in a 6-to-8 fold increase in neutrophils. The pI 5 pyrogen did not significantly alter neutrophils at any dose between 0.1 and 1.0 pyrogenic unit.

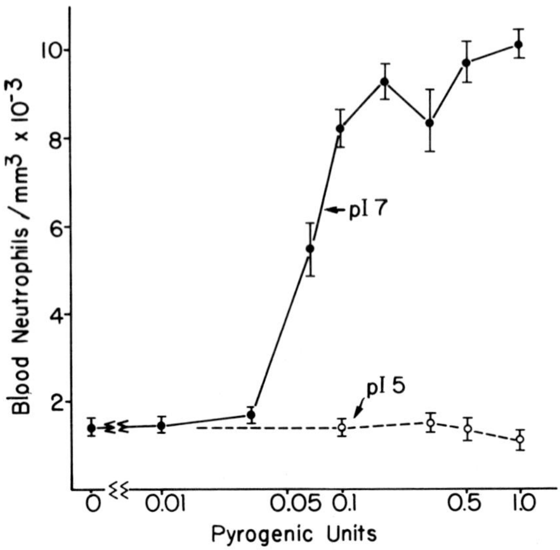

FIGURE 1. Effect of varying doses of pI 7 and pI 5 endogenous pyrogen on the blood neutrophils of rats. The method used to determine pyrogenic units has been described (3).

II. INTERLEUKIN-1 ACTIVITY OF pI 7 AND pI 5 RABBIT ENDOGENOUS PYROGEN

There has been a number of investigations showing that highly purified EP from a number of different species will stimulate the uptake of tritiated thymidine (^{3}H-TdR) by mouse thymocytes (7)(16)(17)(18). A comparison of thymocyte proliferation resulting from pI 5 and pI 7 EP at various pyrogenic doses is presented in Figure 2. The pI 7 EP was more active than the 5 at the lower, but not at higher pyrogenic doses.

III. THE EFFECTS OF SILICA ON THE PRODUCTION OF pI 5 AND pI 7 EP BY OIL ELICITED RABBIT PERITONEAL MACROPHAGES

Gery and coworkers (19) recently demonstrated that silica increased the production and release of IL-1 from cultured murine peritoneal macrophages. We found that the

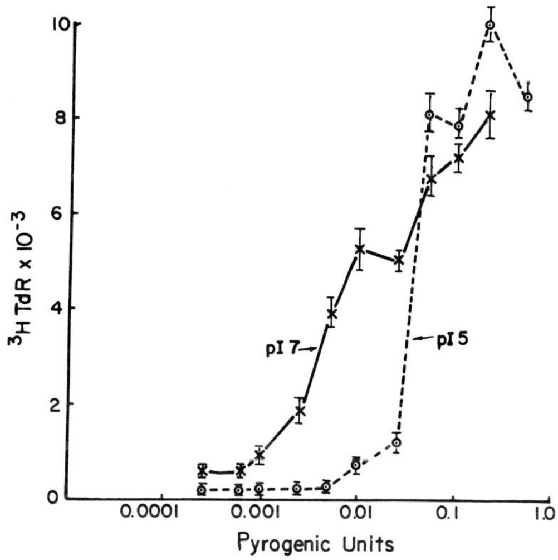

FIGURE 2. A comparison of varying doses of pI 7 and pI 5 endogenous pyrogen upon the uptake of ^{3}HTdR by mouse thymocytes.

addition of 50 μg/ml of silica to the activation and incubation media also increased IL-1 release from oil elicited rabbit peritoneal macrophages. Fractionation of this activity by Sephadex G-50 and isoelectric focussing indicated that there was an increase of several fold in the amount of pI 5 and pI 7 EP produced. In addition there was a release of a higher molecular weight IL-1 which was not pyrogenic and which had no effect on plasma iron, fibrinogen or blood neutrophils.

IV. DISCUSSION

Even though the pI 5 and pI 7 EP are both believed to be produced by rabbit peritoneal macrophages they have different physical properties (12)(20)(21). The results of these studies suggest that they also have different biological activities. The pI 7 EP will decrease plasma iron, cause neutrophilia and increase plasma fibrinogen in either rabbits, rats or mice (9-11). The pI 5 will only reproduce a small portion of these activities. We have made frequent attempts to shift rabbit peritoneal macrophages from the production of one isoelectric point EP to the other. Although these attempts have been moderately successful it is still not possible to attribute these shifts to a particular lymphocyte or granulocyte product.

ACKNOWLEDGMENTS

We wish to thank Mrs. Eva Franks and Mr. Martin Worthington for expert technical assistance.

REFERENCES

1. Atkins, E., and Bodel, P. in "The Inflammatory Process" 2nd Edition (B. W. Zweifach, L. Grant, and R. T. McCluskey eds.) Vol. 3, p. 467. Academic Press, New York, (1974).
2. Dinarello, C. A., *Fed. Proc. 38,* 52 (1979).
3. Murphy, P. A., Chesney, P. J., and Wood, W. R. Jr., *J. Lab. Clin. Med. 83,* 310 (1974).
4. Dinarello, C. A., Renfer, L., and Wolff, S. *Proc. Natl. Acad. Sci. USA, 74,* 4623 (1977).
5. Merriman, C. R., Pulliam, L. A., and Kampschmidt, R. F., *Proc. Soc. Exp. Biol. Med. 154,* 224 (1977).
6. Bornstein, D. L. and Walsh, E. C., *J. Lab. Clin. Med. 91,* 236 (1978).
7. Sztein, M. B., Vogel, S. N., Sipe, J. D., Murphy, P. A., Mizel, S. B., Oppenheim, J. J., and Rosenstreich, D. L., *Cell. Immunol. 63,* 164 (1981).
8. Kampschmidt, R. F. in "Infection: The Physiologic and Metabolic Responses of the Host" (M. C. Powanda and P. G. Canonico, eds.) p. 55. Elsvier Biomedical Press, Amsterdam (1981).
9. Merriman, C. R., Pulliam, L. A., and Kampschmidt, R. F., *Proc. Soc. Exp. Biol. Med. 149,* 782 (1975).
10. Kampschmidt, R. F., *J. Reticuloendothelial Soc. 23,* 287 (1978).
11. Kampschmidt, R. F., Pulliam, L. A., and Upchurch, H. F., *J. Lab. Clin. Med. 95,* 616 (1980).
12. Cebula, T. A., Hanson, D. F., Moore, D. M., and Murphy, P. A., *J. Lab. Clin. Med. 94,* 95 (1979).
13. Kampschmidt, R. F., Pulliam, L. A., and Merriman, C. R., *Am. J. Physiol. 235,* C118 (1978).

14. Kampschmidt, R. F., and Upchurch, H. F., *Am. J. Physiol. 216,* 1287 (1969).
15. Kampschmidt, R. F., and Upchurch, H. F., *J. Reticloendothelial Soc. 28,* 191 (1980).
16. Rosenwasser, L. J., Dinarello, C. A., and Rosenthal, A. S., *J. Exp. Med. 150,* 709 (1979).
17. Murphy, P. A., Simon, P. L., and Willoughby, W. F., *J. Immunol. 124,* 2498 (1980).
18. Kampschmidt, R. F., Upchurch, H. F., and Pulliam, L. A., *Annals N.Y. Acad. Sci. 389,* 338 (1982).
19. Gery, I., Davies, P., Derr, J., Krett, N., and Barranger, J. A., *Cell. Immunol. 64,* 293 (1981).
20. Murphy, P. A., Cebula, T. A., Levin, J., and Bradford, E. W., *Infect. Immun. 34,* 177 (1981).
21. Murphy, P. A., Cebula, T. A., and Windle, B. E., *Infect. Immun. 34,* 184 (1981).

DISCUSSION

ROSENWASSER: With the pI5 LAF from rabbit macrophates, that manifests no pyrogenic activity, does it express such activity when concentrated?

KAMPSCHMIDT: The pyrogenic activity is *much* lower than we would expect from the LAF activity. It is of course possible that concentrated material would be pyrogenic. This is the high MW fraction from a Sephadex G50 column. We presently have no information about its pI or other properties.

LANDY: I am impressed with Kampschmidt's demonstration of the striking effects of silica on the release by macrophages of endogenous pyrogen (and other biologic activities). This is noteworthy; we would do well to recall the injurious and toxic effects of silica on these cells. Our attention should be directed to a most relevant report by Gery et al., Cellular Immunol. *64*:293(1981) in which they explored in detail the factors and situations that lead to *production* by macrophages of IL-1 and the quite separate and at times bizarre situations that make for its *release*. These mechanisms are worthy of continued and more intensive investigation if we are to attain a real understanding of the complex interactions between "active" agents and macrophages that lead to evocation of monokines.

KAMPSCHMIDT: I agree, and that is why we are looking for other possibly related biological activities which are also released. Our report certainly acknowledges the important work by Gery and his associates.

WOOD: Would Kampschmidt elaborate on his claim that he can switch pI 7 IL-1 to pI 5 and vice versa?

KAMPSCHMIDT: We cannot change the proteins *after* they are released by macrophages. We alter the incubation medium for the macrophages by adding sonicates of granulocytes or lymphocytes. Under *these* conditions, when the macrophages are activated, they frequently switch the type of pI released.

DUMONDE: Can endogenous pyrogen be demonstrated in the circulation of animals undergoing febrile reactions?

KAMPSCHMIDT: Yes, it can. A heat-labile pyrogen can be transferred with the plasma of endotoxin-injected rabbits. This activity has been shown not to be due to his residual endotoxin (see also the comment by Rosenwasser) -- that in these circumstances IL-1 cannot be measured because of an inhibitor).

ROSENWASSER: Dinarello, Wolff, and I have examined the sera of endotoxin-treated normal volunteers for LP/IL-1 activities at the peak of fever by utilizing a rabbit anti-LP, IL-1 anti-serum and the thymocyte proliferation assay for IL-1. We were unable to demonstrate circulating LAF/LP/IL-1 but were able to identify a competitive inhibitor of human IL/LAF/LP (J. Immunol. *127*:2517, 1981).

PRODUCTION OF LYMPHOCYTE-ACTIVATING FACTOR (LAF) IN THE ABSENCE OF ENDOGENOUS PYROGENS (EP) BY RABBIT OR HUMAN LEUKOCYTES STIMULATED BY A MURAMYL DIPEPTIDE DERIVATIVE[1]

C. DAMAIS
G. RIVEAU
M. PARANT
L. CHEDID

Immunothérapie expérimentale
Institut Pasteur
Paris France

I. INTRODUCTION

N-acetylmuramyl-L-alanyl-D-isoglutamine (MDP) is a synthetic adjuvant known to stimulate monocyte/macrophage *in vitro* as well as *in vivo* (reviewed in Leclerc and Chedid, 1982). In 1980, Oppenheim *et al.* described that MDP stimulated human monocytes and peritoneal macrophages from BCG-treated mice to produce LAF(IL-1). On the other hand, Dinarello *et al.*(1978) demonstrated the presence of endogenous pyrogens (EP) in the supernatant of human peripheral blood mononuclear cells or rabbit phagocytes cultivated in the presence of MDP. According to certain physical similarities between LAF(IL-1) and EP, Murphy *et al.*(1980) and Rosenwasser and Dinarello (1981) have suggested that the same molecule was responsible for both activities. A MDP derivative, MDP(Gln)-O*n*Bu, as adjuvant active as MDP but lacking of pyrogenic activity and unable to produce EP *in vivo* and *in vitro* (Chedid *et al.*, 1982) had been selected to evaluate its ability to induce the production of LAF(IL-1) by human or rabbit cells.

[1]*This work was in part supported by grant 81/306 from DRET*

ISBN 0-12-527540-4

II. RESULTS AND DISCUSSION

To ascertain the absence of pyrogenicity of MDP(Gln)-O*n*Bu (prepared by Lefrancier *et al.*, 1982) it was also administered to the rabbit by the much more sensitive intracerebroventricular (i.c.v.) route (Table I).

TABLE I. In Vivo Properties of MDP and MDP(Gln)-OnBu

Assay	*MDP*	*MDP(Gln)-OnBu*
Adjuvant effect . in oily emulsion	+	+
. in saline	+	+
Nonspecific resistance to infection	+	+
Pyrogenic effect . i.v.	+	0
. i.c.v.	+++	0
Transferable circulating EP	+	0
Granulocytopenia	+	0
Toxic synergism with LPS or Pseudomonas	+	0

Detailed results in Chedid et al.*, 1982.*

The dialyzed supernatants prepared with human or rabbit cells were tested simultaneously in EP and LAF(IL-1) assays. The presence of a pyrogenic factor in supernatants was determined either by i.v. or i.c.v. injection in the rabbit since the latter route has been extensively used as a sensitive assay method for testing EP (Adler and Joy, 1965; Turchik and Bornstein, 1980). Obviously, to perform these experiments, we carefully prepared pyrogen-free reagents and equipment.

A. *Production of Human LAF(IL-1)*

Experiments were carried out with the peripheral blood mononuclear cells (PBMC) from 15 human donors. Even under endotoxin-free conditions about half of the normal donor cells released detectable level of preformed LAF(IL-1) (high background donors). Thus, typical example of a low and high background donors are reported on Table II. In both cases, even with a high background, MDP and MDP(Gln)-O*n*Bu stimulated significantly the release of LAF(IL-1). To evaluate their pyrogenic activity, the supernatants were injected to rabbit by i.v. or i.c.v. route and the results show clearly that MDP(Gln)-OnBu-supernatant did not produce any detectable febrile response.

TABLE II. LAF(IL-1) and EP Activities in Supernatant of Human PBMC Stimulated with MDP or its Derivative

Compound	Donor with a Low bg[b] LAF activity				Donor with a High bg LAF activity			
	LAF[c]		EP[d]		LAF		EP	
(μg/ml)[a]	1:8	1:4	5	10	1:8	1:4	5	10
	0.2	0.1	-	0.2	12	12	-	0.2
MDP 10	4	2	-	-	16	17	1.6	1.6
MDP(Gln)-OnBu 10	15	12	-	-	29	27	0.3	0.2
MDP 100	7	4	1.5	1.8	15	18	-	-
MDP(Gln)-OnBu 100	15	12	0.1	0.1	20	20	-	-

[a]*Concentration of stimulant incubated for 18-24hrs with $25x10^6$ PBMC.*

[b]*Background.*

[c]*LAF(IL-1) activity was evaluated on C3H/HeJ thymocytes ($1.5x10^6$ cells/well) in the presence of PHA-P Wellcome (1μg/ml) and supernatants diluted 1:8 and 1:4 (final dilution); the results are expressed as mean cpm $x10^{-3}$ of ^{3}HTdR incorporated by triplicate cultures. The SEM varied by <15% from the mean.*

[d]*EP activity was evaluated as the increment of rectal temperature in a rabbit receiving 5 or 10 μl/kg of supernatant by i.c.v. route. Results are expressed as maximal change in temperature recorded for at least 5 hours.*

The stimulants MDP and MDP(Gln)-OnBu did not interfere in the assays since they were eliminated from the supernatants by extensive dialysis; moreover, the pyrogenic activity was abolished by heating at 56°C for 30 minutes, method known to destroy the effect of EP. It is noteworthy that the high LAF(IL-1) background activity found in control supernatants was never correlated with a pyrogenic effect; however, MDP elicited always comparable levels of pyrogenicity independently of LAF(IL-1) responses (Table II). The synthetic adjuvants, by themselves, did not enhance thymidine uptake by thymocytes and were devoid of any comitogenic activity. Finally, the results obtained in the LAF(IL-1) assay were not related to the presence of IL-2 in the supernatant: 1) our supernatants do not contain any IL-2 activity (D. Fradelizi, personal communication); 2) MDP and its derivative cannot directly induce IL-2 production (see Leclerc and Chedid, this volume).

B. *Production of Rabbit LAF(IL-1)*

Rabbit oil-induced peritoneal cells (at least 90% macrophages as evaluated by non-specific esterase staining) were incubated for 18 hours with MDP or MDP(Gln)-O*n*Bu. Like with human cells, both compounds induced a markedly elevated LAF(IL-1) secretion by rabbit peritoneal exudate cells (PEC); again, pyrogenic activity was only detected in MDP-supernatant (Table III). However, MDP(Gln)-O*n*Bu which failed to generate EP release did not block the stimulatory effect of MDP when added simultaneously to the culture nor did produce an inhibitory factor. *In vivo* MDP(Gln)-O*n*Bu showed no antipyretic activity.

TABLE III. LAF(IL-1) and EP Activities in the Supernatants of Rabbit Peritoneal Exudate Cells (PEC) Stimulated by MDP or MDP(Gln)-OnBu with or without Colchicine

Compound ($\mu g/ml$)[a]		-- LAF[c]	-- EP[d]	Colchicine (10^{-6}M)[b] LAF	Colchicine (10^{-6}M)[b] EP
Control		0	0	±	0
MDP	1	+	+	-[e]	-
	10	++	++	+++	++
	100	+++	+++	-	-
MDP(Gln)-OnBu	1	++	0	-	-
	10	+++	0	+++++	0
	100	++++	0	-	-

[a]*Concentration of adjuvants incubated for 18 hours with $5x10^7$ PEC.*

[b]*Colchicine was added at the beginning of the culture.*

[c]*LAF(IL-1) activity was evaluated as mentionned in the footnotes of Table II.*

[d]*EP activity was evaluated either i.c.v. as mentionned in the footnotes of Table II or i.v. by administering 0.5 to 1 ml/kg of crude dialyzed supernatant.*

[e]*Not tested.*

C. *Influence of Colchicine*

Oppenheim *et al.* (1978) has reported that colchicine alone can stimulate the release of LAF(IL-1). Since Bodel (1976) pointed out that colchicine had no activity on EP release with rabbit cells this prompted us to study the influence of this anti-inflammatory agent on the MDP-induced release of LAF(IL-1) and EP. After incubation of rabbit PEC with MDP in the presence of colchicine, results summarized in Table III show that the release of EP was unchanged. There was no detectable EP activity in the colchicine control supernatants; however, colchicine alone increased the production of LAF(IL-1) and in the presence of MDP or MDP(Gln)-O*n*Bu, an additive effect was always found and often a synergism. Again, in the case of supernatants prepared in the presence of colchicine and MDP(Gln)-O*n*Bu no pyrogenic effect was revealed even after i.c.v. administration into the rabbit.

III. CONCLUSION

By using a non pyrogenic MDP derivative, we succeeded to obtain LAF(IL-1) without pyrogenic activity in the supernatants of human or rabbit cells (Damais *et al.*, 1982). A separation of both activities has already been suggested by Rocklin *et al.* (1980) after reviewing some discrepancies in reports concerning these monokines. Differences in their susceptibility to heating have also been noted (Rosenwasser and Dinarello, 1981).

REFERENCES

1. Adler, R.D., and Joy, R.J.T., *Proc. Soc. Exp. Biol. Med., 119,* 660, (1965).
2. Bodel, P., *J. Exp. Med., 143,* 1015, (1976).
3. Chedid, L., Parant, M., Audibert, F., Riveau, G., Parant, F., Lederer, E., Choay, J., and Lefrancier, P., *Infect. Immun., 35,* 417, (1982).
4. Damais, C., Riveau, G., Parant, M., Gerota, J., and Chedid L., *Int. J. Immunopharmacol., 4,* 451, (1982).
5. Dinarello, C., Elin, R.J., Chedid, L., and Wolff, S.M., *J. Infect. Dis., 138,* 760, (1978).
6. Leclerc, C., and Chedid, L., *in* "Lymphokines" (E. Pick, ed.), vol. 7, p. 1, Academic Press, New York, (1982).
7. Lefrancier, P., Derrien, M., Jamet, X., Choay, J., Lederer E., Audibert, F., Parant, M., Parant, F., and Chedid, L., *J. Med. Chem., 25,* 87, (1982).

8. Murphy, P.A., Simon, P.L., and Willoughby, W.F., *J. Immunol., 124,* 2498, (1980).
9. Oppenheim, J.J., Mizel, S.B., and Meltzer, M.S., *in* "Biology of the lymphokines" (S. Cohen, E. Pick and J.J. Oppenheim, eds.), p. 291, Academic Press, New York, (1979).
10. Oppenheim, J.J., Togawa, A., Chedid, L., and Mizel, S., *Cell. Immunol., 50,* 71, (1980).
11. Rocklin, R.E., Bendtzen, K., and Greineder, D., *Adv. Immunol., 29,* 55, (1980).
12. Rosenwasser, L.J., and Dinarello, C.A., *Cell. Immunol., 63,* 134, (1981).
13. Turchik, J.B., and Bornstein, D.L., *Infect. Immun., 30,* 439, (1980).

DISCUSSION

DAVID: Has Damais characterized the "LAF" and the "pyrogenic activity" induced by these MDP analogues?

DAMAIS: This has not yet been done.

LANDY: Especially in the presence of our visitor from France, reporting these remarkable findings with butyl ester of muramyldipeptide (MDP), I would like to pay tribute to our friend and colleague, Professor Edgar Lederer. It was his genius and infinite persistance that led to the progressive isolation, identification and structural analysis of the biologically active components of the cell walls of Mycobacteria and many other microbial genera. Most recently it was he who established MDP as the minimal structure accountable for the adjuvant attributes of microbial cell walls and their peptidoglycans.

Now, I would ask Damais if she could tell us a bit about the background events or rationale that led to the discovery that the butyl ester of MDP could so neatly elicit IL-1, but not endogenous pyrogen responses in the host and, *in vitro,* the *release* of the former but not the latter activity. I assume that there *was* some rationale, and that it was not purely fortuitous that this derivative was somehow selected from an extensive series of compounds that had been prepared for no especially compelling reason.

DAMAIS: We elected to test MDP (Glu)O_2Bu in our *in vitro* study because it was the first *non-pyrogenic* MDP derivative found that was as adjuvant-active as MDP.

McADAM: I would like to add some further evidence for the capacity of the butyl ester of MDP to "discriminate" between different functions of IL-1. The native MDP (L-D) is 100-fold more active than the butyl ester of MDP in provoking mice to elaborate the acute phase protein, SAA.

PICK: Could Damais tell us what was the rationale for the use of colchicine as an enhancer of IL-1 production by macrophages? Does colchicine also enhance IL-1 production as elicited by other stimulants?

DAMAIS: We wanted to know the importance of the cytoskeleton in the release of IL-1 and endogenous pyrogen. We knew from the work of Oppenheim that colchicine could stimulate the production of IL-1. Also, Bodel had reported that rabbit peritonel cells, stimulated by colchicine, were no longer able to produce EP and that there was no synergistic effect with EP-inducers. Therefore we added colchicine and MDP derivatives to the cultures to determine whether there would be differences in the *release* of EP and IL-1 activities.

TEMPERATURE SENSITIVITY OF INTERLEUKIN-DRIVEN MITOGENESIS[1]

D.F. Hanson
P.A. Murphy

Department of Medicine
The Johns Hopkins University
Baltimore, Maryland

I. INTRODUCTION

Endogenous pyrogen (EP) and Interleukin-1 (IL-1) appear to be identical (1). (Unpublished observations in our laboratory support this hypothesis.) IL-1 has been implicated in the initiation of many immune cellular reactions (2). The precise interactions of interleukins in all immune responses remain controversial. However, it is clear that one of the principal consequences of the action of IL-1 is the production of a second mitogenic factor, Interleukin-2 (IL-2), which actually drives the proliferation observed in many interleukin assay systems (3).

Kluger and coworkers have shown that artificial fever markedly increases the survival of infected lizards; similar conclusions using mammals are less certain (4,5). One principal action of EP is to raise body temperature. Therefore, we performed experiments to determine if the central effect of EP/IL-1 (fever, ie. changing temperature) could influence its local effects (IL-1-induced and/or IL-2-induced mitogenesis).

II. RESULTS

A. *Temperature dependence of the IL-1 assay*

A standard IL-1 assay was performed at 32, 34, 36, 38 and 40 ^{o}C using CBA thymocytes in the presence of 5 ug/ml PHA-P

[1]*This work was supported by U.S.Public Health Service Grant* # AI 03772

ISBN 0-12-527540-4

to which were added serial dilutions of IL-1. Mitogenesis was assessed by pulsing the cells for 18 hrs. with tritiated thymidine (^{3}H-TdR), 72 hrs. after the initiation of culture. The preparations of rabbit IL-1 used in these experiments contained greater than 2 X 10^4 EP units and 6 X 10^7 IL-1 units per mg. of protein (1).

The results, shown in Figure 1, indicated that the IL-1 assay was strongly temperature sensitive. Calculation of a Q_{10} value for the reaction gave a figure of 178 (6). However, cells in medium alone or 5 ug/ml PHA alone were not strikingly temperature sensitive and Q_{10} values for these conditions were between 2 and 5.

B. *Temperature sensitivity of the IL-1 assay is not due to trivial effects of temperature*

When the temperature experiment shown in Figure 1 was repeated but with all ^{3}H-TdR pulses performed at 37 ^{o}C, the result was the same and the Q_{10} value 180, indicating that the effect of temperature was not on the rate of transport of ^{3}H-TdR into dividing cells. Likewise, repetition of the experiment using ^{3}H-TdR pulses at 24, 48, 72 and 96 hrs. demonstrated that different incubation temperatures did not shift the time of peak label incorporation. All responses were maximal at 72 hrs.

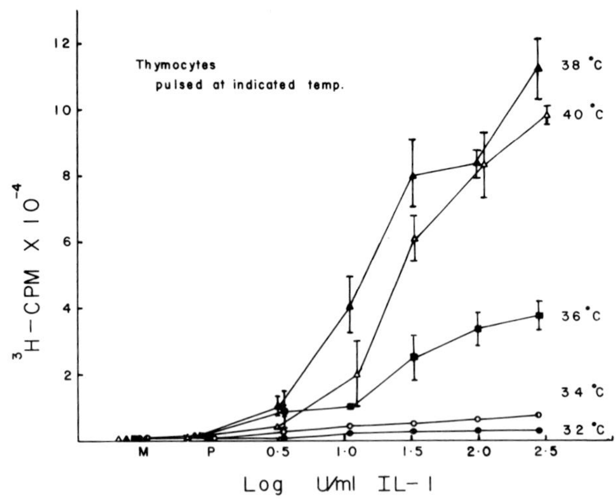

FIGURE 1 - IL-1 assay performed in the presence of 5 ug/ml PHA and at various temperatures; ● = *32,* ○ = *34,* ■ = *36,* ▲ = *38 and* △ = *40°C; M = cells in medium; P = cells in PHA alone*

Thymocytes proliferate rapidly *in vivo* and continue to do so for the first 24 hrs. in culture. Therefore, freshly isolated thymocytes were cultured alone or in the presence of various dilutions of PHA, IL-1 or IL-2. These added factors were used either alone or in combination with one another. The cells were pulsed immediately upon initiation of culture, incubated at various temperatures and harvested 18 hrs. later. Proliferation was vigorous at all temperatures (5-15 X 10^4 CPM per 10^6 thymocytes) and increased as a function of temperature. However, no combination of temperatures and factors revealed large temperature sensitivity (data not shown) and no Q_{10} values were greater than 5. Therefore, temperature sensitivity was not an inherent property of rapidly dividing thymocytes.

C. *High temperature sensitivity is not associated with the action of IL-2*

IL-2 production is a consequence of the action of IL-1. Therefore, we performed temperature experiments on the mitogenic action of IL-2 on either whole thymocytes or on CTL-15H cells, a cell line which requires IL-2 for its proliferation. These experiments utilized partially purified mouse Concanavalin A (Con A) induced IL-2 which contained 10^4 50% maximal IL-2 units/ml as assayed on the CTL-15H cell line (7).

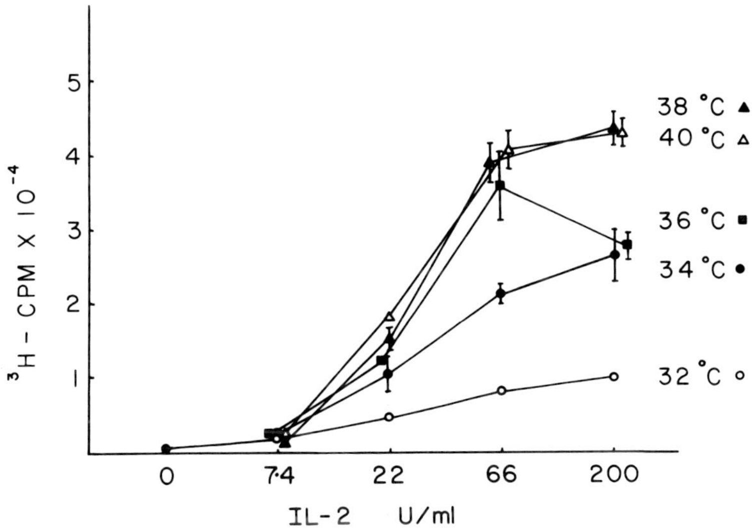

FIGURE 2 - Thymocytes cultured at various temperatures and in the presence of various dilutions of mouse IL-2; standard assay, pulse time and legend as in Figure 1

Figure 2 shows the results of the assay of mouse IL-2 on thymocytes at various temperatures, under the conditions of the standard IL-1 assay but without PHA. The Q_{10} of this reaction was 5-10 although proliferation was vigorous. This result was unchanged if the IL-2 was assayed on thymocytes in the presence of 5 ug/ml PHA or if assayed on CTL-15H cells (data not shown).

D. Peanut agglutinin negative thymocytes respond to IL-1 in a temperature sensitive fashion in the absence of PHA

The thymocyte response to 5 ug/ml PHA alone appeared to possess low temperature sensitivity (Figure 1). In order to rule out the possible contribution of PHA to the temperature sensitivity of the IL-1 assay, we isolated the mature population of thymocytes using peanut agglutinin, a non-mitogenic lectin (8). Peanut agglutinin negative (PNA^-) thymocytes responded to IL-1 alone in a temperature sensitive fashion which did not depend upon the presence of PHA (Figure 3). The Q_{10} of this response was 100.

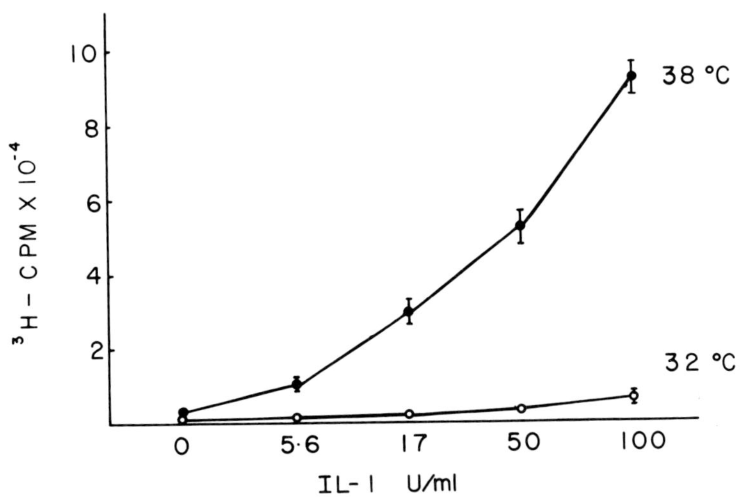

FIGURE 3 - PNA^- *thymocytes cultured at 32°C (   ) or at 38°C (   ) in the presence of various dilutions of IL-1; standard assay and pulse time as in Figure 1*

TABLE 1. *Summary of interleukin Q_{10} values*

Cell type	*Factors*	Avg. Q_{10}[a]	n[b]
Thymocytes	*Medium*	2.7 ± 1.0	18
"	5 ug/ml PHA	3.8 ± 1.7	17
"	50 ug/ml PHA	262 ----	2
"	IL-1	100 ± 133	13
"	IL-1 + 5 ug/ml PHA	189 ± 90	12
"	IL-2	10.7 ± 5.1	18
"	IL-2 + 5 ug/ml PHA	0.6 ± 0.2	12
PNA^- *Thymocytes*	*Medium*	16.3 ± 15.7	4
"	IL-1	72 ± 22	9
CTL-15H	IL-2	5.5 ± 0.6	12

[a] *Q_{10} values reported ± 95% confidence limits*

[b] *n = number of determinations*

E. *Comparison of the temperature sensitivities of interleukin reactions*

Table 1 summarizes our findings with respect to the temperature sensitivities of interleukin reactions. High temperature sensitivity was associated with the effect of IL-1 plus 5 ug/ml PHA or with 50 ug/ml of PHA alone on whole thymocytes. The same result was found for the action of IL-1 alone on PNA^- thymocytes and, less reliably, on whole thymocytes. Low temperature sensitivity was found for thymocytes in medium or 5 ug/ml PHA alone as well as for the action of IL-2 on whole thymocytes or CTL-15H cells. Moreover, the Q_{10} associated with the rate of IL-1 production by rabbit blood mononuclear cells stimulated with heat-killed *Staphylococcus epidermidis* was between 1 and 4 (data not shown).

III. DISCUSSION

Ashman and Nahmias showed that reletively high doses of PHA and Con A induce temperature sensitive mitogenesis and our findings confirm this (Table 1)(9). Our data indicate that IL-1-induced mitogenesis is very temperature sensitive, revealing Q_{10} values of 50 to 200. This is, itself, a signifigant finding since the Q_{10} values of almost all biological reactions are less than 6.0 (6).

High doses of IL-1 or PHA alone, or low doses of both factors acting in concert, can produce highly temperature sensitive responses in whole thymocytes. Moreover, the action of IL-1 alone on PNA^- thymocytes is highly and reproducibly temperature sensitive. Resting cells, minimally stimulated cells or proliferating, unstimulated cells show no such effect.

Temperature sensitivity does not appear to be a property of events prior to or subsequent to the action of IL-1 in its pathway. The production of IL-1 by mononuclear cells is quite insensitive to temperature. Likewise, the mitogenic effect of IL-2, which is produced as a consequence of IL-1 action, is similarly insensitive (Table 1). These observations suggest that the temperature sensitivity observed in IL-1-induced mitogenesis emanates from the action of IL-1 and PHA on the IL-1 target cell in the PNA^- thymocyte population.

It seems likely that temperature sensitivity is a property of the IL-1 target cell itself, rather than a characteristic of either IL-1 or PHA. This is suggested by the facts that high doses of either of these dissimilar factors can produce the temperature effect and that unstimulated PNA^- thymocytes are somewhat temperature sensitive themselves (Table 1).

Some evidence suggests that cell mediated immunity is inefficient at temperatures below 35°C (10,11) and in the human approximately 50% of the body mass is normally at temperatures below 37°C (12). Moreover, a large fraction of natural infections occur within this mass of cooler peripheral tissue. Such infections usually result in local inflammation and/or regional lymph node inflammation, conditions which greatly increase local and lymph node blood supply and, therefore, tend to elevate the temperatures of these sites to values approaching core temperature.

These observations and the data presented in this paper lead us to speculate that local inflammation, draining peripheral lymph node inflammation and fever may all represent means of elevating local temperatures and thus accelerating local IL-1-dependent cellular events. Such a case would serve to explain the apparent conservation of febrogenic and mitogenic functions in the same family of molecules (EP/IL-1). Furthermore, such a process may contribute to the survival value of the febrile response.

REFERENCES

1) Murphy, P.A., P.L.Simon and W.F.Willoughby (1980); *J. Immunol.* *124*;2498
2) Letter to the Editor (1979); *J. Immunol.* *123*;2928
3) Smith K.A., L.B.Lachman, J.J.Oppenheim and M.F.Favata (1980); *J. Exp. Med.* *151*;1551
4) Kluger, M.J. (1978); *Am. Sci.* *66*;38
5) Vaughn, L.K. and M.J.Kluger (1977); *Fed. Proc.* *36*;511
6) Giese, A.C. (1968) Cell Physiology, 3rd ed.; Ch. 10; W.B.Saunders, Philadelphia
7) Gillis, S. P.S.Schwarz and J.M.Brown (1980); *J.Immunol.* *120*; 2027
8) Reisner, Y., I.Linker-Israeli and N.Sharon (1976); *Cell.Immunol.* *25*;129
9) Ashman, R.B. and A.J.Nahmias (1977); *Clin. Exp. Immunol.* *29*;464
10) Purtillo, D.T., G.P.Walsh, E.E.Storrs and I.S.Banks (1974); *Nature* *248*;450
11) Lauwasser, M. and J.W.Shands (1979); *Infect. and Immun.* *24*;454
12) Dubois, E.F. (1939); *Bull.N.Y.Acad.Med.* *15*;143

SENSITIVITY OF T CELLS TO INTERLEUKIN 1 (IL-1) IN RELATION TO THE CELL CYCLE

Miroslav M. Simić
Stanislava Stošić-Grujičić
Milena Kataranovski

Institute of Microbiology and Immunology
School of Medicine
Belgrade, Yugoslavia

Lectin-induced proliferation of at least some subpopulations of T cells, including macrophage-depleted nonadherent thymocytes (NATs), requires participation of both the macrophage-derived IL-1 and the T cell derived IL-2 (1). While it is well documented that resting T cells are unresponsive to IL-2 and that it acts only on cells shifted from G_0 into G_1 phase when they express receptors for IL-2 (2), the timing of IL-1 action in relation to the cell cycle of IL-1 responsive cells is less clear. It has been postulated (3-5) that IL-1 responsive cells are not sensitive to IL-1 when in resting stage and that they become responsive only after activation by lectins when shifted from G_0 into G_1. This postulation appears to be widely accepted (6) although it has primarily been deduced from experiments in which, on one side, cell supernatants were used that were not assayed for IL-2 activity and, on the other hand, the IL-1 effect was measured by enhancement of lectin-induced proliferation. Since both IL-1 and IL-2 have the same capacity to enhance proliferation of T cells in response to lectins, it is conceivable that the effect of contaminating IL-2 in such supernatants could mask the proper effect of IL-1. Therefore, we have reexamined the relationship of the cell cycle to sensitivity of T cells to the action of IL-1 by use of IL-1 containing macrophage supernatants that were devoid of IL-2 activity.

ISBN 0-12-527540-4

Receptivity of T cells to IL-1 was evaluated by measuring its action in enhancing the proliferative response of NATs to PHA and Con A (7). IL-1 containing macrophage supernatant was prepared as previously described (7). As estimated from gel filtration on Sephadex G-100, the major of the two components with comitogenic activity present in the supernatant, had the m.w. 15,000. Supernatant was found free of IL-2 by measuring its growth-promoting activity for T cell blasts. Thus in experiments to be detailed, enhancement of NATs response to lectins reflected the proper action of IL-1. Results were expressed as "IL-1 effect", i.e. Δcpm ± S.E./culture (cpm of cultures with lectins + IL-1 minus cpm of cultures with lectins + medium).

RESULTS AND DISCUSSION

A. *Effects of pretreatment with IL-1 on lectin-responsiveness of NATs*

NATs consist entirely of cells in G_0 which poorly proliferate in response to lectins and do not proliferate at all when exposed to IL-1, alone (7). We pretreated such cells with IL-1 in order to explore whether IL-1 is able to act on cells in G_0 by changing their reactivity to subsequent lectin stimulation. As shown in Fig. 1, a brief pretreatment of NATS with IL-1 enhanced in a dose-dependent way the proliferative response of these cells to lectins in subsequent cultures, thus indicating that IL-1 responsive cells in G_0 are receptive to the action of IL-1. Whether these results demonstrate an irreversible effect of IL-1 on T cells or whether they are merely the result of IL-1 binding to the cells, remains to be established.

Our findings are consistent, on one side, with the observed capacity of IL-1 to convert, following a brief exposure, IL-2 non-producer lymphocyte cell line 1A5 cells to a state in which subsequent PHA stimulation can trigger production of IL-2 (8). On the other hand, they are in agreement with the findings that IL-1 is bound by unstimulated normal T cells (9) or by LBRM 33 cells (8).

Enhancement of lectin-induced proliferation of NATs (or of other macrophage-depleted T cells) is used as a bioassay for IL-1, but it is not specific since IL-2 has similar effects. However, while IL-2 can act only on T cells being shifted from G_0 into G_1, IL-1, as our results demonstrate, is able to act on T cells in G_0, i.e. before they had been activated by the lectins. This makes it possible that these differential pro-

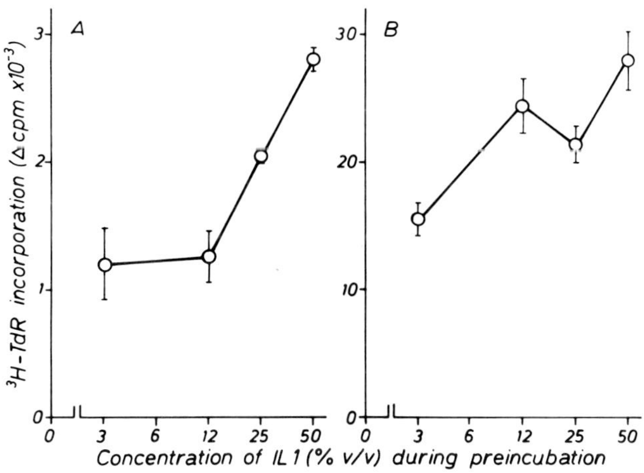

FIGURE 1. Pretreatment with IL-1 enhances the responsiveness of NATs to lectins. NATs were preincubated for 2 hr at 37°C with or without various concentrations of IL-1 as indicated. Following incubation, cells were extensively washed with medium and cultured in the presence of PHA (A) or Con A (B). ^{3}H-TdR incorporation was determined 3 d after lectin stimulation. Data are expressed as "IL-1 effect".

perties of IL-1 versus IL-2 can be used as the basis for a specific assay for IL-1.

B. Kinetics of IL-1 action on lectin-activated NATs

Next group of experiments was directed to the question of sensitivity of NATs to IL-1 as a function of incubation time following lectin stimulation, i.e. when lectin-driven time-dependent G_0-G_1 shift of NATs occurs (5). Three types of experiments were carried out. In the first, we determined the effect of IL-1 added at various intervals after NATs stimulation with lectins (Fig. 2). In the second, the effects of pulsing NATs with IL-1 at various times after lectin stimulation were examined (Fig. 3). Finally, in the third type of experiments, IL-1 was added at the same time as lectins and was removed at various intervals after the initiation of culture (Fig. 4).

Fig. 2 shows the variations in IL-1 comitogenic effect with time when IL-1 was added to cultures at various intervals after the lectins. Enhancement of NATs proliferative response

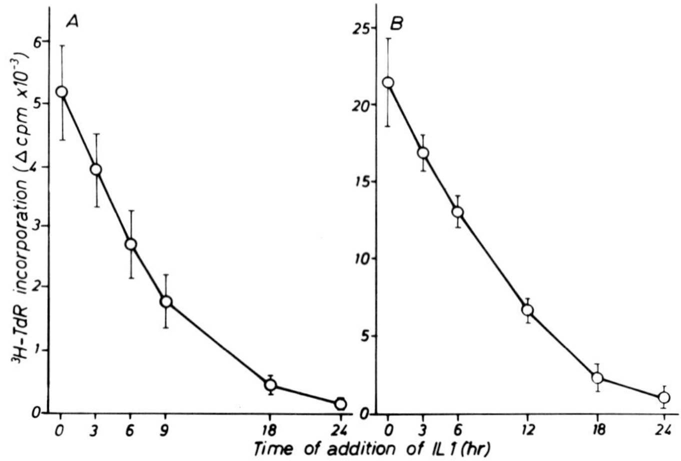

FIGURE 2. Continuous decrease of NATs sensitivity to comitogenic action of IL-1 with time after lectin stimulation. NATs were stimulated with PHA (A) or Con A (B) at the initiation of culture. At the times indicated, IL-1 or medium were added. ^{3}H-TdR incorporation was determined 3 d after the initiation of culture. Data are expressed as "IL-1 effect".

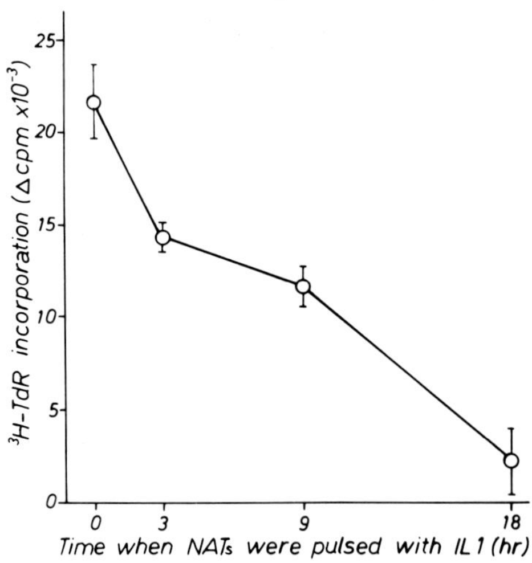

FIGURE 3. Effect of pulsing NATs with IL-1 at various intervals after lectin stimulation. NATs were stimulated with Con A at the initiation of culture. At the times indicated, IL-1 or medium were added. Two hr later cells were extensively washed with medium containing Con A, and were recultured in presence of Con A. ^{3}H-TdR incorporation was determined 3 d after the initiation of culture. Data are expressed as "IL-1 effect"

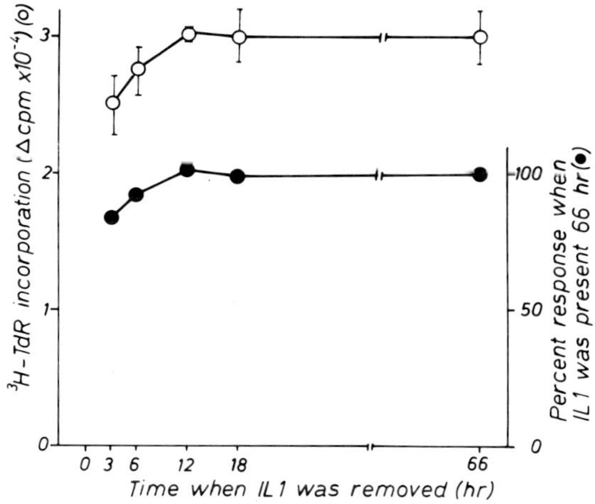

FIGURE 4. Effect of timed removal of IL-1. NATs were incubated with Con A and IL-1 or medium at the initiation of culture. At the times indicated, cells were extensively washed with medium containing Con A and were then recultured in the presence of Con A. ^{3}H-TdR incorporation was determined 3 d after the initiation of culture. Data are expressed as "IL-1 effect" and as percentages of the "IL-1 effect" obtained when IL-1 was present 66 hours.

was maximal when IL-1 was introduced at the same time as lectins, while additions at later times resulted in a continuously progressive decrease of enhancement. The rate of decrease was faster with higher lectin-concentrations (data not shown). Similar time-dependent continuous decrease of IL-1 comitogenic effect was also observed with activated NATs, instead of being continuously cultured in the presence of IL-1, were only pulsed, briefly, with IL-1 (Fig. 3). In these experiments the number of recultured, IL-1 pulsed NATs was corrected for viable cells after washing. It appeared that the decreased IL-1 effect, when added later, could not be attributed to a loss of viability of NATs after lectin stimulation

Experiments in which IL-1 added at the start together with lectins, was removed at various intervals after the initiation of culture, showed that only small differences in NATs proliferation occurred, whether IL-1 was present for 66 or 3-6 hr. This indicates that a major part, if not almost all IL-1 comitogenic action on NATs is irreversibly accomplished within 3-6 hours of exposure to lectins, i.e. within a period when NATs are still in G_0 (5).

Taken together, the presented data indicate that receptivity of IL-1-responsive T cells to IL-1 does not develop following a latent period of insensitivity nor increases within a specific period after lectin stimulation (3). On the contrary, responsiveness is maximal at the very beginning of T cell interaction with lectins and continuously decreases thereafter in a clearly time-dependent way. It has been shown (5) that murine NATs are shifted from G_0 to G_1 by lectins and that the total number of G_1 cells continuously increases with time after lectin stimulation. Since the ability of NATs to respond to IL-1 is lost in a temporal pattern that coincides with this increase in proportion of G_1 cells, our data indicate (a) that IL-1 exerts its action in coordination with the lectins on IL-1 responsive T cells in G_0 by converting them to a state in which lectins can trigger production of IL-2, and (b) that cells shifted by lectins from G_0 into G_1 become unresponsive to IL-1.

REFERENCES

1. Smith, K.A. and Ruscetti, F.W., *Adv. Immunol.* *31*,137 (1981).
2. Larson, E.-L., Iscove, N.N. and Coutinho, A., *Nature 283*, 664 (1980).
3. Waksman, B.H. and Wagshal, A.B., *Cell. Immunol. 36*,180 (1978).
4. Beller, D.I. and Unanue, E.R., *J. Immunol. 123*,2890 (1979).
5. Stadler, B.M., Kristensen, F. and de Weck, A.L., *Cell. Immunol. 55*,436 (1980).
6. Oppenheim, J.J. and Gery, I., *Immunology Today 3*,113 (1982).
7. Stošić-Grujičić, S. and Simić, M.M., *Cell. Immunol. 69*, 235 (1982).
8. Gillis, S. and Mizel, S.B., *Proc. Natl. Acad. Sci. U.S.A. 78*,1133 (1981).
9. Katz, S.P., Shimamura, T., Dessaint, J.-P., Braverman, D. and Waksman, B.H., *Cell. Immunol. 56*,68 (1980).

INHIBITION OF T-CELL ACTIVATION BY ALPHA-1-ANTIPROTEASE IS REVERSED BY PURIFIED RABBIT INTERLEUKIN-1

Philip L. Simon

Departments of Ophthalmology and Microbiology
University of Pittsburgh School of Medicine
Pittsburgh, Pennsylvania

Judith B. Willoughby
William F. Willoughby

Department of Pathology
The Johns Hopkins University School of Medicine
Baltimore, Maryland

I. INTRODUCTION

Alpha-1-antiprotease (a1AP, formerly termed alpha-1-antitrypsin) is a major serum constituent normally present at concentrations ranging from 2-4 mg/ml. Although it is known to inhibit certain protease-mediated reactions such as blood clotting and the kinin system, the nature and extent of its physiological role is not completely understood. Several recent reports have proposed that a1AP may be a regulator for immune responses, producing supression of T and B cell responses both *in vivo* and *in vitro* (1). In fact, those reactions that are supressed by a1AP are principally the same reactions that are enhanced by the interleukins (2). In view of the reports of protease activity of certain lymphokines (3-5), we investigated the relationship between interleukin-1-dependent and independent *in vitro* lymphocyte responses to a

Interleukins, Lymphokines, and Cytokines

ISBN 0-12-527540-4

T cell mitogen, and their suppression by a1AP. Our results show that (a) a1AP, at physiological concentrations, can suppress IL-1-dependent PHA-induced thymus cell DNA synthesis, and (b) a1AP-mediated suppression of PHA-induced spleen cell DNA synthesis can be reversed by the addition of exogenous purified IL-1.

II. MATERIALS AND METHODS

A. *Reagents*

Purified human a1AP was purchased from Sigma Chemical Company, St. Louis, MO. PHA-P was purchased from Difco Labs, Detroit, MI. All tissue culture reagents were purchased from Gibco, Grand Island, NY. IL-1 was prepared from culture fluids of bacterial endotoxin-stimulated rabbit alveolar macrophages, by sequential ammonium sulfate precipitation, gel filtration, and isoelectric focusing as described previously (2).

B. *Cell Culture*

All lymphocyte stimulation procedures were performed using thymus and spleen cells obtained from C3H/HeJ mice purchased from the Jackson Labs, Bar Harbor, ME. The methods for cell culture used in this study have been described previously (2). IL-1 activity was measured by determination of the amount of incorporation of ^{3}H-TdR in mouse thymus cells stimulated with 1 μg/ml phytohemagglutinin (PHA). One unit of IL-1 will produce approximately 50% maximum CPM.

III. RESULTS

A. *Effect of a1AP on IL-1-dependent thymus cell proliferation*

Table 1 shows the results of adding a1AP at time 0 to cultures of mouse thymus cells stimulated with PHA in the presence of IL-1, in a standard 72 hr IL-1 assay system. The results demonstrate that alAP can suppress, in a dose dependent fashion, IL-1 dependent PHA responses of thymus cells. The doses of alAP used in this and subsequent experiments were found (by trypan blue dye exclusion) not to be toxic for the cells (data not shown).

TABLE I. alAP Suppresses Thymus Cell Proliferation[a]

PHA	*IL-1*	*alAP (mg/ml)*	*CPM*
-	-	-	225±76
+	-	-	2,701±289
+	+	-	25,999±3,464
+	+	0.3	19,941±3,711
+	+	1.3	11,398±1,475
+	+	2.5	8,701±969

[a]*Quadruplicate cultures of 1.5 x* 10^6 *thymus cells/0.2 ml culture; PHA = 1 μg/ml; IL-1 = 3 units/ml.*

Table II shows the effects of increasing the dose of IL-1 in an alAP-suppressed PHA-stimulated thymus cell cultures. The results demonstrate that addition of more IL-1 to the suppressed thymus cells nearly restores the level of DNA synthesis to that of unsuppressed cultures.

TABLE II. Effect of Increased Amounts of IL-1 on alAP-Suppressed Thymus Cell Proliferation[a]

IL-1 (Units/ml)	*alAP (2 mg/ml)*	*CPM*
0	-	463±29
3	-	10,273±141
6	-	12,161±129
12	-	12,346±762
3	+	1,275±83
6	+	5,601±246
12	+	9,376±125

[a]*Culture conditions same as for Table I. All cultures received 1 μg/ml PHA. Note that at 5 Units/ml or above, maximum CPM is achieved with unsuppressed cultures.*

B. *Effects of IL-1 and alAP on PHA-Induced Spleen Cell Proliferation*

Table III shows the results of adding IL-1 to spleen cell cultures stimulated with PHA. Spleen cells presumably do not respond with increased proliferation to exogenous IL-1 because sufficient levels of IL-1 are produced by the cells themselves.

TABLE III. IL-1 Does Not Affect Normal Spleen Cell Proliferation[a]

PHA	*IL-1 (Units/ml)*	*CPM*
-	-	2,261±126
+	-	23,128±5011
+	5	23,095±2435
+	10	20,694±2925

[a]*Quadruplicate cultures of 1 x* 10^6 *spleen cells: PHA = 2 μg/ml.*

Table IV shows the results of adding alAP to cultures of PHA-stimulated spleen cells. As was seen with the thymus cells, alAP inhibits the response of spleen cells to PHA in a dose dependent-manner.

TABLE IV. alAP Suppresses Spleen Cell Responses to PHA[a]

alAP (mg/ml)	*CPM*
0.0	54,086±3,662
0.3	45,017±2,440
1.3	28,514±2,142
2.5	23,659±3,566

[a]*Culture conditions same as for Table III.*

If, as is generally believed, all lymphocyte responses to T-cell mitogens are IL-1 dependent, and if alAP inhibits spleen cell responses by inhibiting the action of endogenously produced IL-1, then, this suppression should also be reversible by addition of exogenous IL-1, as seen with the thymus cell response. The results of adding IL-1 to alAP-inhibited (and normal unhibited) spleen cells(with

PHA) are presented in Figure 1. At relatively high concentrations, IL-1 restores fully the response of the inhibited spleen cells, while as noted previously, it does not enhance the response of normal (uninhibited) spleen cells. Thus, a1AP, in effect, renders spleen cells dependent upon exogenous IL-1. It should be noted, however, that the amount of IL-1 required to restore the response was significantly greater than that required to restore the thymus cell response.

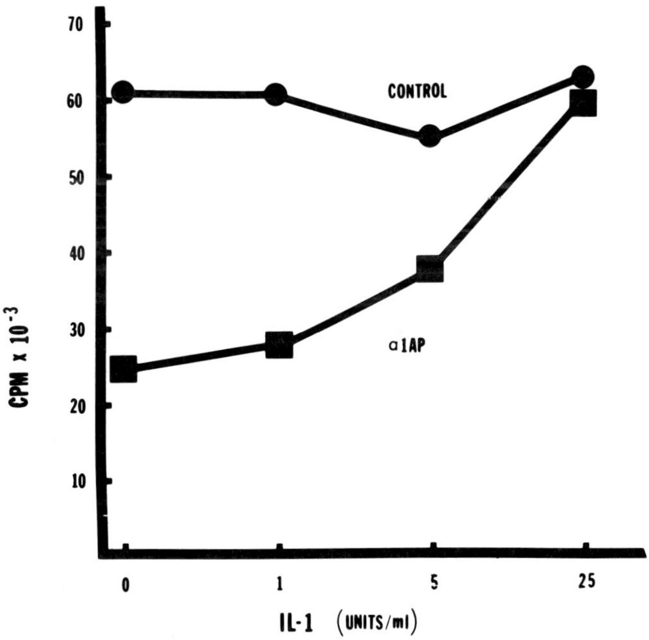

FIGURE 1. IL-1 reverses a1AP-mediated suppression of PHA-induced spleen cell proliferation. Spleen cells (1 x 10 /0.2 ml culture) were cultured for 3 days with 1 µl/ml PHA. All cultures received either a1AP (final concentration, 1.25 mg/ml) or culture medium alone as control on day 0.

IV. DISCUSSION

Our data demonstrate that a1AP inhibits IL-1 activity, as measured by the ability of IL-1 to enhance lymphocyte 3H-TdR incorporation following mitogen stimulation, and thus suggest that a1AP contributes to immunologic homeostasis *in vivo* by regulating the biologic activity of IL-1. We believe that a1AP inhibits thymus cell proliferation by inhibiting, in a reversible fashion, the exogenously-added IL-1. On the other hand, spleen cell responses (where addition of exogenous IL-1 normally has no effect on proliferation) are presumably inhibited by the blocking of endogenously-produced IL-1 by the added a1AP. Support for this hypothesis is derived from the fact that, as with thymus cells, inhibition of spleen cell responses could be reversed by the addition of exogenous IL-1. This present study not only provides evidence for the requirement for endogenously produced IL-1 during the proliferative response of lymphoid cells to T cell mitogens, but also demonstrates that such IL-1-dependent responses are inhibited, and presumably modulated, by a1AP at or near normal serum concentrations.

The role of inherited a1AP deficiencies in the pathogenesis of several diseases has become increasingly suspected (6). Interestingly, many of these diseases, including fibrosing alveolitis, acute anterior uveitis, rheumatoid arthritis, juvenile arthritis, and systemic lupus erythematosis are characterized by uncontrolled or excessive immune and inflammatory reactions. It is possible that at least some of the excess immune reactivity is due to an inability to adequately control IL-1 mediated reactions because of a deficiency of a1AP.

REFERENCES

1. Arora, P. K., Miller, H. C., and Aronson, L.D. *Nature 274*, 590 (1978).
2. Simon, P. L., and Willoughby, W. F. *J. Immuno. 126*, 1534 (1981).
3. Geczy, C. L. *Mol. Immunol. 17*, 601 (1980).
4. Rocklin, R. *Fed. Proc. 37*, 2743 (1978).
5. Dessaint, J. P., Katz, S. P., Waksman, B. H. *J. Immunopharmacol. 1*, 399 (1979).
6. Brelt, S. N., and Penny, R. *Aust. N.Z. J. Med. 10*, 449 (1980).

DISCUSSION

SALVIN: Where does the α-1-antiprotease arise as regards the specific producing cell? How does the α-1-antiprotease relate to IL-1 *in vivo*?

SIMON: The majority of the α-1-antiprotease is produced by hepatocytes, although there are some reports that macrophages and/or lymphocytes may also produce it. We presently have no information on the *in vivo* relationships between IL-1 and α-1-antiprotease.

KURNICK: Why is it that Simon can obtain a response in the presence of normal serum, which contains large amounts of α-1-antiprotease, if this substance is inhibitory? Does its removal increase proliferative responses?

SIMON: Most *in vitro* proliferative responses are performed with serum concentrations of 10% or less, which would effectively lower the α-1-antiprotease to a level below that which is inhibitory. I have not yet determined whether removing it from serum has an effect on proliferative responses.

LINKER-ISRAELI: As a comment to the preceding query I would say that studies done with Horwitz at USC show that PBM-adherent cells of SLE patients produce less IL-1 than those derived from normal controls. I do not know how to connect this finding with the deficiency of α-1 antiprotease found in SLE.

SIMON: Nor do we have any information yet as regards the effect of α-1-antiprotease on the production of IL-1.

ROLE OF ARACHIDONATE METABOLISM IN THE IMMUNOREGULATORY FUNCTION OF HUMAN LEUKOCYTIC PYROGEN/LYMPHOCYTE ACTIVATING FACTOR/INTERLEUKIN-1

Charles A. Dinarello, Scott O. Marnoy and Lanny J. Rosenwasser

Department of Medicine
Divisions of Experimental Medicine and Allergy
Tufts University School of Medicine and
New England Medical Center Hospital
Boston, Massachusetts

Leukocytic pyrogen (LP) is the polypeptide mediator of fever and shares several characteristics with lymphocyte activating factor (LAF). Pyrogenic and LAF containing fractions stimulate prostaglandin E production from cultured monocytes and both the fever and PGE release are blocked by ibuprofen. However concentrations of ibuprofen which prevent LP-induced PGE production have no effect on LP/LAF-induced enhancement of thymocyte proliferation. These studies suggest that products of cyclooxygenase are not involved with the proliferative effects of LP. However, eicosa-5, 8, 11, 14-tetraynoic acid (ETYA) at concentrations which do not affect PHA responses reduced LP-induced enhancement of proliferation by 70%. Similar results were also obtained using the water soluble inhibitor of lipoxygenation, BW755C. These studies demonstrate that LP/LAF stimulates arachidonic acid release from several cells but that the lipoxygenase pathway or its products are critical for thymocyte activation.

I. INTRODUCTION

Recent studies from this and other laboratories have demonstrated that human leukocytic pyrogen (LP), the polypeptide mediator of fever, possesses lymphocyte stimulatory properties and that LP may be identical to lymphocyte activating factor (LAF) (1-5). The implication of these findings is that during fever, LP/LAF may play a direct role in the immunostimulation seen in certain infectious, inflammatory

ISBN 0-12-527540-4

and other febrile conditions and could, in part, contribute to the hypergammaglobulinemia seen in many febrile or immunologically mediated diseases. In addition, LP may indirectly affect immune regulation by its ability to increase body temperature to the same levels which have been shown to augment human lymphocyte responses to mitogens (6,7). Because most patients with severe infections or chronic inflammatory diseases receive antipyretics and/or nonsteroidal anti-inflammatory drugs, the role of these agents on the action of LP is of considerable importance.

The ability of LP to initiate fever have focused on the production of hypothalamic prostaglandins (PG). Both LP and PGE injected into the pre-optic anterior hypothalamus produce rapid onset of fever (8,9). LP injected intravenously produces fever and results in increased third cerebroventricular fluid PGE levels (10); furthermore, these levels decrease during defervescence with antipyretics of the cyclooxygenase-inhibitor class (10,11). LP incubated with hypothalamic tissue in vitro was also shown to induce a 4-fold increase in PGE production (12). The ability of LP to increase PGE in non-thermoregulatory tissue may be important. Most studies have shown that PG of the E series suppress proliferative responses (reviewed in 13), presumably through their ability to alter intracellular levels of cyclic nucleotides (14,15). More recent studies have demonstrated that thymus-derived (T) cells are heterogeneous with regard to the effects of PGE in that some populations are suppressed while others show enhanced mitogen-induced blastogenic responses (16). The production of certain lymphokines including suppressor factors is also enhanced by PGE (17,18). Thus, increased production of arachidonate metabolites during fever is of considerable importance to the mechanism by which LP affects immunocytes.

II. MATERIALS AND METHODS

A. *LP Production and LP Purification*

Two methods of LP fractionation were used: gel-filtration and immunoadsorption followed by gel-filtration. Gel-filtration took place over Sephadex G-50 (fine), 165 x 5.6 cm, 4^oC, in 0.9% unbuffered NaCl, pH 5.6-5.9. Immunoadsorption using immobilized rabbit anti-human LP has been described elsewhere (19,20). LP eluting from the anti-LP immunnoadsorbent was concentrated in 1% lysine HCl and applied to Sephadex G-50 as above (21).

B. LP Assays

The rabbit pyrogen test was used to assay LP using a two-point dose response as described previously (21,22). Rabbit pyrogen dose (RPD) is fever between 0.6 and 0.9°C.

C. Monocyte Monolayers

Fresh human blood mononuclear cells were obtained from Ficoll-Hypaque gradients, washed twice in physiologic saline and resuspended in Hanks' Balanced Salt Solution (HBSS) at 5 x 10^6 cells/ml with 5% heat-inactivated (56°C, 30 min) human AB serum. Two ml of suspension were plated into 16 mm diameter plastic wells (Linbro Costar plates, Hamden, CT). The plates were placed on a slowly rocking platform at 37°C for one hour to facilitate monocyte adherence. Following incubation, the wells were washed twice with 37°C HBSS to remove non-adherent cells and microscopic examination revealed monolayers of spreading cells. When opsonized Staph. albus was added, greater than 90% of cells revealed ingested bacteria. After washing the wells, 1 ml of replacement HBSS or LP fractions diluted in HBSS was added to adherent monocytes. The plates were incubated for one hour in a stationary incubator at 37°C. Supernates from duplicate wells were aspirated, pooled, immediately centrifuged at 2,000 x g and frozen at -70°C until assayed.

D. PG Assays

The monocyte supernates were thawed, diluted and assayed for PGE using a radioimmunoassay (Clinical Assays, Cambridge, MA). No extractions were done since protein content of supernates was less than 1 mg/ml.

E. Thymocyte Assay

A murine thymocyte assay for LAF based on enhancement of phytohemagglutinin responses was used as described previously (3). Concentrations of LP added to thymocytes are expressed as RPD per well. Inhibitor concentrations are expressed as final molarity in the wells.

F. *Inhibitors*

Ibuprofen was provided by the Upjohn Co., Kalamazoo, MI. It was weighed, dissolved in 100 μl 95% ethanol and diluted in warm media. Eicosa - 5, 8, 11, 14-tetraynoic acid (ETYA) was a gift from Hoffman La Roche Laboratories. It was weighed, dissolved in 100 μl dimethyl sulfoxide (DMSO) and diluted in warm media to a concentration of 100 μM. 3-amino-1-(3-trifluoromethylphenyl)-2-pyrazoline (BW755C) was a gift from Dr. S. Moncada, the Wellcome Research Laboratories, Kent, England. It was weighed, dissolved in 100 μl ethanol and diluted in warm media.

III. RESULTS

A. *Effect of Human LP on PGE Release*

Initial studies were carried out using LP-induced PGE release from murine thymocytes during the first hour of the LAF assay. Various concentrations of semi-purified LP (from 10^{-1} to 10^{-4} RPD/well) were incubated with thymocytes in the presence or absence of PHA for one hour, centrifuged in the microtiter plates and the supernate from triplicate wells were pooled and assayed for PGE. At all concentrations of LP used, these supernates did not contain detectable levels of PGE (data not shown); however, significant proliferative responses were observed when these thymocytes were subsequently cultured with replacement RPMI and PHA.

Using a highly concentrated batch of crude human LP, chromatographed fractions were pooled every 5 fractions, concentrated ten-fold and dialyzed against RPMI. Fractions were assayed for pyrogenic and LAF activities as well as the ability to stimulate PGE production from adherent monocytes. As depicted in Figure 1, there was a concurrence of pyrogenic, LAF and PGE stimulating activities.

Further purification of LP took place using a combination of gel-filtration and immunoadsorption. In previous studies, SDS polyacrylamide gel electrophoresis of these LP preparations revealed significant purification (20). When this highly purified LP was incubated with monocyte monolayers for one hour, significant PGE was measured in the supernate (cells + HBSS = no detectable PGE; cells + LP at 2 RPD/ml = 138 ± 16 pg PGE/ml; LP without cells = no detectable PGE). Thus, this experiment supports the preceding studies that fractions containing LP and LAF activity also stimulate PGE production _in vitro_ from monocyte layers.

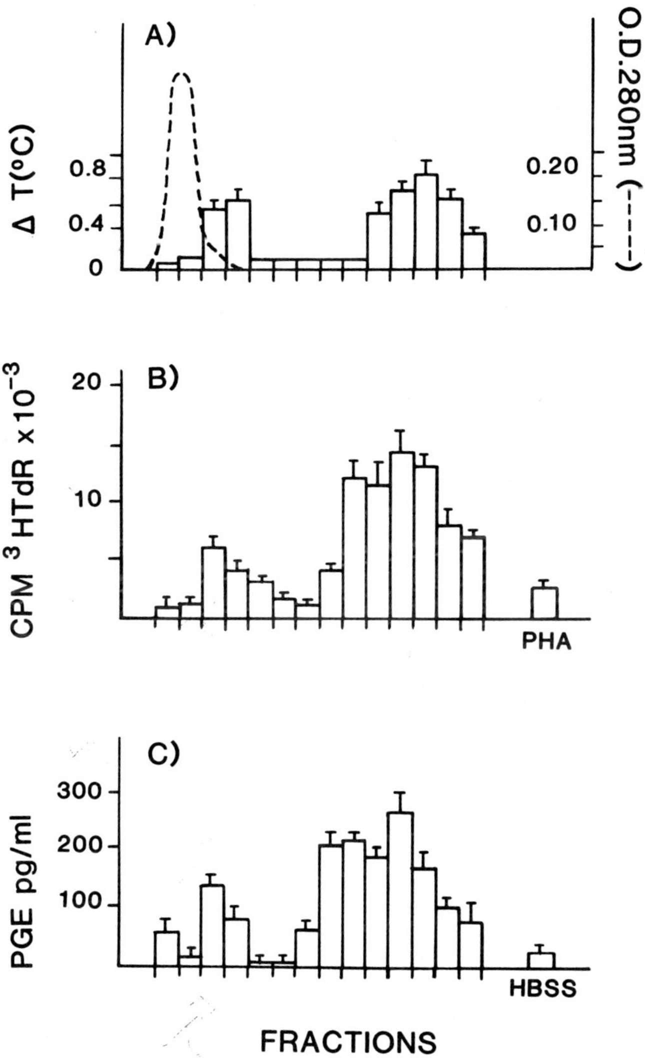

FIGURE 1. Figure 1A shows the mean (± SEM) of three rabbits injected with 1.5 ml of each fraction pool. Figure 1B illustrates thymocyte proliferation in the LAF assay at dilution of 1:20 and Figure 1C represents the ability of each pool to induce PGE production from adherent monocytes.

B. *Effect of Ibuprofen on PG Production and Thymocyte Proliferation*

Ibuprofen was studied in the human monocyte incubations. Monocyte monolayers were prepared, washed and incubated with LP in the presence or absence of ibuprofen. Ibuprofen completely blocked the LP-induced increases in PGE. Thus, it is likely that LP does not stimulate release of PGE but rather induces new PG synthesis. These studies are similar to those reported for LP-induced PGE synthesis from rabbit brain slices (12). However, ibuprofen incubated with murine thymocytes had no effect on the proliferative response to PHA nor on the ability of LP to enhance this response. This was observed at in vitro concentrations from 1-100 μM.

C. *Effect of Lipoxygenase Inhibitors ETYA and BW 755C on LP-induced thymocyte proliferation*

From the preceding experiments it was clear that ibuprofen at concentrations which prevented PGE synthesis from human monocytes has no effect on LP-induced thymocyte proliferation. We next studied the effect of agents which, in addition to inhibiting PG synthesis, are also known to inhibit lipoxygenase pathways (23). As shown in Table I, the DMSO control used to dissolve the ETYA had no effect on thymocyte responses to either PHA or LP. However, concentrations of 10 and 50 μM ETYA, while not affecting the blastogenic effect of PHA markedly reduced 70% of LP-induced enhancement of proliferation at all three concentrations of LP tested. Concentrations of 100 μM ETYA also reduced PHA responses as well as LP enhancement and hence do not reflect a specific effect on LP-mediated responses.

The ability of ETYA to suppress the effect of LP on thymocytes suggested that LP induces an arachidonic acid metabolite which plays a role in the proliferative signal. These studies were also carried out using BW755C, a water soluble agent which, like ETYA, blocks lipoxygenase activity (24).

In Table II, the results of five separate experiments are shown. BW755C at concentrations of 10 and 25 μM had no effect on thymocyte proliferation to PHA but at 50 μM, this agent significantly reduced the ability of all three doses of LP to enhance the PHA response. In addition, there was also a 60% reduction with 25 μM at the lowest LP dose tested. However, a concentration of 100 μM suppressed both PHA and LP enhancement of thymocyte proliferation and hence cannot be considered a specific inhibitor of LP proliferation. Thus, these studies support the data from the specific inhibitor of LP

TABLE I. *Effect of ETYA on Ability of LP to Enhance Thymocyte Proliferation to PHA*[a]

ETYA	PHA	PHA + LP (LP^{-2})[b]	PHA + LP (10^{-3})[b]	PHA + LP (10^{-4})[b]
None (control)	2,997 ± 619	12,680 ± 2,457	8,337 ± 1,925	6,742 ± 1,794
DMSO	2,843 ± 367 (94.9)[c]	11,465 ± 2,172 (90.4)	9,171 ± 1,380 (110.0)	N.D.
10 μM	3,277 ± 165 (109.3)	4,945 ± 864 (38.9)	4,315 ± 385 (51.7)	2,254 ± 518 (33.4)
50 μM	2,999 ± 178 (100.0)	3,911 ± 741 (30.8)	3,854 ± (46.2)	2,791 ± 643 (41.3)
100 μM	967 ± 136 (32.2)	672 ± 216 (5.2)	1,683 ± 362 (20.1)	1,329 ± 489 (19.7)

[a] Mean (± SEM) of four experiments expressed as Δ cpm ^{3}HTdR.

[b] LP concentration expressed as RPD/well.

[c] Numbers in parentheses indicate percent of control.

TABLE II. *Effect of BW755C on Ability of LP to Enhance Thymocyte Proliferation to PHA*[a]

BW755C	PHA	PHA + LP (10^{-1})[b]	PHA + LP (10^{-2})[b]	PHA + LP (10^{-3})[b]
None (control)	1,785 ± 193	17,157 ± 1,817	12,197 ± 3,317	3,348 ± 364
10 μM	2,351 ± 342 (131.7)[c]	20,899 ± 133 (121.8)	13,094 ± 3,646 (106.4)	3,505 ± 655 (104.6)
25 μM	1,883 ± 313 (105.4)	18,092 ± 2,313(105.4)	12,986 ± 1,375 (105.6)	1,458 ± 255 (43.5)
50 μM	1,627 ± 275 (91.4)	10,505 ± 2,053(61.2)	7,073 ± 1,585 (57.5)	2,437 ± 560 (77.7)
100 μM	220 ± 102 (12.3)	2,365 ± 383 (13.7)	1,790 ± 617 (14.5)	624 ± 189 (18.6)

[a]Mean (± SEM) of 5 experiments expressed as Δ cpm ^{3}HTdR incorporation.

[b]LP concentration expressed as RPD/well.

[c]Numbers in parenthesis indicate percent of control.

proliferation. Thus, these studies support the data from the ETYA experiments that the ability of LP to stimulate thymocytes responses to PHA may be dependent on products of the lipoxygenase rather than cyclooxygenase pathway.

IV. DISCUSSION

These studies have focused on certain aspects of arachidonic acid metabolism involved in the biological activity of human LP. The results expand previous knowledge that LP stimulates PGE synthesis in cells or tissues which have significant cyclooxygenase activity (10,12,25). In the present experiments we have shown that LP induces PGE synthesis in adherent human monocytes. Thymocytes, on the other hand, did not produce measurable levels of PGE when stimulated by high doses of LP, similar to the findings of other workers that lymphocytes release little or no PGE (26,27). Others have reported a 15,000 dalton macrophage polypeptide which increases PGE production from adherent synovial cells (28). This factor, mononuclear cell factor, has recently been reported to be similar to LAF (29). These studies together with the data in the present report indicate that LP/LAF stimulation of PGE production may be an important mechanism by which this molecule affects immunoregulation. There is ample data which illustrate the suppressive effects of PGE on lymphocyte responses (reviewed in 13). Therefore, the use of cyclooxygenase inhibitors are generally associated with increased responses to mitogens or antigens (30).

LP also induces neutrophil specific granule release and superoxide production (31,32). These studies indicated that LP is capable of membrane pertubation. However, as in the present studies in thymocytes, LP-induced neutrophil specific granule release was not affected by inhibitors of cyclooxygenase. Thus, an interpretation of these data is that a non-cyclooxygenase pathway may play a role in the biologic activity of LP.

The lipoxygenase pathway may play an important role in T-cell function (33-35). The rate limiting step is the release of arachidonic acid from membrane phospholipids; therefore, LP-induced PGE synthesis implicates a membrane pertubation which results in the release of arachidonic acid with subsequent cyclooxygenation. Some of the arachidonic acid released by the action of LP undergoes lipoxygenation. In the present studies two lipoxygenase inhibitors, ETYA and BW755C suppressed thymocyte responses to LP. It should be emphasized that the concentration of inhibitor which effectively reduced the augmentation of proliferation by 40-70% had no effect on PHA responses. Thus, these results suggest that the concentration of drug was not toxic but rather effective for inhibiting LP-induced responses. The concentrations of ETYA and BW755C used in these studies block cyclooxygenase activity (24,36) but since ibuprofen has no effect on proliferative responses to LP/LAF, we conclude that inhibition was due to the ability of these agents to block lipoxygenase rather than cyclooxygenase pathways. Others have shown that ETYA inhibits mast cell histamine release in vitro in a dose-related response while inhibition of cycloxygenase by indomethacin has no effect (23). ETYA at concentrations of 16 - 33 μM suppresses human T-cell proliferation to mitogens (33). BW755C is a water soluble inhibitor of lipoxygenation. The concentration of BW755C which blocks most of the 5-lipoxygenation step (30 - 100 μM) was effective in preventing LP/LAF-induced proliferation (37). Thus, the concentrations of BW755C used to suppress LP/LAF-induced proliferation are consistent with inhibiting the 5-lipoxygenase and perhaps 12-lipoxygenation step

The mechanism by which LP/LAF enhances lymphocyte proliferative responses remains unknown. LP/LAF has little or no direct mitogenic property itself but rather augments lymphocyte responses to PHA, Concanavalin A and histamine (3,38). In this sense, LP/LAF acts as a co-factor or permissive agent. LP/LAF may function in these systems by influencing turnover of relevant membrane receptors leading to an augmented response (39). Another possible mechanism for the activity of LP/LAF relates to T-cell growth factor (TCGF. IL-2) which has been shown to be dependent on the participation of LAF/Il-1. Inhibition of lipoxygenase pathways may decrease PHA-induced TCGF production (40) or diminish TCGF receptor binding (39). Concentrations of BW755C and ETYA which suppress LP/LAF-induced proliferation have been shown to inhibit lipoxygenase activity associated with leukotrience formation (41,42). Inhibition of the 5-lipoxygenase step in human lymphocytes prevents migration and it has been demonstrated that leukotriene B_4 mediates this response (34). Thus, continued investigations in this field may elucidate the precise metabolite(s) generated in thymocytes exposed to LP/LAF and could lead to a more specific approach to inhibition of ligand-induced T-cell responses.

REFERENCES

1. Dinarello, C.A., and Wolff, S.M., *New Engl. J. Med. 298,* 607 (1978).
2. Rosenwasser, L.J., Dinarello, C.A., and Rosenthal, A.S., *J. Exp. Med. 150,* 709 (1979).
3. Rosenwasser, L.J., and Dinarello, C.D., *Cell Immunol. 63,* 134 (1981).
4. Murphy, P.A., Simon, P.L., and Willoughby, W.F., *J. Immunol. 124,* 2498 (1980).
5. Sztein, M.B., Vogel, S.N., Sipe, J.D., Murphy, P.A., Mizel, S.B., Oppenheim, J.J., and Rosenstreich, D.L., *Cell Immunol. 63,* 164 (1981).
6. Roberts, N.J., Jr., and Steigbigel, R.T., *Infect. Immunol. 18,* 673 (1977).
7. Roberts, N.J., Jr., *Microbiol. Rev. 43,* 241 (1979.
8. Rosendorff, C., and Mooney, J.J., *Am. J. Physiol. 220,* 597, (1971).
9. Stitt, J.T., *J. Physiol. 232,* 163 (1973).
10. Bernheim, H.A., Gilbert, T.M., and Stitt, J.T., *J. Physiol. 301,* 69 (1980).
11. Milton, A.S., *Biochem. Soc. Trans. 6,* 727 (1978).
12. Dinarello, C.A., and Bernheim, H.A., *J. Neurochem. 37,* 702 (1981).
13. Ellner, J.J., *Clin. Immunol. Rev. 1,* 119 (1981).
14. Koopman, W., Gillis, M., and David, J., *J. Immunol. 110,* 1609 (1973).
15. Goodwin, J., Bankhurst, A., and Messner, R., *J. Exp. Med. 146,* 1719 (1977).
16. Stobo, J.D., Kennedy, M.S., and Goldyne, M.E., *J. Clin. Invest. 64,* 1188 (1979).
17. Yoneda, T., and Mundy, G., *J. Exp. Med. 149,* 279 (1979).
18. Rogers, T.J., Nowowiejski, I., and Webb, D.R., *Cell Immunol. 50,* 82, (1980).
19. Dinarello, C.A., Renfer, L., and Wolff, S.M., *J. Clin. Invest. 60,* 465, (1977).
20. Dinarello, C.A., Renfer, L., and Wolff, S.M., *Proc. Natl. Acad. Sci. 74,* 4624 (1977).
21. Dinarello, C.A., Goldin, N.P., and Wolff, S.M., *J. Exp. Med. 39,* 1369 (1974).
22. Dinarello, C.A., and Wolff, S.M., *Inflammation 2,* 179 (1977).
23. Sullivan, T.J., and Parker, C.W., *J. Immunol. 122,* 431 (1979).
24. Eakins, K.E., Higgs, G.A., and Whittle, B.J.R., *Brit. J. Pharm. 70,* 82P (1980).

25. Cranston, W.I., Hellon, R.F., and Mitchell, D., *J. Physiol. 253,* 583 (1975).
26. Lederman, M.M., Ellner, J.J., and Rodman, H.M., *Clin. Exp. Immunol. 45,* 191 (1981).
27. Goodwin, J.S., DeHoratius, R., Israel, H., Peake, G.T., and Messner, R.P., *Ann. Int. Med. 90,* 169 (1979).
28. Dayer, J-M, Stephenson, M.L., Schmidt, E, Karge, W., and Krane, S.M., *FEBS Lett. 124,* 253 (1981).
29. Mizel, S.B., Dayer, J-M., Krane S.M., Mergenhagen, S.E., *Proc. Natl. Acad. Sci. USA 78,* 78 (1981).
30. Goodwin, J.S., Selinger, D.S., Messner, R.P., and Reed, W.P., *Infect. and Immun. 19,* 430 (1978).
31. Klempner, M.S., Dinarello, C.A., and Gallin, J.I., *J. Clin. Invest. 61,* 1330 (1978).
32. Klempner, M.S., Dinarello, C.A., Henderson, W.R., and Gallin, J.I., *J. Clin. Invest. 64,* 996 (1979).
33. Kelley, J.P., Johnson, W.C., and Parker, C.W., *J. Immunol. 122,* 1563 (1979).
34. Payan, D.G., and Goetzl, E.J., *J. Clin. Immunol. 1,* 266 (1981).
35. Goodman, M.G., and Weigle, W.O., *J. Immunol. 125,* 593 (1980).
36. Ahern, D.G., and Downing, D.T., *Biochem. Biophys. Arch. 210,* 456 (1970).
37. Radmark, O., Malmsten, C., and Samuelsson, B., *FEBS Lett. 110,* 213 (1980).
38. Beer, D.J., Dinarello, C.A., Rosenwasser, L.J., and Rocklin, R.E., *Clin. Res. 29,* 363A, (1981).
39. Puri, J., Shinitzky, M., and Lonai, P., *J. Immunol. 124,* 1937 (1980).
40. Gillis, S., and Mizel, S.B., *Proc. Natl. Acad. Sci. USA 78,* 1133 (1981).
41. Narumiya, S., Salamon, J.A., Cottee, F.H., Weatherley, B.C., and Flower, R.J., *J. Biol. Chem. 256,* 4583 (1981).

DISCUSSION

OPPENHEIM: The observations that resting monocytes produce more PGE, whereas activated macrophages produce more IL-1 and less PGE, in conjunction with Rosenwasser's data, suggest that there may be bidirectional feedback interactions occurring between macrophage subsets. Does he have any information from his studies, as to the subsets that produce PGE and EP.

ROSENWASSER: We have no data on reciprocal monocyte subset production of EP/IL-1 vs PGE, but the suggestion Oppenheim makes is intriguing and warrants further experimentation.

DUMONDE: Is Rosenwasser perhaps thinking that mononuclear cells in the CNS are generating PG-like mediators of fever?

ROSENWASSER: Yes, possibly so.

PICK: We should make it clear that ETYA is in no way a specific inhibitor of lipoxygenase - in fact at 100 μM it blocks both lipoxygenase and cycloxygenase. Secondly, I think that the effect of ETYA on the response to PHA (in the absence of IL-1) is in good agreement with data that lipoxygenase products are obligatory participants in the mitotic response of T cells to mitogens.

ROSENWASSER: I agree that the depressed response of thymocytes to PHA on exposure to 100μM ETYA is due to the obligatory participation of lipoxygenase products in T cell mitogenesis.

VADAS: Does Simon know whether the enzymatic function of his protein is responsible for its effect in his assay?

SIMON: No, we do not have such information.

VADAS: Since some proteins are toxic by virtue of their charge or other properties has Simon considered including another, similar protein, lacking α-1-antiprotease activity in his assay, as a control.

SIMON: That is a good point, although our studies indicate that under our culture condition α-1-antiprotease is not toxic.

ASHERSON: How well established is the association between unusual allotypes of α-1-protease inhibition and diseases such as asthma, systemic lupus erythematosus and rheumatoid arthritis?

SIMON: Probably not as good an association as with certain HLA profiles. But it should be pointed out that some of the associations *are* in the range of 40 to 50%.

SECTION VII:
ACTIONS OF EFFECTOR CYTOKINES

MACROPHAGE CYTOTOXIN

Frederick C. Kull, Jr.
Pedro Cuatrecasas

Molecular Biology Department
Wellcome Research Laboratories
Research Triangle Park, North Carolina

I. INTRODUCTION

Tumor necrosis serum (TNS) is cytotoxic to a number of unrelated tumor cell lines in culture (1-3). The cytotoxic activity is ascribed to a tumor necrosis factor of suspected (4) and alleged (5) macrophage origin. Since we have quantitated a number of physiochemical (6) and mechanistic (7) properties of the TNS cytotoxin, and since an established cell line might be an improved source (8), we examined a number of readily-available macrophage-like lines for their capacity to make cytotoxic activity similar to TNS. The murine lines J774.1 and RAW 264.7, when challenged with lipopolysaccharide (LPS), were found to release the largest quantity of activity. Specific activities were typically 100-200X those found in TNS. The RAW 264.7 activity, examined in more detail, exhibited properties identical to the TNS cytotoxin. Polyacrylamide gel electrophoresis analysis indicated the macrophage cytotoxin is a minor component of the proteins released during LPS-induced cell death.

II. MATERIALS AND METHODS

The murine cell lines RAW 264.7, J774.1 and P388D1 were generous gifts from Dr. Yancy Gillespie. The murine line PU-5-1.R was obtained from the Cell Distribution Center, San Diego, CA. The human promyelocytic leukemia HL60 was a generous gift from Dr. Robert Gallo. The human histiocytic line U937 was a generous gift from Dr. Hillel Koren. The KB (human epidermoid carcinoma) and L-M (murine tumorigenic fibroblast) lines were obtained from the American Type

Interleukins, Lymphokines, and Cytokines

ISBN 0-12-527540-4

Culture Collection. The cell lines were maintained with the media formulations and conditions recommended by each source. Serum for culture was fetal bovine (FBS) from Sterile Systems, Logan, UT. Macrophage-like lines were grown in Falcon 1058 dishes, Becton Dickinson Labware, Oxnard, CA. *S. typhimurium* LT-2 LPS was prepared by Dr. Paul Ray. *E. coli* B 026:B6 LPS was obtained from Difco, Detroit, MI. Other materials and methods are described elsewhere (6,7).

III. RESULTS

Table I lists a number of cell lines known to exhibit macrophage-like properties. They were examined for their capacity to release cytotoxic activity into serum-free culture medium in response to challenge with LPS. Two time points and two LPS preparations were examined. These conditions were chosen based on preliminary experiments which indicated that activity release was related to time and LPS concentration. All control supernatants (not challenged with LPS) lacked detectable activity. The murine lines RAW 264.7 and J774.1 consistently gave high titer responses. They also differed in their time course and in their responsiveness to the two preparations of LPS. In contrast, PU-5-1.R and the human lines HL60 and U937, which can be induced to demonstrate macrophage properties (9,10), were induced but did not yield detectable activity. The parent human lines gave no detectable activity (not shown). Human peripheral blood adherent leucocytes released an activity (not shown), but it was not rigorously examined.

Non-denaturing polyacrylamide gel electrophoresis of concentrated supernatants from 3 murine lines is shown in Figure 1. The staining profile of the supernatants is remarkably similar. Control supernatants lacking cytotoxic activity also appear similar. Thus, no LPS-specific bands are readily discernable; however, activity can be detected from the gel slice eluants of LPS-treated supernatants (Figure 1). Two regions of cytotoxic activity are found in RAW 264.7 supernatants. By gel filtration analysis (not shown), these correspond well with the high (>250,000) and low (60,000-45,000) molecular weight forms found in TNS (6).

RAW 264.7 supernatant cytotoxin was partially purified by ion exchange and gel filtration chromatography such that its specific activity equaled that of a partially purified preparation of TNS cytotoxin (200 units/μg, Ref. 7). These preparations were quantitatively compared in a variety of tests which measured physicochemical and mechanistic properties. For example, the heat inactivation and pH optima profiles were identical (not shown). Both toxins were inactivated by 15 min at 70°. Both were pH labile with optima at pH 8.0.

Enzymatic inactivation is compared in Table II. Serine proteases inactivated both preparations equivalently. Glycosidases and nucleases did not affect activity. The effect of some chemical treatments is compared in Table III. Again, both behaved similarly.

TABLE I. Cell Line Sources of Macrophage Cytotoxin[a]

Line	Cytotoxic Units/ml Supernatant 6 hrs Sal.	6 hrs E. coli	18 hrs Sal.	18 hrs E. Coli
RAW 264.7	69 ± 0	75 ± 0	198 ± 12	65 ± 4
J774.1	204 ± 7	40 ± 16	34 ± 2	25 ± 0
P388D1	<4	7	37 ± 4	41 ± 15
PU-5-1-R	<4	<4	<4	<4
HL60 PMA	ND	ND	<4	<4
U937 + MLC CM	ND	ND	<4	<4

[a]Cells in mid to end log phase growth were harvested, washed 2X and seeded in Falcon 3003 dishes at 10^7 cells/10 ml/dish in culture media devoid of serum. HL60 + PMA had been previously cultured for 3 days in 14 nM phorbol myristate acetate. U937 + MLC CM had been previously cultured for 6 days in 25% mixed lymphocyte culture conditioned medium (10). S. typhimurium LT-2 (Sal.) or E. coli LPS was added to concentration of 1 μg/ml (mouse lines) or 10 μg/ml (human lines). Supernatants were harvested at the indicated times and frozen. Quantitative cytotoxic activity measurements were determined on L-M cells as described elsewhere (6,7). Briefly, 1 unit of cytotoxic activity is the ability to cause 50% survival (kill 35,000 cells) in a 20-hour period. Shown is the average of two experiments. ND = not done.

When the target assay system was modified by a number of treatments, either the toxic capacity of the preparations (shown in Table IV) or the sensitivity of the targets to both preparations (shown in Table V) was similarly affected. For example, no killing was observed when the assay temperature was lowered to 32°. Media supplements (with the exception of FBS) failed to influence killing. Inhibitors of protein synthesis increased the sensitivity of targets, whereas tertiary amines and cytoskeletal disrupting agents decreased sensitivity.

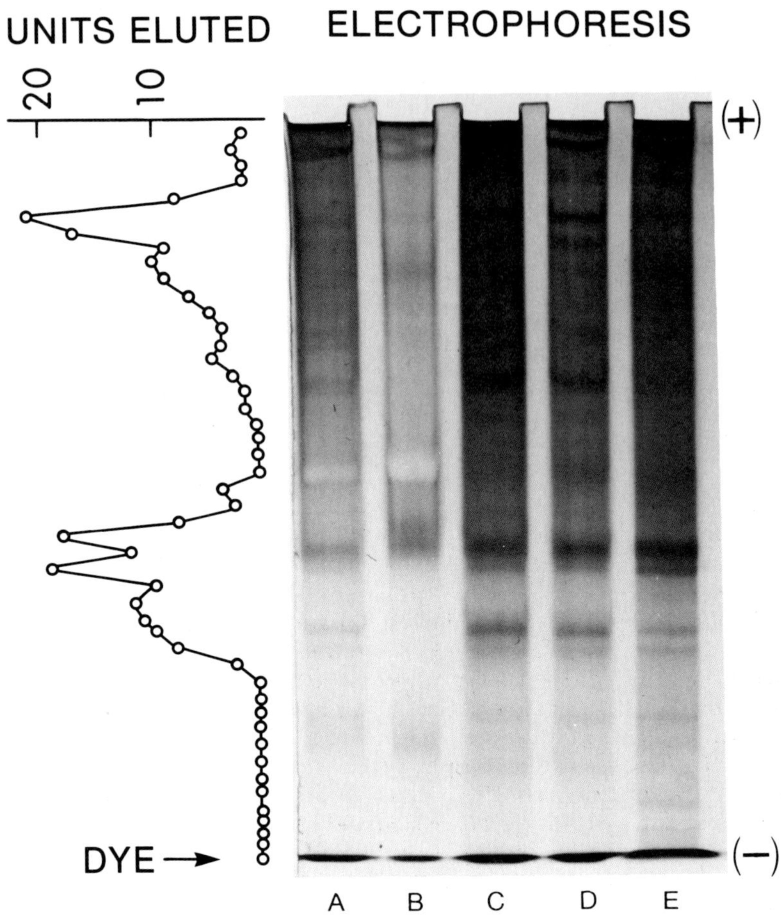

FIGURE 1. Polyacrylamide gel electrophoresis of supernatants. Washed macrophage-like cells were incubated with or without 1 µg S. typhimurium LPS/ml at 2.5 x 10^5 cells/cm^2/ml minimum essential medium for 18 hrs at 37°. Cell-free supernatants were concentrated 20-fold by ultrafiltration (PM10 membranes, Amicon Corp., Lexington, KY). 50 µl was applied to a non-denaturing 1.5 mm thick 7.5% slab gel prepared and run as described (6). The gel was stained by the silver stain method (15). 2.23 mm slices were cut from an equivalent RAW 264.7 + LPS slot, and the segments were eluted overnight with 200 µl 140 mM NaCl, 20 mM NaP_1, pH 7.4. Gel slice eluants were assayed for cytotoxic activity as described in the legend to Table I, except that 1 µg cycloheximide/ml was included in the culture medium to enhance sensitivity. A, RAW 264.7 + LPS; B, RAW 264.7 - LPS; C, J774.1 + LPS; D, J774.1 - LPS; E, P388D1 + LPS.

TABLE II. Enzymatic Inactivation of Cytotoxin[a]

Enzyme	Concentration μg/ml	% Inhibition TNS	MØ
Trypsin	2.5	100	100
Protease K	0.25	92	88
Neuraminidase	2.5	<10	<10
Mixed glycosidases	200	<10	<10
DNAse I	50	<10	<10
RNAse A	50	<10	<10

[a]1 μg partially purified cytotoxin (200 units/μg/0.1 ml) from TNS ad RAW 264.7 supernatant (MØ) was incubated for 1 hr at 37° with the indicated concentration of enzyme. The reaction was quenched with an equal volume FBS, and the samples were measured for inhibition of activity as described in detail elsewhere (6). Briefly, % inhibition is the ratio of the change in the amount of treated activity divided by the change in control X 100.

TABLE III. Chemical Inactivation of Cytotoxin[a]

Inhibitor	Concentration	% Inhibition TNS	MØ
SDS	0.1%	78	82
BPTI	1 mg/ml	< 10	< 10
DFP	1 mM	< 10	< 10
PMSF	1 mM	< 10	< 10

[a]1 μg partially purified cytotoxin (200 units/μg/0.1 ml) from TNS or RAW 264.7 supernatant (MØ) was mixed with the indicated concentration of reagent for 1 hr at 20°. Samples were dialized overnight versus 140 mM NaCl, 20 mM NaP_1, pH 7.4. Samples were then measured for inhibition of activity as described in the legend to Table II. Abbreviations are as follows: SDS, sodium dodecyl sulfate; BPTI, bovine pancreatic trypsin inhibitor; DFP, diisopropyl fluorophosphate; PMSF, phenylmethylsulfonylfluoride.

TABLE IV. Influence of Culture Conditions on Cytotoxicity[a]

Treatment	% Inhibition TNS	MØ
FBS 10%	60	35
32°	>90	>90
Catalase 200 U/ml	<10	<10
MEM amino acids 5X	<10	<10
MEM vitamins 10X	<10	<10

[a]The two cytotoxic preparations were assayed in the presence or absence of altered assay conditions: dec increased concentration of FBS, decreased temperature, the presence of capalase, increased concentration of vitamins amino acids or vitamins. Cytotoxicity was quantitated and the % inhibition was determined as described in the legend to Table II.

TABLE V. Influence of Metabolic Inhibitors on Sensitivity[a]

Treatment	Concentration	Sensitivity TNS	MØ
Chloroquine	10 μM	0.25	0.28
5-dimethyldansylcadaverine	10 μM	ND	0.54
Methylamine	10 mM	0.71	0.68
Colchicine	10 μM	0.34	0.30
Dinitrophenol	2 mM	0.12	0.08
Cycloheximide	4 μM	3.5	3.5
Acetinomycin D	1 μM	5.6	7.6

[a]The two cytotoxic preparations were assayed in the presence and absence of the indicated concentrations of drugs. Sensitivity was determined as detailed elsewhere (7). Briefly, sensitivity is defined as the ratio (control/treated) of the amount of activity required to cause 50% survival (kill 35,000 cells) in a 20-hour period. Numbers less that 1 indicate decreased sensitivity. Numbers greater than 1 indicate increased sensitivity. ND = not done.

Further evidence for the similarity of the preparations is shown in Figure 2. The capacity of the two preparations to kill KB and L-M cells is compared. The differential sensitivities are identical. That is, KB cells are less sensitive than L-M cells; however, the increased amount of either preparation required to kill KB targets is identical.

IV. DISCUSSION

Our evidence indicates that some macrophage-like cell lines may secrete a cytotoxic substance(s) when stimulated with LPS. Time, LPS concentration and type of LPS are important variables. The activity obtained from RAW 264.7 cells is identical to that found in TNS in a number of physicochemical and mechanistic properties. These properties include susceptibility to serine proteases, size, electrophoretic mobility, sensitivities of target cell lines, and the capacity of various metabolic inhibitors and culture conditions to influence target cell sensitivity. Thus, our work confirms recent observations on the cell source of the cytotoxic activity in TNS (5,8,11). Furthermore, our studies discriminate this cytotoxic activity from other macrophage-related cytotoxic capabilities: neutral proteases (12), which are quenched by serine esterase inhibitors, hydrogen peroxide (13), which is quenched by catalase, and essential nutrient deprivation such as arginase (14) which is quenched by excess nutrient. Other evidence also refutes nutrient deprivation. For example, when 200 units of the macrophage cytotoxin were dialized for 72 hrs at 37° versus 10 ml minimum essential medium, the medium was as capable as control medium in supporting L-M cell growth, while the dialized toxin was fully active (data not shown). Thus, nutrient deprivation is not a likely explanation for the mechanism of action.

It appears from polyacrylamide gel analysis that the cytotoxic components are minor constituents in the supernatants. The major bands appear to be related to normal cell death, since they are found in end log phase control cultures; whereas the cytotoxin is specific to LPS-induced changes. LPS-specific bands cannot be detected. Also, activity appears to peak and plateau with major bands. Thus, it may be swept along with the major contaminants. The bands in the cytotoxic region tend to copurify with the cytotoxic activity by conventional separation techniques. In the absence of an affinity handle, we have found it difficult to separate the activity from the major contaminants. The relation of the cytotoxin to known macrophage functions in immunity and inflammation is speculative.

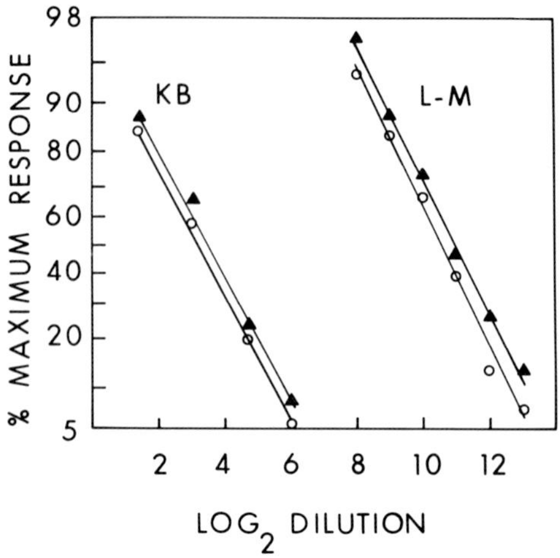

FIGURE 2. Differential sensitivity of TNS and MØ cytotoxin. L-M and KB cells were seeded at 1.5 x 10^4 cells/200 μl minimum essential medium, 5% FBS/well into microtiter trays and incubated 24 hrs. Each line was simultaneously tested for cytotoxicity with the preparations described in the text, except that the toxic incubation period was extended to 48 hrs. Trays were scored for cytotoxicity as described in the legend to Table I. Shown are probit transformations of the dilution curves (see Ref. 6) where maximum response is complete killing. $Log_2 0$ = 50 units/ml. Δ, TNS; O, MØ.

V. REFERENCES

1. Matthews, N., and Watkins, J. F. Br. J. Cancer 38, 302 (1978).
2. Helson, L., Green, S., Carswell, E., and Old, L. J., Nature (Lond.) 258, 731 (1975).
3. Haranaka, K. and Satomi, N., Japan. J. Exp. Med. 51, 191 (1981).
4. Old, L. J., Clin. Bull. 6, 118 (1976).
5. Matthews, N., BR. J. Cancer 44, 418 (1981).
6. Kull, F. C., Jr., and Cuatrecasas, P., J. Immunol. 126, 1279 (1981).
7. Kull, F. C., Jr., and Cuatrecasas, P., Ca. Res. 41, 4885 (1981).
8. Männel, D., Moore, R. N., and Mergenhagen, S. E., Infect. Immun. 30, 523 (1980).

9. Rovera, G., O'Brien, T. G., and Diamond, L., Science 204, 868 (1979).
10. Larrick, J. W., Fisher, D. G., Anderson, S. J., and Koren, H. S., J. Immunol. 125, 6 (1980).
11. Satomi, N., Haranaka, K., and Kuni, O., Japan. J. Exp. Med. 51, 317 (1981).
12. Adams, D. O., Kao, K.-J., Farb, R., and Pizzo, S., J. Immunol. 124, 293 (1980).
13. Nathan, C. F., Silverstein, S. C., Bruckner, L. H., and Cohn, Z. A., J. Exp. Med. 149, 100 (1979).
14. Currie, G. A., Nature (Lond.) 273, 758 (1978).
15. Merril, C. R., Goldman, D., Sedman, S. A., and Ebert, M. H., Science 211, 1437 (1980).

DISCUSSION

MANNEL: Did Kull test the macrophage line-derived (more purified) cytotoxin for tumor necrosis in an *in vivo* model?

KULL: Yes, partially purified cytotoxin from RAW 264 supernatant did not induce tumor necrosis in the *in vivo* model.

OPPENHEIM: I have a number of questions for Kull: Does his approach yield sufficient pure material so he can produce antibodies? Although we identify this factor as a cytotoxin what does he think its real functions may be? Is the cytotoxin also cytostatic?

KULL: Although cell line-derived cytotoxin is substantially more pure than serum-derived material, it is still a minor component of the supernatant. Our attempts to prepare xenogeneic antisera have not been successful thus far. Our goal is to prepare antibodies to help answer this question. The cytotoxin has a dynamic effect on all cell lines in culture. It does not appear to have a cytostatic effect.

DUMONDE: If Kull adds small amounts of endotoxin to the macrophage product -- is it active *in vivo*?

KULL: Possible synergistic effects between either LPS or the multiple forms of the cytotoxin were not examined.

PICK: Would Kull tell us the reason for using PMA *and* LPA when stimulating cytotoxin production by the HL-60 line?

KULL: The rationale is described in the text of my report.

PURIFICATION AND CHARACTERIZATION OF LYMPHOTOXIN FROM HUMAN LYMPHOBLASTOID CELL LINE 1788

Bharat B. Aggarwal
Barbara Moffat
Richard N. Harkins

Department of Protein Biochemistry
Genentech, Inc.
South San Francisco, California

Lymphotoxin is a type of lymphokine secreted by activated lymphocytes *in vitro* (Hiserodt *et al.*, 1979; Rundell and Evans, 1979; Namba and Waksman, 1975; Trivers *et al.*, 1976) and *in vivo* (Granger *et al.*, 1978). It has been demonstrated by several workers that lymphotoxin induces both cytostatic and cytolytic responses in tumor cells (Evans and Heinbaugh, 1981; Rosenau, 1981; Swada *et al.*, 1976) and it has also been shown to be an anticarcinogenic and antitumor agent (see review, Evans, 1982). Most of the biological studies have been carried out with lymphotoxin preparations which are relatively crude, thus making it difficult to interpret precisely the effects of lymphotoxin on transformed cells. Numerous reports (Gately and Mayer, 1974; Granger *et al.*, 1978; Russell *et al.*, 1972; Klostergaard *et al.*, 1981; Amino *et al.*, 1974) have appeared on the purification of this protein but it has been a difficult task due to the minute amount of material produced by lymphocytes.

We have isolated lymphotoxin from several hundred liters of a lymphoblastoid B-cell culture medium and the protein has been purified to homogeneity and partially characterized. The bioassay of lymphotoxin during purification consisted of a semiautomated assay involving the lysis of mouse L-929 fibroblasts (Spofford *et al.*, 1974). One unit of lymphotoxin was defined as the amount required for 50 percent lysis of 12,000 cells plated in each microtiter well. The lymphotoxin was isolated according to the purification scheme as shown below:

Interleukins, Lymphokines, and Cytokines

ISBN 0-12-527540-4

Human Lymphoblastoid 1788 Cell Culture

↓

Cell Removal by Filtration

↓

Concentration and Dialysis

↓

DE-52 Cellulose Chromatography

↓

Preparative Isoelectric Focusing

↓

Lentil Lectin Chromatography

↓

High Pressure Liquid Chromatography (Reverse Phase)

Lymphotoxin purified by this scheme was characterized by several methods. The molecular size was determined both by gel filtration and by sodium dodecyl sulfate polyacrylamide

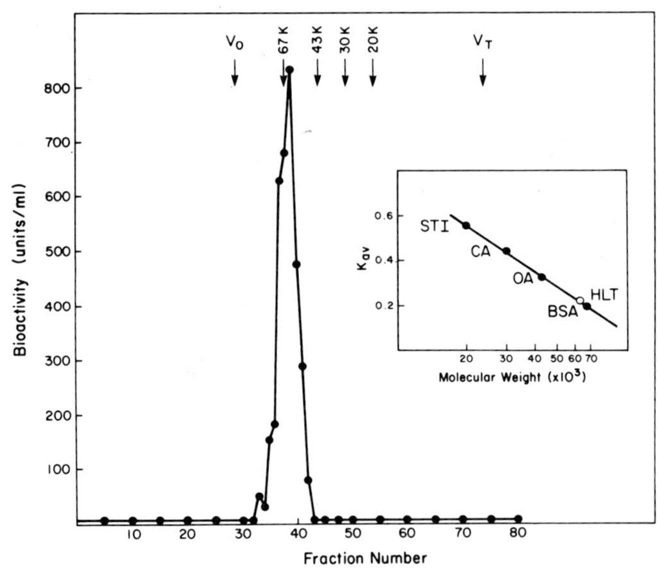

FIGURE 1. The elution profile of lymphotoxin from Sephadex G-100. The column (1 cm x 100 cm) was equilibrated in 10 mM phosphate buffer pH 7.8 containing 0.5 M sodium chloride. The flow rate of the column was 0.2 ml/min. and the sample volume was one ml. The volume of each fraction was 30 drops.

gel electrophoresis (SDS-PAGE). Sephadex G-100 chromatography data (Fig. 1) reveals that purified human lymphotoxin elutes close to bovine serum albumin. This molecular size is consistent with the alpha form of lymphotoxin from human peripheral blood lymphocytes reported by other workers (Russell et al., 1972; Klostergaard et al., 1981; Amino et al., 1974). The SDS polyacrylamide gel electrophhoresis data (Fig. 2) shows that the lymphotoxin from human lymphoblastoid cell line has a molecular weight of approximately 20,000. This suggests the aggregation of the lymphotoxin

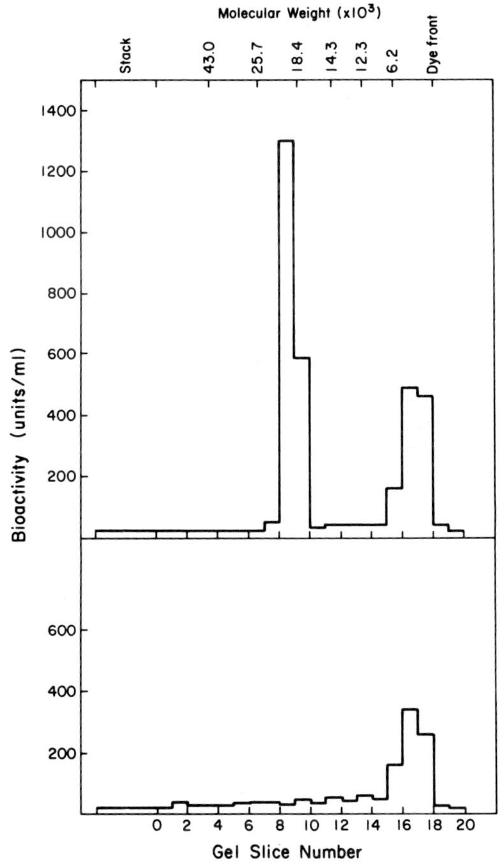

FIGURE 2. The elution profile of purified lymphotoxin (upper panel) from a fifteen percent polyacrylamide gel after electrophoresis in the presence of 0.1 percent SDS. The sample was reduced with 0.1M β-mercaptoethanol. The elution profile of the reduced sample buffer alone is shown in the lower panel.

molecules on the gel filtration column. Immune interferon isolated from peripheral blood lymphocytes has been reported to behave similarly to lymphotoxin with respect to this aggregation phenomenon (Yip et al., 1982).

The purified human lymphotoxin was found to have an isoelectric point (pI) of 5.8 and relative mobility (R_F) of 0.33 on ten percent native polyacrylamide gels at pH 8.0. This suggests that the molecule is acidic in nature which is consistent with data indicating the presence of high number of acidic amino acid residues as determined by amino acid analysis. R_F and pI values close to those observed here were also reported by Klostergaard et al. (1981) for alpha light form of lymphotoxin isolated from human tonsil and adenoid lymphocytes. Furthermore, the lymphotoxin isolated from human lymphoid cells is most probably a glycoprotein since it was found to bind to lentil lectin sepharose and can be eluted by α-methylmannoside. The reverse phase high pressure liquid chromatography on C-18 column was carried out to determine the purity of the material. A major protein peak was observed eluting at approximately fifty percent acetonitrile in 0.1 percent trifluoroacetic acid and it was found to have a molecular weight of 20,000 on SDS-PAGE.

Lymphotoxin was digested with trypsin and the tryptic peptides were isolated by reverse phase HPLC. Only two peptides were observed with molecular weights (as determined by SDS-PAGE) of 15,000 and 5,000. The intact lymphotoxin and the tryptic fragments were partially sequenced by automated amino acid sequencing. The results indicate that the 15,000 molecular weight fragment originates from amino terminus of the lymphotoxin molecule. Further studies are in progress to determine the complete structure of lymphotoxin and also to elucidate the active core of the molecule. By means of recombinant DNA technology we intend to obtain sufficient material to enable its testing in clinical studies.

ACKNOWLEDGMENTS

We would like to thank Rodney Keck, William Kohr, Donald Wheeler, Grace Don, Dennis Tajiri and Lynette Alphonso for their expert technical assistance. We especially appreciate the continuing support and encouragement of Michael Ross and Robert Swanson. A useful consultation with Professor Gale Granger (University of California, Irvine) is also acknowledged.

REFERENCES

1. Hiserodt, J.C., Tiangco, G.J., and Granger, G.A. J. Immunol. 123, 311 (1979).
2. Rundell, J.O., and Evans, C.H. Immunopharm. 2, 19 (1979).
3. Namba, Y., and Waksman, B.H., J. Immunol. 115, 1018 (1975).
4. Trivers, G., Braungart, D., and Leonard, E.G. J. Immunol. 117, 130 (1976).
5. Granger, G.A., Shimizu, I., Harris, L., Anderson, J., and Horn, P. Clin. Immunol. Immunopathol. 10, 104 (1978).
6. Evans, C.H., and Heinbaugh, J.A. Immunopharmacol. 3, 347 (1981).
7. Rosenau, W. Int. J. Immunopharmacol. 3, 1 (1981).
8. Swada, J., Shioiri-Nakano, K., and Osawa, T. Jpn. J. Exp. Med. 46, 263 (1976).
9. Evans, C.H. Cancer Immunol. Immunotherap. 12, 181 (1982).
10. Gately, M.K. and Mayer, M.M. J. Immunol. 112, 168 (1974).
11. Granger, G.A., Yamamoto, R.S., Fair, D.S., and Hiserodt, J.C. Cell Immunol. 38, 388 (1978).
12. Russell, S.W., Rosenau, W., Goldberg, M.L., and Kunitomi, G. J. Immunol. 109, 784 (1972).
13. Klostergaard, J., Long, S., and Granger, G.A. Mole. Immunol. 18, 1049 (1981).
14. Amino, N., Linn, E.S., Pysher, T.J., Mier, R., Moore, G.E., and DeGroot, L.J. J. Immunol. 113, 1334 (1974).
15. Spofford, B., Daynes, R.A., and Granger, G.A. J. Immunol. 112, 2111 (1974).
16. Yip, Y.K., Barrowclough, B.S., Urban, C., and Vilcek, J. Science 215, 411 (1982).

DISCUSSION

FARRAR: Aggarwal seems to have been successful in reducing the tremendous heterogeneity of lymphotoxins, described by Granger, down to a single molecular species. How does he explain the differences between his findings and those of Granger?

AGGARWAL: First of all, one has to remember that we have been working with an LT derived from a cell line, whereas Granger's work was restricted to PBL-derived LT. Secondly, Granger's PBL-LT was induced by a mitogen whereas our 1788-LT is not. We have shown 20K & 64K molecular sizes for 1788 LT by SDS-PAGE and Sephadex G-100 column chromatography, respectively. Whether Granger's multiple forms are derived from a single form or they are distinct species is not yet clear. However, our 1788-LT antibodies *do* neutralize Granger's mitogen-induced crude PBL-LT.

FARRAR: Has Aggarwal tested either of the tryptic fragments for biological activity?

AGGARWAL: No, they have not yet been tested.

PICK: Could Aggarwal increase LT production by the 1788 cell line via the use of mitogens such as Con-A or PHA?

AGGARWAL: We have tested mitogens such as Con-A, PHA, SEB + Thymosine, etc., but we have had no success in inducing the 1788 cell line.

PICK: How does Aggarwal make the choice of conditions that are ideal for HPLC separation of lymphotoxins?

AGGARWAL: I do not know precisely what Pick means by the "ideal conditions" for HPLC; but in the present instance the conditions used for lymphotoxin were those that were already being employed for other proteins in our laboratory. However, there are other organic solvents that one could use for reverse phase HPLC, but we did not find them to be any better than the one we used.

DUMONDE: Does Aggarwal's lymphotoxin possess any other activities we associate with lymphokines?

AGGARWAL: Thus far, the kind of activity associated with IFNγ has been excluded; others are being examined.

STUDIES ON THE ROLE OF ALPHA-HEAVY AND COMPLEX LYMPHOTOXINS HUMAN CELL-MEDIATED CYTOTOXIC REACTIONS *IN VITRO*

Jim Klostergaard[1]
James J. Devlin
Sally L. Orr
Robert S. Yamamoto
Gale A. Granger[2]

Department of Molecular Biology and Biochemistry
University of California, Irvine
Irvine, California

I. INTRODUCTION

Since its initial discovery over a decade ago (1,2), lymphotoxin (LT) has been proposed by some laboratories as the principal cytotoxic mediator involved in a number of cell-mediated cytotoxic (CMC) reactions *in vitro*. While evidence either supporting (3-10) or weakening (5,11,12) this hypothesis has been acquired, ultimate assignment of this molecule(s) or any other to any facet of the molecular mechanism(s) involved in CMC still depends on complete characterization of the moieties potentially employed by the effector lymphocytes. This has, in part, been the impetus to isolate more highly characterized lymphoid cell populations, and to define the LT forms released by them (13-18). These studies have led us to propose that the Complex (Cx) and alpha-heavy (α_H) forms are T-cell derived (17,18). Additional studies have ascribed antigen-receptor function to these forms in some cases (19,20), as well as enhanced cytolytic function compared to the lower molecular weight alpha-light (α_L), beta (β), and gamma (γ) forms (21).

In an attempt to ascertain a role for any of the members of the LT system in CMC reactions, we have employed a strategy based on serology; modulation of the rate or extent of cytolysis in these reactions by reagents specific for certain LT forms has been taken as evidence for the participation of these molecules in the lytic mechanism. Such an approach is critically dependent on the specificity of the immunological

[1]Supported by NCI New Investigator Research Award #1 R23 CA 33090-01
[2]Supported in part by NIAID grant R01 AI-09460

ISBN 0-12-527540-4

reagent employed. To develop such reagents, we have employed as antigens LT preparations of documented purity, which has necessitated purification of these molecules by biochemical techniques. This report will discuss our recent biochemical efforts with the α_H and Cx forms, as well as serological evidence for the participation of these forms in certain CMC reactions *in vitro*.

II. PURIFICATION OF α_H AND CX LYMPHOTOXINS

A. Alpha-heavy (α_H)

The purification of the human α_H LT form from lectin-stimulated lymphocyte culture supernatants has previously been reported (22) and is summarized in Table I. Human tonsil and adenoid lymphocytes were stimulated with the T-cell mitogen Con A in the presence of lactalbumin hydrolysate as a serum-substitute for 5 days in stationary culture. The cell-free supernatants were concentrated and subjected to molecular sieving. The principal lytic species was the

TABLE I. Estimated Specific Activity and Yield of Human Alpha-Heavy Lymphotoxins During Purification

	Specific Activity (units/μg proteins)[a]	Increase Compared to Whole Supernatant	Approximate Net Yield (percent)
Whole Supernatant	<<1	-	-
Molecular Sieving	<1	∿5	80
Con A-Sepharose	∿10	∿250[b]	60
Preparative Isoelectricfocusing	∿200-400	∿5-10,000[b]	40

[a]Based on a unit of activity equalling the destruction of 50% of a monolayer of 10^5 α-L-929 cells; protein determined by fluorescamine assay of whole supernatant and molecular sieving fractions.
[b]Based on distribution of radioiodinated proteins.

α_H; this had the same molecular weight (150,000d) and serological characteristics as the α_H obtained from lectin-stimulated, MLC-driven PBL cultures or SRBC-rosetting PBL populations (17,18). The α_H LT peak was pooled and chromatographed on Con A-Sepharose; bound protein and >80% of the lytic activity was desorbed with methyl-glucopyranoside. The Con A-binding preparation was further purified on a sucrose-stabilized, pH 4-8 ampholine gradient in column isoelectricfocusing. Lytic activity focused in fractions in the pH 6-7 region of the gradient; native polyacrylamide gel electrophoretic analysis demonstrated a single iodinated protein species which migrated in coincidence with the only region in the gel where lytic activity was detectable.

The preparation from isoelectricfocusing was used as a homogeneous antigen to prepare a specific antiserum in rabbits according to previously described procedures (23-25).

B. Complex (Cx)

Previous experiments employing anti-Cx antisera demonstrated blocking of a number of CMC reactions *in vitro* (9,18, and unpublished). Although these results were suggestive, the antisera used were raised against very crude Cx preparations (Table II). We have thus recently pursued the development of a purification scheme which would generate adequate amounts of highly purified Cx preparations for biochemical characterization and use as antigens. These efforts are summarized in Table II.

TABLE II. Estimated Degree of Purification of Human Complex Fractions

Cx Preparation	Separation Techniques	Net Purification Factor
A. Cx-class	Molecular Sieving	∿10
B. Cx-alpha Toxin	Con A-Sepharose (Cx)	∿50
	Phenyl-Sepharose (Cx)	∿150
	Molecular Sieving (Cx)	∿1500
	Preparative Isoelectricfocusing $(Cx - \alpha_L)^a$	∿60,000

[a]Cx lytic activity dissociates into α_L-like toxin during electrofocusing.

Human tonsil and adenoid lymphocytes were stimulated with PHA-P in 5-day stationary culture as previously described (26). The cell-free supernatants were purified on Con A-Sepharose. The methyl-glucopyranoside-eluatable proteins were subjected to negative hydrophobic affinity chromatography on phenyl-Sepharose, and the breakthrough fractions concentrated and further resolved on molecular sieving on Ultrogel AcA 44 (Table II). This highly purified Cx was very significantly further purified ($\sim$40-fold) by isoelectricfocusing (Table II), but this manipulation brought about the dissociation of Cx into an α_L-like lytic subunit (Cx-α-toxin), and, presumably, non-lytic subunits, as well (Devlin, et al., submitted). Although this preparation has still proven to be inhomogeneous, an antiserum has been raised against it and used to probe CMC reactions.

III. SEROLOGICAL EVIDENCE FOR ROLE OF α_H and CX IN CMC REACTIONS

Rabbit anti-human LT sera were generated against a homogeneous α_H preparation (anti-α_H), a crude preparation of Cx LT (anti-Cx class), and against a component of Cx (Cx-alpha-toxin), obtained as described previously (Section I). These sera were screened for their effects on cell-lysis mediated by freshly isolated human peripheral blood lymphocytes (PBL) and by a T-cell growth factor (TCGF) dependent lymphocyte cell line that had been sensitized against a mitomycin-C treated cell line (3158) which was disparate at HLA-A, B, and Dr (Yamamoto et al., manuscript in preparation). The results of multiple experiments employing separate sources of effector cells and different target cell combinations are presented in Tables III and IV. The results indicate that antisera generated against the crude preparation of Cx-LT (anti-Cx class) consistently inhibited cell mediated cytolysis by freshly isolated PBL and TCGF-dependent effectors in all experiments (Table III and Table IV). The anti-α_H could only inhibit lysis of the natural killer sensitive target cell line K-562 and of the sensitizing target cell 3158 mediated by TCGF-dependent effectors (Table IV) and not lysis mediated by the freshly isolated PBL. The anti-Cx-alpha-toxin failed to inhibit any of the cell mediated cytotoxic reactions (Table III and Table IV).

TABLE III. Inhibition of Human Natural Killing by Heterologous Antisera *In Vitro*[a]

Target Cell	NRS (% Lysis)[b]	Percent Inhibition of Lysis in the Presence of:		
		Anti-Cx Class[c]	Anti-α_H[d]	Anti-Cx-α_L-toxin[e]
K-562	48.35 ± 0.65	88%	4%	12%
MOLT-4F	52.98 ± 1.17	53%	0%	0%

[a]Rabbit anti-LT sera were tested at a final concentration of 1:4 for inhibition of natural killing by freshly isolated PBL at a 50:1 effector to target cell ratio in a 4 hr ^{51}Cr release assay.
[b]Percent specific ^{51}Cr release.
[c]Percent inhibition of lysis by antisera generated against preparation A, Table II.
[d]Percent inhibition of lysis by antisera generated against α_H (see Table I).
[e]Percent inhibition of lysis by antisera generated against preparation B, Table II.

TABLE IV. Inhibition of Cytolysis Mediated by TCGF-Dependent MLC-Sensitized Human Lymphocytes by Heterologous Antisera *In Vitro*[a]

Target Cell	NRS (% Lysis)[b]	Percent Inhibition of Lysis in the Presence of:		
		Anti-Cx Class[c]	Anti-α_H[d]	Anti-Cx-α_L-Toxin[e]
3158	60.08 ± 0.20	67	87	5
K-562	30.03 ± 4.22	96	85	11
MOLT-4F	44.27 ± 4.01	84	16	0

[a]Rabbit anti-LT sera were tested at a final concentration of 1:5 for inhibition of direct cell lysis by TCGF-dependent lymphocytes at a 25:1 effector to target cell ratio in a 2.5 hr ^{51}Cr release assay.
[b]Same as b in Table III.
[c]Same as c in Table III.
[d]Same as d in Table III.
[e]Same as e in Table III.

IV. DISCUSSION

The serological analysis employed in these studies is conceptually similar to that previously used by other investigators (3,5-7,10) as well as by this laboratory (4,8,9). Since cell-mediated cytolytic mechanisms clearly depend at least in part on externally disposed membrane structures on both the effector and target cell membranes, antibodies which react with these moieties should perturb the lytic mechanism. The net result of this perturbation is clearly not directly predictable, based on the studies of Cinader with antibodies reactive with soluble enzymes (27). Nevertheless, this approach can provide initial evidence in the determination of the molecular basis of the mechanism(s).

Previous reports from this laboratory have demonstrated that antisera reactive with the smaller LT forms (α_L, β, γ) are rather ineffective in blocking cytolysis mediated by cytotoxic T-lymphocyte (CTL) derived from MLC cultures (8,9). In contrast, Leopardi and Rosenau recently reported that an anti-α_L antiserum could block CTL-mediated cytolysis (10); the basis for this discrepancy is not immediately apparent, but could for example reside in differences in effector cell populations or in the specificity of the antisera.

The current studies indicate that antisera reactive with the highly lytic, receptor-associated α_H and Cx LT forms are generally very effective in blocking both specific, CTL-mediated cytolysis and natural killing (NK) mediated by CTL or fresh PBL's (Tables III and IV). Although the Cx preparation used for development of antisera was very crude and therefore lends ambiguity to the interpretation, the homogeneity of the α_H preparation considerably strengthens the case. These results contrasted with the inability, in our hands, of antisera against α_L to block cytolysis. Since both Cx an α_H can be degraded into the smaller α_L, β, and γ LT forms, and into non-lytic molecules as well, we questioned whether blocking observed with anti-Cx and anti-α_H could be attributed to the recognition of possible unique lytic peptides in these molecules, or to recognition of determinants on non-lytic peptides in the Cx and α_H LT forms. A preparation of α_L obtained from highly-purified Cx LT (Table II) was used as an immunogen in rabbits. The resulting antiserum, although capable of neutralizing soluble Cx and α_L LT activity (data not shown), was far less effective than the anti-Cx or anti-α_H antisera in blocking either CTL- or NK-mediated cytolysis (Tables III and IV). This suggests that the blocking observed with anti-Cx and anti-α_H is likely due to recognition of determinants not associated with the actual lytic peptide(s). The nature of the moieties expressing those important determinants is currently enigmatic, but initial characterization should be possible by immunoprecipitation analysis of the cell-surfaces of effector-target cell conjugates using antisera against the homogeneous α_H and against Cx, should its final purification be achieved. This promises to be one feasible approach to the elucidation of the molecular mechanism(s) of CMC reactions *in vitro*.

REFERENCES

1. Ruddle, N. H., and Waksman, B. H., *J. Exp. Med.* *128*, 1267 (1968).
2. Granger, G. A., and Kolb, W. P. J., *J. Immunol.* *101*, 111 (1968).
3. Walker, S. M., and Lucas, Z., *Trans. Proc.* *5*, 137 (1973).
4. Hiserodt, J. C., and Granger, G. A., *J. Immunol.* *119*, 374 (1977).
5. Gately, M. K., Mayer, M. M., and Henney, C. S., *J. Immunol.* *27*, 82 (1976).
6. Sawada, J. I., and Osawa, T., *Transplant* *26*, 319 (1978).
7. Kondo, L. L., Rosenau, W., and Wara, D. W., *J. Immunol.* *126*, 1131 (1981).
8. Ware, C. F., and Granger, G. A., *J. Immunol.* *126*, 1919 (1981).
9. Ware, C. F., and Granger, G. A., *J. Immunol.* *126*, 1934 (1981).
10. Leopardi, E., and Rosenau, W., *Cell. Immunol.* *70*, 148 (1982).
11. Henney, C. S., Gaffney, J., and Bloom, B. R., *J. Exp. Med.* *140*, 837 (1974).
12. Henney, C. S., *Transpl. Rev.* *17*, 37 (1973).
13. Shacks, S. J., Chiller, J., and Granger, G. A., *Cell. Immunol.* *7*, 313 (1973).
14. Amino, N., Linn, E. S., Pysher, J., Mier, R., Moore, G. E., and Degroot, L. J., *J. Immunol.* *114*, 1334 (1974).
15. Yano, K., and Lucas, Z. J., *J. Immunol.* *120*, 385 (1978).
16. Fair, D. S., Jeffes, E. W. B., III, and Granger, G. A., *Molec. Immunol.* *16*, 186 (1979).
17. Harris, P. C., Yamamoto, R. S., Crane, J., and Granger, G. A., *J. Immunol.* *126*, 2165 (1981).
18. Ware, C. F., Harris, P. C., and Granger, G. A., *J. Immunol.* *126*, 1927 (1981).
19. Hiserodt, J. C., Yamamoto, R. S., and Granger, G. A., *Cell. Immunol.* *38*, 417 (1978).

20. Hiserodt, J. C., Yamamoto, R. S., and Granger, G. A., Cell. Immunol. 41, 380 (1978).
21. Yamamoto, R. S., Hiserodt, J. C., and Granger, G. A., Cell. Immunol. 45, 261 (1979).
22. Klostergaard, J., and Granger, G. A., Molec. Immunol. 18, 455 (1981).
23. Vaitukaitis, J., Robbins, J. B., Nieshlag, E., and Ross, G. T., J. Clin. Endocrinol. 33, 988 (1971).
24. Lewis, J. E., Carmack, C. E., Yamamoto, R. S., and Granger, G. A., J. Immunol. Meth. 14, 163 (1977).
25. Yamamoto, R. S., Hiserodt, J. C., Lewis, J. E., Carmack, C. E., and Granger, G. A., Cell. Immunol. 38, 403 (1978).
26. Lewis, J. E., Yamamoto, R. S., Carmack, C. E., Lundak, R. L., and Granger, G. A., J. Immunol. Meth. 11, 371 (1976).
27. Cinader, B., in "Methods in Immunology and Immunochemistry" (C.A. Williams and M.W. Chase, eds.), p. 313. Academic Press, New York (1977).

LATENT LYMPHOKINES: ISOLATION OF HUMAN LATENT LYMPHOCYTE-DERIVED CHEMOTACTIC FACTOR FOR FIBROBLASTS[1]

Arnold E. Postlethwaite and Andrew H. Kang

Departments of Medicine and Biochemistry
University of Tennessee Center for the Health Sciences
and
Veterans Administration Medical Center
Memphis, Tennessee

Several reports from our laboratory have described the existence of latent lymphokine systems in both guinea pigs and man (J. Immunol. 123:561, 1979, Federation Proc. 40: 1048, 1981 and Federation Proc. 40:1049, 1981). These systems involve conversion in vitro of biologically inactive lymphokines by soluble macrophage-derived factors to active forms. In the present study we have isolated a latent form of human lymphocyte-derived chemotactic factor for fibroblasts (LDCF-F). Isolated latent human LDCF-F has a mol wt of ∿105K as determined by its elution from a BioRad TSK-250-HPLC column. When isolated latent LDCF-F was incubated with monocyte extract and rechromatographed on the TSK-250 column its mol wt had shifted to that of "classical LDCF-F", ∿35K and it was then chemotactically active. A similar conversion of latent LDCF-F to the active 35K form was accomplished by limited exposure of latent LDCF-F to trypsin. These data demonstrate another possible mechanism by which macrophages (monocytes) and lymphocytes may interact to effect certain aspects of cell-mediated immune reactions.

[1] *Supported in part by grants AM16506 and AM26034 from the National Institutes of Health and by research funds from the Veterans Administration.*

ISBN 0-12-527540-4

INTRODUCTION

Interactions between macrophages and T cells are essential for optimal production of many different lymphokines. However, the exact mechanism(s) by which macrophages modulate lymphokine production is (are) not completely understood. Posttranslational modification of proteins by protease cleavage of a large inactive precursor form, or by dissociation of an active protein-inhibitor complex, to produce an active species has been described in a variety of biological systems. Studies conducted in our laboratory have suggested that a similar system may exist for the generation of active forms of some lymphokines (1-3). We described an analogous system in man for LDCF-M (3). In the present study we have observed that human T cells also produce LDCF-F as a latent molecule that can be converted to active form by trypsin or a soluble factor from monocytes.

MATERIALS AND METHODS

Mononuclear leukocytes (MNL) were obtained from peripheral blood of healthy volunteers by the Ficoll-Hypaque technique and were passed over Sephadex G-10 columns to remove adherent cells (1,4). The resulting cell population was found to contain $\leq$ 1% esterase positive cells and $\geq$ 90% T lymphocytes by sheep erythrocyte rosetting (5). Supernatants containing latent LDCF-F were prepared by culturing monocyte-depleted lymphocytes at a concentration of 4×10^6 cells/mol of serum-free RPMI 1640 + HEPES with phytohemagglutinin (PHA, 1 μg/ml) or streptokinase streptodornase (SKSD, 100 U/ml) for 48-72 h. In some experiments supernatant (50 ml) from PHA-stimulated, monocyte-depleted lymphocyte cultures was concentrated 50 X by ultrafiltration and aliquots (0.4 ml) were subjected to high performance liquid chromatography (HPLC) gel filtration on a BioRad TSK-250 column. Monocytes ($\geq$ 95% esterase positive) were prepared by subjecting MNL to counterflow centrifugation (5). Monocytes were disrupted by sonication, and the cell-free extract was sterilized by micropore filtration (1). The monocyte extract or RPMI 1640 medium was added at a 1:10 ratio to aliquots of supernatants or column fractions containing latent LDCF-F and allowed to incubate at 37°C for 16 h. Fibroblast chemotactic activity (CTX) was determined in all samples by measuring the migration of human infant foreskin fibroblasts in Boyden-type blind well chemotaxis chamber (6).

RESULTS

Activation of Latent LDCF-F in Monocyte-Depleted Lymphocyte Culture Supernatants. Monocyte-depleted lymphocytes from three different donors were cultured for 48 h with RPMI 1640, PHA or SKSD (Table I). Aliquots of culture supernatants incubated with media contained no significant CTX, however aliquots of supernatants from PHA- or SKSD-stimulated monocyte-depleted lymphocyte cultures contained significant CTX (Table I).

TABLE 1. Activation of Latent CTX in Supernatants From Antigen and MitogenStimulated Monocyte-Depleted Lymphocyte Cultures

Donor	Culture Stimulant	Chemotactic Activity (Mean ± SEM) Monocyte Sonicate-Treated	RPMI 1640-Treated
1	PHA	47 ± 3	7 ± 1
	SKSD	49 ± 5	7 ± 1
	RPMI 1640	17 ± 1	10 ± 2
2	PHA	76 ± 7	5 ± 1
	SKSD	80 ± 9	4 ± 1
	RPMI 1640	5 ± 1	6 ± 2
3	PHA	48 ± 5	11 ± 1
	SKSD	28 ± 3	10 ± 1
	RPMI 1640	7 ± 2	8 ± 1
Control	PHA	2 ± 1	3 ± 1
	SKSD	4 ± 1	3 ± 1
	RPMI 1640	3 ± 1	4 ± 2

HPLC-Gel Filtration Studies. In order to determine the approximate molecular weight as well as partially purify latent LDCF-F, concentrated supernatant from cultures of PHA-stimulated monocyte-depleted lymphocytes was fractionated on a TSK-250 HPLC column (Figure 1, upper graph). Aliquots of column fractions were incubated for 16 h with monocyte extract or RPMI 1640 medium and tested in the same chemotaxis assay for fibroblast CTX. A single major peak of CTX was detected in monocyte extract-treated fractions (Figure 1, upper graph). This latent chemotactic factor had a mol wt of ∿105K. Media-treated fraction contained no significant CTX (Figure 1, upper graph).

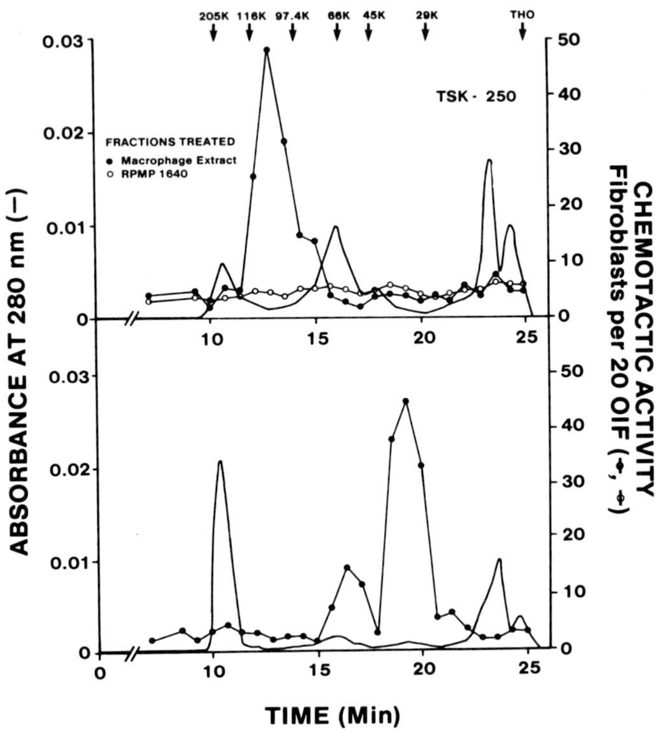

FIGURE 1. Activation of latent LDCF-F and conversion to classic LDCF-F by monocyte extract.

It was of interest to see whether latent LDCF-F activated by monocyte extract treatment would be converted to a smaller molecule similar in size to classic LDCF-F. Therefore, fractions containing latent LDCF-F from three different column runs were activated by monocyte extract, pooled, concentrated, and rechromatographed on the TSK-250 column. Fractions were tested for fibroblast CTX (Figure 1, lower graph). A single major peak of CTX with a mol wt of ∿35K found. This converted chemotactic factor had a molecular weight similar to that of classic LDCF-F produced by PHA-stimulated MNL (Figure 2).

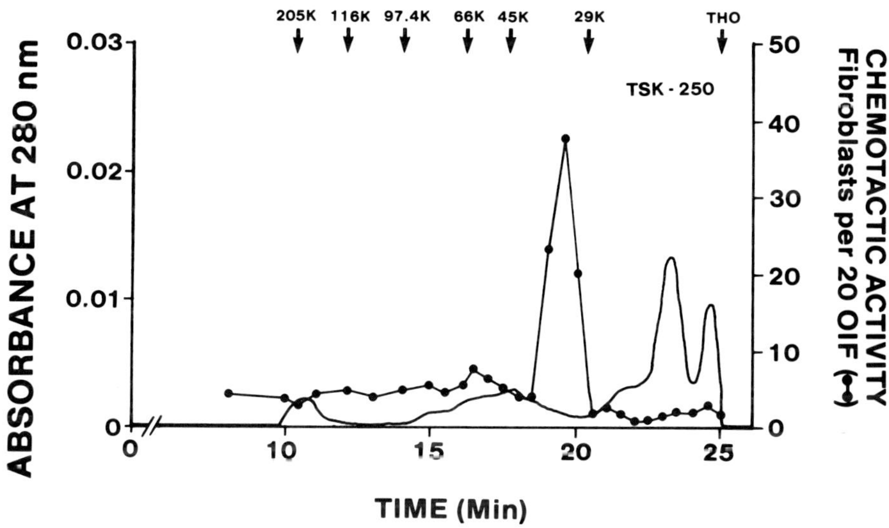

FIGURE 2. Classic Human LDCF-F Produced by PHA-Stimulated MNL.

Since some latent proteins (e.g. interstitial collagenase) can be activated by limited proteolysis, we wanted to determine whether limited trypsinization of latent LDCF-F could convert it to active LDCF-F. To accomplish this, pooled fractions containing latent LDCF-F from three columns (total volume 5 ml) were mixed with 5 ml of RPMI 1640 + HEPES containing 10 mg of insoluble trypsin (ICN Pharmaceuticals, Inc., Cleveland, OH) and incubated at 30°C for 30 min. The insoluble trypsin was removed by centrifugation. The sample was concentrated 50 X and applied to the TSK-250 column (Figure 3).

A single peak of fibroblast CTX was generated by trypsin treatment. This chemotactic factor also had a mol wt of ∿35K, similar to that of classic LDCF-F (Figures 2 and 3).

DISCUSSION

Monocyte-depleted human T cells when stimulated by specific antigen or mitogen, produce a latent form of LDCF-F. Latent LDCF-F has a mol wt of ∿105K and can be converted to an active 35K form (the same size as classic LDCF-F) by a soluble factor from sonicated macrophages or by limited trypsinization.

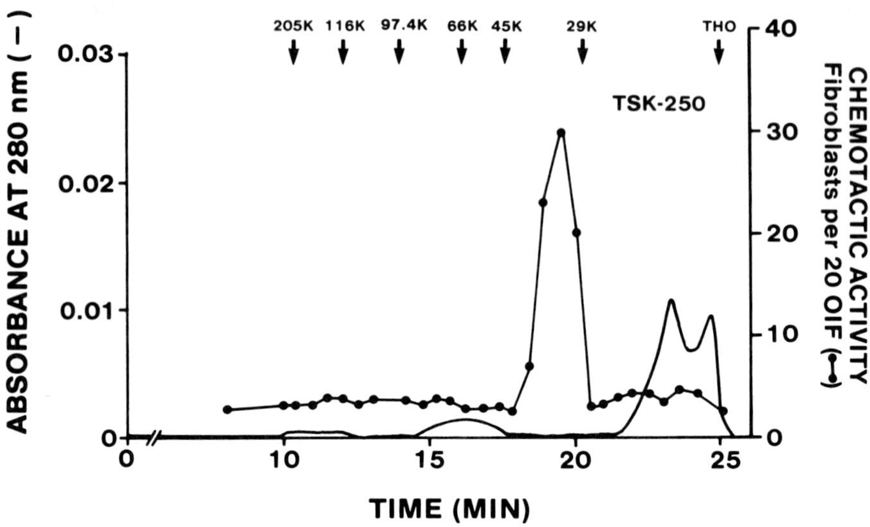

FIGURE 3. Trypsin Treated Latent Human LDCF-F.

These data suggest that the eventual expression of the biologic activity of human LDCF-F in vivo may be dependent upon its being activated by a soluble monocyte-derived factor. We have described a similar mechanism for the eventual expression of the biologic activities of human and guinea pig LDCF-M and guinea pig LDCF-F (1-3).

Latent lymphokines may represent lymphokine-inhibitor complexes, activatable by dissociation of the inhibitor from the active lymphokine or by degradation of the inhibitor by macrophage proteases. They could also represent "pro" forms ("prolymphokines"), activatable by removal of the "pro" piece of the prolymphokine molecule by macrophage proteases. It is not apparent from our data which is the case for latent human LDCF-F.

These studies show that latent lymphokine systems exist for some T cell mediators. These systems illustrate additional mechanisms by which T cells and macrophages interact to effect certain expressions of the cellular immune response.

ACKNOWLEDGMENTS

We thank Cynthia Carter, Patricia Dean, Lisa Lawton, Beth Shaeffer and Susan Turner for their excellent technical assistance and Dorothy K. Davis for typing the manuscript.

REFERENCES

1. Postlethwaite, A.E., and Kang, A.H., *J. Immunol. 123:* 561 (1979).
2. Postlethwaite, A.E., and Kang, A.H., *Federation Proc. 40:*1048 (1981).
3. Hays, R.A., and Postlethwaite, A.E., *Federation Proc. 40:*1049 (1981).
4. Ly, I., and Mishell, R.I., *J. Immunol. Methods. 5:*239 (1974).
5. Postlethwaite, A.E., Jackson, B.K., Beachey, E.H., and Kang, A.H., *J. Exp. Med. 155:*168 (1982).
6. Postlethwaite, A.E., Snyderman, R., and Kang, A.H., *J. Exp. Med. 144:*1188 (1976).

DISCUSSION

DAVID: Can Postlethwaite activate his pro-lymphokine with extracts of other cells, such as PMN? One might expect that the pro-lymphokine involved in chemotaxis of monocytes might be made in the presence of other cells.

POSTLETHWAITE: Only extracts of PMN's have been examined. They are unable to activate latent LDCF-M.

S. COHEN: I have a comment and a question for Postlethwaite. First, I suspect that precursor activities will prove to be a general property of at least some classes of lymphokines, but that monocyte-dependent mechanisms will not necessarily be involved in all instances. For example, we have shown that Eosinophil Chemotactic Factor (ECF) is synthesized in precursor form, and activated by immune complexes. We have found that cells of the C57BL/6 mouse, in serum-free medium, make a precursor MIF that requires serum for activation.

My question is as follows: With the exception of his trypsin data, Postlethwaite's results are also consistent with the possibility that macrophages spontaneously produce inactive forms of molecules we think of as lymphokines, but that they require an activating factor made by stimulated lymphocytes. This is the opposite of his model. Is there any information, other than the trypsin data, to exclude this possibility?

POSTLETHWAITE: We have no additional data at present on the macrophage-derived factor that activates human latent LDCF-F. However, we have previously shown that treatment of the putative macrophage-derived activation for guinea pig LDCF-M with trypsin does not generate monocyte chemotactic activity.

DAYER: If I have correctly understood Postlethwaite, all tests are done in the absence of serum. Will the serum inhibit the action of the monocyte factor on the pro-lymphokine and how does he visualize the operation of such a mechanism *in vivo*?

POSTLETHWAITE: We have used a serum-free system partly because serum contains a specific inhibitor of fibroblast chemotaxis. We have no data relative to the effects of serum factors on the monocyte-derived activator.

RAPPAPORT: Has Postlethwaite investigated the ability of protease inhibitors to alter the activity of monocyte extracts?

POSTLETHWAITE: Yes, but we were not able to classify the active enzyme with a series of inhibitors.

RAPPAPORT: What about trypsin inhibitors?

POSTLETHWAITE: We were not able to inhibit the activity with soybean trypsin inhibitor.

KRAMMER: I wonder whether the different biological activities found in the supernatant of one particular T cell clone could be due to different lymphokines encoded for by separate genes, or whether these activities are attributable to modified products of one or more prototype molecules.

POSTLETHWAITE: Krammer's question addresses a crucial issue. I would guess that both classes of genes would exist within certain T cell clones. We have found that a few lymphokines (e.g. guinea pig lymphocyte mitogenic factor and human T cell-derived fibroblast proliferation factor, also described at this workshop by Wahl, do *not* appear to be secreted by T cells as latent molecules. This would be compatible with the concept that separate genes control production of these lymphokines. The existence of "pro-lymphokines" would suggest the possibility that a prototype gene may encode for a large molecule which by post-translational modification could give rise to several distinct lymphokine molecules, each expressing different biological activities.

PICK: Postlethwaite is dealing with a special form of pro-lymphokine modification, namely by the action of a macrophabe-derived enzyme(s). It is equally possible that in other circumstances the "chopping off" of part of the pro-lymphokine molecule is occurring within the lymphokine-producing cell.

POSTLETHWAITE: I agree. Some pro-lymphokines could be activated intracellularly by proteases and secreted as fully active molecules.

FALK: Has Postlethwaite been able to show that the molecule which he obtains after digestion is the same as the classical factor?

POSTLETHWAITE: No. We do find that the fibroblast chemotactic activity of the converted molecule is not neutralized by anti-fibronectin or anti-C-5 antibodies. This would suggest it is neither C-5 nor fibronectin-derived.

FALK: Is it possible to do distribution experiments?

POSTLETHWAITE: Yes, it is. This would be an excellent way to demonstrate that the converted chemotactic factor is related to classical LDCF-F.

IN VITRO AND IN VIVO EFFECTS OF TMIF[1]

Marion C. Cohen
Helen D'Silva[2]
Takeshi Yoshida
Stanley Cohen

Department of Pathology
University of Connecticut Health Center
Farmington, Connecticut

The possibility that the immune system could play a role in host defense against neoplastic disease has been suggested by the rare but documented phenomenon of spontaneous regression of tumors, the many speculations as to the importance of immunologic reactions in maintaining cellular homeostasis, and analogies with the host response to infectious agents or organ transplants. Evidence for the implication of immune responses in many experimental and clinical situations involving neoplasms is seen in the discovery of a wide variety of tumor-specific antigens and the development of a number of animal models in which immunization against such antigens modified the behavior of transplanted, induced, or spontaneous neoplasms (reviewed in 1). It is generally believed that tumor immunity is a manifestation of cell-mediated immunity (reviewed in 2) although the possibility still exists that humoral antibodies may play a role in counteracting tumor growth by destroying neoplastic cells in the presence of complement.

Most studies concerning the effect of cell-mediated immune mechanisms on tumor immunity have focussed on cytotoxic killing of tumor cells. These mechanisms include cytolysis by T cells, antibody-dependent lymphocyte-mediated cytolysis, "Natural Killer" cells, and activated or armed macrophages (reviewed in 3). Lymphokines appear to play a

[1] *Supported in part by Grants # AI-12477 and CA 32319*
[2] *Supported by CA-09205 Training Grant*

ISBN 0-12-527540-4

role by direct cytotoxic effect (lymphotoxin) or by activating macrophages (macrophage activating factor).

Of equal interest is the possibility that products of the immune system might guard the host from either invasiveness or metastasis of tumor cells without direct cell killing. Thus, we have explored the question of whether mechanisms by which inflammatory cell movement may be modified during cell-mediated immune reactions may be applicable to tumor cell movement as well.

Initially, we determined that an ascites tumor, the P815 mastocytoma maintained in DBA/2 mice was capable of migrating from capillary tubes in the presence of RPMI 1640 medium supplemented with heat-inactivated normal guinea pig serum (4). The pattern was analogous to that observed when inflammatory cells migrate from capillary tubes. This was further extended to other ascites tumors including Ehrlich ascites, Hepatoma 129, and Sarcoma 37 which are of murine origin and Walker carcinosarcoma which is of rat origin (5). This provided an assay system for the analysis of an effect of lymphokines on tumor cell movement. Supernatants with known macrophage migration inhibitory (MIF) activity for macrophages were prepared from antigen or mitogen-activated lymphocytes or from long-term lymphoblastoid T & B cell lines (6). Those of human or murine origin were found to inhibit tumor cell migration. Migration inhibition was not due to cytotoxicity since the effect was reversible. This was confirmed by trypan blue exclusion studies. Of particular interest was that guinea pig MIF-containing supernatants did not inhibit tumor cell migration.

Our initial hypothesis was that the inhibition of tumor cell migration that we were observing was due to conventional MIF. However, data obtained from diaflo ultrafiltration studies indicated that this was not the case. Supernatants prepared from RPMI 8392, one of the long-term lymphoblastoid cell lines, were passed sequentially through PM 30, PM 10, and DM 5 membranes and the retentates concentrated 3-fold. When tested against guinea pig macrophages, the inhibitory activity was found in the PM 10 retentate demonstrating a molecular weight greater than 10,000 daltons. However, inhibitory activity for mastocytoma cells was found in the DM 5 retentate indicating a molecular weight between 5,000 and 10,000 daltons. From this, and other observations summarized in Table 1, we have concluded that the inhibitory activity is distinct from MIF, as well as other inflammatory lymphokines effecting cell migration. We have named this factor, Tumor Migration Inhibition Factor (TMIF) (4,5,7).

TABLE I. PROPERTIES OF TMIF IN RELATION TO MIF

	TMIF	MIF
Sources		
Human	Yes	Yes
Guinea Pig	No	Yes
Mouse[a]	Yes	Yes
C57Bl/6[b]	Yes	No
Molecular Weight	5,000-10,000 daltons	25,000 daltons (man, mouse) 67,000 daltons (guinea pig)
Trypsin Sensitivity	Yes	Yes
Neuraminidase Sensitivity	No	No
Effect of Protease Inhibitors		
SBTI	No	No
Trasylol	No	No
DFP	Yes	No
Inhibition by Monosaccharides	None known	Fucose Rhamnose

[a]All strains tested except C57Bl/6.
[b]C57Bl/6 cells incubated in absence of serum.

We have already seen that TMIF has no effect on macrophage movement. In addition, we were able to show that it can not inhibit the migration of endothelial cells (8) or fibroblasts (unpublished observations). Currently, studies are underway to compare the effect of TMIF on normal and neoplastic cells of comparable cytologic types; for example, fibroblasts and fibrosarcomas.

To determine whether TMIF can be produced *in vivo*, mice were immunized with Ehrlich's ascites tumor and challenged intravenously with a KCl extract of the tumor. Both MIF and TMIF activity were detected in sera collected at 6, 24, and 48 hours following challenge (9). We also studied tumor-bearing mice. Animals were injected with 10^7 tumor cells intraperitoneally and their sera tested for MIF and TMIF activity at 3, 5, 7, 9, 11, and 13 days post-inoculation. Serum MIF and TMIF activity appeared by Day 7 and persisted (9). However, under these conditions the animals generally die by Day 15. Thus, it would appear that the presence of endogenous TMIF is not sufficient to protect the host from death. For TMIF to have an *in vivo* effect on limiting tumor spread, it would probably need to be present early in the course of tumor growth.

To demonstrate that administration of TMIF can produce a measurable effect *in vivo*, we chose a model analogous to the macrophage disappearance reaction. 10^7 tumor cells were incubated for 3 hours at 37°C with an unfractionated lymphokine-containing supernatant, the fraction containing MIF activity, the fraction containing TMIF activity, or control medium. The cells were washed and resuspended in 1 ml of the solution in which they were incubated. They were then injected intraperitoneally into normal mice. Cells were harvested from the peritoneal cavity at various times following injection. The results are shown in Table II. Animals injected with TMIF-treated tumor cells showed significantly fewer recoverable cells 3 hours after injection. While this indicates an *in vivo* effect of TMIF, it is without therapeutic significance since by 24 hours, there was no difference between experimental and control. It is of interest that only the partially purified TMIF had this effect. The whole supernatant, though positive in *in vitro* assays, did not influence the tumor cells in the peritoneal cavity.

To rule out the possibility that the effect observed was due to an influx of inflammatory cells, supernatants prepared as described before were injected into the peritoneal cavities of normal mice. The peritoneal cavity was washed at 0 hours, 3 hours, and 24 hours following injection and the total number of cells counted. Differential cell counts were performed on Giemsa stained smears. These results are also shown in

TABLE II. RECOVERY OF TUMOR AND INFLAMMATORY CELLS FROM THE PERITONEAL CAVITY FOLLOWING TREATMENT WITH TMIF OR MIF

	Cell number[a] ($x10^{-6}$)			
	Tumor cells[b]		Inflammatory cells[c]	
Preparation	0 hours	3 hours	0 hours	3 hours
Supernatant	5.96 ± 0.74	4.12 ± 0.53	3.05 ± 0.31	5.21 ± 0.40
MIF fraction	6.05 ± 0.58	6.78 ± 0.41	ND[d]	5.75 ± 0.32
TMIF fraction	5.90 ± 0.63	1.13 ± 0.09	ND	2.10 ± 0.20
Control Medium	6.40 ± 0.84	5.83 ± 0.53	2.00 ± 0.18	1.90 ± 0.32

[a]Cells recoverable from peritoneal cavity at time indicated.
[b]$1x10^7$ tumor cells were injected intraperitoneally at start of experiments.
[c]Total inflammatory cell content. Increase due to MIF was in macrophage fraction.
[d]ND = not done

Table II. Although MIF-containing preparations caused an influx of inflammatory cells into the peritoneal cavity, the fractions with TMIF activity did not. Thus, the effect of tumor cells in vivo was not due to inflammatory response to that mediator in the peritoneal cavity. In addition, these results provide yet another distinction between TMIF and MIF.

In summary, we have shown that human and murine, but not guinea pig, lymphokine preparations contain a factor (TMIF) that reversibly inhibits tumor cell migration in vitro. TMIF is distinct from other lymphokines affecting cell movement. It can be generated in vivo as well as in vitro, and can influence the distribution of tumor cells in the peritoneal cavity of experimental animals. Thus, although it is not cytotoxic, it has the potential for serving a protective role by immobilizing tumor cells at sites where specific or non-specific killing mechanisms are operative. Further work is required to explore this possibility. It is also possible that TMIF could influence local invasiveness and/or metastasis if administered early in the course of neoplastic disease.

REFERENCES

1. Ristow, S., and McKhann, C. F., *In* "Mechanisms of Tumor Immunity" (I. Green, S. Cohen, and R.T. McCluskey, eds.), p. 109. John Wiley and Sons, New York, (1977).
2. Herberman, R.B., *Adv. Cancer Res. 19*, 207 (1974).
3. Cohen, S., and Cohen, M.C., *Am. J. Path. 93*, 449 (1978).
4. Cohen, M.C., Zeschke, R., Bigazzi, P.E., Yoshida, T., and Cohen, S., *J. Immunol. 114*, 1641 (1975).
5. Cohen, M.C., Goss, A., Yoshida, T., and Cohen, S., *J. Immunol. 121*, 840 (1978).
6. Yoshida, T., Kuratsuji, T., Takada, A., Takada, Y., Minowada, J., and Cohen, S., *J. Immunol. 117*, 548, (1976).
7. Cohen, M.C., *Cancer Research 42*, 2135, (1982).
8. Cohen, M.C., Picciano, P.T., Douglas, W.J., Yoshida, T., Kreutzer, D.L. and Cohen, S., *Science 215*, 301 (1982).
9. D'Silva, H., Cohen, M., Yoshida, T., and Cohen, S., *Clin. Immunol. Immunopath. 23*, 77 (1982).

DISCUSSION

ZSCHIESCHE: Does the T-MIF lead to agglutination of tumor cells, which effect could simulate migration inhibition?

M. COHEN: As you know, migration inhibition assays measure both decreased locomotor activity and secondary consequences of increased cellular adhesiveness. Both play a role in the interaction of MIF and the macrophage, and both could play a role in the interaction of T-MIF and the tumor cell. Thus far, however, we have not been able to observe increased cell sticking nor clumping.

SANDRU: Since Cohen found T-MIF activity in immunized animals, what is the status of T-MIF in tumor-bearing animals?

M. COHEN: We have studied tumor-bearing animals, and find that they too have high levels of circulating TMIF.

SANDRU: Inasmuch as Yoshida showed an MIF-induced densitizing protein, is T-MIF also able to induce desensitization; this could explain the anergy observed in advanced stages of the tumor-bearing state.

M. COHEN: The desensitization model of Yoshida and Cohen involved abolition of chemostatic gradients in the early stages of anergy, thereby inhibiting local accumulation of inflammatory cells. However, T-MIF does not affect these cells.

SANDRU: By utilizing the supernatant of mitogen-stimulated lymphocytes, containing T-MIF, has Cohen observed a variation in the migration inhibition of tumor cells with respect to tumor type or developmental stage, and has she ascertained whether T-MIF can react with specific antibodies raised against MIF?

M. COHEN: Yes, it is likely that tumors with different metastatic potential *in vivo* would react differently to T-MIF *in vitro*. This could be a good diagnostic or prognostic test. Finally, we are in the process of making monoclonal anti-T-MIF, but do not have any definitive results as yet.

GARLAND: Does T-MIF have any effect on the cell cycle by extending G_0. Is there any relationship between T-MIF and fibronectin synthesis? Has anyone utilized electron microscopy to explore fine morphologic changes in inhibited cells?

M. COHEN: We have not had occasion to look into any of these experiments, but they sound like great ideas.

S. COHEN: The problem with morphologic studies on mediator-treated cells is that with the multiplicity of mediators operative in key biologic situations it is hard to pinpoint the cause and significance of the changes observed. For example, the alterations in macrophages, said to be elicited by MIF, are probably due to MAF. The situation for T-MIF would be much simpler as we can readily achieve functional homogeneity, and chemical homogeneity is not so far removed. The excellent experiments Garland suggested, are feasible and well worth doing.

PICK: In reply to Garland's query, the EM findings with MIF-treated cells were in essence two: (a) an increase in the number of macrophage microvilli, and (b) loss of the polysaccharide-containing extracellular matrix (coat).

DUMONDE: When Cohen works with migrating endothelial cells, does she get them to emerge from the capillary tubes?

M. COHEN: No! That does not work, possibly because of contact inhibition. We use the agarose microdroplet technique for endothelial cell studies.

LIU: Is T-MIF produced and secreted by Cohen's tumor target cells (Walker-P815 mastocytoma).

M. COHEN: No, it is not.

LIU: Does T-MIF affect the migration of hepatocytes or mast cells?

M. COHEN: We have no information at this point.

LIU: Does T-MIF render tumor cells more susceptible to cytolysis by macrophages or NK cells?

M. COHEN: Such studies have just been initiated by us.

RECEPTOR BINDING AND ENDOCYTOSIS OF PURIFIED HUMAN INTERFERON-GAMMA (IFN-γ)

Paul Anderson
Y.K. Yip
Jan Vilček

Department of Microbiology
New York University School of Medicine
New York, New York

I. PURIFICATION AND ^{125}I-LABELING OF IFN-γ

Human IFN-γ produced by co-stimulation of plateletpheresis residues with 12-0-tetradecanoylphorbol 13-acetate and phytohemagglutinin was purified to near homogeneity by sequential processing on controlled pore glass, Con A-Sepharose, DEAE-Sephacel and Bio Gel P-150 (1-5). The resulting IFN-γ preparation was iodinated with ^{125}I-Bolton-Hunter reagent to a specific activity of approximately 100 μCi/μg protein (5). Sodium dodecyl sulfate polyacrylamide gel electrophoresis (SDS-PAGE) of the labeled IFN revealed the presence of two major protein bands (Mr=20,000 and 25,000) corresponding to the two IFN-γ subspecies previously described (4). In addition, a relatively minor band of Mr=40-50,000 was observed. Eluates of gel slices from this region contained low levels of IFN activity, suggesting that at least some of this material is IFN-γ in an undissociated form (4).

II. IDENTIFICATION OF CELLULAR RECEPTORS FOR IFN-γ

To analyze binding of IFN-γ to cells and to determine the specificity of this binding, ^{125}I-IFN-γ was incubated

Interleukins, Lymphokines, and Cytokines

ISBN 0-12-527540-4

TABLE I. Competitive Inhibition of ^{125}I-IFN-γ Binding by Various Interferons[a]

Competing IFN (units/ml)	% Specific binding of 125-I-IFN-γ in the presence of IFN-α[b]	IFN-β[c]	IFN-γ[d]
0	100	100	100
10	100	100	100
100	100	100	81
1,000	100	63	50
10,000	100	24	6
100,000	100	16	0

[a] 125-I-IFN-γ (300 units/ml, 100,000 cpm) was added to confluent monolayers of GM-258 fibroblasts together with various concentrations of unlabeled IFNs as indicated. Following a 2 hour incubation at 37° C, cells were washed, solubilized and counted for ^{125}I cpm. Specific binding averaged 65% of total binding which in the absence of unlabeled IFN-γ was 1100 cpm.

[b] IFN-α contained approximately 10^6 units and 1 mg protein per ml.

[c] IFN-β contained approximately 10^7 units and 1 mg protein per ml.

[d] IFN-γ contained approximately 10^5 units and 50 μg protein per ml.

with human GM-258 fibroblasts (trisomic for chromosome 21 and highly sensitive to the antiviral action of IFN) in the presence of increasing concentrations of unlabeled IFN-α, β and γ (Table I). Both IFN-γ and IFN-β competitively inhibited the binding of ^{125}I-IFN-γ to these cells, although at all concentrations tested significantly more IFN-β than IFN-γ was required to produce a given level of binding inhibition. IFN-α had no measurable effect on the binding of ^{125}I-IFN-γ to these cells.

Scatchard analysis using data from competition experiments performed at 4°C revealed the presence of approximately 1000 IFN-γ receptors per cell, each binding with an apparent Kd of 4×10^{-10} M (6).

In order to confirm that the observed binding was in fact due to specific binding of IFN-γ, we measured the effect of anti-IFN-γ sera (raised against IFN eluted from

TABLE II. Internalization of Surface-Bound ^{125}I-IFN-γ

Time (min.)	Cell associated cpm following[a] MEM wash	Acid wash	% acid resistant counts[b]
0	2910	448	15
2	2724	595	22
5	2673	1050	39
10	2010	1240	62
15	1900	1060	56

[a] ^{125}-I-IFN-γ (100 units/ml, 375,000 cpm) was added to confluent monolayers of GM-258 fibroblasts in 35 mm plastic dishes and incubated for 2 hours at 4°C. Cells were washed 4 times with Hanks' balanced salt solution (HBSS) and then shifted to 37°C in pre-warmed MEM plus 1 mg/ml human serum albumin (HSA) for the indicated times. Cells were then washed once with cold HBSS and incubated for 5 min at 4°C with either MEM/HSA, pH 7.4 or 0.2 M acetic acid, 0.5 M NaCl, pH 2.5. Cells were then washed, solubilized in 1% SDS and counted for ^{125}I cpm.

[b] Percentage of acid resistant counts was calculated by dividing the cell-associated cpm following acid wash by the cell-associated cpm following MEM wash.

gel slices corresponding to the isolated 20 K and 25 K regions of SDS-PAGE) on the binding of our ^{125}I-IFN-γ preparation. Pooled antisera to the 20 K and 25 K IFN components reduced the binding of ^{125}I-IFN-γ by 60% relative to a similar dilution of pre-immune control serum. Since specific binding was determined to be 65% of total binding, this represents a reduction of the specific binding by about 90%. In addition, the dilution of anti-IFN-γ sera required to inhibit ^{125}I-IFN-γ binding was, within the error of the assay system, the same as that required to neutralize IFN-γ biological activity (5).

III. INTERNALIZATION OF IFN-γ BY CELLS

The fate of receptor-bound ^{125}I-IFN-γ was examined in the following manner. ^{125}I-IFN-γ was reacted with GM-258

TABLE III. Effect of Inhibitors on Degradation of ^{125}I-IFN-γ in GM-258 Fibroblasts

Time of exposure (hours)	Acid-soluble CPM [a]			
	MEM	Chloroquine (50 μM)	$MeNH_2$ (20 mM)	Azide (15mM)
0	0	0	0	0
2	1800	400	600	0
4	3400	500	700	800
6	4500	400	900	1300
8	6400	ND[b]	ND	ND

[a] ^{125}-I-IFN-γ (200 units/ml, 250,000 cpm) was incubated with cultured fibroblasts in the presence or absence of inhibitors for the indicated exposure times. Culture media were then collected, and precipitated with 10% TCA. Soluble material was then oxidized with hydrogen peroxide and extracted with $CHCl_3$ to remove free iodide (9). The indicated values express TCA-soluble, non-iodide cpm.

[b] Not determined.

fibroblasts at 4°C for 2 hours, and unbound IFN was removed by washing. Cells were then pulsed with minimal essential medium (MEM) pre-warmed to 37°C for times ranging from 0-15 min. At the indicated times, cells were washed with either 0.2 M acetic acid, 0.5 M NaCl, pH 2.5 or MEM, pH 7.4 for 5 min at 4°C. Treatment at pH 2.5 was found to release up to 85% of surface-bound ^{125}I-IFN-γ without causing cell lysis. A similar treatment was used previously to remove surface-bound epidermal growth factor (7) or IFN-α (8). Table II shows the percentage of cell-associated cpm resistant to acid elution after various times of incubation at 37°C. At times up to 10 min, increasing amounts of cell-associated IFN-γ became resistant to acid elution, suggesting that either IFN-γ is being internalized by the cells, or that the nature of the IFN:receptor association changes with time.

If IFN-γ were being internalized, one might expect it to be degraded intracellularly. We therefore measured the appearance of trichloroacetic acid (TCA)-soluble counts in the media of cells exposed to ^{125}I-IFN-γ. After an initial lag of 15-30 min, TCA-soluble counts began appearing in the media and increased linearly over an 8-hour exposure period (Table III). Labeled IFN-γ maintained at 37°C in the absence of cells showed no increase in acid-soluble counts over the 8 hour incubation period. Table III also shows the inhibitory effects of lysosomotropic amines such as chloroquine and methylamine, as well as of a metbolic inhibitor, azide, on the degradation of ^{125}I-IFN-γ, suggesting that the observed degradation occurs intracellularly, probably within lysosomes.

IV. CONCLUSIONS

Our data reveal the presence of specific, high affinity receptors for IFN-γ on the surface of cultured fibroblasts. We also provide evidence that IFN-γ is continuously internalized and degraded over a period of at least 8 hours. Since it is known that the antiviral action of IFN-γ increases as a function of cellular exposure time, it is likely that the observed uptake process contributes in some way to the biological activity of IFN-γ. While it was generally assumed that simple binding of IFN to its cell surface receptor is sufficient to induce the antiviral state (10,11), there is no good evidence that steps distal to cell-surface binding (e.g., internalization) are not also required. We have previously shown that certain primary amines known to inhibit the internalization of various receptor binding ligands also inhibit the antiviral action of IFN (12). It is therefore conceivable that internalization plays some role in the biological action of IFN.

Our data also suggest the existence of separate receptors (or at least separate binding domains on a common receptor) for IFN-γ and IFN-α. These results are consistent with those of Branca and Baglioni (13) showing that IFN-γ fails to compete with the binding of labeled IFN-α. Somewhat unexpectedly we observed that IFN-β inhibited the binding of labeled IFN-γ, suggesting that it too may bind to the IFN-γ receptor, albeit with a somewhat lower affinity. Further experiments are necessary before definitive conclusions regarding the nature of the binding inhibition of IFN-γ by IFN-β can be drawn.

REFERENCES

1. Yip. Y. K., Pang, R. H. L., Oppenheim, J. D., Nachbar, M. S., Henriksen, D., Zerebeckyj-Eckhardt, I., and Vilček, J., Infect. Immun. 34, 131-139 (1981).
2. Yip, Y. K., Pang, R. H. L., Urban, C., and Vilček, J., Proc. Nat. Acad. Sci. 78, 1601-1605 (1981).
3. Yip, Y. K., Barrowclough, B. S., Urban, C., and Vilček, J., Proc. Nat. Acad. Sci. 79, 1820-1824 (1982).
4. Yip, Y. K., Barrowclough, B. S., Urban, C., and Vilček, J., Science 215, 411-413 (1982).
5. Anderson, P., Yip, Y. K., and Vilček, J., J. Biol. Chem. (in press).
6. Anderson, P., Yip, Y. K., and Vilček, J., manuscript in preparation.
7. Haigler, H. T., Maxfield, F. R., Willingham, M. C., and Pastan, I., J. Biol. Chem. 255, 1239-1241 (1980).
8. Zoon, K., zur Nedden, D., and Arnheiter, H., J. Biol. Chem. 257: 4695-4697 (1982).
9. Goldstein, J. L. and Brown, M. S., J. Biol. Chem. 249, 5153-5162 (1974).
10. Ankel, H., Chany, C., Galliot, B., Chevalier, M. J., and Robert, M., Proc. Nat. Acad. Sci. 70, 2360-2363 (1973).
11. Aguet, M., and Blanchard, B., Virology 115, 249-261 (1981).
12. Anderson, P., Tycko, B., Maxfield, F., and Vilček, J., Virology 117, 510-515 (1982).
13. Branca, A. A. and Baglioni, C., Nature 294, 768-770 (1981).

DISCUSSION

ZLOTNICK: Has Anderson tested his semi-purified γ-interferon for other biological activities?

ANDERSON: I have not personally performed such studies.

SCALA: Is interferon β internalized and processed after binding?

ANDERSON: We have not examined the internalization of labeled IFN-β.

OPPENHEIM: Anderson's data reveals an unusually low number of 1000 IFN receptors per cell. How did he deal with the usual problem of controlling the inevitable nonspecific background absorption of IFN? Is there any precedent for such a low number of receptors for any other agents?

ANDERSON: Recent studies have shown that human IFNα also binds specifically to high-affinity cell-surface receptors which are present at a density of 1,000-5,000 receptors per cell. In our studies, specific binding is defined as the difference between total binding and the background nonspecific binding which occurs in the presence of a 1,000-fold excess of unlabelled IFNγ. In these studies, specific binding varied between 65-80% of the total binding, depending on the ^{125}I-IFNγ preparation used.

LINNA: Does Anderson have information on the biochemical events in target cells which follow binding of γ-interferon to its receptor?

ANDERSON: No, we do not.

PAULY: In Anderson's experience, what per cent of the total IFNγ are the 50 K, 25 K and 20 K moieties before and after treatment with SDS?

ANDERSON: Native IFNγ is found only in the 40-50 K form as evidenced by molecular seive chromatography. Upon SDS treatment, approximately 90% of the total IFNγ radioactivity and biological activity is found in the 20 K and 25 K species. The remaining 10% is in the 40-50 K form.

LIU: Do the results of Anderson's competitive binding experiments contradict the data of Ankel et al. which demonstrated binding of α- and β-IFN but not γIFN to gangliosides? Has he tried similar experiments with his human γIFN?

ANDERSON: We have not looked at competition by gangliosides. It is interesting that while the antiviral activity of mouse IFN-α/β seems to be inhibited by gangliosides, and cholera toxin, Agnet finds that these agents do not compete with the specific binding of ^{125}I-IFN-α/β to cells.

FALK: Has Anderson sought to define the cells which carry IFNγ receptors?

ANDERSON: We are presently testing various peripheral blood lymphocyte sub-populations for IFNγ receptors.

GARLAND: Has anyone looked at the binding of interferons to liposomes, or lipoprotein liposomes?

ANDERSON: We have previously shown that IFN-α may be trapped within the aqueous interstices of liposomes. However, in this case we observed no binding of IFN to the liposomal bilayers. While others have studied the association of IFN with liposomes of various composition, to my knowledge no one has yet shown specific binding of purified, labeled, IFN to liposomes.

OXIDATIVE METABOLISM OF LYMPHOKINE ACTIVATED MACROPHAGES - FREE FATTY ACIDS AS SECOND MESSENGERS OF SUPEROXIDE GENERATION

Yael Bromberg
Edgar Pick[1]

Laboratory of Immunopharmacology
Department of Human Microbiology
Sackler School of Medicine
Tel-Aviv University
Tel-Aviv, Israel

I. INTRODUCTION

Progress in the purification and characterization of macrophage (MP) activating lymphokines (LKs) is hampered by the lack of reliable and sufficiently sensitive biochemical markers of the activated state (reviewed in Ref. 1). A number of recent reports have indicated that MPs activated *in vivo* or by LK treatment *in vitro* express an enhanced capacity to respond by an oxidative burst (OB) to membrane stimulation (2-4). This finding does not only offer a reliable indicator for quantitating MP activation but also provides a biochemical basis for at least some of the cytocidal properties of activated MPs. In our studies with guinea pig MPs we found that LK treatment can both enhance and inhibit the stimulated OB depending on the nature of the elicitor, the length of exposure to LK and the concentration of LK (M. Freund and E. Pick, in preparation). It is, therefore, more appropriate to talk about *modulation* by LK of the oxidative metabolism of MPs. We felt that a prerequisite for the study of the modulatory effects of LKs was a better understanding of the molecular mechanisms by which various stimulants activate the OB in MPs.

[1]To whom correspondence should be addressed

ISBN 0-12-527540-4

II. ROLE OF PHOSPHOLIPIDS IN THE ELICITATION OF THE OXIDATIVE BURST

All experiments were performed on adherence-purified MPs derived from the peritoneal cavity of mineral oil-injected guinea pigs. Superoxide (O_2^-) generation in response to a battery of stimulants was assayed by the classical ferricytochrome C reduction method (5) or by its automatic microassay variation (6). We have shown that guinea pig MPs respond by intense O_2^- production to the following stimulants: phorbol myristate acetate (PMA), opsonized zymosan, the lectins concanavalin A (ConA) and wheat germ agglutinin (WGA), the Ca^{2+} ionophore A23187, N-formyl-L-methionyl-L-leucyl-L-phenylalanine (FMLP), phospholipase C from *Cl. welchii* (PLC), NaF, colchicine and sodium nitroprusside (7). While looking for a property shared by this heterogeneous collection of stimulants, we noted the fact that many were found to induce the production of prostaglandins (PGs) by MPs or neutrophils. Since the rate limiting step in PG synthesis is the availability of the arachidonic acid (AA) substrate, we examined the ability of the aforementioned stimulants to release AA from cellular phospholipids. We found that all stimulants, with the exception of colchicine and nitroprusside, induced liberation of AA from MPs prelabeled for 1 h with ^3H-AA, at concentrations identical to those eliciting O_2^- production. The most potent inducer of AA liberation was A23187 followed by PMA and opsonized zymosan. Similar results were obtained when the ability of OB elicitors to induce the production by MPs of thromboxane A_2, was examined. Eight out of 10 stimulants were capable of doing so; again, colchicine and nitroprusside were inactive. There was, however, only partial agreement between the ability of an agent to elicit O_2^- generation and AA liberation. PMA and opsonized zymosan were potent activators in both assays; A23187 was more active as an AA liberator than as an OB inducer, while ConA and WGA were potent OB elicitors but induced only moderate AA release. Since phospholipase mediated AA liberation is a Ca^{++}-dependent process, we examined the requirement for extracellular Ca^{++} for stimulated O_2^- production. With the exception of PMA and nitroprusside, the presence of Ca^{2+} in the medium enhanced O_2^- generation; the responses to A23187 and PLC were absolutely dependent on extracellular Ca^{2+}. We next examined the effect of blocking cellular phospholipase activity on O_2^- production. This was done by using p-bromophenacyl bromide (BPB), an agent which binds to histidine residues in phospholipase A_2(PLA_2). Incubation of MPs with 5 μM BPB totally prevented O_2^- production elicited by ConA and WGA, partially inhibited the OB to most stimulants but did not affect the response to PMA and nitroprusside. The marked inhibitory effect of BPB on ConA and WGA-elicited O_2^- production and its lack of effect on the PMA response were mirrored in the influence of the drug on AA liberation. We also found that AA liberation from MP phospholipids by all but two OB elicitors could be prevented by elevating the cellular cyclic AMP level by incubation of MPs with PGE_2 in the presence of the phosphodiesterase inhibitor RO 20-1724. The same treatment also prevented the induction of an OB by the majority of stimulants, most notably by ConA, WGA, and FMLP. O_2^- production in response to PMA, A23187, NaF and nitroprusside was, however, little affected by elevated cyclic AMP levels. These results suggest that: 1. AA liberation from MPs occurs via more than one pathway, exhibiting different sensitivities to inhibition by BPB and cyclic AMP elevation. The most likely possibilities are activation of a PLA_2 or a PLC - diacyglycerol lipase sequence; 2. BPB probably affects mostly PLA_2 while cyclic AMP elevation apparently blocks both PLA_2 and PLC-mediated AA liberation, most likely via interference with Ca^{2+} transport; 3. O_2^- generation in response to most stimulants appears to be linked to PLA_2 activation. Exceptions are PMA (for which AA liberation and O_2^- generation are parallel but causally unrelated phenomena) and nitroprusside (for which there is no evidence of phospholipase activation). A special type of stimulation is represented by colchicine (and other microtubule disrupting drugs (7)) because it does not cause significant AA liberation under conditions leading to O_2^- production; 4. The higher sensitivity of O_2^- production induced by sugar-binding lectins to BPB could be due to a better accessibility of PLA_2 activated by lectins to the inhibitor, possibly because of proximity of the enzyme to the outer limit of the cell membrane.

We hypothesized that a product of the action of PLA_2 on MP membrane phospholipids acts as a messenger for the activation of the O_2^- forming enzyme (NADPH oxidase) in MPs. We therefore examined the ability of fatty acids, applied exogenously, to elicit O_2^- generation. AA (C20:4), linolenic acid (C18:3), linoleic acid (C18:2) and oleic acid (18:1) induced rapid and pronounced O_2^- formation. The capacity of fatty acids to induce O_2^- production was proportional to the number of double bonds in the molecule (AA being the most potent) and was concentration dependent in the 25-100 μM range. Saturated fatty acids, such as stearic (C18:0), were inactive. Lysophosphoglycerides, representing the remnant of the phospholipid molecule after removal of fatty acid, were incapable of eliciting O_2^- production when added to MPs at concentrations of 1-100 μg/ml. All these data pointed toward free polyunsaturated fatty acid being responsible for activating the OB. MPs oxidize AA by either the cyclooxygenase or the lipoxygenase pathways to yield E, I and $F_{2\alpha}$-type PGs, thromboxane, hydroxy-eicosatetraenoic acids (HETEs) and leukotrienes. The nature of these products and their relative amounts vary considerably with the tissue origin of the MPs, the agent used for elicitation of the MP containing exudate and the state of MP activation (8-10). It was of interest to test whether O_2^- production was induced by the liberated fatty acid prior to oxidation or by an oxidation product derived from one or the other pathway. MPs were exposed to the various stimulants in the presence of either indomethacin (a blocker of cyclooxygenase) or 5,8,11,14-eicosatetraynoic acid (ETYA) (an inhibitor of both cyclooxygenase and lipoxygenase). Neither drug inhibited O_2^- production induced by membrane stimulants or exogenous AA. However, when indomethacin or ETYA were utilized at high concentrations (100-200 μM), the OB to most stimulants was enhanced. The most pronounced enhancement was evident when cells were stimulated with microtubule disrupting agents (colchicine, vinblastine, podophyllotoxin) in combination with 100-200 μM indomethacin. When the mechanism of this effect was examined, it was found that indomethacin alone elicits O_2^- formation in a dose-dependent manner, when present at 50-400 μM. Indomethacin-elicited O_2^- production could be inhibited by BPB and by elevating MP cyclic AMP levels. This phenomenon, as well as the enhancement by indomethacin of the OB elicited by trivial stimulants cannot be explained by the specific effect of indomethacin on cyclooxygenase because it is not evident at drug concentration below 10 μM. It was recently reported that, at high concentrations, indomethacin inhibits the enzyme lysophosphatide acyltransferase which is responsible for the reacylation of deacylated phospholipids (11). Inhibition of acyltransferase results in an increase in the cellular level of unesterified fatty acids and we propose that the enhancing and direct stimulatory effect of indomethacin on O_2^- production is the result of this increase. Although these results allowed us to conclude that AA does not *have* to be oxidized in order to function as a messenger of O_2^- generation, it was still possible that an AA metabolite possesses the ability of stimulating O_2^- production. We examined this question by treating AA with soybean lipoxygenase. The product of this reaction is 15-hydroperoxy-eicosatetraenoic acid (15-HPETE) which is converted to 15-HETE. We found that soybean lipoxygenase-treated AA maintains its ability to induce O_2^- production by MPs. Addition of purified 15-HETE (kindly provided by Dr. J. M. Bailey, G. Washington University) to MPs elicited intense O_2^- production; the dose-response curve of 15-HETE closely resembled that of unoxidized AA. Incubation of MPs with PGE_2 (up to 100 μM) did not elicit an OB. Further information on the structural requirements of fatty acids for the activation of O_2^- production was provided by the finding that the methyl ester of AA was incapable of provoking O_2^- generation.

These experimental data provide the basis for the following proposal concerning the mechanism of stimulation of the O_2^- forming enzyme in MPs: 1. The O_2^- forming, membrane localized NADPH oxidase is activated by unesterified polyunsaturated fatty acid. Considering the predominance of AA in MP phospholipids, it is likely that the enzyme is activated principally by AA; 2. We do not know whether activation is mediated by a direct effect of the fatty acid on the enzyme protein or is secondary to changes in membrane fluidity, structure or ion transport leading to enzyme activation; 3. The common link among OB elicitors is their capacity to activate a membrane-localized PLA_2 leading to the release of free fatty acid; 4. The level of unesterified fatty acid in the cell membrane is determined by the relative extent of phospholipase activity and lysophospholipid reacylation ("*free arachidonate tonus*") and by the consumption of AA by oxidation; 5. The predominance of a certain oxidation pathway and the further metabolism of AA hydroperoxides and endoperoxides is also of significance; our preliminary data suggest that at least some lipoxygenase products can activate O_2^- production while cyclooxygenase-derived products are inactive. A shift towards lipoxygenase will tend to prolong the OB; a predominance of cyclooxygenase products will tend to shorten the OB and also inhibit AA release by the elevation of cyclic AMP levels induced by E type PGs; 6. It is probable that fatty acid release is a highly compartmentalized process and that effective activation of the O_2^- forming enzyme can only occur when membrane receptors for stimulants, phospholipase, substrate phospholipid and NADPH oxidase are in a cluster-like association; 7. A key unanswered question remains the mechanism by which membrane stimulation results in phospholipase activation. One obvious possibility is that stimulants activate a Ca^{2+} gate and that a Ca^{2+}-calmodulin complex is involved in phospholipase activation. There is experimental evidence for calmodulin participation in the activation of PLA_2 and in the stimulation of O_2^- formation by leukocytes. AA or its metabolites, released as a consequence of PLA_2 activation, may act as secondary Ca^{2+} ionophores and accentuate the primary event by further PLA_2 (or PLC?) activation or act as mediators of a negative feedback via an increase in cyclic AMP and the resulting inhibition of phospholipase. An alternative mechanism of phospholipase activation is by phosphorylation of the phospholipase inhibitory protein, lipomodulin (12). Phosphorylation of lipomodulin by a cyclic AMP or a Ca^{2+}-dependent protein kinase results in the loss of enzyme inhibitory activity and consequent phospholipase activation. This mechanism is compatible with our earlier finding of PG-independent cyclic AMP elevation in MPs induced by agents capable of eliciting an OB (13). Further unanswered questions are the type of phospholipid serving as substrate for activated phospholipase (phosphatidylcholine of phosphatidylinositol) and the possibility that various stimulants may activate different phospholipases (A_2 or C). In summary, our results depict the OB of MPs as resembling closely the sequence of events, common to many cell types, known as "stimulus-secretion coupling". The key innovative elements in our proposal are the causal relationship between phospholipase activation and O_2^- generation and the identification of unesterified unsaturated fatty acid as the intracellular messenger for the activation of the O_2^- forming enzyme.

III. CONCLUSION

In the light of these findings, we have recently suggested that the modulatory effect of LK on the oxidative metabolism of MPs is effected by changes in the composition and/or topography of membrane phospholipids (14,15). The evidence for the involvement of phospholipids in LK action on MPs is considerable and ranges from changes in the content of unsaturated fatty acids to the modulation of phospholipid breakdown and AA oxidation pattern following membrane stimulation. Work is under way in several laboratories on changes in phospholipid metabolism exhibited by MPs activated *in vivo*. The increasing availability of purified LK preparations should enable us to analyze the mechanisms of these changes in molecular terms. It is hoped that this will lead to the definition of a new set of biochemical markers for LK action on MPs which, in addition to their purely diagnostic value, will bear a direct relationship to the enhanced cytocidal properties of the activated MP.

REFERENCES

1. Edelson, P.J., *Lymphokines 3,* 57 (1981).
2. Nathan, C.F., Nogueira, N., Juangbhanich, C., Ellis, J., and Cohn, A., *J. Exp. Med. 149,* 1056 (1979).
3. Murray, H.W., and Cohn, Z.A., *J. Exp. Med. 152,* 1596 (1980).
4. Murray, H.W., *J. Exp. Med. 153,* 1690 (1981).
5. Babior, B.M., Kipnes, R.S., and Curnutte, J.T., *J. Clin. Invest. 52, 741 (1973).*
6. Pick, E., and Mizel, D., *J. Immunol. Meth. 46,* 211 (1981).

7. Pick, E., and Keisari, Y., *Cell. Immunol. 59,* 301 (1981).
8. Humes, J.L., Burger, S., Galavage, M., Kuehl, F.A., Wightman, P.D., Dahlgren, M.E., Davies, P., and Bonney, R.J., *J. Immunol. 124,* 2110 (1980).
9. Scott, W.A., Pawlowski, N.A., Andreach, M., and Cohn, Z.A., *J. Exp. Med. 155,* 535 (1982).
10. Scott, W.A., Pawlowski, N.A., Murray, H.W., Andreach, M., Zrike, J., and Cohn, Z.A., *J. Exp. Med. 155,* 1148 (1982).
11. Kroner, E.E., Peskar, B.A., Fischer, H., and Ferber, E., *J. Biol. Chem. 256,* 3690 (1982).
12. Hirata, F., *J. Biol. Chem. 256,* 7730 (1981).
13. Bromberg, Y., and Pick, E., *Cell. Immunol. 61,* 90 (1981).
14. Pick, E., and Bromberg, Y., *Transpl. Proc.,* in press.
15. Pick, E., and Bromberg, Y., *in* "Proceedings of the 2nd International Conference on Immunopharmacology", Pergamon Press, Oxford, in press.

DISCUSSION

ANDERSON: Oxidized fatty acids such as linolenic and linoleic acid have been shown to act as calcium ionophores with potencies similar to that of A23187. Is Pick certain that the unsaturated fatty acids,which he has purchased, are in fact *un*oxidized prior to their addition to cells?

PICK: We have taken every possible precaution to eliminate oxidation of the fatty acids during storage and use in experiments. However, we did not perform an analysis (HBLC) of the percentage of oxidized fatty acid present. However, it seems unlikely that the effect is due to contaminating fatty acid peroxides as purposeful oxidation of arachidonate with lipoxidase did not *increase* the ability of the fatty acid to elicit superoxide generation by macrophages.

PRODUCTION OF MCF(MAF) AND ANTI-MCF(MAF) BY T AND B-CELL HYBRIDOMAS

Deming Sun and Maria-Luise Lohmann-Matthes

Max-Planck-Institut für Immunbiologie
D-7800 Freiburg i. Br.

SUMMARY

A T-cell hybridoma is described, which produces large quantities of the lymphokine MCF (MAF), which activates macrophages to cytotoxicity. We also report on a monoclonal rat anti mouse MCF (MAF) antibody, which inhibits the activation of mouse macrophages by MCF (MAF).

INTRODUCTION

It is now generally agreed that macrophages can be activated to cytotoxicity by the lymphokine MCF, MAF (1). Such activated macrophages kill intracellular parasites and extracellular tumor cells. MCF (MAF) has been partially purified (2,3,4) and separated from several other lymphokines (5). Recent experiments have shown, that in vitro produced MCF containing lymphokine preparations, injected intraperitoneally, protect the animals from tumor growth (8) and infectious disease (6,7).

ABBREVIATIONS
used in this paper:
Il-2 = Interleukin 2; mab = monoclonal antibody;
RIA = radioimmunoassay
MCF (MAF) = macrophage cytotoxicity (activating) factor; ConA = Concanavalin A; FCS = fetal calf serum;
DMEM = Dulbeccos modified Eagle's Medium.

Interleukins, Lymphokines, and Cytokines

ISBN 0-12-527540-4

Envisaging the possibility of MCF (MAF) application in vivo several crucial points have to be solved: a) Methods must be developped, to produce MCF (MAF) in large quantities, b) chemical purification must proceed much further and c) the assay system should be improved and facilitated. In order to approach some of these aspects, we here report on a T cell hybridoma, which is derived from a fusion of BW 5147 and MCF (MAF) producing spleen cell blasts, which produces high amounts of MCF. The present study also reports on monoclonal rat anti mouse MCF antibodies, which inhibit MCF activity, which may be suited for further chemical purification by affinity chromatography.

MATERIALS AND METHODS

A. Establishment of T hybridoma producing macrophage cytotoxicity (activating) factor (MCF / MAF).

1. Cell fusion between lymphoblasts and HAT-sensitive Thymoma cell line BW5147 was done essentially as described before (8) using B-cells and SP2 cells as fusion partners.

2. Screening test for MCF (MAF)-producing T hybrids.
Both Con A-stimulated (2 µg/ml) hybrids supernatants and the original supernatants of the hybrids (without further Con A-stimulation) were screened for their activity of inducing macrophage activation.
Peritoneal exudate cells of thioglycollate-induced C57BL/10 mice were used as effector cells.
The tests were performed in microtiter plates. 1×10^5/well effector cells in 50µl DMEM with 10% FCS were preincubated with 200µl supernatants of hybrids for 18-24 hrs. Thereafter, the supernatants were exchanged by 1×10^4/well ^{51}Cr-labelled target cells (P815) in a volume of 0.25ml medium. The cell free supernatants were collected after 18-24 hrs and measured in a gamma-counter. % cytotoxicity was calculated according to the formula:

$$\% \text{ cytotoxicity} = \frac{\text{cpm supernatant}}{\text{cpm sediment} + \text{cpm supernatant}} \times 100$$

Cloning of the positive wells was performed by limiting dilution in 96 well flat bottom microtiter plates.

B. Production of monoclonal anti-MCF antibodies.

1. Immunization of rats.

Lewis rats were injected s.c. with 2 doses each of 1000 U partially purified MCF (1 U MCF is defined by its ability of activating 2×10^5 macrophages) (2).

MCF partially purified by chromatography on Phenyl-Sepharose was injected in complete Freund's adjuvans at multiple sites at 14-21 day intervals. 4 days after the last immunization (an intravenous injection of 1000 U MCF) the spleen was removed and a single cell suspension was prepared for fusion.

Spleen cells harvested from Lewis rat previously immunized 3 times with Phenyl Sepharose purified mouse MCF were fused with the HAT-sensitive plasmacytoma cell line X63Ag8653.

The supernatants of the HAT-resistant hybrid cell clones were screened as following:

2. Bioassay of inhibition of MCF activity by monoclonal antibodies.

300µl clonal supernatants were preincubated with 50µl MAF-containing medium in 24-well costar plate for 2 hrs.

Thereafter, 150µl of the "MAF-mixed" hybridoma cell supernatants were added to each well of a 96-well microtiter plate containing 1×10^5 thioglycollate-induced C57BL/10 mouse peritoneal macrophages. After 18 hrs incubation the supernatants were exchanged by ^{51}Cr-labelled target cells and the released ^{51}Cr in the supernatants were measured after further 18-24 hrs.

C. Bioassay of IL-2 activity.

Proliferation (^3H-TdR in corporation) of 1×10^4 IL-2 dependent mouse spleen T cell blasts was tested in the presence of IL-2-containing conditioned medium (supernatant from Con A-stimulated rat spleen cells). After 48 hrs in culture 20µl of ^3H-TdR solution (0.5µCi) was added and 18 hrs later cells were harvested on filters. Individual filters were then counted in a beta-counter.

RESULTS

1. Production of MCF (MAF) by the T cell hybridoma M 20/15

Con A blast cells from mouse spleen cells, which had been propagated for 10 days in IL-2 were fused as described in Material and Methods with BW5147. Out of 100 growing clones 4 clones showed high MCF (MAF) activity upon stimulation with Con A. None of the clones

produced MCF (MAF) constitutively in appreciable amounts. One of these four clones proved to be highly stable, whereas the other clones lost gradually their MCF producing ability. Table 1 shows a comparison between MAF activity produced by hybridoma and by spleen cells corrected for the input of MAF producing cell. Thus, 1 hybridoma cell produces about 500-fold more MCF (MAF) than one spleen cell.
Biochemically M20/15 MCF (MAF) seems to correspond to the spleen cell derived MCF(MAF) at least in terms of the parameters tested so far. M20/15 MCF (MAF) elutes from Phenyl-sepharose and from Phenylalanin-sepharose columns in an identical way as spleen cell derived MCF (MAF) (2).
Also with respect to the capacity of *in vitro* prepared MCF (MAF) to activate peritoneal macrophages *in vivo* (9) both MCF (MAF) preparations are identical (Stimpel et al., *in prep.*).

2. Production of monoclonal antibodies against MCF, MAF.

Since the assay system of MCF (MAF) *in vitro* is not only time consuming, but is also strongly influenced by the preactivation conditions of the individual batch of effector macrophages, we started to produce monoclonal antibodies with the aim to simplify the biological assay and also to use such antibodies for further biochemical purification of MAF.
From a fusion of spleen cells of a rat immunized with Phenylsepharose purified MCF (MAF) with X63Ag8 653 cells we obtained several clones, which inhibited the bioassay of MCF (MAF) (Table 2).

Table 1

Production of MCF (MAF) by T-hybridoma cells (M 15) or spleen cells. Comparison on a per cell basis.

Cellular input	Units of MCF produced during 24 h of Con A-stimulation
10×10^6 spleen cells	20 U
10×10^6 M 15 cells	10 000 U

T-hybridoma cells or spleen cells were stimulated under serum-free conditions for 24h in the presence of 2 µg ConA.
The supernatant was collected and tested on thioglycollate-induced peritoneal macrophages. 1 Unit of MCF is defined by its capacity to activate 2×10^5 macrophages to full cytotoxicity against P815 cells (2).

Table 2

Effect of anti-MCF monoclonal antibodies on MCF induced macrophage cytotoxicity

Effector macrophages incubated with	% ^{51}Cr release from P815	%inhibition by monoclonal anti-MCF antibodies
Medium	28 ± 2	-
MCF	75 ± 3	0
MCF + rat serum directed against mouse MCF (M 37 serum)	40 ± 2	74
MCF + mab 37/10-4	29 ± 1	100
MCF + mab 37/88	26 ± 2	104
MCF + mab M 143 [a]	77 ± 3	0

Macrophages were incubated with MCF or with MCF mixed with monoclonal antibodies directed against MCF, as described in Materials and Methods. After 24 h the supernatants were removed and ^{51}Cr labeled P815 target cells were added. The assay was terminated after 18 - 14 hours by removing the supernatants.

[a] M18/143 is a monoclonal rat anti mouse macrophage antibody (10), which does not interfere with macrophage cytotoxicity.

3. Influence of anti MCF (MAF) monoclonal antibody on Il-2 activity.

IL-2 containing supernatants of Con A stimulated mouse spleen cells were incubated in the same way as MCF (MAF) preparations with supernatant of the anti MCF (MAF) clone 37/10. No decrease in Il-2 activity was seen at any given concentration. These results indicate, that the monoclonal antibody 37/10 discriminates between these two lymphokine activities (Table 3).

Table 3

Effect of anti MCF monoclonal antibodies on Il-2 dependent proliferation of T-lymphoblasts.

Il-2 dependent lymphoblasts incubated with	^{3}H TdR uptake cpm
medium	149 ± 53
Il-2	840 ± 55
Il-2 + mab 37/10-4	754 ± 61
Il-2 + mab 37/88	794 ± 82

1 x 10^4/well Il-2 dependent mouse spleen lymphoblasts were incubated for 48h with DMEM + 10% FCS containing Il-2 or Il-2 mixed with monoclonal antibodies directed against MCF. 0,5 µCi (20 µl)/well ^{3}H TdR were then added and 18h later ^{3}H TdR incorporation was determined. Il-2 was used at a dosage, which was just enough to allow significant ^{3}H TdR incorporation.

DISCUSSION

Working with macrophage mediated cytotoxicity poses to major problems: One is the small amount of material available and the other is the bioassay which needs 48 h at least and is strongly influenced by conditions of the macrophage effector cell. The results presented provide some help to deal with these two major points. The T-hybridoma clone M20/15 provides us with large quantities of MCF (MAF) material. The stability of the M20/15 cells is extremely good, although the titer of the MCF (MAF) produced decreases slowly when the cells are kept in tissue culture for several months. The sensitivity for MCF (MAF) activation of every individual batch of effector macrophages varies strongly. Therefore, the quantity of MCF (MAF), which is needed for activation differs within a range of 1:20. As a consequence of this the amount of MCF (MAF) had to be titrated for every set of experiments, where inhibition of MCF (MAF) by anti MCF (MAF) antibodies was tested. A limiting amount of MCF (MAF) had to be used in order to obtain optimal inhibition. Too little MCF (MAF) resulted in no cytotoxicity, thus no inhibition could be demonstrated. To much excess MCF (MAF) resulted in an incomplete inhibition by anti-MCF (MAF) monoclonal antibodies. Thus, every inhibition assay had to be done with 3 MCF (MAF) concentrations.

We tried to approach the question of specificity of our monoclonal antibody 37/10 for MCF by testing it also in an Il-2 dependent system. In the biological Il-2 assay using Con A blasts as proliferating targets there was no condition, were Il-2 was inhibited by anti-MCF, thus indicating, that there is no crossreactivity. However, more experiments will be performed with other lymphokines. We also started to use the monoclonal antibody for further chemical purification. 37/10 was coupled to BrCN Sepharose. In such a column MCF (MAF) activity is retained whereas Il-2 passes without loss of activity (in preparation). A control column coupled with normal rat IgM has no influence on either lymphokine passage.

Our next aim will be to purify MCF (MAF) by affinity chromatography using our monoclonal antibodies as has been done with monoclonal antibodies raised against Il-2 (10,11).

REFERENCES

1. Lymphokine reports No 3.
   E. Pick and M. Landy ed. 1981: Macrophage activation, Academic Press
2. Kniep, E.M., Domzig, W., Lohmann-Matthes, M.L. and Kickhöfen, B.: Partial purification and chemical characterization of macrophage cytotoxicity factor (MCF, MAF) and its separation from migration inhibitory factor (MIF). J. Immunol. 127, 417, 1981.
3. Leonard, E.J., Ruco, L.P. and Meltzer, M.S.: Characterization of macrophage activation factor (MAF) a lymphokine, which causes macrophages to become cytotoxic for tumor cells. Cell. Immunol. 41, 341, 1978.
4. Dy, M.R., Kamoun, P., Dimitriu, A. and Hamburger, J.: Studies on mouse macrophage arming factor and its molecular weight. Transplantation 21, 273, 1976.
5. Hübner, L., Kniep, E.M., Laukel, H., Sorg, C., Fischer, H., Gassel, W.D., Havemann, K., Kickhöfen, B., Lohmann-Matthes, M.-L., Schimpl, A. and Wecker, E.: Chemical characterization of MCF (MAF), MIF, TRF and CSF from culture supernatants of ConA-stimulated murine spleen cells. Immunobiology 157, 169, 1980.
6. Lang, H., Blumenstock, H. and Lohmann-Matthes, M.-L.: Protection of mice against tumor growth and bacterial infection by treatment with the lymphokine MCF (MAF). Immunobiology, 160, 64, 1981.

7. Lang, H. and Lohmann-Matthes, M.-L.: In vivo application of the lymphokine MCF (MAF) in mice. Submitted 1982.
8. Sun, D. and Lohmann-Matthes, M.-L.: Functionally different subpopulations of mouse macrophages recognized by monoclonal antibodies. Eur. J. Immunol. 12, 134, 1982.
9. Lohmann-Matthes, M.-L., Lang, H. and Sun, D.: Macrophages as cytotoxic effector cells. Immunobiology, 161, 401, 1982.
10. Gillis, S. and Henney, C.S.: Biochemical and biological characterization of lymphocyte regulatory molecules VI. Generation of a B-cell hybridoma, whose antibody product inhibits Il-2 activity. J. Immunol. 126, 1978, 1981.
11. Stadler, B.M., Berenstein, E.H., Siraganian, P. and Oppenheim, J.J.: Monoclonal antibody against human Interleukin 2 (Il-2). J. Immunol. 128, 1620, 1982.

This work was supported by the German BMFT.

CHAIRMEN'S COMMENTS

This session provided substantive recent progress concerning the nature of certain effector lymphokines, their interactions, and their possible actions on target cells. Measurement of lymphokine activities is procedding with greater confidence.

There is growing evidence that certain lymphokines can be generated in precursor forms (pro-lymphokines), which may require limited proteolysis to become activated. This observation suggests interesting possibilities as to how lymphokine activities may be controlled within a micro-environment.

The action of effector lymphokines on target cells includes a reduction of their motility. In addition, the action of macrophage activating factors (MAF) involves the release of intracellular fatty acid which might stimulate NADPH oxidase directly or indirectly to produce an oxidative burst. The relationship of effects on mobility or oxidative metabolism to the microbiocidal or tumoricidal activities of macrophages is not yet clear.

Confusion still reigns as to the relationship of interferon to other lymphokines participating in macrophage mediated immunological reactions. With the advent of gene cloning, this problem should become clarified in the next two to three years.

D. Dumonde
E. Pick

SECTION VIII: PRODUCTION OF LYMPHOKINES BY HYBRIDOMAS, LYMPHOCYTE LINES, CLONES, AND SUBPOPULATIONS

MOLECULAR BASIS OF T-CELL MEDIATED REGULATION OF THE GROWTH OF HEMOPOIETIC CELLS AND THEIR EXPRESSION OF THY-1 AND Ia ANTIGENS[1]

John W. Schrader
Ian Clark-Lewis
Grace H.W. Wong
Jennifer McKimm-Breschkin

Immunoregulation Laboratory, The Walter and Eliza Hall Institute of Medical Research, Post Office Royal Melbourne Hospital, Melbourne, Australia.

Previously our laboratory has shown that cloned T cell hybridomas can be induced to produce multiple bioactivities that regulate the growth of lymphoid and hemopoietic cells (1-4). Using biochemical techniques and by analysis of the pattern of activities produced by several different T cell hybridomas, we have established that these activities are due to a multiplicity of molecular species (4-7). Initially we were able to delineate three classes of T-cell derived molecule (Table 1). Properties such as molecular weight, hydrophobicity and mean isolelectric point, differentiated T cell growth factor (TCGF or IL-2) from T-cell derived granulocyte-macrophage colony stimulating factor (GM-CSF), and from a molecule or class of molecules that stimulated both the in vitro growth of persisting cells (P cells) (4,8-10) (probable counterparts of the progenitors of a subset of mast cells), and the in vitro production of multipotential hemopoeitic stem cells (CFU-s) (4,11).

[1] Supported by the NH & MRC, Canberra, Australia, The Bushell Trust and The Windermere Hospital Foundation.

ISBN 0-12-527540-4

Table 1.

Three classes of regulatory molecules from T-hybridoma 123[a]

	MW (PBS)	MW (guanidine)	Hydro-phobicity	Mean pI	Lectin binding
TCGF	35,000	30,000	++++	4.9	Wheat germ agglutinin
GM-CSF	30,000	23,000	++	4.9	Wheat germ agglutinin Con A
PSF	30,000	23,000	++	6.5	Wheat germ agglutinin Con A
CFUs-SA	"	"	++	6.5	ND
TIF	"	ND	++	ND	ND

[a] References 4-7,11.

This list is by no means exhaustive. Recently with Nossal's group we found that the T cell hybridomas 123 and T6 also produce a factor or factors that promotes the response of single B cells to antigen and is distinct from TCGF (12). We have concentrated on the biochemical analysis of PSF with the initial aim of determining whether PSF or closely related molecules affect other cells such as hemopoietic stem cells.

Characterization and Purification of PSF.

The T cell hybridomas we have studied all produce regulators only when induced by concanavalin A. The myelomonocytic tumor WEHI-3B, however, produces large amounts of PSF constitutively and is a more convenient source of large amounts of material for biochemical analysis. The properties of PSF derived from WEHI-3B resemble closely those defined previously for material from concanavalin A-stimulated spleen cells or the T cell hybridoma 123 (Table 2).

Table 2. Properties of PSF

MW 23,000
pI 4 - 8 (mean pI, 6.5 after neuraminidase)
Glycoprotein - sensitive to trypsin
- binds to Con A and Wheat germ agglutinin
Sensitive to 2-mercaptoethanol
Resistant to heat, guanidine HCL
Elutes from C^3 short chain column at 41% acetonitrile
Active at 10^{-11}M or less

HPLC has proven a useful tool in the purification of PSF; to date PSF has been purified to a degree where it is active at about 10^{-11}M or less. The activity stimulating the in vitro production of CFU-s (CFUs-SA) co-purifies with PSF to this stage, suggesting that a single molecule or very similar molecules affect both P cells and multipotential stem cells.

TIF-a factor stimulating the appearance of Thy-1 positive hemopoietic progenitor cells.

Much interest has centered around the possibility that TCGF or other lymphokines regulate the maturation of immature T cells. In many studies the Thy-1 antigen has served as a marker for the appearance of T cells. However, recently we have shown that Thy-1 occurs on a range of hemopoietic stem cells and progenitor cells - including CFU-s, granulocyte-macrophage colony-forming cells and P cell precursors - following activation by a T cell derived factor, Thy-1 inducing factor or TIF (13,14). TIF was produced by T cell hybridomas and the myelomonocytic tumor WEHI-3B and resembles in its properties PSF (Table 3). It seems likely that the expression of Thy-1 is an accompaniment of the activation of murine hemopoietic stem and progenitor cells by regulatory molecules; Thy-1 is thus unsuitable as a marker to distinguish T cells from activated hemopoietic cells.

Table 3. Thy-1 inducing factor (TIF)

Sources - Con A-stimulated spleen cells
- Con A-stimulated T cell hybridomas
- WEHI-3B
M.W. 30,000
Distinct from TCGF - sources
- molecular weight
- hydrophobicity
Co-purifies with PSF and CFUs-SA on HPLC

An IFN-γ-like molecule induces increased expression of Ia and H-2 antigens.

Initially we reported that P cells, grown in medium conditioned by Con A-stimulated spleen cells (CAS), were strongly Ia positive (9). Subsequently, we have found that when cultured in partially purified PSF, P cells were essentially Ia negative (14). When these Ia negative P cells were cultured with CAS, Ia antigens appeared within 24 hrs. Ia-antigens were detected by indirect immunofluorescence using monoclonal anti-Ia antibodies (10-2.16 and 14-4.S) and the FACS (Fig. 1). Control experiments showed that the Ia antigens were coded for by the P cell genome and expression required protein synthesis. Levels of H-2K antigens were also increased, although not those of Ly 5 or receptors for Fc or peanut agglutinin.

Medium conditioned by the con A-stimulated T-cell hybridomas 123, T6 and T19.1 and by WEHI-3B all failed to induce Ia on Ia negative P cells. Of the lymphokines assayed, only IFN-γ, present in CAS, correlated with the Ia-inducing activity. Furthermore, supernatants of antigen-stimulated clones of T cells known to contain high levels of IFN-γ (15) were potent sources of Ia-inducing activity. Gel filtration of CAS showed that the Ia-inducing activity was due to a factor with a molecule weight of 40,000-45,000, which co-chromatographed with antiviral activity, assayed by reduction of Semliki Forest virus plaques. Both the antiviral activity and the Ia-inducing activity were sensitive to exposure to pH 2. Properties of the Ia-inducing factor (IIF) are summarized in Table 4.

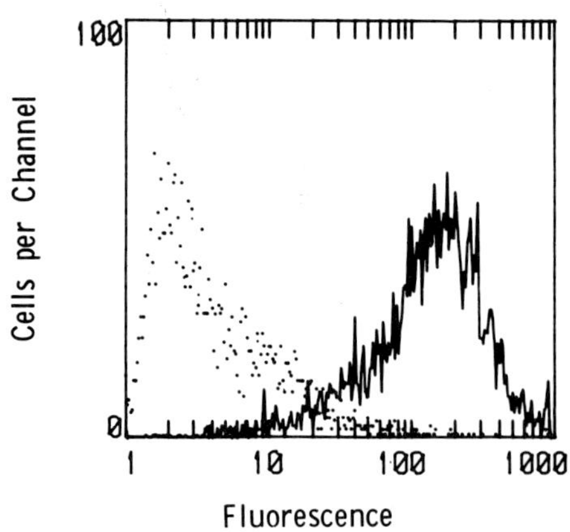

Fig. 1. Levels of Ia-antigens on P cells cultured in the presence (——) or absence (....) of CAS.

Table 4. Properties of Ia-inducing factor (IIF)

Sources - antigen stimulated T-cell clones
- Con A activated spleen cells
MW: 40,000 - 45,000
Sensitive to pH 2
Trypsin sensitive
Heat stable (56°C, 30 min)
Co-chromatographs with IFN-γ

Our studies thus indicate that IFN-γ, or a closely related molecule is the T-cell derived molecule that induces Ia antigens on P cells, and also on certain macrophage and B cell tumor lines (Wong et al, in preparation). IFN-γ (or perhaps a closely related product of a duplicated gene) may exert many of its immunoregulatory effects by enhancing expression of antigens of the major histocompatibility complex, particularly the class II (Ia) antigens.

Conclusion

Our laboratory has established that activated T cells directly synthesize a minimum of four classes of regulatory molecule (Table 5). These four classes may each contain multiple members and there are likely to be other classes (e.g., MAF, MIF). Information about the total number of these T-cell derived regulatory molecules, their interrelationships and the control of their production should rapidly follow cloning of the genes.

Table 5. Four classes of T-cell derived regulatory molecule

Factors	Cloned T cell source	MW	pH 2
TCGF	T cell hybridomas	35,000	Resistant
GM-CSF	T cell hybridomas	30,000	Resistant
PSF CFUs-SA TIF	T cell hybridomas	30,000	Resistant
IFN-γ IIF	Cloned T cell lines	40-45,000	Sensitive

Acknowledgments

We thank Ms. Pat Mottram and Dr. J.F.A.P. Miller for the supernatants of cloned T cell lines and Ms. D. Billingsley and D. Grail for their excellent technical assistance.

References

1. Schrader, J.W., Arnold, B., and Clark-Lewis, I., Nature 283, 197 (1980).
2. Schrader, J.W., Clark-Lewis, I., and Bartlett, P.F., In "Biology of Bone Marrow Transplantation" (R.P. Gale and C.F. Fox, eds) p.443, Academic Press, N.Y. (1980).
3. Schrader, J.W., and Clark-Lewis, I. J. Immunol. 126, 1101 (1981).
4. Schrader, J.W., Clark-Lewis, I., and Bartlett, P.F., Lymphokines 5, 291 (1982).
5. Clark-Lewis, I., and Schrader, J.W., J. Immunol. 127, 1941.
6. Clark-Lewis, I., Schrader, J.W., J.Immunol. 128, 168 (1982).
7. Clark-Lewis, I., and Schrader, J.W., J. Immunol. 128, 175 (1982).
8. Schrader, J.W. and Nossal, G.J.V., Immunol. Rev. 53, 61 (1980).
9. Schrader, J.W., J. Immunol. 126, 452 (1981).
10. Schrader, J.W., Lewis, S.J., Clark-Lewis, I., and Culvenor, J., Proc.Soc.Natl.Acad.Sci.USA. 73, 323 (1981).
11. Schrader, J.W. and Clark-Lewis, I. J.Immunol. 129, 30 (1982).
12. Pike, B.L., Vaux, D., Clark-Lewis, I., Schrader, J.W. and Nossal, G.J.V., Proc.Soc.Natl.Acad.Sci.USA. (in press).
13. Schrader, J.W., Battye, F., and Scollay, R., Proc.Soc. Natl.Acad.Sci.USA. 79, 4161 (1982).
14. Schrader, J.W. and Clark-Lewis, I., Curr.Topic in Microbiol. Immunol. 100, 221 (1982).
15. McKimm-Breschkin, J.L., Mottram, P.L., Thomas, W.R. and Miller, J.F.A.P., J.Exp.Med. 155, 1204 (1982).

DISCUSSION

KRAMMER: Is WEH1-231 an *in vitro* line?

SCHRADER: Yes. The WEH1-231 is a clone line, passaged *in vitro*.

KRAMMER: Do P cells kill WEH1-164?

SCHRADER: We have not tested this. However it should be done, as mast cells have been reported to kill fibrosarcoma cells.

GARLAND: Do P-cell lines display 20αSDH activity; are they inducible for Thyl and for non-specific cytotoxicity by γINF?

SCHRADER: P cell lines manifest no cytotoxicity on targets such as YAC, C118, EL-4; P815 even in the presence of supernatants containing IFNγ. Our continuous P cell lines are Thy-1 negative. We have not tested for 20αSDH activity in P cells.

PAETKAU: Would Schrader comment on the relationship between PSF and Ihle's "IL-3"?

SCHRADER: We have tested highly purified preparations of IL-3 sent us by Ihle and find them to be very active in supporting the growth of P cells. I understand that Ihle is in fact now routinely assaying IL-3 using factor-dependent lines similar to P cells.

SWAIN: A number of investigators have reported the development of Thy-1 on previously Thy-1^- cells. Is it likely that Schrader's TIF was responsible for these inductions?

SCHRADER: It is highly likely that the induction of Thy-1 on non-T hemopoietic or myeloid cells could be occurring in these experiments, especially if crude preparations of "IL-2" are used, as most sources of IL-2 also contain TIF (probably identical with PSF).

SWAIN: Does Schrader know of sources of TIF which do *not* contain IL-2?

SCHRADER: Yes. The WEHI-3B tumor produces large amounts of TIF in the absence of IL-2.

LATTIME: Has Schrader assessed other tumor targets for susceptibility to killing by P cells?

SCHRADER: We have not tried a fibrosarcoma such as WEHI-164 as a target. This would be a logical move as it has been reported that mast cells can kill these cells.

PALLACIOS: Are P cells capable of presenting antigen?

SCHRADER: P cells can stimulate in mixed lymphocyte type reactions. Hapten-conjugated P cells can also stimulate the hapten-specific T cell lines produced by Jacques Miller's group. However, we have not yet determined whether P cells are capable of presenting soluble protein antigens.

PRODUCTION OF LYMPHOKINES BY T CELL HYBRIDOMAS DERIVED FROM A FUSION OF THE AKR T CELL TUMOR BW5147 AND A SELECTED HIGH PRODUCER T CELL CLONE IN LONG-TERM CULTURE

Brigitte Kaltmann

Institut für Immunologie und Genetik
Deutsches Krebsforschungszentrum
Heidelberg, F.R.G.

Diethard Gemsa

Institut für Immunologie und Serologie
Universität Heidelberg
Heidelberg, F.R.G.

Lothar Hültner

Abteilung für Experimentelle Hämatologie
der Gesellschaft für Strahlen- und
Umweltforschung
München, F.R.G.

Ursula Kees

Institut für Immunologie und Genetik
Deutsches Krebsforschungszentrum
Heidelberg, F.R.G.

Fabrizio Marcucci

Institut für Virologie
Deutsches Krebsforschungszentrum
Heidelberg, F.R.G.

Peter H. Krammer

Institut für Immunologie und Genetik
Deutsches Krebsforschungszentrum
Heidelberg, F.R.G.

Murine (AKR/J) T cell clones propagated in tissue culture for > 2 years in medium substituted with T cell growth factors

ISBN 0-12-527540-4

(TCGF) were selected for high production of lymphokines upon T cell mitogen (concanavalin A) stimulation. These clones released immune interferon (IFN-y), macrophage activating factors (MAF), and colony stimulating factors (CSF). T cell hybridomas derived from fusions of the AKR T cell tumor BW5147 and one of these T cell clones produced either: i) all 3 lymphokines simultaneously, ii) a combination of two lymphokines, or iii) only one lymphokine. Thus, production of IFN-y, MAF and CSF by these hybridomas did not only show quantitatively different levels but could also be separated. This indicates that IFN-y, MAF and CSF are molecularly not identical. These T cell hybridomas will be a valuable source for lymphokine purification.

INTRODUCTION

Mitogen (e.g. concanavalin A (Con A)) or antigen induced supernatants from lymphoid cells contain a variety of different lymphokines. T cells are the producer cells of most lymphokines in such supernatants. This has been shown by: 1) Abrogation of production after treatment of the producer cell population with anti-T cell reagents and complement (1), 2) analysis of lymphokine production from mitogen activated, enriched T cells in limiting dilution microcultures (2,3,4), and 3) secretion of lymphokines from selected T cell clones in long-term culture and from T cell hybridomas (3,4,5,6,7).

Experiments with long-term T cell clones reveal that the progeny of a single T cell is capable of releasing several lymphokines simultaneously. This is supported by the fact that the frequency (f) of precursor producer T cells for these lymphokines is high (f = 1/2 for IFN-y, f = 1/3 for CSF, and f = 1/3 for MAF) (2,3,4).

In principle, several strategies or combinations thereof can be envisaged to distinguish between various lymphokines secreted from T cells: 1) differential effects on target cells, 2) absorption of a particular lymphokine on a defined target cell population and removal of the biological effect, 3) biochemical separation, 4) cloning of lymphokine genes and selective production in pro-or eukaryonts, 5) production of monoclonal antibodies against a defined lymphokine, 6) selection of variant T cell clones which secrete a reduced number of lymphokines, and 7) fusion of lymphokine producing normal lymphoid cells or T cell clones with a T cell tumor in order to select hybridomas in which lymphokine production has been separated.

We have chosen to fuse a high producer T cell clone with the AKR T cell tumor BW5147 in order to develop hybridomas which upon induction produce one or a few lymphokines. Our data indicate that IFN-y, CSF and MAF are molecularly not identical, and their Biological effects can be separated.

MATERIAL AND METHODS

T cell clone PC-AKR clone 96 (clone 96). Cells from the draining axillary and inguinal lymph nodes of AKR/J mice, sensitized on the abdominal skin by 2,4,6-trinitrophenylchloride (Picrylchloride, PC) 5 days earlier were cultured for 4 days. Clones were derived by limiting dilution cloning (4,5). Clone 96 has been passaged in tissue culture for > 2 years and grows strictly dependent on the presence of TCGF. Lymphokine release was induced (after washing the cells in TCGF-free medium containing 2% FCS) by pulsing the cells in a large tissue culture flask (Falcon bottles, 3024 F) for 3 h with 10 ug/ml Con A, washing, and culturing for 24 h in medium free of both TCGF and Con A.

Supernatants from hybridomas derived from a fusion of clone 96 with the HAT-sensitive AKR T cell tumor BW5147 were induced in the same way as the supernatant of clone 96 except that cells ($1x10^6$/ml) were induced in 5 ml Falcon tubes (Falcon N° 2005) with Con A containing medium overnight.

Con A induced supernatants from BW5147 were negative for lymphokines.

Fusion of clone 96 with the AKR T cell tumor BW5147. Fusion was carried out according to standard methods previously described (8).

IFN assay. IFN was assayed by measuring reduction of the cytopathic effect (CPE) of vesicular stomatitis virus on L cells (9). The IFN titer was expressed as the reciprocal of the highest serial threefold dilution of the sample showing 50 % CPE reduction. One unit (U) of IFN α/ß in our assay (reference) corresponds to 2 international U of IFNα/ß (NIH stadared G-002-904-511).

The anti-viral activity from supernatants of clone 96 was shown to be IFN-y as previously described (2).

CSF assay. CSF was determined as previously described (3). Briefly, supernatants of all cultures were assayed individually for CSF in 1 ml semi-solid agar cultures containing 75,000 CBA bone marrow cells. After 7 days the bone marrow cultures were scored for colony growth.

MAF assay. MAF derived from clone 96 has been tested in a variety of assays (D. Gemsa et al., in preparation) (Table 1). MAF derived from T cell hybridomas was determined in a tumor growth inhibition assay (GIA). The non-metastasizing variant Eb of the L 5178y lymphoma was either propagated in vivo by weekly intraperitoneal injections in syngeneic DBA/2 mice or maintained in vitro by weekly passages in tissue culture flasks. The in vivo and in vitro passaged tumor cells were free of lactate dehydrogenase elevating virus. The tumor cytostatic activity of macrophages was determined in a GIA. Briefly, 4 x 10^4 resident macrophages were incubated with 2 x 10^4 Eb tumor cells (macrophage to tumor ratio 2:1) in a total volume of 0.2 ml in flat bottom microtiter plates at 37^oC with or without the addtion of MAF containing supernatants. The assay was terminated at 48 h after a 4 h pulse with 2 uCi ^{3}H-thymidine (specific activity 400 mCi/mmol). Cells were harvested by an automatic cell harvester (Flow Laboratories, Rocksville, MD) and incorporation of ^{3}H-thymidine into tumor DNA was measured in a scintillation counter. Growth inhibition of tumor cells by activated macrophages is expressed as per cent growth inhibition compared with tumor cell cultures with resident macrophages. As macrophages alone incorporated negligeable amounts of ^{3}H-thymidine, the incorporation could be accounted for by tumor cells.

RESULTS AND DISCUSSION

Table 1 shows that clone 96 secreted three different lymphokines: MAF, CSF, and IFN-y. MAF was measured by its activity on macrophage metabolism, induction of release of oxygen metabolites and prostaglandin E (data not shown), and most notably by the induction of macrophage killing of schistosomula and tumor cells and by induction of tumor cytostasis. CSF was determined by the formation of colonies from hematopoietic precursor cells in semi-solid agar cultures of bone marrow cells, and IFN-y by reduction of the CPE of vesicular stomatitis virus on L cells.

Thus, at least 3 lymphokines acting on a variety of target cells from different organs could be found in the mitogen

Table 1. Activities measured of lymphokines secreted by clone 96

MAF	CSF	IFN-y
Schistosomula killing	Formation of colonies of hematopoietic cells	Inhibition of virus replication
Tumor cytostasis		
Tumor cytolysis		

(Con A) induced supernatant of clone 96. In order to investigate whether or not these lymphokines are different molecules and exert defined different biological effects, clone 96 was fused to the HAT-sensitive AKR T cell tumor BW5147. After fusion, hybridomas were consecutively cloned 3 times with an input of 0.1 cells per well in limiting dilution microcultures. After the last cloning the hybridomas were grown up to suitable cell numbers and stimulated with Con A to secrete lymphokines into the supernatant (see: Material and Methods). The supernatants of each clone have been tested for IFN, MAF and CSF activities. Data obtained from > 500 clones showed secretion of one, two, or all three lymphokines. The following combinations occured (with decreasing frequency): IFN-y, MAF, CSF > MAF, CSF > CSF > IFN, CSF > MAF > IFN, MAF > IFN. Some clones shown in Table 2 demonstrate that, apart from different combinations of the three lymphokines secreted, the quantity of lymphokines in the supernatant of the individual clones showed a considerable variation.

These data show that hybridomas can be derived from a fusion of a high producer T cell clone with BW5147 which secrete

Table 2 Activity of lymphokines of clone 96 x BW5147 hybridomas

hybridoma clones	IFN[1]	MAF[2]	CSF[3]
E4	-	-	34
E9	-	33.1	22
E11	10	-	35
F2	30	50.9	15
F7	10	-	50
F11	30	67.1	15
G2	-	52.2	11
G6	-	75.1	-

1 IFN titer (see Materials and Methods).

2 growth inhibition (%) of Eb tumor cells (see Materials and Methods). Supernatants were used at a concentration of 30 % (v/v) in medium.

3 colonies per 75,000 bone marrow cells (see Materials and Methods).

IFN, MAF and CSF. These lymphokines are secreted in all possible combinations as well as separately (an IFN-y-only-secretor was found in a different series of clones; data not shown). Although we cannot entirely exclude that in each particular case low contaminating amounts of each of the three lymphokines are present which escaped our detection system, these data indicate that MAF, CSF and IFN-y are molecularly not identical and that the biological effects of macrophage activation, colony formation of hematopoietic cells and inhibition of viral replication are exerted by at least three different lymphokines. The described hybridomas should facilitate the purification of these lymphokines and help to define their biological effects.

Acknowledgments

We thank Ms. W. Kramer, U. Seitz, Mr. W. Müller and P. Genthner for expert technical and Ms. A. Riedl and A. Schygulla for expert secreterial assistance.

References

1) Landolfo,S., Kirchner,H., and Simon M.M. Eur.J.Immunol. 4, 295 (1982).

2) Krammer,P.H., Marcucci,F., Waller,M., and Kirchner,H. Eur. J.Immunol. 3, 200 (1982).

3) Staber,F., Hültner,L., Marcucci,F., and Krammer,P.H. Nature, 298, 79 (1982).

4) Krammer,P.H., Dy,M., Hültner,L., Isakson,P., Kees,U., Lohmann-Mattes,M.-L., Marcucci,F., Michnay,A., Puré,E., Schimpl,A., Staber,F., Vitetta,E., and Waller,M., in "Isolation, Characterization, and Utilization of T Lymphocytes" (Fathman,D., and Fitch,F., ed.), p. 253. Academic Press, New York, (1982).

5) Marcucci,F., Waller,M., Kirchner,H., and Krammer,P.H. Nature 291, 79 (1981).

6) Schrader,J.W., Arnold,B., Clark-Lewis,J. Nature 283, 197 (1980).

7) Ratliff,T.L., Thomasson,D.L., McCool,R.E., Catalona,W.J. Cell.Immunol. 68, 311 (1982).

8) Goldsby,R.A., Osborne,B.A., Simpson,E., Herzenberg,L.A. Nature 267, 707 (1977).

9) Marcucci,F., Nowak,M., Krammer,P., Kirchner,H. J.Gen.Virol. 60, 195 (1982).

DISCUSSION

DAVID: What was the antiserum directed to, which inhibited MAF from clone 96 but not the other? Do antisera to IFNγ block MAF?

KRAMMER: The antiserum was raised in rabbits by immunization with column-enriched lymphokines induced from PL-AKR clone 29 (similar in its lymphokine secretion to PC-AKR clone 96). This antiserum blocks IFNγ and MAF from PC-AKR clone 96 but not MAF from clone AKR anti-B6 PK7.1.2F8.

PICK: What was the stimulus used for the induction of O_2 and H_2O_2 release? It seems to me that this is not a general finding - to demonstrate enhanced O_2 and H_2O_2 production induced by lymphokines; your finding is therefore of particular interest.

KRAMMER: O_2 and H_2O_2 was induced from zymosan-pre-stimulated macrophages (using MAF as indicated).

PICK: The increase in PGE production is also rather exceptional; in the main, macrophage activation has generally been found to be associated with *decreased* PGE synthesis.

KRAMMER: In our case, MAF derived from a PC-AKR clone 96 increased PGE release from macrophages.

WOOD: To follow-up on David's query, has Krammer compared the activities of authentic IFNγ with the activities of the clone 96 MAF?

KRAMMER: No, we have not. Biochemically pure IFNγ was not available to us.

DAYER: I am very interested in Krammer's dissociation of PGE_2 release by the two clones. Guerne and I are reporting to this workshop that $Ly2^+$ supernatant can stimulate PGE_2 release by macrophages, and that in the case of $Ly2^-$ to a lesser extent. Accordingly, I ask Krammer whether he had an opportunity to examine other arachidonic metabolite products.

KRAMMER: No, we did not.

DAYER: Has Krammer identified the Ly markers on his cells?

KRAMMER: Yes. Ly markers were not found -- which is not too surprising for cells that have been maintained in culture for 2 to 3 years.

DUTTON: Does Krammer think that the finding of different factors from different sublines indicates the presence of different gene products -- or could they be variants of the same gene product?

KRAMMER: We have to consider that one lymphokine may exert a number of different biological activities. Alternatively, we might be dealing with many different lymphokine proteins, encoded for by different genes. Still a third possibility is that modified split products of a given molecule exert different activities.

B CELL-SPECIFIC IMMUNOREGULATORY FACTORS: CONSTRUCTION OF PRODUCER T CELL HYBRIDOMA LINES AND DEVELOPMENT OF BIOASSAYS[1]

Amnon Altman, Robin G. Weiner, Karen L. Evans, E. Fernandez-Cruz and David H. Katz

Department of Immunology
Medical Biology Institute
La Jolla, California

I. INTRODUCTION

The biological effects of crude T cell-derived immunoregulatory lymphokines (LK) originally were described in *in vitro* systems of antibody responses (1). In recent years, the study of T cell-specific LK, *e.g.*, T cell growth factor (TCGF) (3,4) has progressed at a somewhat faster speed due to the establishment of: a) TCGF-producing T cell lines and hybridomas (reviewed in 2) and b) a sensitive and rapid TCGF assay, using cloned T cells as targets (4). Studies of B cell-specific factors (BCSF) have been hampered by the lack of precise bioassay methods. Thus, it remained unclear whether factor(s) which stimulate antibody responses act directly on B target cells, what role (if any) TCGF plays in such responses and whether BCSF provide proliferation and/or differentiation signals to their cellular targets. Very recently, methods have been devised to: a) establish lines of normal B lymphocytes in long-term culture (5,6), b) bioassay BCSF using neoplastic B cell targets (7) or limiting dilution analyses of normal B cells (8), and c) distinguish various BCSF from other LK, *e.g.*, from TCGF (9-11). In this communication, we report the construction and identification of cloned T cell hybridoma lines which constitutively produce BCSF, and the development of new bioassays for such BCSF, using the anti-DNP IgA-secreting myeloma line, MOPC-315, as a cellular target.

[1]Supported by USPHS Grants CA-33650 and AI-19476

ISBN 0-12-527540-4

II. MATERIALS AND METHODS

A. T Cell Hybridoma Lines

The construction and properties of various T cell hybridoma lines producing various LK, *i.e.*, allogeneic effect factor (AEF) (12), TCGF (13), macrophage activating factor (MAF) (14), immune interferon (IFNγ) (14) and BCSF(15) was described earlier. Briefly, DBA/2 T cells blasts alloactivated *in vivo* and secondarily *in vitro* to (C3HxDBA/2)F_1 alloantigens were fused with the $HGPRT^-$ mutant of the AKR/J-derived T cell lymphoma, BW5147, using 30% polyethylene glycol. Hybridoma lines were selected for growth in HAT selection medium and screened for constitutive or lectin-induced production of various LK species. T hybridoma lines were phenotyped and cloned by the limiting dilution technique. All T cell hybridoma lines were found to be free of mycoplasma and are maintained routinely in Dulbecco's modified Eagle's medium (DMEM) supplemented with 5-10% fetal calf serum (FCS) or in synthetic, serum-free Iscove's-modified DMEM.

B. MOPC-315 Cells

Three variants of the BALB/c-derived, IgA-secreting MOPC-315 myeloma were used in these studies. The *in vivo*-growing (ascites) variant, which normally fails to grow in culture, was obtained from the laboratory of Dr. J.M. Davie, St. Louis, MO. and maintained by weekly intraperitoneal passage in BALB/c mice. Two variants of the same tumor adapted to growth in culture were obtained from Drs. G.L. Milburn and R.G. Lynch, Iowa City, IA. Both of these, 315/P and 315/J, display TNP-specific cell surface IgA receptors (16). About 40% of the 315/J cells secrete anti-TNP IgA antibody detected by their ability to form plaques against TNP-SRBC while only 0-3% of the 315/P cells secrete antibody (16). These cells were maintained by subculturing them every 2-3 days at an initial concentration of $3x10^5$/ml in RPMI-1640 supplemented with 10% FCS and 25% spent culture medium.

C. Bioassays for BCSF

T cell hybridoma-derived immunoregulatory LK with apparent specificity for murine B cell targets were screened in three systems:

1. Primary Anti-TNP-SRBC Antibody Response. These cultures were performed according to established procedures (17). Briefly, $5x10^5$ twice anti-Thy 1.2+C-treated DBA/2 or (C57BL/6xDBA/2)F_1 hybrid spleen cells were cultured in microtiter plates with 0.01% TNP-SRBC and dilutions of different polyclonal and monoclonal LK preparations. Direct (IgM) plaque responses were determined after 5 days of culture in a modified Jerne's plaque assay.

2. B Cell Growth Assay. One x 10^2 to $1x10^5$ MOPC-315 ascites cells were cultured in 96 well, flat-bottom culture plates in RPMI-1640 plus 5% FCS containing serial dilutions of LK preparations. After 24-96 hours of culture, the cells were pulsed for an additional 6 hr. period with 1μCi ^{3}HTdR, harvested on a Skatron cell harvester and counted in a liquid scintillation counter.

3. B Cell Differentiation Assay. One to $5x10^5$ 315/P cells were cultured for 36-48 hrs in 1 ml RPMI+5% FCS in the absence or presence of LK preparations in 16 mm. wells of a 24 well culture plate. Thereafter, the cells were harvested, counted and assayed for their TNP-specific plaque forming cell (PFC) responses at 250 cells per slide, using TNP-coated SRBC and a rabbit anti-mouse IgA antiserum at a final dilution of 1:500 for developing the plaques (16).

III. RESULTS AND DISCUSSION

A. Initial Screening for BCSF Production

T cell hybridoma lines were initially screened for the constitutive production of LK species which stimulate primary anti-TNP-SRBC antibody responses of T cell-depleted spleen cultures. Several AEF-producing hybridomas were identified by this procedure (12) and one, hybridoma 34, was selected for further studies. This hybridoma was recloned twice by limiting dilution and many of the subclones were found to secrete constitutively BCSF capable of stimulating primary antibody responses *in vitro* (15). We consistently failed to detect TCGF activity in the culture supernatants of all such clones, using a sensitive TCGF bioassay (3) which could detect TCGF activity in conditioned media of Con A-stimulated rat spleen cells or some other murine T cell hybridoma lines constructed by us (13) at dilutions of up to 1:500. These results strongly suggest that hybridoma-derived TCGF does not play a role in these antibody responses. In addition, many of the BCSF-producing T cell hybridoma clones were found to display $I\text{-}A^d$ determinants and receptors for the Fc portion of murine IgG (15).

Although the spleen cells used in these assays were depleted of T cells by two treatments with monoclonal anti-Thy 1.2 antibody + C (resulting in a ≈95% inhibition of their responses to T cell mitogens) it was still possible that residual T cells or pre-T cells could become activated in these cultures and produce their own LK, including TCGF. Furthermore, such cultures still contain macrophages as well as other cell types in addition to B lymphocytes and such cells could respond to exogenous LK or produce their own LK. We focused our efforts, therefore, on the development of new assays using homogeneous populations of B cells. We elected to study initially a population of neoplastic B cells, represented by the IgA-secreting myeloma, MOPC-315. This myeloma was selected since it: a) secretes a defined, DNP-specific idiotype, b) is sensitive to T cell-derived immunoregulatory signals (18), and c) includes several variants distinguished by their growth and secretion properties (16).

B. Stimulation of B Cell Growth by LK Preparations

The ascites variant of MOPC-315 does not grow in culture although variants selected for *in vitro* growth were derived from it (16). We hypothesized that the failure of this ascites line to grow *in vitro* may reflect its requirement for growth factor(s) which is normally provided by the murine host but is missing *in vitro*. In this case, the *in vitro*-adapted lines, 315/P and 315/J may represent selected growth factor-independent variants. To test this hypothesis, we studied the effects of different LK preparations on the growth of ascites MOPC-315 cells *in vitro*. Conventional AEF supernatants, a potent source of various LK, were used initially to investigate the parameters of this system. The parameters studied in these experiments included: a) the number of MOPC-315 cells per well, b) the concentration of AEF supernatants, and c) the time of harvest following a 6 hr. terminal pulse with ^{3}HTdR.

These experiments confirmed that ascites MOPC-315 cells grow poorly in culture in comparison with 315/P cells and, secondly, demonstrated that AEF supernatants can markedly stimulate the growth of the ascites variant *in vitro*. These effects were evident with as few as 100-200 cells per well and as early as after 24 hrs. of culture and were more striking with low cell numbers (200-2,000) which gave very low background responses and, thus, resulted in 20-50 fold stimulation indices. With increasing cell numbers, i.e., J $1x10^4$ cells/well, background responses increased and tended to obscure the response to the added LK. In routine assays, we are now using $1\text{-}5x10^3$ MOPC-315 ascites cells per culture in a 3 day assay. Other LK preparations found to have substantial biological activity in this assay include: conditioned media of Con A-stimulated murine spleen cell cultures, supernatants of allostimulated cloned T cell lines established by Dr. G. Dennert (found before to contain various BCSF; 19) and constitutive supernatants of several subclones of hybridoma 34.

Although the overwhelming majority of cells in a MOPC-315 ascites population are actually myeloma tumor cells, this population is contaminated with some other cells, including T lymphocytes. Therefore, it is possible that the LK does not act directly on the MOPC-315 but rather on other, contaminating cells. Furthermore, such contaminating cells may themselves produce in culture LK which affect the myeloma tumor cells. However, these possibilities appear unlikely since: a) marked effects are observed within 24 hrs. of culture and with as few as 100-200 cells, a time frame and cell dose conditions under which the possible contribution of contaminants is, at most, negligible, b) conditioned media of Con A-stimulated spleen cultures produce an identical growth effect on the MOPC-315 population irrespective of whether α-methyl mannoside is incorporated in the culture medium to block residual Con A activity, and c) LPS, but not Con A, stimulates the growth of the ascites cells *in vitro*. The last two findings demonstrate that the observed effects are not due to activation and/or proliferation of contaminating T cells.

C. Stimulation of B Cell Differentiation by LK Preparations

To ascertain the presence of B cell differentiation-promoting activity in LK preparations, we performed experiments to determine whether such preparations will induce the differentiation of the low IgA secreting variant, 315/P, to a state of high IgA secretion. MOPC-315/P cells cultured for 40 hrs. with or without several LK preparations were subsequently enumerated for the presence of anti-TNP PFC in a modified Jerne's plaque assay. The results (Table I)

TABLE I. Stimulation of MOPC-315/P Differentiation by LK Preparations[a]

Cells	*LK Source*	*% TNP-specific PFC*
MOPC-315/P	*None*	*0*
"	*Murine Con A sup.*	*5.6*
"	*AEF (DBA/2 → C3D2F$_1$)*	*16.4*
"	*C.C3.11.75 T cell line*	*10.8*
MOPC 315/ascites	*None*	*52.8*

[a]*40 hrs. incubation with 5% (v/v) LK preparations.*

indicate that, in fact, MOPC-315/P cells may serve as an adequate cellular target for titrating B cell differentiation factor(s). No PFC could be detected in the control population while incubation with some polyclonal or monoclonal LK preparations increased the fraction of PFC to levels of 5.6-21.2%. The positive control, i.e., the MOPC-315 ascites variant had about 53% PFC, as expected (16). Our efforts are presently focused on optimizing this system and screening our T cell hybridoma clones for the presence of differentiation-promoting activities.

Although neoplastic B cells are known to be responsive to T cell-derived immunoregulatory signals, it remains to be determined whether a similar spectrum of LK species affects both neoplastic and normal B lymphocytes. Furthermore, the physiological significance of such BCSF needs to be studied. However, assays of the type described herein may facilitate the study of B cell-specific immunoregulatory molecules and their roles in B cell activation and proliferation.

ACKNOWLEDGMENTS

We are grateful to Drs. J.M. Davie, G.L. Milburn and R.G. Lynch for MOPC-315 cell lines and to Dr. G. Dennert for generously providing us with LK preparations from his alloreactive T cell lines.

REFERENCES

1. Dutton, R.W., Falkoff, R., Hirst, J.A., Hoffman, M., Kappler, J.W., Kettman, J.R., Lesley, J.F., and Vann, D. Progr. Immunol. 1,355 (1971).
2. Altman, A., and Katz, D.H. Adv. Immunol. 33, in press (1982).
3. Morgan, D.A., Ruscetti, F.W., and Gallo, R.C. Science 193, 1007 (1976).
4. Gillis, S., Ferm, M.M., Ou, W., and Smith, K.A. J. Immunol. 120, 2027(1978).
5. Howard, M., Kessler, S., Chused, T., and Paul, W.E. Proc. Natl. Acad. Sci. USA 78, 5788 (1981).
6. Sredni, B., Sieckmann, D.G., Kumagai, S., House, S., Green, I., and Paul, W.E. J. Exp. Med. 154, 1500 (1981).
7. Pure, E., Isakson, P.C., Takatsu, K., Hamaoka, T., Swain, S.L., Dutton, R.W., Dennert, G., Uhr, J.W., and Vitetta, E.S. J. Immunol. 127, 1953 (1981).
8. Wetzel, G.D., Swain, S.L., and Dutton, R.W. J. Exp. Med. 156, 306 (1982).
9. Schimpl, A., Hübner, L., Wong, C., and Wecker, E. Progr. Immunol. 4, 403 (1980).
10. Takatsu, K., Tanaka, K., Tominaga, A., Kumahara, Y., and Hamaoka, T. J. Immunol. 125, 2646 (1980).
11. Howard, M., Farrar, J., Hilfiker, M., Johnson, B., Takatsu, K., and Paul, W.E. J. Exp. Med. 155, 914 (1982).
12. Katz, D.H., Bechtold, T.E., and Altman, A. J. Exp. Med. 152, 956 (1980).
13. Altman, A., Sferruzza, A., Weiner, R.G., and Katz, D.H. J. Immunol. 128, 1365 (1982).
14. Schreiber, R.D., Altman, A., and Katz, D.H. J. Exp. Med. 156, in press (1982).
15. Altman, A., Schreiber, R.D., and Katz, D.H. in "T Cell Hybridomas", Current Topics in Microbiology and Immunology (H. von Boehmer, W. Haas, G. Köhler, F. Melchers and J. Zeuthen, eds.) 100, 179, Springer Verlag, Berlin, Heidelberg, New York, 1982.
16. Gebel, H.M., Autry, J.R., Rohrer, J.W., and Lynch, R.G. J. Natl. Cancer Inst. 62, 201 (1979).
17. Armerding, D. and Katz, D.H. Activation of T and B lymphocytes in vitro. II. Biological and biochemical properties of an allogeneic effect factor (AEF) active in triggering specific B lymphocytes. J. Exp. Med. 140, 19, 1974.

18. Lynch, R.G., Rohrer, J.W., Odermatt, B., Gebel, H., Autry, J.R. and Hoover, R.G., Immunol. Rev. 48, 45 (1979).
19. Swain, S.L., Dennert, G., Warner, J.F., and Dutton, R.W. Proc. Natl. Acad. Sci. USA 78, 2517 (1981).

DISCUSSION

SWAIN: Altman stated that MOPC315/P might be a growth factor-independent line. Did he directly test this by adding growth factor to low numbers of these cells?

ALTMAN: Yes, we did such an experiment and found that lymphokine preparations had *no* discernible effect on the growth of MOPC-315/P cells *in vitro*, in contrast to their effect on the ascites variant of this tumor.

PRODUCTION OF MURINE IMMUNE INTERFERON BY A T LYMPHOCYTE CELL LINE: REGULATION BY AN EL4 THYMOMA-DERIVED FACTOR(S)

William R. Benjamin
Patricia S. Steeg
John J. Farrar

National Institute of Dental Research
National Institutes of Health
Bethesda, Maryland

I. INTRODUCTION

The identification of continuous cell lines and hybridomas capable of producing lymphokines which appear to be identical to products of normal lymphoid cells has facilitated both biochemical and biological characterization of a number of such factors(1,2). Very little data, however, has been reported concerning the development of cell lines capable of producing immune interferon (IFN-γ).

In this study we describe the isolation of an Interleukin 2-independent T cell line (BFS) which, after stimulation with phorbol 12-myristate 13-acetate (PMA), produced IFN-γ. In addition, we demonstrate that costimulation of the BFS cells with PMA and EL4 thymoma culture supernatant synergistically enhances IFN-γ production.

II. RESULTS AND DISCUSSION

This laboratory has routinely utilized an IL-2 dependent cytotoxic T cell line (CT 6, provided by J. Watson) to measure levels of IL-2 activity. By subculturing CT 6 cells in decreasing levels of IL-2, we isolated a subline (BFS) which no

Interleukins, Lymphokines, and Cytokines

ISBN 0-12-527540-4

longer required exogenous growth factor for maintenance of cell growth. Both the CT 6 and BFS cells are H-2^b and Thy 1.2 positive and Lyt-1.2, Lyt-2.2 and sIg negative by FACS analysis.

Recent studies have demonstrated that PMA either alone or in combination with additional stimulants can induce the production of lymphokines from normal lymphoid cells or cell lines (3,4). Additional reports have demonstrated that Lyt 2^+ lymphocytes are a source of IFN-γ(5). Since CT 6 cells were originally a cloned cytotoxic T cell line we examined whether PMA could induce IFN production by either the CT 6 or BFS cell lines. Cells were suspended to $1x10^6$ per ml in RPMI 1640 medium containing 10% fetal bovine serum, antibiotics and selected doses of PMA. The results in Table I demonstrate that PMA induced IFN production in a dose-dependent manner from BFS but not CT 6 cells. Experiments examining the kinetics of IFN production by the BFS cell line demonstrated that maximal IFN levels were obtained 48 hr after stimulation with PMA.

Interferon-γ can be distinguished from IFN α/β by a number of criteria(6). Since PMA stimulated BFS cells to produce IFN, we next determined whether the IFN produced was IFN-γ or IFN α/β. The results in Table II show that BFS-derived IFN was sensitive to pH 2 treatment but resistant to heating at 56°C for 60 minutes. In addition, the IFN activity was neutralized by antiserum against IFN-γ(7) but not by antiserum specific for Type I (IFN α/β) IFN. Thus, these results suggest that BFS cells produce IFN-γ in response to stimulation with PMA. Further biochemical characterization of BFS-derived IFN-γ demonstrated an apparent mw of 48,000 as determined by AcA Ultrogel gel filtration and an isoelectric point of 5.5-6.0.

TABLE I. PMA-stimulated Induction of IFN Production by the BFS Cell Line

	IFN (Units/ml)[a]	
PMA (ng/ml)	BFS cells	CT 6 cells
0	<5	<5
5	<5	<5
10	7	<5
50	15	<5
100	27	<5
500	5	<5
1000	5	<5

[a]*The IFN activity of the culture supernatants was determined by a viral plaque reduction assay.*

TABLE II. Characteristics of Interferon Produced by the BFS Cell Line

Treatment	*IFN activity remaining, %*[a]
None	*100*
pH 2	*10*
56°C, 60 min	*114*
Goat anti-type I IFN antiserum	*101*
Rabbit anti-IFN-γ antiserum	*4*

[a]*Each IFN preparation initially contained 70-250 units of IFN per ml.*

We have examined BFS supernatants for a variety of additional lymphokine activities and have found them to be negative for IL-1, IL-2, macrophage-granulocyte colony stimulating factor and T cell-replacing factor. However, IFN-γ-containing supernantants obtained from the BFS cell line do induce macrophage Ia antigen expression, which is consistent with previous reports demonstrating this activity for IFN-γ(7,9). Thus, in contrast to the large number of lymphokines produced by cell lines such as the EL4 thymoma cell line(10), which appears to be of a helper cell lineage, the BFS cell line appears to produce a limited number of lymphokine activities.

The titers of IFN-γ obtained from the BFS cell line after stimulation with PMA are lower than those obtained after lectin stimulation of conventional spleen cell preparations. The production of IFN-γ by splenic Lyt-2$^+$ T lymphocytes appears to involve T-T interactions presumeably mediated by soluble factors biochemically similar to IL-2(5,11). However, as stated above the cloned BFS cells do not produce these requisite "helper" signals or factors. We, therefore, examined whether partially purified EL4 thymoma-derived IL-2 could enhance PMA-induced IFN-γ production by the BFS cell line. In these experiments an IL-2-containing sample was prepared by sequential ammonium sulfate precipitation (40-60% saturation), phenyl sepharose chromatography and preparative isoelectric focusing. After preparative isoelectric focusing the fractions were combined into three pools (pH 4.25-6.25, 6.25-8.25 and 8.25-9.50) and examined for their ability to synergize with PMA in the induction of IFN from the BFS cell line. The results in Table III show that the pool of material which contained maximal IL-2 activity (pH 4.25-6.25) synergistically enhanced PMA-induced IFN production by the BFS cell line.

TABLE III. Synergy Between Partially Purified EL4 Thymoma IL-2 and PMA in the Induction of BFS Interferon

IEF pool (Units IL-2/ml)	Dilution	IFN (U/ml) +PMA(40ng/ml)	IFN (U/ml) -PMA
None		-	<3
None		14	-
pH 4.25-6.25(4500)	1:10	334	11
	1:20	370	-
	1:40	315	-
	1:80	96	-
pH 6.25-8.25(350)	1:10	247	11
	1:20	82	-
	1:40	32	-
	1:80	15	-
pH 8.25-9.50(20)	1:10	57	<3
	1:20	21	-
	1:40	13	-
	1:80	13	-

This IFN activity is also labile to treatment at pH 2 but resistant to heating at 56°C and to treatment with anti-Type I IFN antiserum and thus has the characteristics of IFN-γ. These results indicated that although BFS cell line growth is independent of IL-2, the cell line presumeably can still respond to IL-2 and is subject to the regulation of IFN production by IL-2.

III. CONCLUSIONS

The results in this study have demonstrated the isolation of an IL-2-independent T cell line which produces IFN after stimulation with PMA. The IFN is labile to treatment with pH 2 buffer and is neutralized with anti-IFN-γ antiserum but is resistant to heating at 56°C, 60 min and anti-Type I IFN antiserum. Furthermore, the mw of the IFN is 48,000 and the pI is 5.5-6.0. All of these characteristics are similar to those reported previously for spleen cell-derived IFN-γ. We have a also shown that a factor(s) present in EL4 thymoma culture supernatants synergizes with PMA in the induction of IFN-γ production by the BFS cell line. Thus, the BFS cell line appears to have retained responsiveness to regulatory signals associated with IFN-γ production by T lymphocytes.

ACKNOWLEDGMENTS

The authors thank Drs. Joost J. Oppenheim, Tadashi Kasahara and Fran Hakim for their critical reviews of this manuscript. In addition, we thank Dr. Howard M. Johnson for the anti-IFN-γ antiserum. We also thank Mrs. Vickie Benjamin for the expert preparation of this manuscript.

REFERENCES

1. Farrar, J.J., Fuller-Farrar, J., Simon, P.L., Hilfiker, M.L., Stadler, B.M. and Farrar, W.L., *J. Immunol. 125,* 2555(1980).
2. Harwell, L. Skidmore, B., Marrack, P., and Kappler, J., *J. Exp. Med. 152,* 843 (1980).
3. Nathan, I., Groopman, J.E. Quan, S.C., Bersch, N. and Golde, D.W., Nature (London) 292, 842 (1981)
4. Pang, R.A.L., Yip, Y.K. and Vilceck J., *Cell. Immunol. 64,* 304 (1981).
5. Torres, B.A., Farrar, W.L. and Johnson, H.M., *J. Immunol., 128,* 2217 (1982).
6. Stewart, W.E. II., Nature (London) 286, 110, 1980.
7. Osborne, L.C., Georgiades, J. and Johnson, H.M., *ICRS Med. Sci. 1,* 212 (1980).
8. Oppenheim, J.J., Luger, T., Sztein, M.B. and Steeg, P.S., in "Role of Macrophages in Self Defense Mechanisms" (N. Ishida, ed.) Elsevier/North Holland, New York) (1982) in press.
9. Steeg, P.S., Moore, R.N. and Oppenheim, J.J., *Fed. Proc. 40,* 771 (1981).
10. Farrar, J.J., Benjamin, W.R., Hilfiker, M.L., Howard, M. Farrar, W.L. and Fuller-Farrar, J., *Immunological Rev. 63,* 129 (1982).
11. Farrar, W.L. Johnson, H.M. and Farrar, J.J., *J. Immunol. 126,* 1120 (1981).

DISCUSSION

DUMONDE: Can Benjamin absorb out IL-2 with BFS cells?

BENJAMIN: We have performed preliminary experiments concerning this matter, but the results have been inconclusive. We will be pursuing these studies further, both before and after PMA treatment to determine whether IL-2 receptors are increased.

LACHMAN: If the BFS line retains IL-2 receptors, does Benjamin believe that there is one kind of receptor for IL-2-mediated growth, and another one for IL-2-mediated IFN production.

BENJAMIN: We have no information relevant to Lachman's proposition.

LACHMAN: It seems to me that radiolabeled IL-2 would be the ideal reagent to probe these cells.

BENJAMIN: Once we have definitely established that IL-2 is the active lymphokine involved in this process, the use of radiolabeled IL-2 would indeed be one way of ascertaining whether the receptor involved in the induction of IFN has the same characteristics as the one operative in T cell proliferation.

SCHRADER: In Nacy's report earlier in this workshop, we learned that supernatants of EL-4, stimulated with PMA, contained IFNγ. In Benjamin's experiments he reports that under certain conditions EL-4, stimulated with PMA did *not* produce IFNγ; can he account for the opposing findings?

BENJAMIN: El-4 only produces IFNγ when stimulated with PMA in the presence of 10% serum. In the present work supernatants were made in 1% serum. Thus, for the production of IFN by the EL-4 cells there appears to be a differential requirement for FCS, which is in contrast to the production of other lymphokines by the EL-4 cell line.

ZLOTNIK: Benjamin does not find TRF or MAF in his γIFN-containing supernates. Would he explain these assays further?

BENJAMIN: In the TRF assay we measured the effect of the supernatants on the plaque-forming cell response of nude mouse spleen cells, and spleen cells from mice treated 48 hrs earlier with anti-thymocyte serum. In these assays we found no detectable TRF activity.

In collaborative studies with Meltzer we have examined crude and partially purified lymphokines for their capacity to induce tumoricidal macrophages, using the assay system described earlier by Nacy. Based on these preliminary studies we have found no MAF activity. However, in both cases, and in these types of systems generally, we must be mindful of the possibility of inhibitors which could mask any particular activity.

PICK: Does Benjamin know how long his IL-2 has to be present in his BFS cell cultures in order to enhance IFNγ production?

BENJAMIN: Preliminary studies suggest that the IL-2 preparation must be present during the entire culture period in order to obtain maximal enhancement of IFN production.

BERTOGLIO: What is the evidence that the IFN Benjamin measures from the BFS line stimulated with EL-4 supernatants is actually synthesized by BFS? How certain can he be that it does not result from the processing by BFS of an inactive precursor for IFN, present in the EL-4 supernatant?

BENJAMIN: We have not yet performed the internal labeling studies which would definitely answer this question. However, this proposition by Bertaglio is a very interesting possibility and merits further examination.

LYMPHOKINES FROM THE EL-4 T-CELL LINE INDUCE MACROPHAGE MICROBICIDAL AND TUMORICIDAL ACTIVITIES

Carol A. Nacy
Wayne T. Hockmeyer

Walter Reed Army Institute of Research
Washington, District of Columbia

William R. Benjamin
John J. Farrar
Stephanie L. James
Monte S. Meltzer[1]

National Institutes of Health
Bethesda, Maryland

Immune responses during neoplastic or infectious disease induce unique subpopulations of mononuclear phagocyte effector cells. These activated macrophages are functionally different from both differentiated resident macrophages of various tissues or inflammatory cells that accumulate after injection of sterile irritants such as starch, oil, or casein. Immunologically activated macrophages effect a number of cytotoxic reactions: some are dependent upon known endogenous or exogenous mediators, such as antibody, bacterial lipopolysaccharides, or the secretagogue, phorbol myristate acetate (1-3); other cytotoxic reactions either do not require such mediators, or the mediator is not known. An example of this latter reaction was demonstrated in studies by Adams and coworkers that showed activated, but not resident or inflammatory macrophages bound and killed tumor cells directly (4). Among the nonspecific macrophage effector activities, we can define a distinct subset that share a number

[1]Present address: Department of Dermatology, Walter Reed Army Medical Center, Washington, District of Columbia.

ISBN 0-12-527540-4

of common characteristics: (a.) cytotoxicity is a short-lived cell function: activity is maximal 8-12 hr after treatment of macrophages with lymphokines, then declines rapidly to control levels by 24 hr (5-7); (b.) cytotoxicity is the end result of a sequence of reactions, a cascade of events similar in concept to that of complement or clotting systems: each reaction must occur within a defined time, and requires both appropriate signal and responsive macrophage (8,9); and (c.) cytotoxicity fails to develop with macrophages from certain mouse strains: strains with the Lps^d gene (C3H/HeJ), strains derived from A mice (A/J), and the P/J strain (9,10). Each of the defects, however, is controlled by a different genetic locus (11,12). Transient cytotoxic reactions by activated macrophages with each of these characteristics occur against a wide variety of targets. Such targets include a diverse group of tumor cells (5), the intracytoplasmic bacterium, *Rickettsia tsutsugamushi* (6), the intraphagolysosomal protozoan, *Leishmania tropica* (7), and skin stage schistosomula of the extracellular helminth, *Schistosoma mansoni* (13).

Despite broad commonality in induction and expression of transient cytotoxic responses, these reactions can be divided even further by characterization of responsive cell populations or of activation signals. Macrophages that migrate to a site of inflammation are at least 10-fold more responsive than equal numbers of resident cells to lymphokines that induce nonspecific tumoricidal and extracellular microbicidal activities (5,13). In contrast, it is the resident peritoneal macrophage that kills amastigotes of L. tropica after lymphokine treatment (7): lymphokine treated cells from chronic irritant-induced inflammatory exudates fail to destroy this obligate intracellular parasite. Differences between inflammatory and resident cells for development of effector activities after lymphokine treatment may be secondary to acquisition and loss of certain cytotoxic effector mechanisms (production of O_2^-, prostaglandins, etc) during maturation. Superimposed upon this diversity in responsive macrophage effector cells is the heterogeneity within lymphokine mediators. Preliminary analysis of activities in lymphokines from antigen-stimulated spleen cells suggests that factors for regulation of transient cytotoxic reactions against various targets may be different. With the same, but quite heterogenous cell population (resident peritoneal macrophages) we can induce both tumoricidal and microbicidal activities with spleen cell-derived lymphokines (7,13,14). Further characterization of these lymphokines, however, showed that factors for induction of microbicidal activity against intracellular targets (*R. tsutsugamushi, L. tropica*) were completely ineffective for induction of tumoricidal and

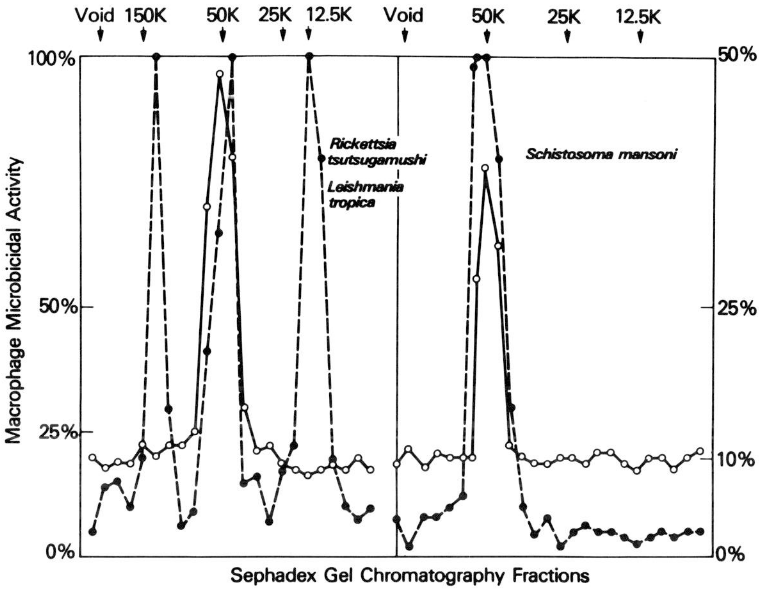

FIGURE 1. Activation of macrophages for microbicidal and tumoricidal activities by spleen cell-derived lymphokines.

microbicidal activities against extracellular targets (tumor cells or schistosomula) (13-15; summarized in Fig. 1). Lymphokines that induced intracellular destruction of rickettsiae and leishmania eluted from Sephadex G-100 in three distinct regions of 130,000, 50,000, and 10,000 mw. In contrast, lymphokines that activate macrophages for extracellular destruction of tumor and larval targets eluted in a single peak of activity of mw 45-50,000.

The recent report of lymphokines from a continuous T cell line allowed us the unique opportunity to examine the question of diverse regulatory signals with another source of activation stimuli (16). Lymphokines from a variant of the EL-4 thymoma cell line, EL-4(FARRAR), were prepared by treating cells with 20 ng/ml PMA for 48 hr; culture fluids were chromatographed on Sephadex G-100 columns (13-15). Cells were treated with EL-4 column fractions as macrophage monolayers (tumoricidal assay) or as unfractionated peritoneal cells (microbicidal assays) for 4 to 72 hr (5,7,13). Macrophages treated with column fractions of culture fluids from PMA-stimulated EL-4 cells developed capacity to kill both the

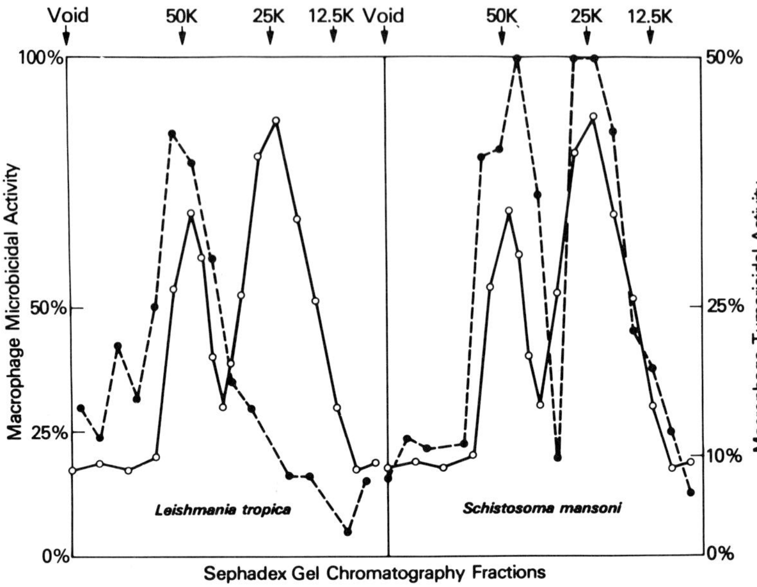

FIGURE 2. Activation of macrophages for tumoricidal and microbicidal activities by factors in culture supernatants of PMA-stimulated EL-4 cells.

intracellular and extracellular targets. Activities for destruction of *S. mansoni* schistosomula cofractionated with activities for nonspecific tumor cytotoxicity: both extracellular microbicidal and tumoricidal activities were induced with factors in the 45,000 and 23,000 mw regions (16; Fig. 2, right panel). In contrast, EL-4-derived factors that activate macrophages for microbicidal activity against the intracellular, amastigote form of *L. tropica* were found only in the 45,000 mw region (Fig. 2, left panel). Interferon (IFN) activity was also detected in fractions of 50,000 mw, and thus co-chromatographed with the larger mw factor(s) for induction of macrophage microbicidal and tumoricidal activities (16). This IFN activity has physicochemical and antigenic properties of IFN, but definite confirmation of a role for IFN in macrophage activation for these effector functions awaits availability of monospecific IFN antibodies, or purified IFN. We did confirm, however, that IFN was not active for induction of any of these microbicidal or tumoricidal activities (Table-I).

TABLE 1. Treatment of Macrophages with Culture Supernatants That Contain Interferon Activities.

Culture fluids from:	*IFN titer (IU/ml):*	*Tumor cytotoxicity:*	*Microbicidal Activity: L. tropica*	*S. mansoni*
EL-4 (45K)	*15*	*60%*	*90%*	*100%*
+ anti IFN	*15*	*80%*	---	---
EL-4 (23K)	*0*	*60%*	*0*	*100%*
L-929	*1000*	*10%*	*0*	*0*
+ anti IFN	*0*	*10%*	---	---
antigen-stimulated spleen cells	*100*	*50%*	*85%*	*100%*

In contrast to lymphokine factors from antigen-stimulated spleen cells (Fig. 1), two distinct activities for extracellular destruction of tumor targets and schistosomula could be detected in EL-4 culture fluids: the larger (45,000 mw) factor may be IFN, but the smaller activity (23,000 mw) has no IFN activity (antiviral or induction of macrophage Ia) and is not observed in spleen cell-derived lymphokines (Fig. 2). Three distinct lymphokines induce microbicidal activities against 2 different intracellular targets in spleen cell lymphokines (Fig.1); only a single activity (45,000 mw) was present in the EL-4 supernatants (Fig. 2). Thus, with lymphokine signals from two different sources, we can completely dissociate extracellular microbicidal and tumoricidal activities from intracellular killing activities of activated macrophages. Destruction of two different extracellular targets, a fibrosarcoma cell and a schistosomulum, is induced by the same lymphokine factors, regardless of the source of these factors. Moreover, destruction of two intracellular microorganisms that reside in different cellular compartments (an intracytoplasmic bacterium, *R. tsutsugamushi*, and an intraphagolysosomal protozoan parasite, *L. tropica*) is also induced by similar lymphokine factors. These factors are different, however, than those that regulate extracellular killing. Control of cytotoxicity is dictated not by the target *per se*, but by the site of the cytotoxic reaction.

REFERENCES

1. Koren, H.S., Anderson, S.J., and Adams, D.O., Cell. Immunol. 57, 51 (1981).
2. Pace, J.L., and Russell, S.W., J. Immunol. 126, 1863 (1981).
3. Nathan, C.F., Brukner, L.H., Silverstein, S.C. and Cohn, Z.A., J. Exp. Med. 149, 84 (1979).
4. Marino, P.A., and Adams, D.O., Cell. Immunol. 54, 11 (1980).
5. Ruco, L.P., and Meltzer, M.S., J. Immunol. 120, 1054 (1978).
6. Nacy, C.A., and Meltzer, M.S., J. Immunol. 123, 2544 (1979).
7. Nacy, C.A., Meltzer, M.S., Leonard, E.J., and Wyler, D.J., J. Immunol. 127, 2381 (1981).
8. Ruco, L.P., and Meltzer, M.S., J. Immunol. 121, 2035 (1978).
9. Pappas, M.G., Oster, C.N., and Nacy, C.A., in: "Host Defenses Against Intracellular Pathogens" (T.K. Eisenstein, H. Friedman, Eds.) Plenum Press, New York (in press, 1982).
10. Boraschi, D., and Meltzer, M.S., Cell. Immunol. 45, 188 (1979).
11. Ruco, L.P., Meltzer, M.S., and Rosenstreich, D.L., J. Immunol. 121, 543 (1978).
12. Boraschi, D., and Meltzer, M.S., J. Immunol. 122, 1592 (1979).
13. James, S.J., Leonard, E.J., and Meltzer, M.S., Cell. Immunol. (in press, 1982).
14. Nakamura, R.M., and Meltzer, M.S., Cell. Immunol. 65, 52 (1981).
15. Nacy, C.A., Leonard, E.J., and Meltzer, M.S., J. Immunol. 126, 204 (1981).
16. Meltzer, M.S., Benjamin, W.R., and Farrar, J.J., J. Immunol. (in press, 1982).

DISCUSSION

SCHRADER: Was Nacy's factor from the EL-4 supernatant that killed Leishmania, pH2-sensitive? Handman and Burgess have reported that purified GM-CSF enhanced killing, and I would expect that the EL-4 supernatant should contain GM-CSF.

NACY: The sensitivity of pH2 is unknown, but the EL-4 activity *is* heat labile. Various CSF's have been shown to be inactive against this form of the parasite.

ZLOTNIK: Nacy showed a protocol that indicates resident macrophages to be more responsive to induction of microbicidal activity than elicited macrophages, while the opposite is true for tumoricidal activity. Is this also the case for the EL-4 supernatant lymphokines?

NACY: We have thus far only used resident cells with the EL-4 factors for analysis of intracellular killing.

ZLOTNIK: Do all eliciting agents work the same way? What agents does Nacy use for eliciting macrophages?

NACY: We have analyzed intracellular destruction of *Leishmania tropica* with murine cell populations elicited by intraperitoneal injection of phosphate-buffered saline, fetal calf serum, casein, proteose-peptone, starch, and thioglycollate. In all cases, intracellular killing activity was less than 30% that of lymphokine-treated resident peritoneal macrophages.

VADAS: Ian-Clark-Lewis, John Schrader and I also have a 23,000 MW EL-4 fraction that powerfully enhances both eosinophil and neutrophil-mediated, antibody-dependent killing of tumor cells, that also has strong CSF activity. Therefore, unless the population of macrophages killing parasites is devoid of granulocytes, Nacy's EL-4 fraction may work via such contaminating cells.

NACY: The cells routinely used for analysis of extracellular killing reactions are 3 day, casein-elicited peritoneal exudate cells that are comprised of ~85% macrophages, 10^{+}% lymphocytes, and $\leq$ 5% other cell types (mast cells, neutrophils). We rarely observe eosinophils in these cell populations.

NETA: γIFN has been reported to be resistant to heating, yet Nacy suggests that her EL-4 activity is heat labile.

BENJAMIN: Not really. The high molecular weight activity reported by Nacy is stable at 56°C for 30 min. but is destroyed at 100°C.

OPPENHEIM: Does Nacy have any data indicating whether the two factors that she has isolated act through the pathway described by Pick, to stimulate the production of oxygen intermediates, which may then function as the actual mediators of the macrophage bactericidal effects?

NACY: We have not yet been successful in demonstrating a significant increase in secretion of O_2 by resident peritoneal macrophages treated with lymphokines. Whether these toxic oxygen products participate in the killing reaction within the cell is presently unknown.

DUMONDE: Are there other differences in the response of mouse macrophages to other lymphokines in the *L. tropica* system?

NACY: Strain C3H/HeJ and A/J murine macrophages do *not* respond to lymphokine signals for intracellular killing unless exogenous signals such as LPS are added. Several other mouse strains have completely unresponsive macrophages (P/J, NZW/N) and these macrophage defects seem to correlate with the natural susceptibility of these strains to this infectious disease.

BIOCHEMICAL CHARACTERIZATION OF LYMPHOBLASTOID MACROPHAGE-ACTIVATING FACTOR (MAF)[1]

J. E. McEntire
P. A. Dunn
B. W. Papermaster

Cancer Research Center
P.O.Box 1268
115 Business Loop 70 W
Columbia, Missouri 65205

A macrophage activating factor (MAF) has been isolated from the long-term human B-lymphoblastoid cell line RPMI-1788. MAF is purified from an acid-soluble fraction of the concentrated culture supernatant that has been shown to contain MAF, MIF, LIF, chemotactic factor, and skin reactive factor and to enhance survival in mouse tumor systems when used as an adjuvant to chemotherapy. MAF has been purified by high performance liquid chromatography (HPLC) on molecular sieving, ion exchange and reversed phase supports. The product is shown to be a polypeptide of 8500 daltons when dissociated from carrier proteins and appears to be homogeneous by analytical electrophoresis and HPLC using both exogenously-labeled samples and post-column derivatization with fluorescamine. Analysis of the factor for other potential biologically-active components shows no polyamines, nucleotides, or fatty acids.

I. INTRODUCTION AND STATEMENT OF THE PROBLEM

Intensive clinical and laboratory investigations with interferon during the past several years have led to rapid progress in our knowledge of this group of active glycoproteins (1-5). Provided that an interferon is made by a cell of lymphoid origin, it would fit within the class of substances known as *lymphokines* (6). However, from a structural standpoint, much greater information is known about interferons, and the amino acid sequences of fibroblast and lymphoblast interferons have been published (4,5). Such data are not

[1]*Supported by grant No. CA-29145 from the National Cancer Institute, National Institutes of Health, DHHS.*

ISBN 0-12-527540-4

available for other lymphokines due to the lack of sufficient quantities of purified materials.

There is still much controversy about the molecular properties of MAF and its mode of action, and neither MIF nor MAF has been purified to chemical homogeneity with identification of N- or C-terminal amino acids. For example, are MIF and MAF the same or different molecules? Among the many points of disagreement, one about which most workers agree so far is that both MIF and MAF are heterodisperse. Studies by Sorg *et al.* have suggested that multiple molecular species exist for MIF in human and mouse (7-9). McDaniel has reported multiple molecular species for human MAF derived from the NAMALWA cell line (10). Using gel filtration and preparative isotachaphoresis, Block *et al.* separated three components or fractions with MIF activity from Con A-stimulated human lymphocytes with 80% of the activity in pool B3 (11-13). The molecular weight of this pool was 23,000, with a pI between 3.5-4.0. Recently, Weiser *et al.* have reported three species from human Con A-treated lymphocytes (14). These MIF species ranged in size from 23,000 to 85,000 daltons, and pI ranged from 2.4-5.6. These distributed as two charge groups, pH 3 and pH 5 MIFs. A pH 3 guinea pig MIF isolated from biosynthetically-labeled (3H leucine) culture medium by Remold *et al.* yielded a component of 30,000 daltons purified 30,000 fold (15). We have reported repeatedly that lymphotoxin, MIF, and MAF activities bound to purified serum albumin and alpha-2-macroglobulin from culture medium were dissociable in SDS or guanidine (16-18). Possanza *et al.* have reported a subunit structure for human pheripheral blood lymphocytes and cell line MIF with a small subunit, <10,000 daltons, dialyzable from a larger, non-dialyzable subunit (19). Yoshida and Reisfeld (20) and Papageorgiou (21) also have found MIF activity for thymosin β4, a sequenced peptide of 4892 daltons.

Macrophage activation factors have been summarized recently. Characterization has been hampered by even greater assay problems than with MIF. Assays for MAF are more complex and variable in readout than those for MIF. Less agreement exists on their acceptance and standardization. However, nearly all investigators in the field agree that one assay necessary to test MAF is the macrophage-mediated cytotoxic assay, various types of which have been discussed by Nathan *et al.* (22), David (23), Piessens *et al.* (24), Raz *et al.* (25), and Fidler *et al.* (26).

Recently, Kniep *et al.* have claimed to have separated MAF from MIF on the basis of isoelectric focusing in urea and ultragel chromatography (27). McDaniel has described multiple molecular species of human MAF in 6M guanidine (10). Raz *et al.* have described MAF as heat-stable to boiling (25). Pace *et al.* have pointed out the role of endotoxin and prostagladins as factors in macrophage activation and expression (28). Both Sorg *et al.* (9) and Ruco and Meltzer (29) have stressed the qualitatively different stages of differentiation necessary for macrophage-mediated tumor cell killings.

The problem involves characterizing polypeptide chains of complex lymphokine molecules which, like toxins and hormones, may have different subunits for attachment and induction of response. In addition, it appears that the target cells must be at an appropriate stage of differentiation to carry receptors or that they must be encapsulated in liposomes in order to bypass receptors (26).

II. CURRENT STATUS OF THE BIOCHEMICAL PROGRESS ON MAF IN OUR LABORATORY.

Isolation efforts have involved, in summary, concentration and preparation of starting material (Fig. 1) and bulk extraction into TCA, followed by a number of analytical and preparative protein chemical purification steps. These have included gel filtration on Sephadex G-25, G-200, BioGel A1.5M, and affinity chromatography on Con-A and BSA-conjugated Sepharose, preparative electrophoresis in agarose and polyacrylamide gels with and without urea and/or SDS, and HPLC using gel filtration, ion exchange, and reversed phase columns. The process of TCA extraction results in active lymphokine peptides in the soluble phase of 10% TCA. The solvent phase is neutralized, TCA is removed by G-25 gel filtration chromatography, and the protein-containing peak is lyophilized. The salient features of acid extraction in TCA are 1) removal of more than 99% of the total protein, 2) reduction in volume, and 3) solubilization of water-soluble, polar peptides. An example of the heterogeneity of the acid-soluble fraction is included in the following description and should be compared with similar procedures used by Todaro *et al.* for extraction of tumor growth factors with similar properties (30).

The lymphokine prepared as in Figure 1 was monitored by analytical HPLC on various columns as described previously (31). Fractions were assayed for biological activity in a number of assays as noted below and biosynthetically-labeled protein fractions were pooled, rechromatographed, and dissociated with 6M guanidine HCl and purified further. The Waters I-125 column has an 80,000 dalton nominal exclusion limit, and the separation of fractions consisted of several peaks eluting in a size range comparable to the markers used to callibrate the column. The material applied to the column was biosynthetically-labeled with 3H amino acids as reported previously (31). Phagocytic activity was found in peaks corresponsing to albumin and another peak at lower weight range.

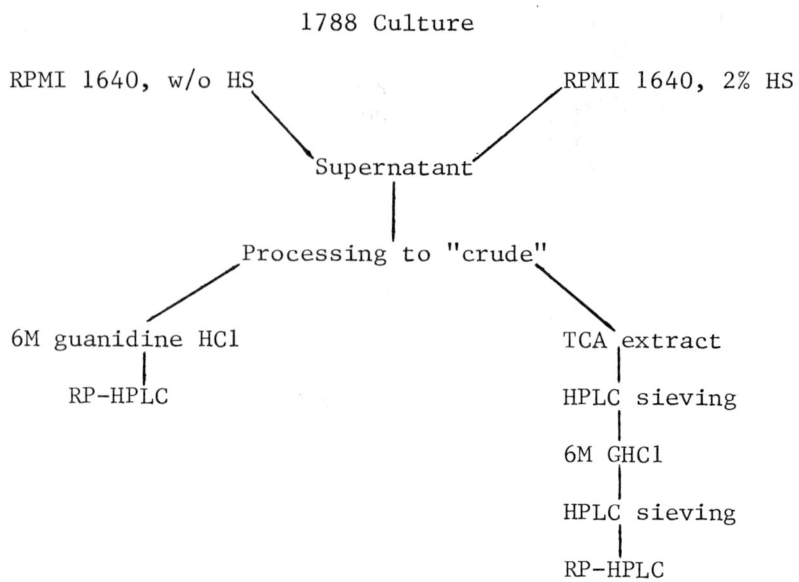

FIGURE 1. Preparation of lymphokine-containing supernatants.

The albumin-associated peak was dissociated with 6M guanidine HCl and rechromatographed. The predominance of biosynthetic protein elutes as a peak having a molecular weight of 8500 daltons. MAF activity also chromatographs in this same fraction. The molecule is purified further by repeated HPLC and has been shown to be an acidic, polar protein.

Bioassays are carried out as described by Dunn *et al.* (32,33). At present, various human tumor target lines are being studied for utility as target cells in cytotoxic assay systems, among them the A375 human melanoma cell line.

Analysis for non-protein components that might be activating macrophages showed no nucleotides/nucleosides (34), polyamines (35), or saponifiable fatty acids (36) of biological importance to be detectable within the sensitivity limits employed. Therefore, all MAF activity appears to be associated with this small protein that we have purified. Analytical HPLC using a number of procedures, conventional gel filtration and ion exchange, and electrophoretic analysis show only a single band of protein in the final preparation.

Since this polar molecule is not retained on reversed phase columns in unmodified solvent systems (31), we are attempting currently to use ion-paired reversed phase chromatography on C18 and C8 columns in which the solvents are modified with trifluoroacetic acid to allow separation (Fig. 2).

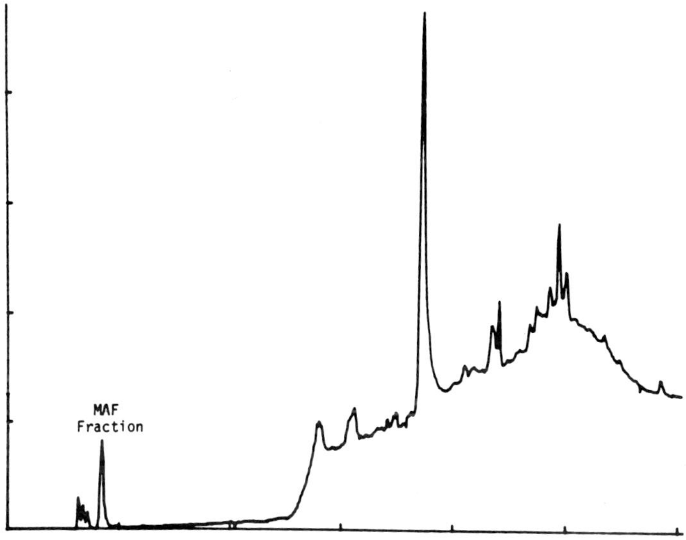

FIGURE 2. Ion-paired reverse phase HPLC on a TCA-soluble lymphokine extract. Conditions: RP 300 (C-8) column; buffer A: 0.1% trifluoroacetic acid; buffer B: 0.1% TFA in acetonitrile; chromatography was isocratic for first 5 min., then in a 0-100% linear gradient of buffer B; flow rate 2.0ml/min., and detection at 214nm.

These solvent systems should allow us to separate directly guanidine-containing fractions without prior removal of the dissociating solvent. This technique appears to be adaptable to large-scale preparative separations that should enable us to produce amounts of material adequate for sequence analysis in the near future.

REFERENCES

1. Pestka, S., Maeda, S., Hobbs, D. S., Levy, W. P., McCandliss, R., Stein, S., Moshera, J. A., and Staehlin, T., *in* "Cellular Responses to Molecular Modulators" (L. W. Mozes, W. A. Scott, J. Schultz, and R. Werner, eds.), p. 455. Academic Press, New York (1981).

2. Stewart, W. E., II, "The Interferon System." Springer-Verlag, Vienna (1979).
3. Maeyer, B. W., Glasso, G., and Schellekers, H. (eds.), "Biology of the Interferon System." Elsevier, N. Holland (1981).
4. Knight, E., Jr., Hunkapiller, M. W., Dorani, B. D., Hardy, R. W. F., and Hood, L. E., *Science 207*, 525 (1980).
5. Zoon, K. C., Smith, M. E., Bridgen, P. J., Anfinsen, C. B., Hunkapiller, M. W., and Hood, L. E., *Science 207*, 527 (1980).
6. Dumonde, D. C., Wolstencraft, R. A., Panayi, G. S., Matthew, W., Morley, J., and Howson, W. T., *Nature 224*, 38 (1969).
7. Sorg, C., and Bloom, B. R., *J. Exp. Med. 137*, 148 (1973).
8. Klinkert, W., and Sorg, C., *Molec. Immunol. 17*, 555 (1980).
9. Sorg, C., *in* "Lymphokines and Thymic Hormones" (A. L. Goldstein, and M. A. Chirigos), p. 135. Raven Press, New York (1981).
10. McDaniel, M. C., *Inflammation 4*, 125 (1980).
11. Block, L. H., and Jaksche, H., *in* "The Lymphokines: Biochemistry and Biological Activity" (J. H. Hadden, and W. E. Stewart, II, eds.), p. 55. Humana Press, Clifton, New Jersey (1981).
12. Block, L. H., Aloni, B., Biemesderfer, D., Kashgarian, M., and Betensky, M. W., *J. Immunol. 121*, 1416 (1978).
13. Block, L. H., Jaksche, H., Bamberger, S., and Ruhenstroth-Bauer, G., *J. Exp. Med. 147*, 541 (1978).
14. Weiser, W. V., Greineder, D. K., Remold, H. G., and David, J. R., *J. Immunol. 126*, 1958 (1981).
15. Remold, H. G., McCarthy, P. L., and Mednis, A. D., *Proc. Nat. Acad. Sci. USA 78*, 4088 (1981).
16. Papermaster, B. W., McEntire, J. E., Gilliland, C. D., Buchok, S. J., and Smith, M. E., *in* "Biochemical Characterization of Lymphokines" (A. L. de Weck, F. Kirstensen, and M. Landy, eds.), p. 193. Academic Press, New York (1979).
17. Papermaster, B. W., Gilliland, C. D., McEntire, J. E., Smith, M. E., and Buchok, S. J., *Cancer 45*, 1248 (1980).
18. Papermaster, B. W., Smith, M. E., and McEntire, J. E., *in* "The Lymphokines: Biochemical and Biological Activity" (J. W. Hadden, and W. E. Stewart, II, eds.), p. 149. Humana Press, Clifton, New Jersey (1981).
19. Possanza, G., Cohen, M. C., Yoshida, T., and Cohen, S., *Science 205*, 300 (1979).
20. Yoshida, T., and Reisfeld, R. A., *Nature 226*, 857 (1970).
21. Papageorgiou, P. S., Henley, W. L., and Glade, P. R., *J. Immunol. 108*, 494 (1972).

22. Nathan, C. F., Karnovsky, M. L., and David, J. R., *J. Exp. Med. 133,* 1356 (1971).
23. David, J. R., *Fed. Proc. 34,* 1730 (1975).
24. Piessens, W. F., Churchill, W. H., Jr., and David, J. R., *J. Immunol. 114*, 293 (1975).
25. Raz, A., Fogler, W. E., and Fidler, I. J., *Cancer Immunol. Immunother. 7,* 157 (1979).
26. Fidler, I. J., and Fogler, W. E., *in* "Lymphokines and Thymic Hormones" (A. L. Goldstein, and M. A. Chirigos, eds.), p. 277. Raven Press, New York (1981).
27. Kniep, E. M., Domzis, W., Lohmann-Matthews, M. L., and Kickhofen, B., *J. Immunol. 127,* 417 (1981).
28. Pace, J. L., Taffet, S. M., and Russell, S. W., *J. Reticuloendothel. Soc. 30,* 15 (1981).
29. Ruco, L. P., and Meltzer, M. S., *Cell. Immunol. 41,* 35 (1978).
30. Todaro, G. J., Marquardt, H., De Larco, J. E., Fryling, C. M., Reynolds, F. H., Jr., and Stephenson, J. R., *in* "Cellular Responses to Molecular Modulators" (L. W. Mozes, W. A. Scott, J. Schultz, and R. Werner, eds.), p. 183. Academic Press, New York (1981).
31. McEntire, J. E., Dunn, P. A., Gehrke, C. W., and Papermaster, B. W., *in* "Lymphokines and Thymic Hormones" Raven Press, New York (1981).
32. Dunn, P. A., and Tyrer, H. W., *J. Lab. Clin. Med. 98,* 374 (1981).
33. Dunn, P. A., Eaton, W. R., Lopatin, E., Tyrer, H., McEntire, J. E., Papermaster, B. W., Miller, A., and Vosika, G., *in* "Human Lymphokines: The Biological Immune Response Modifiers" (A. Khan, and N. O. Hill, eds.), p. 67. Academic Press, New York (1982).
34. Gehrke, C. W., Juo, K. C., and Zumwalt, R. W., *J. Chromatogr. 188,* 128 (1980).
35. Gehrke, C. W., Tuo, K. C., and Ellis, R. L., *J. Chromatogr. 143,* 345 (1977).
36. Gehrke, C. W., *J. Chromatogr. 143,* 335 (1977).

DISCUSSION

ASHERSON: Solubility in 10% trichloroacetic acid is a frequent property of glycoproteins. Does McEntire's product contain carbohydrate?

McENTIRE: We have not yet determined the carbohydrate content of our preparation, but have plans to do so.

VILCEK: McEntire mentioned that 1788 cells do not produce interferon. Some years ago we examined supernatants of 1788 cultures for interferon activity and did find interferon-α present. Therefore I wonder if the absence of interferon has been confirmed by other laboratories.

McENTIRE: To my knowledge, no one has reported consistant production of interferon by the 1788 cell line. The Organon preparations described earlier at this workshop contained interferon only when processed under certain conditions and, even then, not in all batches. Although the RPMI 1788 cell line certainly might produce interferon with appropriate stimulus, all of our preparations have been assayed in a number of laboratories including those of Baron, Johnson, Anfinsen and at the Bureau of Biologics and have always been completely negative. Additionally, our cells have been regularly typed and show no changes from the originally established cell line. I would suggest that any putative 1788 interferon producers be typed, in order to establish its proper designation as 1788.

VILCEK: Does McEntire have other evidence for the purity of his MAF preparation? Has he analyzed his material on SDS-PAGE? Has he been able to produce antibodies to the purified protein?

McENTIRE: Our evidence for purity includes a single band on SDS--PAGE, by silver staining and exogenous ^{125}I labelling and single peaks on various molecular sieving, HPLC columns, as determined by post column fluorescein derivatization. We have not yet attempted to produce monoclonal antibodies but consider it a high priority project.

LANDY: The past decade has seen great emphasis given to the study of all facets of macrophage functions in immunobiology. The literature abounds in reports on macrophage activation - but activation for *what*? All too often there is ambiguity in the numerous attempts to establish correlates between agents that elicit *heightened performance* by these cells (*quantitative* changes) and specific changes (*qualitative* effects) or sets of alterations these cells have undergone in response to exposure or interaction with any of a wide spectrum of agents. The collection of such natural and synthetic products examined and found active would seem to have no discernible property or effect in common to warrant reliable correlation.

Now McEntire has presented us with data that are straightforward, to the effect that certain lymphokines also can lead to enhanced functional activities of macrophages in one regard or another. The problem will be how these latest findings are to be integrated into a larger framework as represented by the mountain of reports with findings on macrophage activities that *do not* fit any recognizable pattern. I, for one, am especially mindful of the fact that the purely physical act of phagocytosis is one of the most effective means of augmenting macrophage manifold capabilities. Yet here McEntire has utilized phagocytosis itself as his *assay* parameter!

There is simply an embarrassment, an overabundance, of means by which this generally amorphous endpoint can be achieved. It will take some considerable thought and effort to avoid having lymphokines be regarded as merely one more operational means of provoking macrophages to a level of higher performance.

McENTIRE: I agree that there are certainly large confusing masses of data concerning macrophage activation by many types of molecules. However, my approach to understanding or integrating these molecules requires purification and chemical characterization of the active molecules as a prerequisite to attempting to understand their complex biological roles. I certainly do not feel that my chemical definition of an activator adds to the confusion in the field, but instead, more clearly defines a molecule and its function which is important to the understanding of macrophage activation.

With respect to use of phagocytic enhancement as a screen for MAF-containing fractions, this assay offers a quick and convenient method with which to identify certain samples for further analysis. I would suggest that other lymphokines which have been chemically characterized have had the advantage of such screening assays which allow analysis of hundreds of samples daily. With respect to correlation of phagocytic enhancement to more conventional measurements of MAF, I have presented data which shows this highly significant correlation in well controlled studies in which our MAF "classically" activates macrophages. These macrophages had never been in contact with beads, such as we use for phagocytic studies.

OLDHAM: Is McEntire's homogenous material (MAF-MW 8,500) active in MIF assays?

McENTIRE: We have not assayed this purified material for MIF activity.

OLDHAM: Does it have MAF activity on tumor target cells in both mouse and human systems?

McENTIRE: The MAF does have activity on both mouse L1210 and human cultured melanoma cells.

OLDHAM: How would McEntire characterize the heat stability of this MAF?

McENTIRE: We have not performed adequate heat stability studies.

CORRELATION BETWEEN RESISTANCE MARKERS AND INTERLEUKIN-2 PRODUCTION OF THE T CELL LINE HSB-2[1]

Jürg Liechti
Beda M. Stadler
Alain L. de Weck

Institute of Clinical Immunology
University of Bern
Bern, Switzerland

INTRODUCTION

T cell lines of mouse (1) or human origin (2,3) have been described to generate the T cell growth factor activity Interleukin 2 (IL 2). Such T cell lines are also of interest because of their production of other lymphokine activities (4). The production of certain lymphokines may also be increased as compared to normal T cells or can be increased by cloning of cell line cells with an increased capacity to produce a certain lymphokine. We have cloned the HSB-2 line, but so far, we have not been able to detect clones that were able to produce more IL 2 activity than the parental line. Therefore, we have used selective growth conditions using growth inhibitory drugs, such as glucocorticoids, or we have irradiated cells, in order to obtain variant cells which might be different from the parental line in their capacity to generate lymphokines. Here we report the correlation of resistance markers with the production of IL 2.

[1]*This work was supported by the Swiss National Science Foundation. Grant NF 3.165-0.81*

ISBN 0-12-527540-4

MATERIAL AND METHODS

Conconavalin A (Con A) was obtained from Calbiochem (San Diego, CA) while 8-azaguanine, bromodeoxyuridine (BUdR) and phorbol myristate acetate were supplied by Sigma (St. Louis, MD). Dexamethasone was obtained from Roussel UCLAF (France). Culture of the HSB-2 cells, IL 2 assays, as well as proliferation assays have been described recently (3,5). Cells were cloned by micromanipulation. Single cell cultures were maintained in 50% conditioned medium obtained from HSB-2 cells in logarithmic growth phase.

RESULTS

We have shown earlier that the human T cell line HSB-2 produced the T cell growth factor activity IL 2 (3). Supernatant IL 2 activity was detected after lectin and PMA stimulation at about the same level as IL 2 produced by normal human mononuclear cells. Dexamethasone has been reported to inhibit IL 2-mediated growth of normal T cells (6). However, HSB-2 cells which were not able to absorb IL 2 (7), and seemed not to be dependent on IL 2, were also inhibited in their proliferation by 10^{-8} M dexamethasone. Therefore, we generated dexamethasone resistant variant lines of the HSB-2 line by stepwise growth adaptation to increasing dexamethasone concentrations. A variant line was considered to be growth adapted as soon as the cell cultures contained less than 3% dead cells after several passages at the same dexamethasone concentration. Then, for a part of the culture, the dexamethasone concentration was increased for further growth adaptation, whereas the residual cells were further propagated in the absence of dexamethasone. Subsequently, cells in dexamethasone free medium were stimulated with Con A and PMA and supernatant IL 2 activity was determined. Table I shows that dexamethasone resistance (up to 1×10^{-4} M) was paralelled by an increased production of IL 2.

Next we generated radioresistant variant lines of HSB-2. Table 2 shows that the subline HSB-2-Rad 2 derived from an irradiation treatment with 500 rad was already slightly increased in its capacity to produce IL 2. When this subline was further irradiated with 500 rad or 100 rad, sublines were generated with even greater IL 2 production. However, all the generated sublines had to be stimulated with Con A and PMA for IL 2 production. Thus, the introduction of resistance markers did

TABLE I. Dexamethasone Resistance and IL 2 Production of HSB-2 cell line variants

HSB-2 variant cell designation	Dexamethasone concentration used for growth adaptation[a] (M)	IL 2 production after stimulation[b] (units/ml)
HSB-2	0	122 ± 36
HSB-2-Dx I	6×10^{-5}	355 ± 28
HSB-2-Dx II	8×10^{-5}	827 ± 112
HSB-2-Dx III	1×10^{-4}	912 ± 98
HSB-2-Dx IV	$1,5 \times 10^{-4}$	871 ± 164

[a]HSB-2 cells were continuously grown in the presence of dexamethasone. After growth adaptation to the lowest dexamethasone concentration cells were split into dexamethasone-free culture medium for IL 2 production or into medium containing a higher dexamethasone concentration.

[b]$1x10^6$ cells/ml were stimulated with 10 ng/ml PMA and 10 µg/ml Con A in RPMI 1640 with 1% FCS for 48 h. Mean units/ml are shown of 3-6 independent experiments.

not generate variant cell lines with constitutive IL 2 production.

Furthermore, we generated from the high producer line HSB-2-Dx III sublines resistant for BUdR and 8-azaguanine. BUdR resistant cells produced only 200 units/ml of IL 2 and the 8-azaguanine resistant variant ceased IL 2 production. It seems therefore, that some resistance markers might have a negative correlation with the capacity to produce IL 2.

Subsequently we cloned parental HSB-2 cells as well as cells from the HSB-2-Dx III line. None of the tested clones produced more IL 2 than the parental HSB-2 cells. This finding suggested that the cloning procedure selected for cells with low IL 2 production. Table 3 shows that the growth characteristics of the sublines might indeed be correlated with the IL 2 production, because all the generated high IL 2-producing sublines had longer generation times as documented here by the decreased ^{3}H-thymidine (^{3}H-TdR)-uptake compated to the parental line or the cloned lines. Further, Table 3 shows that the sublines producing more IL 2 were also more sensitive to the growth-inhibitory effect of PMA. In contrast, the low IL 2-producing clones were not affected by 1 ng/ml of PMA and were

TABLE II. Radioresistance and IL 2 Production of HSB-2 Variant Cell Lines

HSB-2 variant line designation	*Irradiation[a] (rad)*	*IL 2 production of stimulated cells[b] (units/ml)*
HSB-2	-	*122 ± 36*
HSB-2	*500*	*205 ± 45*
HSB-22	*+500*	*367 ± 72*
HSB-23	*+1000*	*502 ± 31*

[a]*HSB-2 cells were irradiated with a ceasium source in tissue culture flasks. The surviving cells were propagated and again irradiated with either 500 or 100 rad.*

[b]*$1x10^6$ cells/ml were stimulated with 10 ng/ml PMA and 10 μg/ml Con A in 1% FCS for 48 h. Mean values of 3 independent experiments are shown.*

always less then 40% inhibited in their proliferation by 10 ng/ml of PMA (Table 3).

The HSB-2-Dx III subline was subsequently made radioresistant, which resulted in a further increase in IL 2 production. Table 3 shows that the dexamethasone resistance was maintained in these two sublines (HSB-2-Dx III-Rad 2 and HSB-2-Dx III-Rad 3). This result and the fact that the radioresistant lines did not become dexamethasone-resistant, suggested that the increased IL 2 production was not directly caused by the two resistance markers.

DISCUSSION

Our finding that glucocorticoid resistance in HSB-2 cells was paralelled with an increased capacity to produce IL 2 is in contrast to the described effect of dexamethasone on the production of IL 2 by normal T cells. Gillis et al. have reported that dexamethasone inhibited the production of IL 2 (6) as well as the generation of cytolytic T cells (8). It is, therefore, likely that glucocorticoids have different effects on normal T cells or on T cell lines. The HSB-2 cells were inhibited in their proliferation by dexamethasone. The dexamethasone-resistant cells had prolonged generation times. The fact that radioresistant cells had also increased generation times

TABLE III. Proliferation of HSB-2 Cells in the Presence of PMA and Dexamethasone

Variant of HSB-2 line[a]	^{3}H-TdR incorporation[b] (cpm)	% inhibition of ^{3}H-TdR incorporation by: PMA (10ng/ml)	Dexamethasone (10^{-4}M)
HSB-2	45,405	63.1	50.2
HSB-2-C2	156,208	35.2	67.5
HSB-2-C3	36,609	16.9	58.2
HSB-2-C5	95,620	20.9	33.8
HSB-2-Rad 2	4,284	44.6	46.0
HSB-2-Rad 22	3,546	63.6	40.3
HSB-2-Rad 23	4,077	50.6	42.7
HSB-2-Dx III	6,404	60.9	0.0
HSB-2-Dx III-Rad 2	3,738	48.4	0.0
HSB-2-Dx III-Rad 3	2,796	53.1	0.0

[a]Variant lines of the HSB-2 lines were cloned (designated with C) or selected by radioresistance (Rad) or/and by dexamethasone resistance (Dx).

[b]^{3}H-Thymidine incorporation of triplicate cultures. $2x10^5$ cells/well in microtiter tissue culture plates were cultured and pulsed for 24 hours.

leads us to hypothesize that the duration of the G_1 phase may determine the amount of IL 2 produced. Prolongation of the G_1 phase has also lead to increased IL 2 production in normal T cells (3). This phenomenon could also explain the failure to obtain clones with increased IL 2 production. The cloning procedure might favor cells with shorter generation times, therefore, we might have selected for low IL 2-producing cells. The BUdR- as well as the 8-azaguanine-resistant variants produced less than expected or ceased IL 2 production respectively. These sublines in conjunction with the high producer lines might be suitable tools for understanding the intracellular regulation of IL 2 production. Furthermore, the selection of variant lines might also be useful for studies on the production of other lymphokines.

ACKNOWLEDGMENTS

We would like to thank Ms. H. Lüthi for her outstanding technical assistance and Ms. M. Bader for her excellent secretarial assistance. We also express our appreciation to Dr. G. D. Bonnard for his critical review of this manuscript.

REFERENCES

1. Farrar, J. J., Fuller-Farrar, Y., Simon, P. L., Hilfiker, M. L., Stadler, B. M., and Farrar W. L., *J. Immunol. 125,* 2555 (1980).
2. Gillis, S., and Watson, J., *J. Exp. Med. 152,* 1709 (1980).
3. Stadler, B. M., Dougherty, S. F., Farrar, J. J., and Oppenheim, J. J., *J. Immunol. 127,* 1936 (1981).
4. Farrar, J. J., Paetkau, V., Fuller-Farrar, J., Moore, R.N., Hilfiker, M. L., Stadler, B. M., and Farrar, W. L., *Lymphokines, 5,* 353 (1982).
5. Stadler, B. M., and Oppenheim, J. J., *Lymphokines, 6,* 117 (1982)
6. Gillis, S., Crabtree, G. R., and Smith, K. A., *J. Immunol. 123,* 1624 (1979).
7. Stadler, B. M., Berenstein, E. H., Siraganian, R. P., Oppenheim, J. J., *J. Immunol. 128,* 1620 (1982).
8. Gillis, S., Crabtree, G. R., and Smith, K. A., *J. Immunol. 123,* 1632 (1979).

DISCUSSION

PAETKAU: The observation that H2B2 variants, which produce high IL-2, are also PMA-resistant correlates with our observations on the Farrar EL-4 line. When we cloned the original EL-4 cells, obtained from Farrar, we found that the high IL-2-producing clones stopped growing in (i.e. were highly sensitive to) PMA, while low-Il-2-producers were relatively insensitive to PMA.

STADLER: I would say, additionally, that our finding correlates also with our earlier observation on normal T cells -- that IL-2 seems to be a product of cells in the G_1 phase of the cell cycle. Thus, a longer generation time for an IL-2-producing tumor cell line would, at a given constant rate of protein synthesis, mean that the cells have a longer interval in which to produce the IL-2.

ORSON: Did Stadler look at the IL-2 receptors in the lines "overtime" to see if there was, in fact, any correlation with the changes in IL-2 production?

STADLER: Our initial working hypothesis was that IL-2 also plays an important role in tumor cells. Therefore, inhibition of IL-2 production by dexamethasone, as it has been shown for normal T cells, should result in variant HSB-2 cells that produce more IL-2 and therefore would be able to bypass the dexamethasone-mediated growth inhibition. However, we have never been able to absorb supernatant IL-2 with HSB-2 cells.

FITCH: In view of the apparent relationship between IL-2 release and duration of the cell cycle, what is the effect of hydroxyurea on other agents which interrup the cell cycle?

STADLER: Hydroxyurea prevents only further utilization of IL-2, but not absorption. Hydroxyurea did not affect the production of IL-2 by HSB-2 cells, suggesting that they also are not utilizing this growth factor.

MULTIPLE LYMPHOKINE ACTIVITIES ARE SECRETED BY A CLONED MURINE T HELPER CELL[1]

Frank W. Fitch
Michael B. Prystowsky
John M. Ely
David Wilde
Jose Quintans
Eugene Goldwasser
John Goldman
Margaret Goldman

The University of Chicago
Chicago, Illinois

David I. Beller
Heinz Remold

Harvard Medical School
Boston, Massachusetts

James N. Ihle

Frederick Cancer Center
Frederick, Maryland

Stefanie N. Vogel

Uniformed Services University of Health Sciences
Bethesda, Maryland

[1]Supported in part by USPHS Grants AI-04197, AI-12110, AI-14530, AI-14732, AI-15869, AI-18797, CA-14723, CA-19266, CA-18375, USPHS Protocol R07338, an American Cancer Society Grant - Ill. Div. (82-27), a grant from the Leonard S. Florsheim, Jr. Fund, a March of Dimes Birth Defect Foundation Grant (1-736), and Contract N01-CO-75380 with Litton Bionetics, Inc. from National Cancer Institute.

ISBN 0-12-527540-4

Many of the regulatory functions of T lymphocytes are carried out by soluble mediators - lymphokines - released by the cells after appropriate stimulation. Lymphokines usually have been defined operationally in terms of the effects that they produce. Nearly one hundred different biological activities have been attributed to lymphokines (1). Frequently, cell suspensions obtained from lymphoid tissues have been used to produce lymphokines. Stimulation of such cells yields culture supernatants containing multiple lymphokine activities. We have used cloned murine T lymphocytes to address two questions for which answers have been difficult to obtain with heterogeneous cell populations:

1. How many lymphokine activities are produced by a single T cell? 2. How many different molecules account for these activities?

A murine alloreactive T cell clone, designated L2 (2,3), secretes ten different lymphokine activities when stimulated with appropriate alloantigen or with the T cell mitogen concanavalin A (Con A) (4). These lymphokines act on five different cell types (Table I). Several of these lymphokines [interleukin 2 (IL 2), interleukin 3 (IL 3), colony stimulating factor (CSF), and B cell stimulating factor (BCSF)] are "growth factors" which stimulate cell division. Others [macrophage activation factor (MAF), migration inhibitory factor (MIF), macrophage Ia-inducing factor (MIRF), interferon (γ-IFN), and factors enhancing production of complement components and inducing Fc-mediated phagocytosis] modify the phenotypic characteristics of the

Table I

Summary of Lymphokine Production by Murine T Lymphocyte Clone L2

Target Cells	Lymphokine	Ref.
T Lymphocytes	Interleukin 2 (IL 2)	5
	Interleukin 3 (IL 3)	6
B Lymphocytes	B cell stimulating factor (BCSF)	7
Hemopoietic Stem Cells	Colony stimulating factor (CSF)	8
Macrophages	Induction of Ia expression (MIRF)	9
	Macrophage activation factor (MAF)	10
	Migration inhibitory factor (MIF)	11
	Enhancement of C4/C2 production	12
	Induction of Fc mediated-phagocytosis	13
Fibroblasts	Interferon (γ-IFN)	14

Table II

Array of Lymphokine Activities Produced by T Lymphocyte Clones

Lymphokine Activity	T Lymphocyte Clones				
	non-cytolytic				cytolytic
	L2	L2V	HK16	A-10	L3
IL 2	++	V	++	+	-
BCSF	+	+++	nt	nt	-
CSF	+++	+++	+++	+++	-
MIRF	++	-	++	++++	++
γ-IFN	++	V	nt	++++	++

+ to ++++ = relative lymphokine levels in culture supernatants
V = variable lymphokine production by different subclones
nt = not tested

target cells. This same array of lymphokines as well as the same relative amounts are produced after stimulation with either alloantigen or Con A (4). However, another T cell mitogen, phytohemagglutinin, does not stimulate lymphokine secretion by cloned L2 cells (15).

Not all T cell clones produce the same array or relative amounts of lymphokines (Table II). Cells studied include non-cytolytic helper T cell clones reactive with $Mls^{a,d}$ alloantigens (L2) and with alloantigens specified by the $\underline{I\text{-}A}^k$ region of the major histocompatibility complex (HK-16). Clone A-10 is a helper T cell clone which has dual antigen recognition: ovalbumin in the presence of syngeneic ($H\text{-}2^a$) spleen cells or $H\text{-}2^s$ alloantigen. These clones appear to secrete the same array but different relative amounts of the several lymphokine activities tested. Variant clone L2V arose spontaneously during long term passage of L2 cells (16). This clone fails to secrete MAF, MIF, and MIRF; subclones of L2V produce low but variable amounts of IL 2 and γ-IFN. A cytolytic clone, L3 which appears to be reactive toward $H\text{-}2D^d$ (2,3), does not secrete any of the lymphokines which stimulate cell division but produces most but not all of the lymphokines which modify phenotypic characteristics of target cells. Multiple patterns of lymphokine production by cloned T cells have been observed by other investigators (17), and some cytolytic clones have been found to produce IL 2 (18).

The multiple patterns of lymphokine production suggest that different molecules are responsible for several of the lymphokine activities. The time course of lymphokine production also suggests that this is the case (4,16). IL 2 is secreted soon after stimulation of cloned L2 T cells with alloantigen, reaches peak levels by 24 hours, and decreases rather rapidly to undetectable levels after 8 days of culture. Other lymphokines appear to be secreted somewhat later, possibly at differing rates. CSF and IL 3 reach peak levels 24 to 48 hours after stimulation; MAF, MIRF, and γ-IFN do not attain peak levels until 48 hours of culture. These lymphokine activities remain elevated throughout the 8 day culture period. Collectively, the patterns of secretion and time course of lymphokine production indicate that <u>at least</u> 4 distinct molecules account for the 10 lymphokine activities produced by the cloned murine helper T lymphocyte L2.

Results obtained using physical separation methods are consistent with this interpretation. IL 2 has been separated from BCSF and CSF (19). Using a combination of phenyl-Separose and hydroxylapatite chromatography, selective enrichment of IL 2, CSF, and γ-IFN has been achieved (20). Thus, several different molecules are responsible for the lymphokine activities which are found in culture supernatants from a cloned murine T cell.

However, it also appears that a single molecule can have more than one lymphokine activity. Highly purified IL3 both induces 20 α-steroid dehydrogenase, a characteristic activity of IL 3, <u>and</u> has CSF activity (unpublished observations). Highly purified interferon appears to induce Fc-mediated phagocytosis in addition to having anti-viral activity (13). It is probable that other molecules can mediate more than one lymphokine activity, but materials must be purified to biochemical homogeneity to permit adequate evaluation of this possibility.

An additional complexity has been recognized recently: several different molecules can mediate the same lymphokine activity. CSF activity present in culture supernatants from cloned L2 helper T cells can be separated into two fractions based on affinity for Con A-Sepharose (20). The fraction which binds contains CSF activity but no detectable IL 2 or IL 3; the colonies produced by this CSF consist predominantly of macrophages. The fraction which does not bind to Con A-Sepharose contains CSF, IL 2, and IL 3; mixed granulocyte/macrophage colonies are produced by this CSF. Preliminary experiments indicate that ion exchange chromatography can yield further separation of CSF activities. Thus, at least two and probably 3 different molecules secreted by cloned L2 helper T cells have CSF activity. Staber and Burgess also have separated two distinct types of CSF (21). It seems possible that colonies which consist prediminantly of different cell types may be produced by distinctive CSF molecules (22).

An additional cautionary note is in order. Culture supernatants from stimulated cloned T cells may also contain inhibitory materials which influence some lymphokine activities. Effluents from phenyl-Sepharose columns were found to contain greater than 150% of the CSF activity present in the starting culture supernatant suggesting that inhibitors of CSF were bound to the phenyl-Sepharose gel (20). Also, the slopes of dose-response curves for crude and partially purified CSF differ, while those for crude and partially purified IL 2 are indistinguishable. These findings are consistent with the presence of inhibitory materials for CSF in the crude supernatant (unpublished observations).

Collectively, these results lead to several conclusions: Cloned murine T cells secrete multiple lymphokine activities, but varied arrays of lymphokines are produced by different cloned T cells. As yet, there appears to be no correlation between particular phenotypic characteristics of the T cell and the kinds of lymphokines produced. Several different molecules account for these multiple lymphokine activities, but at least some molecules can produce several different biological effects. However, the same biological activity can be produced by several different molecules. The availability of cloned T lymphocytes should facilitate the characterization of the cellular and molecular events associated with lymphokine production and with the regulatory processes mediated by lymphokines.

REFERENCES

1. Waksman, B.H., <u>in</u> "Biology of the Lymphokines" (S. Cohen, E. Pick, and J.J. Oppenheim, eds.), p. 585. Academic Press, New York, (1979).
2. Glasebrook, A.L. and Fitch, F.W., <u>Nature</u> <u>278</u>, 171 (1979).
3. Glasebrook, A.L. and Fitch, F.W., <u>J. Exp. Med.</u> <u>151</u>, 876 (1980).
4. Prystowsky, M.B., Ely, J.M., Beller, D.I., Eisenberg, L., Goldman, J., Goldman, M., Goldwasser, E., Ihle, J., Quintans, J., Remold, H., Vogel, S., and Fitch, F.W., <u>J. Immunol.</u> (In press) (1982).
5. Gillis, S., Ferm, M.M., Ou, W., and Smith, K.W., <u>J. Immunol.</u> <u>120</u>, 2027 (1978).

6. Ihle, J.N., Pepersack, L., and Rebar, L., J. Immunol. 126, 2184 (1981).
7. Glasebrook, A.L., Quintans, J., Eisenberg, L., and Fitch, F.W., J. Immunol. 126, 240 (1981).
8. Stanley, E.R., Metcalf, D., Marite, J.S., and Yeo, G.F., J. Lab. Clin. Med. 79, 657 (1972).
9. Scher, M.G., Beller, D.I., and Unanue, E.R., J. Exp. Med. 152, 1684 (1980).
10. Ruco, L.P., and Meltzer, M.S., J. Immunol. 119, 889 (1977).
11. David, J.R. and David, R.A., in "In Vitro Methods in Cell-Mediated Immunity" (B.R. Bloom and P.R. Glade, eds.), p. 249. Academic Press, New York, (1971).
12. Goldman, M.B., O'Rourke, K.S., and Goldman, J.N., Cell. Immunol. (In press) (1982).
13. Vogel, S.N., Weedon, L.L., Moore, R.N., and Rosenstreich, D.L., J. Immunol. 128, 380 (1982).
14. Rubinstein, S., Familletti, P.C., and Pestka, S., J. Virol. 37, 755 (1981).
15. Ely, J.M., Ph.D. Thesis, University of Chicago (1982).
16. Ely, J.M., Prystowsky, M.B., Eisenberg, L., Quintans, J., Goldwasser, E., Glasebrook, A.L., and Fitch, F.W., J. Immunol. 127, 2345 (1981).
17. Glasebrook, A.L., Kelso, A., Zubler, R.H., Ely, J.M., Prystowsky, M.B., and Fitch, F.W., in "Isolation, Characterization, and Utilization of T Lymphocyte Clones" (C.G. Fathman and F.W. Fitch, eds.), p.341. Academic Press, New York, (1982).
18. Widmer, M.B. and Bach, F.H., Nature 294, 750 (1982).
19. Nabel, G., Greenberger, J.S., Sakakeeny, M.A., and Cantor, H., Proc. Natl. Acad. Sci. 78, 1157 (1981).
20. Prystowsky, M.B., Ely, J.M., Vogel, S.N., Goldwasser, E., and Fitch, F.W., Fed. Proc. (In press) (1982).
21. Staber, F.G. and Burgess, A.W., J. Cell. Physiol. 102, 1 (1980).
22. Burgess, A.W., Metcalf, D., Russell, S.H.M., and Nicola, N.A., Biochem. J. 185, 301 (1980).

DISCUSSION

ALTMAN: One of Fitch's protocols showed that the target for IL-3 is a T cell; while earlier in this workshop we heard from Ihle that the target for IL-3 is *not* a T cell. Can Fitch clarify this issue?

FITCH: IL-3 may well be active on different target cells. We listed the T cell as the target cell since Ihle, who performed the IL-3 assays for us, used induction of 20α-steroid dehydrogenase in immature thymocytes as the measure of IL-3 activity.

KURNICK: If Fithc's cell lines will replicate in the presence of IL-2 alone, and they produce IL-2 upon Con-A stimulation, do they continue to divide after Con-A addition in the absence of exogenous IL-2? If not, why not?

FITCH: The situation is somewhat complicated. Cloned L2 cells are stimulated to release IL-2 as well as other lymphokines, and to proliferate if exposed to Con-A for 6 hours and then washed repeatedly with α methyl-mannoside to remove residual Con-A. However, if Con-A remains in the culture, lymphokine release occurs, but the cells do *not* proliferate. If syngeneic T-cell-depleted spleen cells are present, as well as the cloned T cells, lymphokine release *and* cell proliferation occurs. We have no satisfactory explanation for these observations. It is noteworthy that L2 cells are also stimulated to release lymphokines by pokeweed mitogen, but *not* by PHA.

GARLAND: I note that in Fitch's study the S.N "IL-3" did not stick to Sepharose-Con-A. In our experience, "IL-3", which is heavily glycosylated, *does* stick to Sepharose-Con-A. The antibody used was presumably against Ihle's IL-3 which also appears to adhere.

FITCH: The IL-3 analyses were performed by Ihle, and the IL-3 activity present in the culture supernatants of our T cell clone L2 did not bind to Con-A Sepharose columns.

OPPENHEIM: Fitch hinted that factors or agents in addition to interferon may induce Ia. Would he care to elaborate on that?

FITCH: Culture supernatants obtained soon after L2V variant clone was derived *did* contain interferon activity, but did *not* have Ia-inducing activity. Some recently obtained subclones produce little or no interferon *or* Ia-inducing activity. Thus, our results indicate that interferon activity and Ia-inducing activity are not invariably correlated.

PALACIOS: Has Fitch tested whether an antibody against Lyt2 blocks the production of IL-2?

FITCH: We do not have Lyt-2^+, IL-2 producing clones. However, I believe that Glasebrook has found that anti-Lyt-2 inhibits release of IL-2, and other lymphokines by Lyt-2^+, IL-2-producing clones.

MacDONALD: In response to Palacio's query I can add that Glasebrook has indeed shown that the production of IL-2 by a Lyt-2^+ CTL clone *can* be blocked by monoclonal anti-Lyt-2 antibodies.

FREQUENCY AND LYT PHENOTYPE OF PRECURSORS OF LYMPHOKINE-SECRETING ALLOREACTIVE T LYMPHOCYTES

Anne Kelso

Department of Immunology,
Swiss Institute for Experimental Cancer Research,
Epalinges, Switzerland

H. Robson MacDonald

Ludwig Institute for Cancer Research, Lausanne Branch,
Epalinges, Switzerland

INTRODUCTION

Antigen- or mitogen-stimulated T lymphocytes release a variety of soluble mediators which enhance or inhibit the activities of other lymphoid cells participating in the immune response (reviewed in ref. 1). With the development of cloned T cell lines and hybridomas, it has recently become possible to examine many aspects of lymphokine production at the clonal level (this volume). However, although long-term lines have many advantages for characterizing lymphokines and the cells that secrete them, it is not clear whether such lines are representative of the total T lymphocyte pool which can respond to an antigen. We have therefore used an alternative approach, the limiting dilution microculture system for estimating frequencies of antigen-reactive normal T cells (2). By combining this system with cell separation on a fluorescence-activated cell sorter (FACS) and sensitive lymphokine microassays, we have simultaneously assessed the

[1] *A.K. was supported by a grant from the Swiss National Foundation for Scientific Research*

ISBN 0-12-527540-4

frequencies of alloreactive, Lyt-2+ and Lyt-2- precursors of cytolytic T lymphocytes (CTL) and T lymphocytes secreting the three independent lymphokines, Interleukin-2 (IL-2, T Cell Growth Factor), Macrophage Activating Factor (MAF) and Granulocyte-Macrophage Colony Stimulating Factor (GM-CSF).

MATERIALS AND METHODS

As detailed elsewhere (3), positive selection of Lyt-2+ and Lyt-2- lymphocytes was performed by staining nylon wool column-purified C57BL/6 ($H-2^b$, Mls^b) spleen cells with monoclonal rat anti-Lyt-2 antibody (53-6.7, provided by Dr J.A. Ledbetter) and fluoresceinated mouse anti-rat Ig ; cells were sorted on the basis of fluorescence using a FACS II. Graded numbers of sorted or control responder cells were cultured with irradiated DBA/2 ($H-2^d$, Mls^a) stimulator spleen cells and IL-2-containing supernatant from an EL-4 thymoma subline (provided by Dr J.J. Farrar) (4) in limiting dilution microcultures (2). After 7-9 days incubation, an aliquot of each culture was assayed for cytolytic activity against ^{51}Cr-labeled P815 ($H-2^d$) tumor cells. The remaining cells were washed to remove residual EL-4 supernatant and restimulated by addition of irradiated T cell-depleted DBA/2 spleen cells; supernatants were removed after 24 hr for lymphokine assays. Positive cultures were defined as those in which activity exceeded that in control wells (no responder cells) by more than 3 standard deviations above the mean. Minimal estimates of precursor frequency for each activity were obtained by the minimum Chi^2 method from the Poisson distribution relationship between the number of responder cells cultured and the percentage of nonresponding cultures (5).

Lymphokines were assayed by the following methods (3,6): IL-2 by the induction of ^3H-thymidine incorporation by a cloned IL-2-dependent T cell line, MAF in a ^{51}Cr-release assay for tumoricidal activity of cultured bone marrow-derived macrophages against P815 tumor cells, and GM-CSF by the differentiation and proliferation of granulocytes and macrophages from bone marrow in liquid microcultures.

RESULTS AND DISCUSSION

Precursor frequencies of CTL and cells secreting MAF, IL-2 and GM-CSF were estimated in a modification of the limiting dilution microculture system used previously in this laboratory for frequency analysis of cytolytic and proliferating T lymphocytes (2). An example of the data obtained using this protocol is shown in Figure 1 for microcultures established from unseparated splenic T cells, and assayed for cytolytic activity (panel A) and, after restimulation, supernatant IL-2 (panel B) and MAF (panel C) activities. Preliminary experiments indicated that restimulation of primary microcultures significantly increased the apparent frequency of MAF-secreting cells, and that 24 hr of restimulation allowed detection of maximal lymphokine activities (3). It can be seen from Figure 1 that for each activity, increasing the responder cell dose increased the percentage of positive microcultures and produced cultures where the activities reached 80-100% of maximal levels.

Table I summarizes 2 experiments comparing precursor frequency estimates for CTL and lymphokine-secreting cells in FACS-passed unselected, Lyt-2^+ and Lyt-2^- lymphocytes. Equivalent results have been obtained with populations depleted of Lyt-2^+ cells by treatment with antibody and complement (3). The following conclusions can be drawn.

(1) Essentially all precursors of CTL were found in the Lyt-2^+ population, where the frequency was 50-fold higher than in the Lyt-2^- population. This is in agreement with

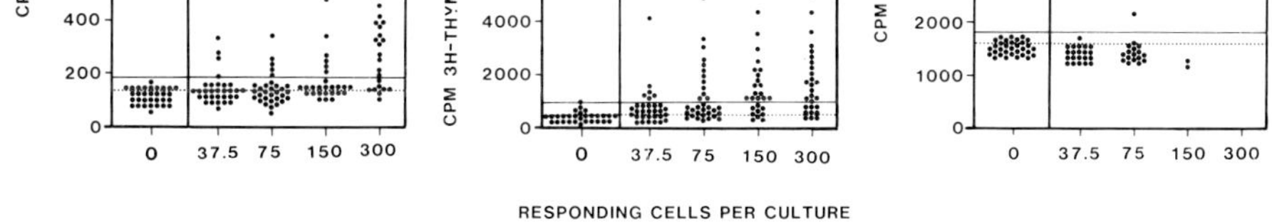

FIGURE 1. Cytolytic, IL-2 and MAF activity in limiting dilution microcultures of nylon wool column-purified C57BL/6 spleen cells stimulated with DBA/2 alloantigens. Each dot represents the activity in a single microculture. Positive cultures were defined as those where activity exceeded the mean activity in control wells (no responders, dotted line) by more than 3 standard deviations (solid line).

TABLE I. Precursor frequencies of CTL and lymphokine-secreting cells in spleen cells separated according to Lyt-2 phenotype[a]

Activity	*Precursor frequency (per 10^4 cells)*					
	Unselected		*Lyt-2*$^+$		*Lyt-2*$^-$	
	1[b]	*2*	*1*	*2*	*1*	*2*
CTL	*37*	*51*	*99*	*46*	*1.8*	*0.7*
IL-2	*58*	*38*	*2.5*	*2.0*	*93*	*98*
MAF	*123*	*66*	*38*	*25*	*135*	*87*
GM-CSF	*185*	*36*	*56*	*18*	*217*	*57*

[a]*Limiting dilution microcultures were established from nylon wool-purified C57BL/6 spleen separated on the basis of Lyt-2 phenotype as in Figure 1 (33% Lyt-2*$^+$ *cells in experiment 1, 35% in experiment 2).*

[b]*Experiment number.*

previous reports that precursor and effector CTL bear the Lyt-2 marker (7-9).

(2) Conversely, the majority of precursors of IL-2-secreting cells were found in the Lyt-2$^-$ fraction, where the frequency was 40-fold higher than in the Lyt-2$^+$ population. Although conflicting results have been obtained in bulk cultures (10-12), the present results and those of other workers (13-15) suggest that the frequency of IL-2-secreting, H-2K/D-reactive Lyt-2$^+$ cells is much lower than that of IL-2-secreting cells in Lyt-2$^-$ H-2I- or Mls-reactive populations.

(3) In contrast to IL-2, precursors of cells secreting MAF and GM-CSF were present in both the Lyt-2$^+$ and Lyt-2$^-$ populations. Whereas both these assays detected almost all the responding Lyt-2$^-$ cells, only about one-third of responding Lyt-2$^+$ cells produced these factors. This observation contrasts with our analysis of 60 long-term clones, in which approximately 90% of both cytolytic H-2K/D-reactive and noncytolytic H-2I- or Mls-reactive clones were found to secrete MAF (6; unpublished observations). However, the average titers of MAF released by cytolytic clones were 2- to 3-fold lower than those produced by noncytolytic clones, suggesting that the lower precursor frequencies of MAF- and GM-CSF-secreting cells in Lyt-2$^+$ as compared to Lyt-2$^-$

populations noted here may reflect subthreshold production by some Lyt-2+ cells.

The latter possibility was addressed in the experiment shown in Figure 2, which compares the amounts of IL-2, MAF and GM-CSF present in pooled supernatants from restimulated microcultures of unselected, Lyt-2+ and Lyt-2- cells. In panel A, it can be seen that at least 30 times more IL-2 was detected in supernatants from Lyt-2- than from Lyt-2+ cells, which is consistent with the higher precursor frequency estimates obtained for the Lyt-2- population (Table I). In addition, panels B and C show that MAF and GM-CSF titers were approximately 10-fold greater in cultures of Lyt-2- cells than of Lyt-2+ cells. This weaker total production by Lyt-2+ cells can be attributed in part to the 3-fold lower precursor frequency of cells secreting MAF and GM-CSF in this population (Table I), but also suggests that average production by the cells which do release these factors may be lower than in the Lyt-2- population. Preliminary experiments showing that MAF release by both cytolytic and noncytolytic clones is enhanced in the presence of an IL-2-containing supernatant (K.T. Brunner and A. Kelso, unpublished observations) raise

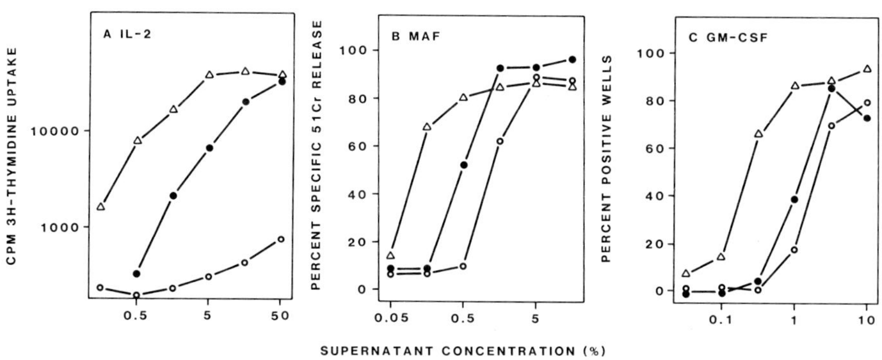

FIGURE 2. Lymphokine activities in supernatants from microcultures of lymphocytes separated according to Lyt-2 phenotype. FACS-passed unselected (●), Lyt2+ (○) and Lyt-2- (△) responder cells were cultured at 1000 cells/well under limiting dilution microculture conditions. After 7 days incubation, cultures were restimulated and 24 hr supernatants pooled from groups of 12 wells and titrated in the lymphokine assays. No activity was detected in supernatants from cultures without responder cells.

the possibility that the capacity of many noncytolytic Lyt-2$^-$ clones to secrete IL-2 accounts for the higher MAF production by both short-term and long-term clones of this phenotype.

In conclusion, the finding that most Lyt-2$^-$ and at least one-third of Lyt-2$^+$ alloreactive T lymphocytes are precursors of MAF- and GM-CSF-secreting cells points to an important role of both these populations in amplifying the immune response by the release of factors which induce differentiation, proliferation and activation of cells of the granulocyte and macrophage lineages.

REFERENCES

1. Biology of the Lymphokines. S. Cohen, E. Pick, and J.J. Oppenheim, eds. Academic Press, New York (1979).
2. Ryser, J.-E., and MacDonald, H.R., *J. Immunol. 122*, 1961 (1979).
3. Kelso, A., and MacDonald, H.R., *J. Exp. Med. 156*, In press (1982).
4. Farrar, J.J., Fuller-Farrar, J., Simon, P.L., Hilfiker, M.L., Stadler, B.M., and Farrar, W.L., *J. Immunol. 125*, 2555 (1980).
5. Taswell, C., *J. Immunol. 126*, 1614 (1981).
6. Kelso, A., Glasebrook, A.L., Kanagawa, O., and Brunner, K.T., *J. Immunol. 129*, 550 (1982).
7. Cantor, H., and Boyse, E.A., *J. Exp. Med. 141*, 1390 (1975).
8. Teh, H.-S., and Teh, S.-J., *J. Immunol. 125*, 1977 (1980).
9. Cerottini, J.-C., and MacDonald, H.R., *J. Immunol. 126*, 490 (1981).
10. Swain, S.L., *Proc. Natl. Acad. Sci. 78*, 7101 (1981).
11. Kern, D.E., Gillis, S., Okada, M., and Henney, C.S., *J. Immunol. 127*, 1323 (1981).
12. Miller, R.A., and Stutman, O., *Cell. Immunol. 68*, 114 (1982).
12. Lutz, C.T., Glasebrook, A.L., and Fitch, F.W., *Eur. J. Immunol. 11*, 726 (1981).
14. Miller, R.A., and Stutman, O., *J. Immunol. 128*, 2258 (1982).
15. Krönke, M., Scheurich, P., Pfizenmaier, K., Röllinghoff, M., and Wagner, H., *J. Exp. Med. 156*, 41 (1982).

DISCUSSION

PAETKAU: Was the Lyt-2 phenotype determined after induction or had the cells been separated by Lyt-2 phenotype beforehand?

MacDONALD: The cells were separated before induction.

PAULY: What is the Lyt phenotype of the IFN-producing precursor, and the cloned lymphocyte populations?

MacDONALD: We find IFN-producing precursor and effector (cloned) T cells of both the Lyt-2^+ and Lyt-2^- surface phenotype.

PAULY: What is the precursor frequency of the IFN-producing cell in the non-cloned populations?

MacDONALD: The frequency is of the order of 1:100 in the C57BL/6 anti-DBA/2 strains combination (using nylon wool-purified responder cells).

GARLAND: In the case of alloreactive T cell clones can the secretion of lymphokines be correlated with Ir gene control?

MacDONALD: This question has not been directly addressed.

DUTTON: I think that a comment might now be in order about the reliability of comparisons based on the frequencies determined by limiting dilution assays. It is possible that not all the cells present are induced under the conditions of the response; hence the finding of a low frequency could represent a failure of induction rather than the apparent absence of such cells (specifically IL-2-producing Lyt 2^+ cells in the MLR).

MacDONALD: I agree fully. In fact, all frequency estimates in this (or any other) limiting dilution system represent *minimal estimates* as long as the plating efficiency is not known.

PALACIOS: Has MacDonald carried out the same frequency experiments using as stimulator cells, T-cell-depleted irradiated spleen cells?

MacDONALD: We always utilized T-cell-depleted spleen cells to induce lymphokine production. Our control for T cell depletion was to stimulate the treated cells with Con-A and assay for lymphokine production.

LYT 2$^+$ AND LYT 2$^-$ T LYMPHOCYTES ARE BOTH CAPABLE OF PRODUCING LYMPHOKINES ACTING ON T LYMPHOCYTES (IL_2), B LYMPHOCYTES (TRF), MACROPHAGES (IPA,MAF) AND MASTOCYTES (MCGF)[1]

P.-A. Guerne, P.-F. Piguet, J.-M. Dayer[2] *and P. Vassalli*

Department of Pathology
University of Geneva
Geneva, Switzerland

I. INTRODUCTION

This work explores the synthesis of various lymphokines and of interferon (IFN) by subpopulations of Lyt 2$^+$ and 2$^-$ T lymphocytes purified by a panning procedure and submitted to polyclonal activation by Con A.

II. METHODS

A. *Cells*

Lymph node and spleen cells were prepared from CBA mice. Enrichment for T cells was performed by passage through nylon

[1] *This work was supported by grant No 3.053.81 from the Swiss National Science Foundation.*
[2] *Division of Immunology and Allergology, Department of Medicine, Hôpital cantonal, Geneva, Switzerland.*

ISBN 0-12-527540-4

wool columns (1). Lyt 2^+ lymphocytes were separated from Lyt 2^- by the panning procedure (2), using monoclonal rat anti-Lyt 2 antibody (clone 53-6-7, The Salk Cell Distribution Center, San Diego) and affinity-purified mouse anti-rat Ig. The purity of the separated cells was monitored after each separation by double immunofluorescence (3); the frequency of Thy 1^+ cells bearing the Lyt 2 antigen was 96% (92-98) for the Lyt 2^+ subpopulation and 3% (1-6) for the Lyt 2^- (3).

B. *Culture*

Cultures were performed in DMEM supplemented with amino acids and 2% FCS (3). Supernatant (SN) production was performed with various doses of T cells in 1 ml wells (Nucleon, Danemark) with a dose of Con A (Sigma chemical, St-Louis) producing the optimal mitogenic transformation (1 ug/ml). In order to provide non limiting number of adherent cells, the plate had been previously incubated with 4 x 10^4 peritoneal cells for 60 min, followed by extensive washing. In these conditions, Lyt 2^+ and 2^- cells responded equally to Con A, when the response was measured by ^3H TdR incorporation. After 24 hrs of incubation, the SN were collected and replaced by fresh medium and Con A for 24 more hrs. The SN were centrifuged and filtered on millipore (0.45 mcm) and supplemented with alpha-methyl-mannoside, 0.1 M final.

III. RESULTS

A. *IL_2*

The proliferative response of an IL_2 dependant T lymphocyte clone (4) to dilution (1/4 final) of the SN produced by various numbers of Lyt 2^+ and 2^- lymphocytes is shown (representative experiment).

Comparable activity is recovered from the SN of either the Lyt 2^+ or 2^- populations. T cell titration suggests that optimal IL_2 production per producing cell occurs at the lowest T cell concentration. This is particularly obvious for the 24-48 hrs collection, where activity is detected in the SN of 0.125, but not of 2 x 10^6 cells/culture.

TABLE I. ^{3}H *TdR Incorporation (cpm)*

SN of		*nb of T cells/ml x* 10^{-6}					
		5	*2*	*0.5*	*0.25*	*0.125*	*0*[a]
Lyt 2+	*0-24 hrs*	*1800*	*1600*	*900*	*600*	*450*	*180*
Lyt 2-	*"*	*2200*	*2300*	*1000*	*650*	*300*	
Lyt 2+	*24-48 hrs*	*250*	*650*	*2000*	*2300*	*1500*	*250*
Lyt 2-	*"*	*100*	*300*	*2000*	*1800*	*800*	

[a]*SN of the adherent cells + Con A*

B. T Replacing Factor (TRF)

The table shows the response to SRBC of T depleted spleen cells after addition of the SN (1/4) produced by various numbers of T lymphocytes.

TRF was produced of the same time and T cell concentration conditions as IL_2. It is not excluded that IL_2 plays a role in this assay, by activation of the residual T cells (as suggested by Gibson et al. (6)). This seems however unlikely since examination by immunofluorescence on day 4 of a B cell culture with SN added on day 2 showed only few Thy 1^+ cells (2-5%).

TABLE II. *PFC/culture*[a]

SN of		*nb of T cells/ml x* 10^{-6}				
		5	*2*	*0.5*	*0.125*	*0*
Lyt 2+	*0-24 hrs*	*70*	*177*	*50*	*24*	*0*
Lyt 2-	*"*	*150*	*129*	*123*	*33*	
Lyt 2+	*24-48 hrs*	*0*	*22*	*100*	*0*	*0*
Lyt 2-	*"*	*0*	*35*	*70*	*9*	

[a]*Addition of SN on the 2nd day. PFC determined on the 4th day of culture*

C. Macrophage Activation

1. Induction of secretion of plasminogen (Pmg) activator (IPA). Peritoneal exudate cells (5 x 10^3) were incubated with SN produced by various numbers of T cells. After 20 hrs, they were overlayed with a solution made of agar (1.2%), milk (1.6%) and purified human plasminogen (40 ug/ml), according to J.D. Vassalli et al. (7).

TABLE III. Pmg Dependant Caseinolytic Plaque[a]

SN of		nb of T cells/ml x 10^{-6}					
		5	2	0.5	0.25	0.125	0[a]
Lyt 2⁺	*0-24 hrs*	200	210	100	100	65	2
Lyt 2⁻	"	200	250	190	100	51	

[a]*These were scored after a 10 hr incubation.*

Similar activity was produced by the Lyt 2+ and 2⁻ subpopulations. Activity recovered during the 0-24 hrs and 24-48 hrs period was similar; this lymphokine is therefore not produced with the same kinetics and at the same T cell concentration as TRF and IL_2.

2. Induction of the cytolytic function of macrophages (MAF). This activity was explored according to Kelso et al. (8). T cell titration for MAF was identical to that of IPA, the Lyt 2^+ or 2^- subpopulations producing similar activity (not shown).

3. Induction of PGE_2 secretion by macrophages. The SN of Lyt 2^+ lymphocytes were in general more active inducers of PGE_2 secretion than Lyt 2^-, but only when cultured at low T cell concentration (not shown).

D. Mastocyte Growth Factor (MCGF)

A mast cell line, raised in our laboratory as described elsewhere (9) was used as responding cell in a proliferative assay (representative experiment).

TABLE V

SN of		^{3}H TdR Incorporation (cpm) nb of T cells/ml x 10^{-6}		
		2	0.25	0
Lyt 2^+	0-24 hrs	2050	1250	220
Lyt 2^-	"	1720	1150	220

Thus similar MCGF activity was produced by the Lyt 2^+ or 2^- lymphocytes.

E. Interferon (IFN)

The SN were titrated for their inhibition of the cytopathic effect of the vesicular stomatitis virus on a transformed fibroblast LF_2 (10).

TABLE VI

SN of		IFN Units/ml nb of T cells/ml x 10^{-6}					
		5	2	0.5	0.25	0.125	0
Lyt 2^+	0-24 hrs	405	270	30	10	10	0
Lyt 2^-	"	45	30	10	0	0	0

IFN seems to be produced predominantly, if not exclusively by the Lyt 2^+ lymphocytes.

IV. DISCUSSION

Mouse T lymphocytes can be divided into two main classes with the aid of the anti-Lyt antibodies; the Lyt 1^+2^- and the Lyt 1^+2^+ which represent about 40 and 60% of the mouse peripheral T cells respectively (11). It is presently controversial whether these two subpopulations are restricted in their functional capacity or their antigenic repertoire (12). An essential action of the helper T cell is to induce, by the secretion of lymphokine, the function of other effector cells. The present study leads to the conclusion that both the Lyt 2^+ and 2^- subpopulations are capable of producing lymphokines which induce the proliferation and/or differentiation of cytolytic T lymphocytes (IL_2), B lymphocytes (TRF), macrophages (IPA, MAF) and mastocytes (MCGF). These results are therefore not compatible with the limitation of the "helper" phenotype to the Lyt 2^- cells, but they are in accord with the possibility that, depending upon the antigen, either the Lyt 2^+ or 2^- will be the major source of lymphokine.

In contrast to the lymphokines described above, IFN is released mainly, if not exclusively by Lyt 2^+ lymphocytes. Indeed the small activity produced by the high concentration of Lyt 2^- might represent either a very low frequency of Lyt 2^- producing cells or, more likely, a production by the Lyt 2^+ contaminants.

REFERENCES

1. Julius, M.H., Simpson, E., and Herzenberg, L.A., *Eur. J. Immunol.* 3, 645 (1973).
2. Mage, M., Mathieson, B., Sharrow, S., McHugh, L., Hämmerling, V., Kamellopoulos-Langerin, C., Brideau, D., and Thomas, Ch.A., *Eur. J. Immunol.* 11, 228 (1981).
3. Piguet, P.-F., Irlé, C., Kollatte, E. and Vassalli, P., *J. Exp. Med.* 154, 581 (1981).
4. Gillis, S., Ferm, M.M., Ou, W., and Smith, K.A., *J. Immunol.* 120, 2027 (1978).
5. Schimpl, A., and Wecker, E., *in* "Biology of the lymphokines" (S. Cohen, E. Pick, and J.J. Oppenheim, ed.), p. 369. Academic Press, New York, (1979).
6. Leibson, H.J., Marrack, P., and Kappler, J.W., *J. Exp. Med.* 154, 1681 (1981).
7. Vassalli, J.-D., Hamilton, J., and Reich, E. *Cell* 11, 695 (1977).
8. Kelso, A., Glasebrock, A.L., Kanagawa, O., and Brunner,K.T. *J. Immunol.* (in press).
9. Razin, E., Cordon-Cardo, C., and Good, R.A., *Proc. Nat. Acad. Sci. USA*, 78, 2559 (1981).
10. Dahl, H., and Degré, M., *Acta Path.Microbiol.Scand. Section B*, 80, 863 (1972).
11. Ledbetter, J.A., Rouse, R.V., Spedding Micklew, H., and Herzenberg, L.A., *J. Exp. Med.* 152, 280 (1980).
12. Simpson, E., *Nature* 295, 366 (1982).

DISCUSSION

VADAS: Could Guerne tell us the composition of Lyt-2$^-$ cell populations and the contribution of Lyt-2$^+$ cells to the overall production of the various lymphokines.

GUERNE: The makeup of the Lyt-2$^-$ cell population was 96% T cells, 97% of these T cells were Lyt-2$^-$. For all the lymphokines we have tested, Lyt-2$^+$ supernatants had activity comparable (slightly lower) to supernatants of the same number of non-separated T cells.

OPPENHEIM: In general, data indicating that a cell produces something is difficult to refute. However, negative data such as the failure of Lyt-2$^-$ lymphocytes to produce interferon may be due to inadequate stimulation. Did Guerne try to stimulate with anything more than 1 mg/ml of Con-A?

GUERNE: No, we did not try anything other than 1 mg/ml of Con-A, inasmuch as this amount stimulated lymphocytes perfectly well for a great range of other lymphokine activities.

PALACIOS: Does Guerne's panning procedure for recovery of Lyt-2$^+$ cells render these cells sensitive to IL-2?

GUERNE: We have not explored that possibility.

INTERLEUKIN-2 PRODUCTION BY INDIVIDUAL HUMAN T LYMPHOCYTE SUBPOPULATIONS

Stefan C. Meuer
John F. Daley
Stuart F. Schlossman
Ellis L. Reinherz

Division of Tumor Immunology
Sidney Farber Cancer Institute
Boston, Massachusetts

I. INTRODUCTION

Recent studies have indicated that clonal expansion of T lymphocytes following antigen or mitogen triggering requires the presence of a soluble mediator (1-5). In addition, such a factor also appears to be necessary for differentiation of pre-cytotoxic T lymphocytes into cytotoxic effector cells (6-8). Several lines of evidence support the notion that a single T cell derived glycoprotein, termed IL-2, possesses both activities in humans as well as experimental animals (9,10). Although requirements for IL-2 production have been extensively investigated, the T cell subset origin of IL-2 has been less well defined, particularly in man.

The development of antibodies to human T lymphocyte surface determinants led to the identification of functionally distinct human T cell subpopulations that express unique cell surface molecules (11). Thus, the T4+ subset predominantly provides inducer/helper function for T-T, T-B and T-macrophage interactions whereas the T8+ subset principally functions in a suppressive mode and contains the majority of cytotoxic effector T cell activity following MLC (12-15).

In the present study, we utilized monoclonal antibodies in conjunction with cell sorting techniques to obtain highly purified human T cell subsets. Subsequently, the capacity of T4+ and T8+ T cells to secrete IL-2 upon stimulation with mitogens or alloantigens was examined.

ISBN 0-12-527540-4

II. RESULTS

A. IL-2 Secretion by Separated T Cell Subpopulations upon Stimulation with Mitogens

In order to determine whether IL-2 secretion was restricted to the T4+ or the T8+ T cell subset, fractionated T8+ or T4+ (2.5 x 10^6/ml) were stimulated in vitro with a combination of phytohemagglutinin (PHA, 5 μg/ml) and phorbol myristate acetate (PMA, 4 μg/ml) known to generate high amounts of IL-2 from unfractionated cells (16,17). In order to remove free PHA and PMA, the cultures were washed twice after 2 hours of incubation. Following 48 hours of in vitro culture, supernatants were harvested and passed through 0.45 μm filters. Serial dilutions of culture supernatant were then assayed for IL-2 activity utilizing a long term human T cell line (SML) (18). As demonstrated in Fig. 1, supernatants from both mitogen stimulated T4+ and T8+ cells induced proliferation of SML to a similar extent. Moreover, these same supernatants contained proliferative activity for 2° MLR blasts as well as the IL-2 dependent mouse T cell line, Lyt23.4 (data not shown). In contrast, the control supernatant derived from autologous monocytes stimulated in an identical fashion had no effect in the various test systems.

PRODUCTION OF IL-2 FOLLOWING STIMULATION WITH MITOGEN

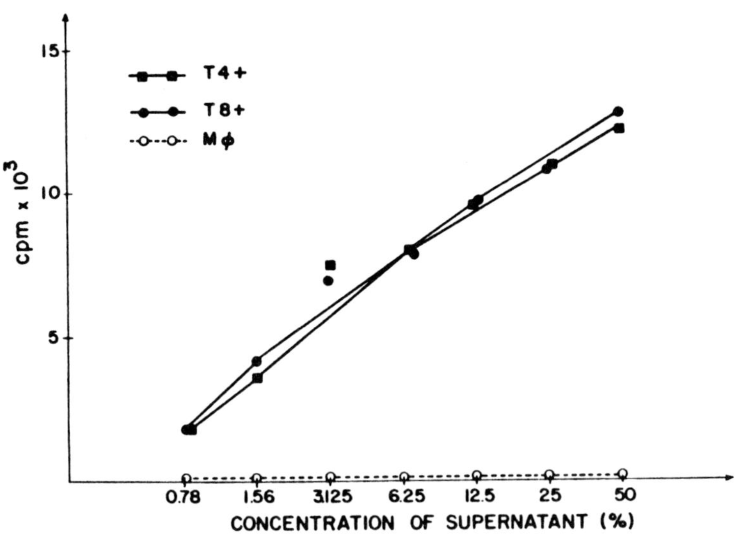

B. *IL-2 Secretion by Separated Human T Cell Subpopulations upon Stimulation with Alloantigens*

It is well established that supernatants of lymphocytes undergoing a mixed lymphocyte reaction (MLR) contain IL-2 activity (5). To characterize the source of IL-2 following alloactivation, fractionated T4+ and T8+ cells were incubated for 48 hours in MLR and then the supernatants were harvested and processed as above. To avoid the possibility of back stimulation, an irradiated B lymphoblastoid line, Laz 156, was used as alloantigen. Again, the amount of IL-2 produced by the various responder cell populations was judged by measuring proliferative responses of SML to serial dilutions of supernatants. In contrast to mitogen stimulation (compare Fig. 1), Fig. 2 demonstrates that the major source of IL-2 after alloactivation is the T4+ population. Under identical conditions, the T8+ lymphocyte secreted about 10 times less of the lymphokine than the T4+ cells. Supernatants derived from T8+ lymphocytes displayed significant IL-2 activity only at final concentrations of >25%. It should be noted, however, that in 2 responder/stimulator combinations, T8+ T cells, as judged by this assay system, reproducibly secreted an amount of IL-2 which was approximately half of that made by the T4+ cells. Again, proliferative assays with human 2° MLR blasts and Lyt23.4 to the various supernatants gave comparable results (data not shown).

PRODUCTION OF IL-2 FOLLOWING STIMULATION WITH ALLOANTIGEN

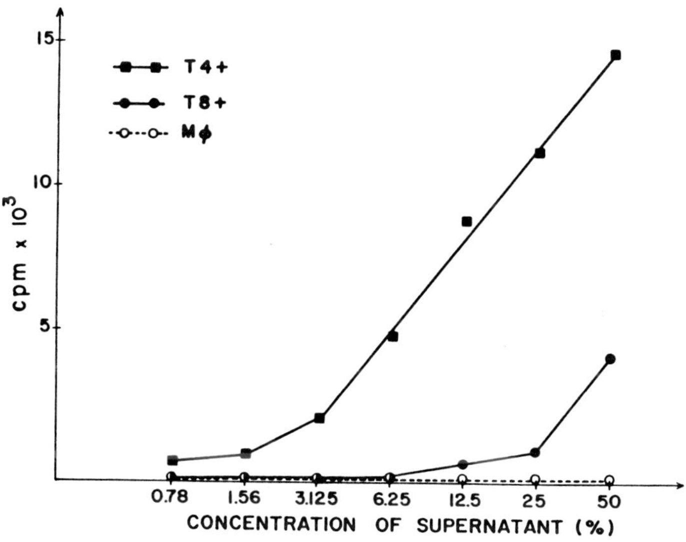

C. In vitro Differentiation of CTL with Supernatants from Separated Human T Cell Subpopulations

The differentiation of precytotoxic lymphocytes into CTL effector cells has been shown to require the presence of the T cell product IL-2 (6-8). Consequently, if the proliferative activity of the above supernatants was due to IL-2, then the supernatants should be capable of inducing CTL effector differentiation from resting T8+ T lymphocytes (9-12). To test this possibility directly, unstimulated cell sorter purified T8+ cells were incubated with alloantigens and 3% autologous macrophages for 6 days in the presence of media, or supernatants derived from mitogen or alloantigen stimulated T4+ or T8+ cells (compare Fig. 1 and 2). Subsequently, cells were harvested and examined for cytotoxic activity against ^{51}Cr labelled Laz 156 targets at different E/T ratios.

Table 1 demonstrates the results of 10 representative independent experiments performed. As shown, stimulation of T8+ T cells with alloantigen in the absence of IL-2 containing supernatants resulted in minimal or low levels of CML. More importantly, supernatants derived from mitogen stimulated T4+ or T8+ cells induced the T8+ subpopulation to kill at all E/T ratios tested. The cytotoxicity as determined on ^{51}Cr labelled stimulator cells was specific since unrelated allogeneic tar-

TABLE I

EFFECTS OF SUPERNATANTS FROM MITOGEN OR ALLOANTIGEN STIMULATED T CELL SUBSETS ON INDUCTION OF CELL MEDIATED LYMPHOLYSIS BY T8+ LYMPHOCYTES

Derivation of supernatant		% specific lysis		
	E/T ratio:	40/1	20/1	10/1
Exp. 1				
Medium alone		13	9	8
Mitogen stimulated MØ		10	11	5
Mitogen stimulated T4+ cells		37	25	24
Mitogen stimulated T8+ cells		31	29	22
Exp. 2				
Medium alone		2	2	-
Alloantigen + MØ		2	2	-
Alloantigen stimulated T4+ cells		37	33	28
Alloantigen stimulated T8+ cells		8	5	5

gets were not killed by these effectors (data not shown). In marked contrast, only the supernatant derived from the alloactivated T4+ population but not the alloactivated T8+ population provided significant help for CTL differentiation. These results are consistent with the present findings indicating that alloantigen stimulated T4+ but not T8+ cells induced major proliferation of SML and previous observations that although both T4+ and T8+ T cells are capable of proliferating to alloantigen, T8+ cytotoxic precursor cells require a T4+ T lymphocyte or its soluble product for optimal development of cytotoxicity (12). It should be noted that in 2 additional experiments, however, significant CTL generation occurred in the T8+ alloantigen stimulated population alone which correlated with the production of IL-2 by the same individual's T8 subset. Thus, the capacity of alloactivated T8 cells to become cytotoxic appears to be directly linked to their ability to produce sufficient quantities of IL-2 which seems to be critically dependent on the stimulus used for activation.

III. DISCUSSION

In the present study the capacity of individual human T cell populations to produce IL-2 was examined. For this purpose, cell sorter separated T4+ and T8+ human T cell subsets were stimulated with either mitogens or alloantigens. After 48 hours of in vitro culture, supernatants were analyzed for their IL-2 activity in each of two test systems: 1) induction of proliferation of long term T cell lines; or 2) induction of cytotoxic effector T cells by stimulation of resting T8+ lymphocytes with an allogeneic B lymphocyte population. In both systems, the results indicate that although mitogens triggered equivalent amounts of IL-2 from both T4+ and T8+ cells, in the majority of responder/stimulator combinations used for MLC, secretion of IL-2 was almost restricted to the T4+ subset.

It is very unlikely that the effects observed with supernatants derived from lectin stimulated T cell subsets were due to mitogen effects alone for several reasons. First, these supernatants were not mitogenic for resting T lymphocytes. Second, neither SML, the long term human T cell line, nor Lyt 23.4, an IL-2 dependent mouse cell line, responded to mitogens. Third, supernatants derived from mitogen treated autologous monocytes failed to exhibit any effects in either system. The latter control would also rule out the possibility that proliferation by SML to the various supernatants was initiated by Interleukin 1 (IL-1) with subsequent triggering of IL-2 production from SML cells themselves (19).

IL-2 appears to represent an essential second signal for T lymphocyte proliferation (3). In view of the above data, the finding that both regulatory T cell subsets proliferate to PHA and Con A is not surprising (11,18). Moreover, the production of IL-2 in both T4+ and T8+ subsets after allogeneic stimulation is consistent with the finding that those subsets proliferate to alloantigen (11,18). However, the major quantitative difference in the amount of IL-2 obtained from the T8+ and T4+ alloantigen stimulated populations suggests that under these circumstances, either T8 production of IL-2 is much less than that of T4 production or that the T8+ subset utilizes IL-2 to a greater extent than the T4+ subset. Alternatively, several recent reports suggest the possibility that the signals required for lymphocyte proliferation and CTL differentiation might be mediated by two distinct lymphokines (20,21). The present studies do not address this question. It should be noted, however, that isolated T8+ cells show brisk proliferative responses to alloantigens but, in general, do not mount significant cytotoxic responses. Such generation of maximal cytotoxicity in CML has been shown to require collaborative interaction between two subsets of T lymphocytes in human and experimental systems (6,8,12,22). In man, both with monoclonal antibodies and heteroantisera, it has been found that although T8+ (TH_2+) T cells contain the majority of cytotoxic effector cells for CML, a T4+ (TH_2-) T cell was required to facilitate differentiation of cytotoxic effector cells from the T5/8 population (23). In the absence of this T4+ population, cytotoxicity after primary MLC stimulation is markedly reduced (12). Generation of CTL is also dependent on IL-2 (6-8). Given the present results indicating that in most MLC responder/stimulator combinations the vast majority of IL-2 activity was derived from the T4+ and not the T8+ population, it is obvious how the IL-2 requirement and T-T interactions may be linked.

The observation that in several experiments the T8+ population generated cytotoxic effector cells and produced IL-2 activity by itself is of great interest. Recent studies in the mouse system with defined stimulator/responder combinations may shed light in this apparent paradox. In those studies, it was found that when stimulator and responder cells differed at only I region encloded MHC antigens, Lyt1+ and not Lyt2+ T cells produced IL-2. In contrast, when the MHC combination differened singularly at the K/D region of MHC, the Lyt2+ and not the Lyt2- cells were responsible for IL-2 production (24,25). Such studies emphasize the importance of the stimulus in activation of individual T cell subpopulations. This is also suggested in human studies because the induction of proliferation of human T cell subsets and their surface expression of activation antigens was found to be dependent on the individual stimulus (26). Furthermore, more recent studies performed with both polyclonal and monoclonal human T cell populations demonstrated that the human T4+ inducer and T8+ cytotoxic/suppressor population responds to products of different MHC loci (27,28).

Therefore, the differential activation of T cell subsets by various stimuli implies that their individual functional repertoires may be linked to unique surface receptors and/or interaction molecules involved in T cell triggering. This difference rather than an absolute subset restricted production of a nonspecific molecule such as IL-2 may dictate the regulatory capacity of T cells after an antigenic pertubation. Analysis of the various stimli, including HLA and Ia antigens, which could preferentially induce IL-2 secretion from T8+ or T4+ T cells will need to be determined.

REFERENCES

1. Gillis, S., Baker, P.E., Ruscetti, F.W. and Smith, K.S. *J. Exp. Med. 148*:1093 (1978)
2. Gillis, S. and Smith, K.S. *Nature 268*:154 (1977)
3. Watson, J. and Mochizuki, D. *Immunolog. Rev.51*:257 (1980)
4. Smith, K.A., Lachman, L.B., Oppenheim, J.J. and Favata, M.F. *J. Exp. Med. 151*:1551 (1980)
5. Smith, K.A. *Immunolog. Rev. 51*:337 (1980)
6. Wagner, H. and Rollinghoff, M. *J. Exp. Med. 148*:1523 (1978)
7. Wagner, H., Hardt, C., Heeg, K., Rollinghoff, M. and Pfizenmaier, K. *Nature 284*:278 (1980)
8. Friedman, S.M., Hunter, S.B., Irigoyen, O.H., Kung, P.C., Goldstein, G. and Chess, L. *J. Immunol. 126*:1702 (1981)
9. Gillis, S., Gillis, A.E. and Henney, C.S. *J. Exp. Med. 154*:983 (1981)
10. Kern, D.E., Gillis, S., Okada, M. and Henney, C.S. *J. Immunol. 127*:1323 (1981)
11. Reinherz, E.L. and Schlossman, S.F. *Cell 19*:82 (1980)
12. Reinherz, E.L., Kung, P.C., Goldstein, G. and Schlossman, S.F. *Proc. Natl. Acad. Sci. USA 76*:4061 (1979)
13. Reinherz, E.L., Kung, P.C.,Goldstein, G. and Schlossman, S.F. *J. Immunol. 123*:2894 (1979)
14. Reinherz, E.L., Kung, P.C., Breard, J.M., Goldstein, G. and Schlossman, S.F. *J. Immunol. 124*:1883 (1980)
15. Reinherz, E.L., Kung, P.C., Goldstein, G. and Schlossman, S.F. *J. Immunol. 124*:1300 (1980)
16. Stadler, B.M., Dougherty, S.F., Farrar, J.J. and Oppenheim, J.J. *J. Immunol. 127*:1936 (1981)

17. Sando, J.J., Hilfiker, M.L., Salomon, D.S. and Farrar, J. *Proc. Natl. Acad. Sci. USA* *78*:1189 (1981)
18. Meuer, S.C., Hussey, R.E., Penta, A.C., Fitzgerald, K.A., Stadler, B.M., Schlossman, S.F. and Reinherz, E.L. *J. Immunol. 129* (1982) in press
19. Farrar, J.J., Mizel, S.B., Fuller-Farrar, J., Farrar, W. and Hilfiker, M.L. *J. Immunol. 125*:793 (1980)
20. Okada, M., Yoshimura, N., Kaieda, T., Yamamura, Y. and Kishimoto, T. *Proc. Natl. Acad. Sci. USA 78*:7717 (1981)
21. Wagner, H., Hardt, C., Rouse, B.T., Rollinghoff, M., Scheurich, P. and Pfizenmaier, K. *J. Exp. Med. 155*: 1876 (1982)
22. Cantor, H. and Boyse, E.A. *J. Exp. Med. 141*:1390 (1975)
23. Evans, R.L., Lazarus, H., Penta, A.C. and Schlossman,S.F. *J. Immunol. 120*:1423 (1978)
24. Okada, M. and Henney, C.S. *J. Immunol. 125*:300 (1980)
25. Okada, M., Klimpel, G.R., Kupper, R.C. and Henney, C.S. *J. Immunol. 122*:2527 (1979)
26. Reinherz, E.L., Kung, P.C., Pesando, J.M., Ritz, J., Goldstein, G. and Schlossman, S.F. *J. Exp. Med. 150*: 1472 (1979)
27. Engleman, E.G., Benike, C.J., Grumet, F.C. and Evans,R.L. *J. Immunol. 127*:2124 (1981)
28. Meuer, S.C., Schlossman, S.F. and Reinherz, E.L. *Proc. Natl. Acad. Sci. USA 79* (1982) in press

DISCUSSION

KASAHARA: What proportion of macrophages or OKMI$^+$ cells are present as contaminants in Meuer's T4$^+$ or T8$^+$ cell populations? Has he looked into a possible requirement for macrophages, or OKMI$^+$ cells, or macrophage products?

MEUER: All our experiments were performed in the presnece of 3% autologous monocytes.

FAHEY: Of the several T4 and T8 clone groups, some make and some do not make IL-2. Do they all have IL-2 receptors, and do they respond to "purified" IL-2?

MEUER: We have not yet investigated IL-2 receptors on our clones and we have had no purified IL-2 available. However, these clones all respond to culture supernatants obtained from stimulated human T cells.

PFIZENMAIER: What is the evidence that IL-2 *per se* drives differentiation of CTL precursors into active CTL in the human allo-MLC?

MEUER: This study does not address the issue whether one or more lymphokines are required for proliferation *versus* CTL induction. The fact, however, that isolated T8$^+$ cells *do* proliferate in response to alloantigens but generally do not become cytotoxic, may suggest that more than one kind of molecule is necessary to generate cytotoxic effector cells.

PALACIOS: Monetta *et al.* and I have different observations at the cl-c cell level with regard to the correlation of Ty or T8 antigenic determinants and function. Some time ago we reported that Ty$^+$ cells produce IL-2 after stimulation by autologous or allogeneic non-T cells, PPD- or tetanus toxoid-pulsed macrophages. We found no evidence that T8$^+$ cells produce IL-2 in these four systems. With regard to the Con-A system, when I examined the capacity of T8$^+$ cells to produce IL-2, I found that T8 cells cultured with macrophages did not produce under the stimulus of Con-A,. However, Luger and colleagues as well as Meuer observed that Con-A-stimulated T8$^+$ cells, in the presence of IL-1 in PMA, produce IL-2. Some months ago, I reported that Ty$^+$ but not T8$^+$ cells help macrophages to produce IL-1 under the stimulus of Con-A. Therefore, our varied observations would seem to be explained by the inability of T8$^+$ cells to assist macrophages with production of IL-1. If one bypasses this stage by adding IL-1 or PMA, T8$^+$ cells can produce IL-2. Because of the inability of T8$^+$ cells to help macrophages to produce IL-1, I would think that the physiological relevance of $T8^+$ cells as producers of IL-2 must be minimal, if any. Has Meuer ascertained whether his highly purified $T8^+$ cells, stimulated by Con-A and IL-1, generate cytotoxic activity?

MEUER: No, we have not done such an experiment.

VADAS: Did Meuer check the phenotype of his cells after Con-A stimulation?

MEUER: All separated T cell populations were re-analyzed by FACS analysis and had a stable phenotype. Moreover, separated $T8^+$ cells never showed a proliferative response to autologous macrophages pulsed with soluble antigens (Tetanus-Toxoid; our donors were Tetanus-responders).

LYMPHOKINE PRODUCTION IN COMMON VARIED IMMUNODEFICIENCY, POLYCLONAL HYPER-GAMMAGLOBULINEMIA AND GRANULOMATOUS DISEASES

Hans H. Peter[1)]
Petra Gronewold

Division of Clinical Immunology
Department of Medicine, Med. Hochschule
D-300 Hannover 61, F.R.G.

Holger Kirchner

Institute of Virology, German Cancer Center
D-6900 Heidelberg, F.R.G.

Hinnak Northoff

Institut of Immunology and Serology,
University of Heidelberg D-6900 Heidelberg, F.R.G.

I. INTRODUCTION

The capacity of the immune system to mount a specific immune response involves a complicated regulatory network in which antigen-specific signals preceed a non-specific amplification process. While abundant experimental evidence suggests a key role of interleukin 2 (IL-2) in this amplification process little is known about alterations of the lymphokine production in human diseases. Notably clinical conditions with signs of an abnormally regulated immune system were of interest to us to study the capacity of blood lymphocytes to secrete IL-2 and interferons (IFN). We report in this preliminary study on lymphokine production in patients with common varied immunodeficiency (CVID), polyclonal hyper-gammaglobulinemia (PHIg) associated with chronic immune complex disease and various granulomatous diseases (GD).

---

1) Supported by DFG grant Pe 151/7-4

ISBN 0-12-527540-4

Compared to a group of 18 healthy control persons lymphocytes from patients with PHIg showed a greatly reduced capacity to produce IL-2 and IFN under the chosen experimental conditions. Lymphocytes from CVID patients tended to produce higher IFN but lower IL-2 titers than the control group.

II. MATERIALS AND METHODS

A. <u>Patients and Control Persons:</u> Some clinical features of the examined patients are listed in Table I.

TABLE I. Clinical Features of the Examined Patients

Pat,Sex,Age			Diagnosis[b)]		IC[d]	Serum-Ig-Levels		
						IgG	IgA	IgM
1.	f,	47[a]	CVID	(27)[c]	-	7.0[e]	0.8	0.5
2.	f,	38	CVID	(5)	-	10.3	0.7	<0.3
3.	f,	45	CVID	(40)	-	1.0	<0.6	<0.3
4.	f,	42	CVID	(6)	-	2.8	<0.6	<0.3
5.	f,	62	CVID	(15)	-	15.0	<0.6	0.3
6.	f,	42	CVID	(12)	-	2.1	<0.6	<0.3
7.	m,	19	CVID	(3)	-	4.9	<0.6	3.8
8.	m,	54	CVID	(4)	-	2.6	<0.6	<0.3
9.	f,	38	SLE	(1)	++	20.1	3.8	3.2
10.	m,	62	AILAP	(2)	+++	21.0	1.9	1.6
11.	m,	55	AILAP	(2)	+	18.9	2.9	3.7
12.	m,	55	AILAP	(2)	++	14.5	1.9	1.2
13.	m,	60	NHL	(3)	+++	21.4	3.5	0.7
14.	f,	28	TB	(13)	++	14.9	2.3	3.5
15.	m,	39	Behcet	(5)	-	11.7	1.5	3.2
16.	m,	37	Behcet	(3)	+	12.6	1.6	1.7
17.	m,	40	Behcet	(3)	+	15.6	3.0	1.9
18.	f,	29	cGD	(8)	-	10.6	1.9	1.5
19.	m,	24	cGD	(22)	-	9.2	2.1	0.9

a) Age in years; f, female; m, male.
b) For details see below.
c) Years of disease duration in paranthesis.
d) Presence of C1q binding immune complexes.
e) Normal ranges: IgG: 8.0 - 18.0 g/l;
IgA: 0.9 - 4.5 g/l;
IgM: 0.6 - 2.8 g/l.

The CVID patients had an increased susceptibility to respiratory and/or gastro-intestinal infections. They received regular intravenous gamma-globulin substitution and antibiotics when needed. The five patients with PHIg presented all with SLE-like symptoms such as recurrent arthralgias, Raynaud's phenomena, lymphadenopathy, fever, skin rush, oedema but no renal disease. They had elevated circulating immune complexes and an increased proportion of polyclonal plasmablasts in their circulation. Only one patient (No9) had a typical SLE as judged by the presence of high titered anti-DNA antibodies. Three patients (No 11,12,13) were classified as angio-immunoblastic lymphadenopathy (AILAP) based on a typical lymphnode histology. One patient (No 13) developed a low-grade non-Hodgkin lymphoma (NHL) with hepato-splenomegaly after a three year history of recurrent fever, allergic skin rushes and arthralgias. Among the patients classified as granulomatous diseases (GD) three (No 16,17,18) presented with typical symptoms of Behcet's syndrome such as recurrent aphthous lesions of the oral and ano-genital mucosal membranes, granulomatous gut lesions, arthralgias, vasculitis and uveitis. Patient No 14, a young woman, had a severe pulmonary tuberculosis that resisted to treatment with classical tuberculastatic drugs. Two patients (No 18,19) presented with chronic granulomatous disease (cGD) characterized by recurrent furunculosis and skin abscesses. Both the HPIg and GD patients were tested prior to any steroid treatment. The healthy control individuals were recruted from the blood bank and from our laboratory personal.

B. <u>Lymphokine Production Cultures:</u> Human peripheral blood mononuclear cells (MNC) were isolated by Ficoll Hypaque gradient centrifugation from 30-40 ml of defibrinated blood. The isolated cells were washed twice and resuspended to $2x10^6$ or $1x10^6$ cells/ml in medium RPMI 1640 supplemented either with 10% fetal calf serum (FCS) or 10% autologous serum (AS). Aliquots of $2x10^6$ cells/ml were stimulated for 24 hrs in round-bottom tissue culture tubes with either no stimulant, 5ug/ml of PHA (Burroughs Welcome), or loug/ml of Con A (Pharmacia). Maximal lymphokine production was induced by Co-stimulation with 5ug/ml of Con A, 5ng/ml of phorbol myristic acetate (PMA, Serva Lab.) and 2 mM of hydroxyurea for 72 hrs (1,2). MNC were also preincubated for three days to reduce the effects of suppressor cells and exposed to mitogens for another 24 hrs (2).

IFN induction cultures were set up with 10^6 cells/ml and exposed to previously established optimal concentrations of the following aIFN inducers (3): Mycoplasma orale 1 infected L1210 cells, Corynebacterium parvum (Cp) and Herpes simplex virus type I antigen (HSV). The cultures were incubated for 24 hrs at 37°C cell-free supernatants were harvested and stored at -70°C until testing for lymphokine activity.

C. Assays for IFN and IL-2 Activities: Supernatants from test and control cultures were sent as coded samples to Drs.H.Kirchner and H.Northoff in Heidelberg. IFN was measured in a virus plaque reduction assay as previously described (4). IL-2 activity was assayed according to the technique described by GILLIS et al (5). Briefly, serial dilutions of the culture supernatants were added to the IL-2 dependent murine T cell line M2. The cells were cultured for 22 hrs in medium RPMI 1640+10% FCS and then pulsed for 4 hrs with 1 µCi of ^{3}H-Thymidine. IL-2 activity was expressed as units/ml based on a reference rat IL-2 preparation, which was arbitratily set 100 U/ml.

III. RESULTS

The results of IFN and IL-2 production cultures are summarized in Table II and Figure 1. In contrast to the negative IL-2 medium controls most patient and control lymphocytes released low levels of IFN in the presence of 10% FCS or AS. In stimulated cultures the different serum supplements did not appear to influence IL-2 and IFN production. No IL-2 was elicited in lymphocyte cultures exposed to the -IFN inducers L1210, HSV and Cp. In mitogen stimulated cultures, on the other hand, IL-2 and IFN production occured simultaneously with the exception of one CVID patient (No 1), who presented with a severely impaired cellular immunity and showed a dissociation of high IFN and absent IL-2 production. Only after preincubation for 3 days to reduce suppressor cell activity a low IL-2 production was detectable (Table III). Preincucubation resulted also in most other lymphocyte cultures in a moderate enhancement of mitogen induced IL-2 but did not increase IFN production (data not shown here). By comparing lymphokine production of patients and controls the most striking finding was the low IFN and IL-2 titers in the 5 PHIg patients. Notably the 3 patients with AILAP failed to produce any significant IL-2 and secreted only low IFN activities (Figure 1).

Table II: IFN and IL-2 Production in 24 hrs Cultures of Human Lymphocytes

Lymphokine	Source of MNL	Serum Suppl	n	Medium	Con A	PHA	ConA+PMA	L1210	HSV	Cp
				Stimulants						
IFN (U/ml)	Control	FCS	16	26±15[&]	314± 96	294± 79	417± 72	325± 80	339± 80	147±60
		AS	12	20± 8	162± 57	282± 71	n.t.	275± 91	238± 97	223±98
	CVID	FCS	7	39±34	346± 62	284±153	589±140	263±124	282±118	52±18
		AS	5	8± 5	314±170	243±162	n.t.	310±172	198±177	165±73
	PHIg	FCS	5	7± 5	9± 9	21± 6	81	7± 5	8± 5	16±16
		AS	4	10± 6	11± 8	13± 7	n.t.	6± 2	10± 6	68±59
	GD	FCS	5	11± 5	n.t.	n.t.	n.t.	81± 42	362±154	n.t.
		AS	5	4± 2	n.t.	70± 11	n.t.	57± 24	18± 5	n.t.
IL-2 (U/ml)	Control	FCS	18	1	23± 8	165± 45	1099±242	3± 2	1	0
		AS	14	1± 1	25± 11	130± 47	n.t.	0	0	0
	CVID	FCS	8	1	7± 6	40± 14	391±100	0	0	0
		AS	6	1	9± 7	65± 31	n.t.	0	1	0
	PHIg	FCS	5	0	1	8± 5	70± 44	3± 2	0	0
		AS	4	0	0	4± 2	n.t.	0	0	0
	GD	FCS	6	0	n.t.	46± 15	361±139	0	1± 1	n.t.
		AS	5	0	n.t.	53± 21	n.t.	0	0	n.t.

&)Arithmetic mean of n patients ± SE; nt, not tested or technical failure.

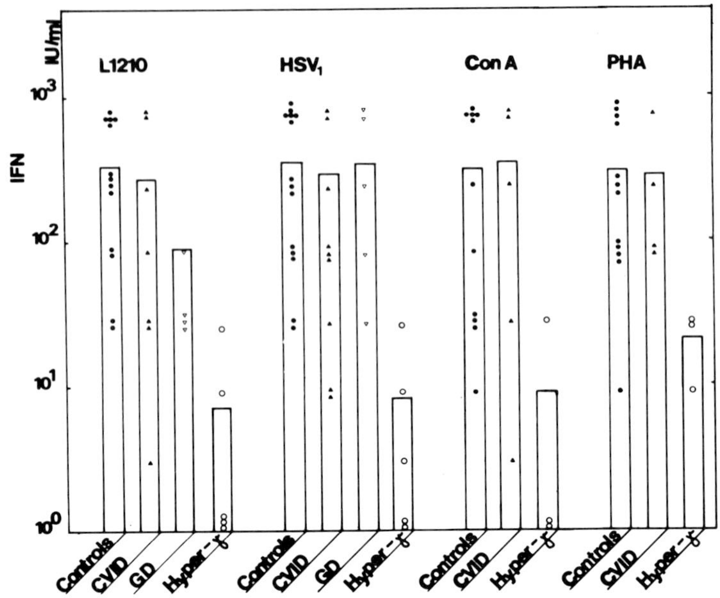

FIGURE 1 a-IFN and y-IFN production of MNC from patients with common varied immunodeficiency (CVID) granulomatous diseases (GD) and polyclonal hypergammaglobulinemia (HPIg).

TABLE III: Dissociation of IL-2 and IFN Production in Lymphocyte Cultures from a Patient with CVID

Stimulant	72 hrs Pre-Culture	IFN (IU/ml) FCS	IFN (IU/ml) AS	IL-2 (U/ml) FCS	IL-2 (U/ml) AS
Medium	no	9	0	0	0
	yes	81	0	0	0
PHA	no	243	81	0	0
	yes	243	27	13	15
Con A	no	243	81	0	0
	yes	243	27	75	50
Con A + PMA	no	729	nt	0	nt
L1210	no	243	81	0	0
HSV	no	243	27	0	0
C.parvum	no	81	243	0	0

CVID lymphocytes produced normal to slightly elevated IFN titers but had a tendence to release less IL-2 than the controls. A similar pattern was noted for the GD patients although this group is still too heterogeneous and incomplete to draw any meaningful conclusions.

IV CONCLUSIONS

The results of the present study suggest that lymphokine production in vitro reflects a functional state of the immune system. MNC from patients with an up or down regulated specific immunity exhibit different patterns of IFN and IL-2 production. Compared to a group of healthy individuals lymphocytes from patients with CVID secreted normal IFN titers in response to mitogens tumor cells, viral and bacterial stimuli, whereas their IL-2 production was moderately reduced confirming a recent report (6). Lymphocytes from patients with a polyclonally activated immune system were poor producers of both a- and y-IFN (Figure 1) and of IL-2. Notably the 3 cases of AILAP could not be induced to any significant lymphokine production under our experimental conditions. GD patients' lymphocytes had a reduced capacity of lymphokine production whithout however, reaching the low levels of PHIg patients. In the literature a defective IFN production has been noted in a variety of chronic viral infections (7,8,9) and IL-2 deficiency has been reported besides in primary immunodeficiency (6) in autoimmune mice (10). Our data indicate that the hyporesponsiveness of polyclonally activated lymphocytes to various IFN inducers extends also to the induction of IL-2. Since recently pokeweed induced polyclonal Ig synthesis has been shown to be suppressed in lymphocyte cultures from patients with active SLE (11), a certain refractoriness to non-specific amplification of effector mechanisms appears to be a common feature of in vivo activated human lymphocytes.

V. REFERENCES

1. Stadler, R.M.et al, Behring Inst.Mitt.67,245(1980)
2. Northoff, H.et al, Behring Inst.Mitt.67, 90 (1980)
3. Peter, H.H.et al, Eur.J.Immunol. 10, 547 (1980)
4. Kirchner, H.et al, J.Immunol.Meth. 48, 213 (1982)
5. Gillis, S.et al, J.Immunol. 120, 2027 (1978)
6. Lopez-Botet, M.et al, J.Immunol. 128, 679 (1982)
7. Emödi, G., Just, M.,Acta Paed.Scand. 63,183 (1974)
8. Tolentino, P.et al, J.Inf.Dis. 132, 459 (1975)
9. Virelizier, J.L.et al, Lancet,ii, 231 (1978)
10. Dauphinée, M.J. et al, J.Immunol. 127, 2883 (1981)
11. Beale, M.G. et al, J.Immunol. 128, 486 (1982)

DISCUSSION

RODE: What is the percentage of T cells and the ratio of T cell subsets in Peter's patients with immune abnormalities?

PETER: The percentage of T cells for all patients tested was within the normal range. In the common variable immunodeficiency group, 6 of the 8 patients have been tested for T cell subsets; three of these had an OKT_4/OKT_8 ratio of <1, the other 3 patients had normal ratios. Similarly, in the hyper-γ globulinemic group, 3 of the 5 patients tested had an OKT_3/OKT_8 ratio of <1 and 1 out of 4 in the GD group, respectively. As regards Ia expression, it was significantly augmented in the hyper-γ-globulinemic group ranging from 20-34% positive cells. We have also done mitogen responsiveness and NK testing in most patients but have not seen any clear-cut correlation with lymphokine production. All patients responded to PHA and Con-A; subnormal responses were only seen in 2 CVID patients and in 1 hyper-γ-globulinemic patient.

WOOD: Has Peter added exogenous IL-1 to his cultures in order to test whether the defective IL-2 production is secondary to defective IL-1 production?

PETER: No, this has not been done so far.

DAYER: As a followup on Wood's query, what was the monocyte repertoire or the ratio between monocytes and lymphocytes after Ficoll-Hypaque separation, and its relationship to the immune category of the patient?

PETER: As far as I can tell, the monocyte counts were within the normal range of Ficoll-Hypaque preparations; that is 10-30%.

PICK: Since the defect in cGD is limited to the mono-myelocytic series, it is rather difficult to understand why IL-2 and IFN production would not be decreased. One possible explanation would be that the changes seen are due to the presence of infection in both immunodeficient and cGD patients.

PETER: Pick is right with respect to the 2 cGD patients. Their IFN and IL-2 production was in the normal range. I can also not exclude the possibility that bacterial infections in the CVID and GD patients are responsible for their slight decrease in lymphokine production. However, we were struck by the fact that the most impressive reduction in lymphokine levels was seen in the patients with a polyclonally activated immune system and clinical symptoms resembling chronic GvH disease.

ASHERSON: Could the relatively small changes in IL-2 production found in patients with hypogammaglobulinemia be secondary to their infections? Has Peter, for instance, any data on patients with bronchiectasis or chronic bronchitis not due, in an obvious sense, to immune deficiency?

PETER: I really do not know. So far we have no data on patients with chronic bronchitis and normal serum-Ig levels.

ASHERSON: It is known that a minority of patients with late onset hypogammaglobulinemia (common variable immune deficiency) have T suppressor cells which block pokeweed-mitogen-induced immunoglobulin production, when present in *small* numbers. This is seen in patients with hypogammaglobulinemia due to thymoma and in some patients with hypogammaglobulinemia and hemolytic anemia or pure red cell aplasia. Did Peter have the opportunity of examining any of these patients? Would Peter consider the hypothesis that the T suppressor cell in patients with hypogammaglobulinemia due to thymoma inhibits IL-2 production? Did this particular patient give a normal response to Con-A?

PETER: Yes, one CVID patient (No 8) developed his hypo-γ-globulinemia together with the appearance of a benign thymoma. He had an OKT4/OKT8 ratio of <1, responded normally to PHA and Con-A and his cells produced background levels of Ig when stimulated with pokeweed mitogen. We have so far not determined whether his lymphocytes suppress the Ig synthesis of normal B cells.

GERARD: Has Peter looked at the concomitant secretion of a specific or non-specific inhibitor of IL-2 activity that could explain an apparent dimunition of IL-2 production? For example by mixing partially purified IL-2 (with no inhibitor activity) and Peter's samples.

PETER: This is an interesting proposal, but I cannot really comment on it since Nordhoff has performed the IL-2 tests in his laboratory in Heidelberg. I remember however that he noted on the IL-2 values of a few culture supernatants "plus inhibitor". I will have to check this with him.

CHAIRMEN'S COMMENTS

This session was highlighted by information gained from analysis of lymphokines produced by cell lines, as follows:

1) Culture supernatants of stimulated monoclonal T cell lines and hybridomas often contain many different T Cell activities.

2) Some investigators suggested that groups of these activities could be attributed to a single molecule, but most of the evidence suggested that these activities were due to distinct molecules. For example, Swain and Schimpl both agreed that antibody to interferon did not block TRF activity.

3) Furthermore, Krammer provided evidence for considerable heterogeneity even within a single activity. His evidence suggested that supernatants with MAF activity derived from two separate clones was attributable to two separate factors.

4) On the other hand, Altman described some promising results obtained by using an in vivo passaged MOPC 315 line as a bioassay for BCGF and TRF.

5) Another approach was presented by Stadler who selected for variant high-producer lines using irradiation, corticosteroids, BUdR or Azaguanine. However, the success of this approach was limited since the more productive lines were those that grew least well.

6) T Cell hybridomas have been developed by Butler that secrete the desired B cell growth promoting factor, without concommitantly producing any IL2 or TRF.

These examples abundantly illustrate the variety of problems that can be attacked in an advantageous manner with the use of cell lines and hybridomas.

R. Dutton
F. W. Fitch

SECTION IX:
IMMUNOPHARMACOLOGICAL STUDIES OF LYMPHOKINES

ROLE OF INTERLEUKIN-1 IN THE ADJUVANTICITY OF MURAMYL DIPEPTIDE *IN VIVO*

D.D. Wood
M.J. Staruch
*P.L. Durette**
W.V. Melvin III
B.K. Graham

The Department of Immunology and
*Department of Membrane and Arthritis Research
Merck, Sharp & Dohme Research Laboratories
Rahway, N.J. 07065 USA

Many of the immunological adjuvants have been shown to interact with macrophages and modify their behavior. In several cases, it has been shown that macrophages secrete interleukin-1 (IL-1) in response to adjuvants (1,2) raising the possibility that IL-1 is the common mediator of adjuvanticity. Consistent with this possibility, it is known that *in vivo* many adjuvants also induce fever and alter the serum concentrations of the acute phase proteins as does IL-1.

METHODS

Chemicals. Muramyl dipeptide (MDP), the D-alanine and the L-isoglutamine analogues, as well as the 6-0-stearoyl derivative, were obtained from Calbiochem-Behring (La Jolla, CA). 3-0-(D-2-propionyl-L-alanyl-D-isoglutamine)-D-glucopyranose was prepared in a 4-step sequence from 1,2:5,6-di-0-isopropylidene-α-D-glucofuranose (3). 2-Acetamido-2-deoxy-4-0-(D-2-propionyl-L-alanyl-D-isoglutamine)-D-glucopyranose was obtained in 3 steps (4 and Durette and Meitzner, unpublished results) from benzyl 2-acetamido-3,6-di-0-benzyl-2-deoxy-α-D-glucopyranoside (5). 6-Deoxy-6-stearoylamino-muramyl dipeptide was obtained by acylation of 6-amino-6-deoxy-MDP (6).

ISBN 0-12-527540-4

Mice. C57BL/10Sn, C57BL/6J, and C3H/HeJ female mice, 10-12 weeks old, were obtained from Jackson Laboratories (Bar Harbor, ME). C57BL/10ScN Sch female mice were obtained from Harlan Labs (Madison, WI).

Determination of SAA and BSA. The increase in circulating levels of serum amyloid A (SAA) in mice was measured by a micro enzyme-linked immunosorbant (ELISA) inhibition test as described elsewhere (7). A unit of SAA was defined as the amount of serum needed to inhibit the binding of approximately 16 ng of alkaline phosphatase labeled amyloid A by 50%.

Serum antibody specific for BSA was assayed by a micro ELISA method on BSA coated plates. The titres for anti-BSA antibody were calculated on the basis of a standard curve of affinity purified mouse anti-BSA immunoglobulin.

Immunization with BSA and Adjuvants. Mice were immunized subcutaneously with 250 ug of bovine serum albumin (Pentex Fraction V BSA; Miles Laboratories, Inc., Kankakee, IL) dissolved in pyrogen-free saline. MDP was mixed with the BSA solution while complete Freund's Adjuvant (CFA) was emulsified with an equal volume of BSA in saline. IL-1 was given intraperitoneally 2 hours after BSA, while IL-2 was given 4 hours after BSA. All mice were boosted with 100 ug of BSA in saline 21 days after the priming and bled 9-11 days later.

Preparation of Interleukin -1 and -2. Interleukin-1 was prepared from LSP stimulated leukocytes obtained from a single chronic myelomonocytic leukemia patient and purified by gel filtration as a 12-24,000 dalton pool and by isoelectric focusing as a pH 6.9-7.4 pool, free of T cell growth factor (IL-2) activity (8). Partially purified IL-2 was prepared as the 20-30,000 dalton fraction from EL-4 cells stimulated with 10 ng/ml phorbol myristate acetate (9).

RESULTS AND DISCUSSION

Structural Studies. Oppenheim, et al (2) have shown that MDP and its adjuvant active 6-0-stearoyl derivative stimulate IL-1 release from human monocytes *in vitro* whereas the adjuvant inactive D-alanyl analogue does not. In order to extend these observations *in vivo*, we took advantage of the observation that IL-1 induces the production of the acute phase protein known as soluble amyloid A (SAA) *in vivo* (10) and assumed that the SAA titre could be used as an indirect measure of IL-1 production.

C3H/HeJ mice were injected with 100 ug of MDP or selected analogues and the effects measured upon the SAA titre at 6 hours or upon the secondary antibody response to bovine albumin (BSA). The results, shown in Table I demonstrate that there is a fair correlation between the two responses. MDP itself stimulates an increase in SAA from 6 to 90 units/ul and increases the

Table I. Structure-Function Correlations

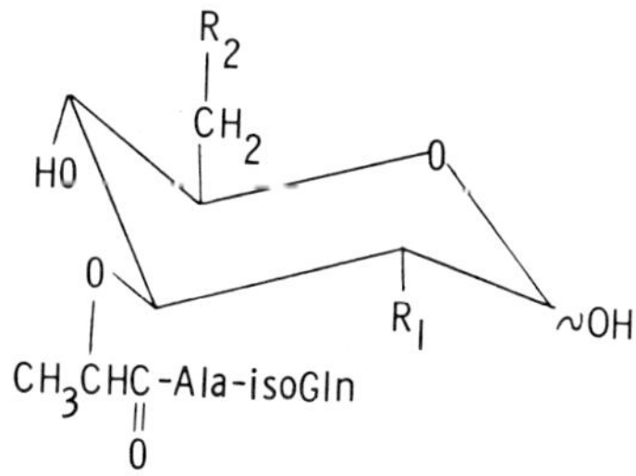

Substitution						% of MDP	
Carbon 3	Carbon 4	ala	isoGln	R_1	R_2	SAA	anti-BSA
β-DP[1/]	α-OH	L	D	-NHAc	-OH	(100)	(100)
α-DP	α-OH	L	D	-NHAc	-OH	3	-2
β-DP	β-OH	L	D	-NHAc	-OH	2	4
β-DP	α-OH	D	D	-NHAc	-OH	4	3
β-DP	α-OH	L	L	-NHAc	-OH	4	26
β-DP	α-OH	L	D	-OH	-OH	0	18
β-OH	α-DP	L	D	-NHAc	-OH	1	1
β-DP	α-OH	L	D	-NHAc	$-O-\overset{O}{\overset{\Vert}{C}}-(CH_2)_{16}CH_3$	143	406
β-DP	α-OH	L	D	-NHAc	$-\overset{H}{N}-\overset{O}{\overset{\Vert}{C}}-(CH_2)_{16}CH_3$	2	-7

[1/] Lactyl-Ala-IsoGln

anti-BSA response from 617 to 5130 ug/ml. In contrast, the configurational isomers of MDP at carbon 3 or 4 of the muramic acid or the D-alanine or L-isoglutamine isomers were much less active in either assay. Similarly, replacement of the acetamido group at carbon 2 with hydroxyl or transposition of the lactyl dipeptide fragment from carbon 3 to carbon 4 also abolished both activities. The acylation of 6-hydroxyl with stearic acid has been shown to substantially increase adjuvant activity (11), and we find that it also increases the SAA response. In contrast, both responses are lost with the 6-deoxy-6-stearoylamino analogue.

These results show that the response to MDP depends upon a highly constrained set of steric and chemical recognition requirements. The fact that the SAA and adjuvant responses showed the same recognition requirements is consistent with the hypothesis that the adjuvanticity of MDP depends upon its ability to stimulate macrophages into a state characterized by the secretion of the SAA inducer (presumably IL-1).

Adjuvanticity of Exogenous IL-1. If IL-1 is really the mediator of adjuvanticity, IL-1 should be able to replace MDP as a stimulant of the humoral response. The selection of an IL-1 dose for testing was based upon the observation that 300-3,000 units of IL-1 were needed to induce an SAA response which was at least as great as that induced by the adjuvant dose of MDP (Table II). A lag period of 2 hours was selected because the SAA response to IL-1 began and peaked approximately 2 hours earlier than the response to MDP. Based upon these observations, the adjuvanticity of IL-1 was tested by injecting 1-2,000 units of IL-1 into C3H/HeJ mice 2 hours after the primary immunization with BSA. Under these conditions, it was found that IL-1 induced a significant enhancement in the antibody response to a secondary challenge with BSA (Table III). Although the response induced by a single injection of IL-1 is never as great as the response induced by MDP, these results clearly show that IL-1 itself is an adjuvant.

Table II. Effect of IL-1 on SAA Induction in C3H/HeJ Mice

Sample (i.p.)	SAA Titre[a] (units/ul)
Saline	6 + 0.2
IL-1 30 units	34 + 14
IL-1 300 units	164 + 24
IL-1 3000 units	302 + 16
MDP 100 ug	115 + 16

[a] Titre measured 6 hours after an i.p. injection of the test material in 0.2 ml saline.

Table III. Adjuvant Effect of IL-1 C3H/HeJ Mice

Immunization		Anti-BSA
Primary	Secondary	(ug/ml Serum)
BSA	BSA	43 ± 17
BSA + 1000 units IL-1	BSA	429 ± 97
BSA + 2000 units IL-1	BSA	704 ± 72
BSA +10 ug MDP	BSA	921 ± 165

Strain Dependence of IL-1 Adjuvanticity. Some mouse strains, such as the BALB/cJ and C3H/HeJ, respond much more strongly to the adjuvant effects of MDP than do others, such as the C57BL strains (12). If IL-1 is indeed involved in the adjuvanticity of MDP, the two groups might be expected to differ in their response to IL-1.

Both the high and low responding strains respond to MDP with a similar enhancement in carbon clearance (54% increase in clearance rate for the C3H/HeJ; 84% for C57BL/6J) showing that the genetic defect is not a failure of the macrophage to recognize MDP. Similarly, both give similar SAA responses to 100 ug MDP (60± 6 units/ul SI=29 for C3H/HeJ; 52 ± 2 units/ul SI=26 for C57BL/10Sn) suggesting that the macrophages *in vivo* probably release nearly equivalent amounts of IL-1. In contrast, the two groups respond quite differently to the adjuvant effects of IL-1. A typical experiment is shown in Table IV. In this case, the secondary antibody response of C3H/HeJ mice was stimulated 13 fold by MDP and 8.7 fold by IL-1. In contrast, the response of C57BL/6J mice was stimulated only 2 fold by MDP and 1.8 fold by IL-1.

The data are consistent with the hypothesis that IL-1 is one of the mediators of the adjuvant response to MDP. Specifically, it appears that the macrophages of both C3H and C57BL mice respond to MDP with a similar IL-1 production but that their immune systems respond quite differently to the IL-1 which is presumably induced by MDP. It seems unlikely that the defect is caused by a defective receptor for IL-1 since the livers of the two strains respond to exogenous IL-1 with similar SAA responses [300 units of IL-1 induce 145 ± 44 units (SI=57) of SAA in the C3H/HeJ and 124 ± 61 units (SI=63) in C57BL/10ScNSch]. Indeed, when partially purified IL-2 is used as an adjuvant, the C3H/HeJ again responds with a marked stimulation of the antibody response while the C57BL/6J is again a poor responder (Table IV).

Table IV. Strain Dependence of the Adjuvant Response

Immunization		Antibody Response (ug/ml)	
Primary	Secondary	C3H/HeJ	C57BL/6J
BSA	BSA	63 ± 20	53 ± 6
BSA + 3000 units IL-1	BSA	548 ± 136	94 ± 17
BSA + 100 units IL-2	BSA	475 ± 144	76 ± 12
BSA + 100 ug MDP	BSA	839 ± 105	106 ± 26
BSA in CFA	BSA	2802 ± 257	1542 ± 62

It is evident that experiments of this type can never conclusively prove that IL-1 mediates the immunological response to adjuvants. Proof will require the development of a method to selectively inhibit IL-1 production or recognition. Nevertheless, these data are consistent with a fairly simple model of adjuvanticity. It is likely that MDP is recognized by macrophages through a highly sterically constrained recognition mechanism. The macrophage then responds with the production of IL-1 which probably stimulates SAA production by the liver and IL-2 production by T cells. The IL-2, in turn, probably expands the T helper population accounting for the enhanced secondary antibody response. It is likely that the genetic defect of the C57BL/6J lies somewhere distal to the production of IL-2.

ACKNOWLEDGMENTS

We wish to thank Marilyn M. Serson for her effort in the preparation of this paper.

REFERENCES

1. Wood, D.D., and Cameron, P.M., J. Immunol 121: 53 (1978).
2. Oppenheim, J.J., Togawa, A., Chedid, L., and Mizel, S., Cell. Immunol. 50: 71 (1980).
3. Hasegawa, A., Kaneda, Y., Goh, Y., Nishibori, K., Kiso, M., and Azuma, I., Carbohydr. Res. 94: 143 (1981).
4. Kiso, M., Kaneda, Y., Goh, Y., Hasegawa, A., and Azuma, I., Agric. Biol. Chem. 44: 1971 (1980).
5. Jacquinet, J.-C. and Sinay, P., Carbohydro. Res. 46: 138 (1976).
6. Hasegawa, A., Okumura, H., Kiso, M., Azuma, I., and Yamamura, Y., Carbohydr. Res. 79: C20 (1980).
7. Wood, D.D., Cameron, M., and Staruch, M.J. J. Immunol. Methods, in press.
8. Wood, D.D., J. Immunol. 123, 2395 (1979).
9. Farrar, J.J., Fuller-Farrar, J., Simon, P.L., Hilfiker, M.L., Stadler, B.M., and Farrar, W.L., J. Immunol. 125: 2555 (1980).
10. Sztein, M.D., Vogel, S.N., Sipe, J.D., Murphy, P.A., Mizel, S.B., Oppenheim, J.J., and Rosenstreich, D.L. Cell. Immunol. 63, 164 (1981).
11. Azuma, I., Sugimura, K., Taniyama, T., Yamamura, Y., Kusumato, S., Okada, S. and Shiba, T., Infect. Immun. 14, 18 (1976).
12. Staruch, M.J. and Wood, D.D., J. Immunol. 128, 155 (1982).

DISCUSSION

LANDY: Has Wood considered the likelihood that the magnitude and character of the host macrophage populations would likely be major determinants of responsiveness to MDP and even IL-1 production by different mouse strains? Such divergencies could conceivably be rendered even more complex by the expectation that at any given time the host peritoneal macrophage population - or even his fixed histiocytes -- would reflect different stages of maturation and/or differentiation and, consequently, different reactivity.

WOOD: I have no doubt that there are strain differences in macrophage maturation and function; nevertheless, our studies have provided no evidence that these differences affect the overall response to MDP. In our hands, MDP caused a similar increase in carbon clearance from the blood and a similar induction of SAA in the high and low responder strains.

WEBB: Would Wood comment on the purity of his IL-1 preparations?

WOOD: The IL-1 was prepared from a single chronic myelomonocytic leukemia patient by LPS stimulation of his monocytes. The IL-1 was used prepared from a pool of the 15,000-20,000 dalton fractions from Sephadex G75, pH6.8-7.4, and isoelectric focusing. It manifests no IL-2 activity.

ROSENWASSER: We have a finding with B6 thymocytes and human IL-1/LD effects, that Wood might want to comment on. We find that unlike other strains tested (C3H, CBA) and PNA-negative set of thymocytes in the B6 mouse will not respond directly to IL-1 with mitogenesis in the absence of lectin costimulation -- perhaps reflecting differences in a thymocyte subset of C3H vs B6 mice. Does Wood feel that this finding fits with the differences in adjuvanticity he described between C3H and B6 mouse strains? Or not at all?

WOOD: I agree that there seems to be little similarity between the genetic constrains seen *in vitro* and those which we see *in vivo*. For example, cells of C57BL mice respond quite well to IL-2 *in vitro* despite their poor response *in vivo*. Further, Damais et al. have shown that *in vitro* C3H/HeJ macrophages make less IL-1 than do C57BL macrophages.

PICK: Considering the fact that Wood's IL-1 is derived from an abnormal human cell line, Wood is in the situation of not being able to construct an appropriate control experiment, i.e. using as control a sample originating from the same cells and run through the same purification procedure but manifesting no IL-1 activity *in vitro* (LAF only). I doubt that we are permitted to draw far reaching conclusions from experiments based on the injection of human material into mice.

WOOD: Indeed, we do not have a "control" preparation; however, the adjuvant activity of 2000 LAF units is quite similar whether we test samples purified by gel filtration and isoelectric focusing or only by gel filtration. Therefore, if the adjuvant effect were to involve a contaminant, it must be one with an isoelectric point similar to that of IL-1.

SUPPRESSION OF IL2 PRODUCTION BY MURAMYL DIPEPTIDE AND DERIVATIVES

Claude Leclerc
Louis Chedid

Service d'Immunothérapie Expérimentale
Institut Pasteur
Paris, France

I. INTRODUCTION

The synthetic glycopeptide MDP(muramyl dipeptide, N acetyl muramyl-L-alanyl-D isoglutamine) has been shown to be the minimal unit required to replace Mycobacteria in Freund's complete adjuvant (1-3). Administered in saline, MDP increases immune responses against a variety of antigens (4,5). In most of these studies, a single dosage of 100 μg or less was injected simultaneously with the antigen. When higher and repeated dosage of MDP were injected prior to the antigen, a decrease in humoral and cellular immune responses was observed (6). This MDP-induced immunosuppression has recently been shown to be mediated by suppressor T cells (7). In this report, we have further investigated the mechanisms of MDP-induced immunosuppression and have demonstrated that the inhibition of IL2 production constitute a major mechanism by which MDP regulate the immune responses.

II. MATERIALS AND METHODS

A. *Mice*

DBA_2 mice, 8 to 12 weeks old, were obtained from Iffa Credo (L'Arbresle, France).

Interleukins, Lymphokines, and Cytokines

ISBN 0-12-527540-4

B. *Chemicals*

MDP was MDP-Pasteur (Institut Pasteur Production, France). MDP(D-D) and MDP-GnBu were synthesized by P. Lefrancier (Choay Laboratories).

C. *Production of IL2*

Spleen cells from PBS or MDP-treated mice were cultured in RPMI 1640 supplemented with 5 x 10^{-5}M 2 mercaptoethenol, 2 mM L-glutamine, antibiotics, and 5% heat-inactivated fetal calf serum (FCS, Gibco Laboratories, Grand Island, N.Y.). After 2-48 hours stimulation in the presence of ConA (5 μg/ml), the supernatants were harvested and stored at - 20°C until used.

D. *Assay for IL2 activity*

IL2 activity was quantitated in a bioassay using cells of a long term IL2 dependent CTL line, which was kindly donated by F. Denizot (Marseille-Luminy, France). These cells (2.10^4/0.2 ml) were cultured in flat-bottomed microtiter plates in medium containing serial dilutions of culture supernates to be tested for their IL2 activity. A standard rat IL2 preparation was used as a reference. After 48 h, cultures were pulsed overnight with 0.5 μCi of tritiated thymidine (1 Ci/m mol, Saclay, France) and cells were harvested with a multiple sample device (Skatron, Flow Laboratories). The responses were quantitated by probit analysis as previously described (8). One hundred units of IL2-activity were defined as the amount of IL2-containing crude culture supernate that produced 50% of the maximal proliferative response generated by the reference IL2 preparation.

III. RESULTS

Five hundred μg of MDP were injected i.p. into DBA_2 mice on days -3, -2, -1. Control mice received only PBS. This *in vivo* treatment by MDP has been previously shown to inhibit the generation of cytotoxic T lymphocytes and to suppress the PFC response against sheep red blood cells (SRBC)(6). As can be seen on table 1, the ability of spleen cells from MDP-treated mice to produce IL2 after 24 hours ConA- stimulation was almost completely abolished (97.5 % inhibition). This suppression of IL2 production was also observed after 6 hours ConA stimulation, which ruled out the possibility of an earlier synthesis of IL2 by spleen cells from MDP treated-mice.

This inhibition of IL2 production could also be observed after an *in vivo* treatment by 500 µg of MDP-GnBu (an apyrogenic, adjuvant active MDP derivative, 9) on days -3, -2, -1. In contrast, the adjuvant inactive MDP stereoisomer, MDP (D-D) exhibited no effect on IL2 production, when injected under the same experimental conditions (Results not shown).

ConA-activated spleen cells from normal mice absorbed IL2 with the same efficiency as ConA-activated spleen cells from MDP-treated mice (Table II). This clearly indicates that the weak IL2 activity is due to an inhibition of production rather than to an increased in situ absorption of IL2 by MDP treated mice splenocytes.

It has been recently demonstrated that prostaglandins of the E type (PGE) inhibits the production of Interleukin 2 (10). Since MDP has been shown to trigger PGE synthesis when incubated with macrophages (11), it could be hypothesized that the inhibition of IL2 production was mediated by PGE. However, *in vitro* addition of indomethacin, a prostaglandin synthetase inhibitor, to spleen cells from MDP-treated mice had no effect on the observed IL2 suppression (Results not shown). Moreover, administration of 100 µg of indomethacin to mice, 1 h before treatment by MDP, did not prevent the induction of suppression (Table III).

To determine whether IL2 deficiency in MDP-treated mice reflects a macrophage defect, an IL1 containing supernatant was added to the spleen cell cultures. Under these experimental conditions, IL1 did not stimulate IL2 production by ConA-activated spleen cells of MDP-treated mice (Results not shown). This observation suggests that IL2 deficiency is not due to a defective production of this monokine. It has to be noticed that this IL1 preparation can restore the production of IL2 by normal purified T cells activated by ConA. Moreover, highly purified T cells prepared from spleen of MDP-treated mice also produced normal levels of IL2 when stimulated with ConA and IL1. Taken together, these results indicate that the inhibition of IL2 production observed with spleen cells of MDP-treated mice is due neither to a macrophage defect nor to a primary T cell defect.

TABLE I. Suppression of IL2 production by MDP

In vivo treatment on days -3,-2,-1	*Units IL2 / ml*			
	Hours following ConA stimulation			
	2	*6*	*14*	*48*
PBS	*0*	*8*	*80*	*0*
MDP 500 µg	*0*	*0*	*2*	*0*

Table II. Absorption of IL2 by ConA-stimulated spleen cells from PBS or MDP-treated mice

Absorption prior to test		Units IL2 / ml
Medium		50
ConA-stimulated spleen cells from PBS	10^7	45.5
	$2.5\ 10^7$	41
	5.10^7	27
ConA-stimulated spleen cells from MDP-treated mice	10^7	50
	$2.5\ 10^7$	43
	5.10^7	28

Table III. The suppression induced by MDP is not dependent upon prostaglandin synthesis

	Days -3,-2,-1	Units IL2/ml	Percent inhibition
	PBS	60	87
	MDP 500µg	8	
Indomethacin 100µg 1h before treatment	PBS	75	85
	MDP 500µg	11	

We have previously demonstrated that MDP induced-immunosuppression is mediated by the induction of T suppressor cells (7). Thus, the inhibition of IL2 production could be related to the induction by MDP of a cell-mediated suppressive mechanism. In order to investigate this possibility, spleen cells from MDP-treated mice were cultured with normal spleen cells at a 1:1 cell ratio. The total cell density was from 3.75 10^6 cell/ml to 1.5 10^7 cell/ml. As can be seen on table IV, there was no evidence for suppressive cells inhibiting IL2 production when ConA stimulation was performed at 3.75; 5; or 7.5 10^6 cell/ml. However, the IL2 production by normal spleen cells was almost completely abolished when cocultured with MDP-treated mice spleen cells, at a final density of 1.10^7 or 1.5 10^7 cell/ml. These results support the idea that suppression of IL2 production is an active process mediated by a suppressor cell population. The exact nature of this population is still unknown but the suppression is not mediated by Lyt-2 positive cells since the treatment of MDP-treated mice spleen cells by a anti-Lyt-2 plus complement did not restore IL2 production (Results not shown). Thus, the mechanisms of IL2 inhibition induced by MDP seems to be different from the one recently described by Gullberg and Larson (12). In this report, it has been described the ConA induction *in vitro* of suppressor T cells, Lyt-2 positive, which inhibit the initiation of IL2 production by normal ConA-stimulated spleen cells (12). In contrast, we failed to detect any IL2 suppression after *in vitro* incubation of normal spleen cells with MDP, whatever the experimental conditions were (Results not shown). The induction of IL2 suppression by this molecule could be induced only by *in vivo* treatment of mice. Moreover, the evidence for the mediation of suppression by suppressor cells could be obtained only at high cell concentrations, which can favour the idea of a low frequency of these cells in the spleen of MDP treated animals. Alternatively, these data could indicate that cell-to-cell contact is required for the establishment of suppression. However, these results suggests that the cells inducing or mediating the MDP-induced immunosuppression are not present in normal spleen cells.

Table IV. Suppressive effect of MDP-treated mice spleen cells on IL2 production

	Units IL2 / ml		
Cell Concentration / ml	*Spleen cells cultured*		
	PBS treated	*MDP treated*	*PBS + MDP treated **
$1.5\ 10^{7}$	39	4 (90)	10 (74)
1.10^{7}	42	3 (93)	17 (60)
$7.5\ 10^{6}$	47	15 (68)	34 (28)
5.10^{6}	58	10 (83)	47 (19)
$3.75\ 10^{6}$	44	11 (75)	43 (2)

* *cell ratio 1:1*

( ) *percent inhibition*

IV. DISCUSSION

It has been repeatedly demonstrated that the proliferation of cytotoxic T lymphocytes (CTL) is totally dependent upon IL2. Therefore, this interleukin represents the ultimate mechanism of the regulation of CTL responses. The different cell-types and mediators controlling the initiation of IL2 production have been extensively studied and described in the past few years. However, only little is actually known of the mechanisms underlying the amplitude and the duration of IL2 production.

Our present data demonstrate that synthetic adjuvants like MDP, when injected under certain experimental conditions, induced the suppression of IL2 production after lectin-stimulation. This results correlate well with the previously described inhibition of generation of CTL by MDP injected under the same experimental conditions (7). Thus, the immunosuppressive property of MDP could likely be explained by its capacity to induce an inhibition of IL2 production. Moreover, such molecules which mimick bacterial-cell wall structures represents interesting biological probes for the study of IL2 regulation.

REFERENCES

1. Adam, A., Ciorbaru, R., Ellouz, F., Petit, J.F., and Lederer, E., *Biochem. Biophys.Res.Commun. 56,* 561 (1974).
2. Ellouz, F., Adam, A., Ciorbaru, R. and Lederer, E., *Biochem.Biophys.Res.Commun.59,* 1317 (1974).
3. Merser, C., Sinaÿ, P. and Adam, A., *Biochem.Biophys.Res. Commun. 66,* 1316 (1975).
4. Audibert, F., Chedid, L., Lefrancier, P. and Choay, J., *Cell. Immunol. 21,* 243 (1976).
5. Chedid, L., Audibert, F., Lefrancier, P., Choay, J. and Lederer, E., *Proc. Natl. Acad. Sci. USA. 73,* 2472 (1976).
6. Leclerc, C., Juy, D., Bourgeois, E. and Chedid, L. *Cell. Immunol. 45,* 199 (1979).
7. Leclerc, C., Bourgeois, E. and Chedid, L., *Eur.J. Immunol. 12,* 249 (1982).
8. Gillis, S., Ferm, M.M., Ou, W. and Smith, K.A., *J. of Immunol. 120,* 2027 (1978).
9. Chedid, L., Parant, M., Audibert, F., Riveau, G., Parant, F., Lederer, E., Choay, J., and Lefrancier, P., *Infect. Immun. 35,* 417 (1982).
10. Rappaport, R.S. and Dodge, G.R., *J.Exp.Med. 155,*943 (1982).
11. Wahl, S.M., Wahl, L.M., McCarthy, J.B., Chedid, L. and Mergenhagen, S.E., *J. Immunol., 122,* 2226 (1976).
12. Gullberg, M. and Larsson, E.L., *J. of Immunol.,128,* 746 (1982).

DISCUSSION

OPPENHEIM: Is it clear to LeClerc that the protocol she utilized caused an *in vivo* suppression of the immune response as well as the depression of *in vitro* IL-2 production?

LeCLERC: Yes. Using this protocol (MDP 500 µg on days -3, -2, -1), we have demonstrated an inhibition of PFC response against SRBC, or of induction of cytotoxic lymphocytes.

OPPENHEIM: Does the paradoxical effect of these immunoadjuvants involve increased absorption or increased turnover of IL-2?

LeCLERC: No. We failed to detect any increased adsorption or utilization of IL-2 by spleen cells of MDP-treated mice.

WOOD: To extend LeClerc's answer to Oppenheim's query, I would point out that the phenomenon of immunosuppression by *pretreatment* with adjuvants is not at all unique to MDP, but is observed with *most* adjuvants. Has LeClerc considered that the suppression which she observed is mediated by the immunosuppressive acute phase proteins evoked by MDP?

LeCLERC: I agree that this phenomenon is a general one and can be observed with most of the immunostimulants, depending upon experimental conditions. This IL-2 suppression could surely be mediated by the immunosuppressive acute phase proteins. However, the butyl ester derivative of MDP, which induced a comparable suppression to the one induced by MDP, seems rather inactive as regards induction of SAA synthesis. Thus, additional data are needed to determine whether there is a clear correlation between suppression and elaboration of acute phase proteins.

PETER: One suggestion on how to resolve this problem would be to do limiting dilution experiments in order to determine the real frequency of IL-2 producing cells in the MDP-treated mice. Does LeClerc know if this kind of experiment has been done in her system?

LeCLERC: No, this type of experiment has not yet been done. However, the response of purified T cells, isolated from MDP-treated mouse spleen cells, was comparable to the response of normal purified cells. Thus, these data are opposed to the idea of a suppression of IL-2 production due to decrease in the frequency of IL-2 producing cells.

INTERLEUKIN-1 RELEASE BY CANCER PATIENTS' MONOCYTES

Michael Hoffmann
Simon Pollack

Memorial Sloan-Kettering Cancer Center
New York, New York

Release of IL-1 by macrophages can be induced not only by the physiological T cell signal but also by the cell wall lipopolysaccharide (LPS) of gram-negative bacteria (1,2,3), and this has provided the basis on which we have developed a test for this aspect of macrophage function in the mouse (2) and in man (3). The latter takes advantage of the fact that the activity of IL-1 is not species-restricted; IL-1 released by human peripheral blood monocytes (PBM) can be assayed on mouse target cells (1,3). Using such target cells derived from inbred strains of mice, the assay is easily standardized. the use of C3H/HeJ mice (which are unresponsive to LPS) provides added assurance that the activity of IL-1 and not LPS is measured in the assay. Initially, we used antibody production by splenic B-lymphocytes as the indicator of IL-1 release (3). More recently we have found thymocyte proliferation as measured by thymidine incorporation (4) to provide an equally reliable and technically simpler assay which permits more precise quantitative estimates of IL-1 (1,5).

In an initial survey, peripheral blood monocytes of many patients with various types of cancer were found to release less IL-1 in response to LPS than PBM of normal individuals (3). Extending this observation, we have now examined patients with breast cancer, colon cancer and malignant melanoma at various stages of disease.

METHODS AND PATIENTS

Of the 159 patients studied, 36 had breast cancer, 29 benign breast lesions, 38 colorectal cancer, 6 benign colorectal lesions, and 50 malignant melanoma. The following staging criteria were used.

Interleukins, Lymphokines, and Cytokines

ISBN 0-12-527540-4

Breast cancer: Stage I - Tumor confined to the breast; Stage II - Metastases in skin regional soft tissues or lymph nodes; Stages III - Metastases in bones viscera or opposite breast.

Colorectal cancer: Dukes Stage A - Tumor confined to the bowel wall; Dukes Stage B - Invasion through the full thickness of the bowel wall without nodal metastases; Dukes Stage C - metastases in regional lymph nodes; Stage D - Distant metastases.

Malignant melanoma: Stage I - Tumor confined to the primary site or its immediate (5 cm) vicinity: Stage II - Metastases in regional lymph nodes or skin; Stage III - Metastases in distant skin, lymph nodes or one viscus; Stage IV - Metastases in more than one viscus.

In patients with resectable disease, blood for study was generally drawn before surgery. In patients with non-resectable disease, blood was drawn no sooner than 4 weeks after chemotherapy or radiation. In no case was there evidence of therapy-related bone marrow depression at the time of study.

Thirty-five normal individuals served as controls.

INTERLEUKIN-1 PRODUCTION: The method used to measure IL-1 release by monocytes has been described (3,4). Briefly, mononuclear leukocytes were isolated from heparinized blood by density gradient centrifugation on sodium metrizoate/ficoll (lymphoprep, Nyegard, Oslo, Norway). The cells were incubated 5 x 10^6 per 35 mm plastic Petri dish, in 1 ml of RPMI 1640 containing 20% fetal calf serum, at 37°C for 1 hours. Nonadherent cells were washed off, and the residual adherent cells were incubated for another 16-18 hours in 1 ml of serum-free Mishell-Dutton culture medium (6) containing LPS (1 μg/ml prepared from S. abortus by C. Galanos, Freilburg, Germany). The supernatant was centrifuged, passed through a Millipore filter (0.45 uM) and stored at -70°C until tested.

IL-1 activity was determined by measuring the proliferation effect of the supernatant on mouse thymocytes (1,4). Thymocytes from 6-8 week-old C3H/HeJ mice were cultured medium containing 2-meracaptoethanol 5 x 10^{-5}M, to which 0.05 ml of serially diluted (1:1, 1:5, 1:25, 1:125) supernatant was added, at a final concentration of 10^6 cells/0.1 ml, in flat-bottom Falcon II microtest tissue culture plates (no. 3040, Falcon Plastics, Oxnard, Calif.) for 3 days at 37°C. Cultures were performed in triplicate. For the last 4 hours of incubation, ^{3}H-thymidine 0.2 μCi was added. The cultures were harvested on glass fiber filters with an automatic harvester, and the radioactivity of the filters was measured in a scintillation counter.

Results (mean CPM of triplicate cultures) were expressed as percent of the response induced by the supernatant prepared from LPS-stimulated monocytes of three normal individuals. These standard supernatants were stored in small alliquots at -70°C and used for the entire study. The response induced by the standard supernatants was close to the response induced by the supernatants of LPS-stimuatled monocytes from 35 normal controls.

RESULTS

The overall results in 124 cancer patients, 35 patients with benign tumors and 35 normal controls are shown in Figure 1.

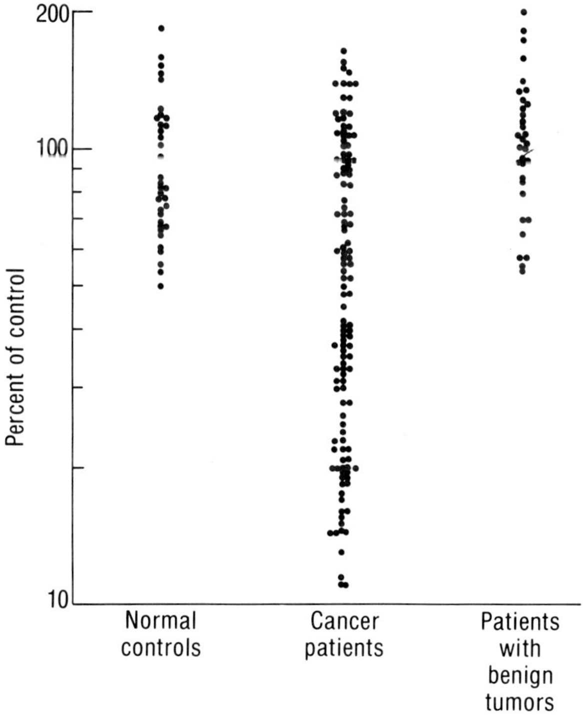

FIGURE 1. Endotoxin-induced in vitro release of IL-1 by monocytes of normal individuals, cancer patients and patients with benign tumors.

IL-1 release by monocytes of patients with benign tumors was found to be within the normal range. By contrast, a wide spread was seen in cancer patients, IL-1 release falling within the normal range in some patients but being lower than normal in others. Subsequently, we related normal or subnormal IL-1 release to the type or stage of cancer.

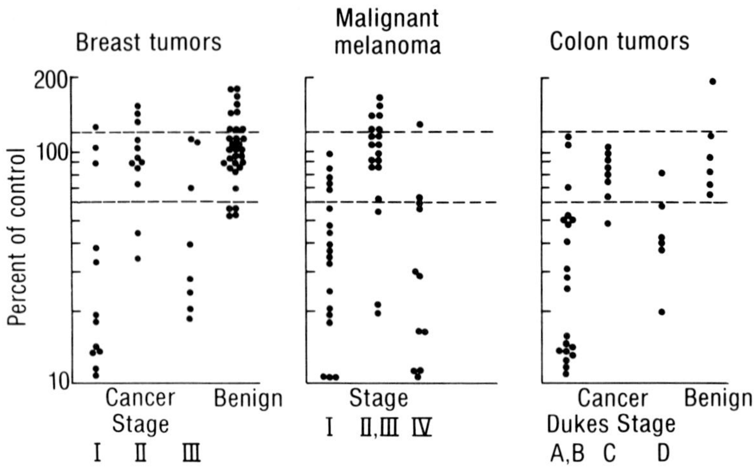

FIGURE 2. Endotoxin-induced in vitro release of IL-1 by monocytes of patients with breast cancer, malignant melanoma, or colorectal cancer at different stages of disease.

The results obtained in patients with breast cancer or benign lesions are shown in Figure 2.

With few exceptions, patients in two groups - Stage II breast cancer and benign lesions - fall in the normal range. By contrast, IL-1 results with monocytes for patients with Stage I or Stage III breast cancer were found to be signficantly lower than normal. Similar results were seen in patients with colorectal cancer or benign colorectal lesions (Figure 2). While patients with Dukes Stage C cancer or with benign lesions showed IL-1 release within the normal range, significantly lower values were found in many patients with Dukes Stage A or B cancer and in patients with disseminated cancer showed subnormal IL-1 values. Analysis of test results in patients with malignant melanoma revealed with same general pattern (Figure 2). Again, IL-1 release was normal in patients with moderately advanced disease - Stage II and III, but significantly lower in patients with Stage I or Stage IV disease.

While the test results reflect IL-1 released by a standardized number of plated peripheral blood leukocytes the possibility had to be considered that difference in IL-1 release paralled differences in the number of circulating moncytes. In Figure 3, plots of monocyte number vs. log of IL-1 release are shown for the melanoma patients, along with 95% confidence band.

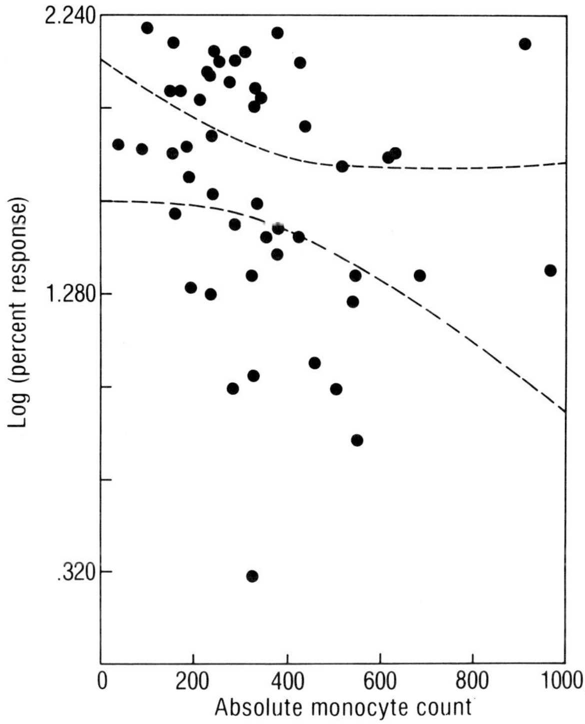

FIGURE 3. Lack of correlation between endotoxin-induced in vitro release of IL-1 and the number of circulating monocytes. Correlation coefficient between monocyte count and logarithm (base 10) of IL-1 release is -0.25 (p=0.088). The band indicates the 95% confidence interval for a linear relationship between monocyte number and log IL-1.

For a given monocyte number, the log of IL-1 release would be predicted to fall within the band with probability .95 if the relationship between monocyte numbers and log IL-1 were linear. As many data points fall outside the band, prediction of IL-1 release by monocyte number would be inaccurate. The results were examined in the same way in the groups of patients with breast cancer or colon cancer (not shown here). No correlation between monocyte number and IL-1 release was found.

DISCUSSION

The results of this study confirm and extend our earlier finding (3) that the release of IL-1 by peripheral blood monocytes in response to LPS is decrease in many but not all cancer patients. We have now found that this change is related to the stage of the disease in patients with breast cancer, colorectal cancer and malignant melanoma. IL-1 values were low in patients with regional extension of cancer, and decreased in patients with disseminated disease. While the results require extension and refinement, the pattern seem sufficiently clear at this point to warrant consideration.

If we assume that these observations made in groups of different patients reflect a sequence of events in each individual patient (an assumption requiring confirmation by longitudinal studies), it seem unlikely the cancer in its earliest stage would cause a general macrophage defect, and that this defect would be repaired as cancer begins to spread regionally. It seems more likely that the decrease in in vitro release of IL-1 by monocytes from patient's with early cancer is the consequence of an in vivo engagement of monocytes (e.g. exhaustion of monocytes or their sequestration to sites other then the blood stream) and that the normal results obtained in patients with regional spread of cancer indicate the breakdown of a recognitive mechanism that is associated with the progression of cancer. In debilitated patients with advanced disseminated disease, on the other hand, a decrease in the Il-1 activity in monocyte cultures supernatants more likley reflects a more complex disturbance which parallels defects in other immunologic functions.

We have previously shown that in patients with cancer in an early stage low IL-1 values are not the result of an inhibitor which interferes with the IL-1 assay (3). More recent experiments indicate a different situation in patients with advanced cancer. Monocytes obtained from these patients and incubated with LPS for 16 hours release an inhibitor whose production can be blocked with indomethacin. This block suggests that the inhibitor released by cancer patient's monocytes belong to the family of prostaglandins. Increased and accelerated prostaglandin synthesis by macrophages has been observed in mice as a consequence of a tumor burden and has been attributed to a state of increase macrophage activity in tumor bearing hosts (7).

REFERENCE

1. Wood, D.W., (1979) *J Immunol 123:* 2400.
2. Hoffmann, M.K., Koenig, S., Mittler, R.S., Oettgen, H.F., Ralph, P., Galanos, C., Hammerling, U. (1979) *J Immunol 122;* 497.

3. Finelt, M., Hoffmann, M.K. (1979) *Clin Immunopathol 12;* 281.
4. Gery, I., Waksman, B.H. (1972) *J Exp Med 136;* 143.
5. Aarden, L.A., Brunner, T.K., Cerottini, J.C., et at, (1979) *J Immunol 123;* 2928.
6. Mishell, R.I., Dutton, R.W. (1967) *J Exp Med 126;* 423.
7. Pelus, L.M. and Bockman, R.S. (1982) *J Immunol 123*, 2118.

DISCUSSION

GILMAN: It is well known that certain cellular immune functions decline with ageing. Recently, in collaboration with Jonathan Rosenberg and Joseph Feldman, I reported that at least part of this age-related decline in T cell function is due to deficient production of IL-2 by cells of aged rats. Similar results have been obtained by Gillis in man and Weigle et al. in the mouse system. My question is, did Hoffmann observe any difference in the capacity for IL-1 synthesis by cells from young healthy individuals compared to healthy elderly individuals?

HOFFMANN: No. We have tested IL-1 release by human monocytes from individuals ranging in age from 12 to 86 years and found no siginficant difference in this regard. We also found no difference in the ability of monocytes from young or aged rats to produce IL-1.

DAYER: When doing Ficoll-Hypaque adhesion, it is a common observation that depending on the diseases under study, or even between normal donors, the percentage of monocytes which adhere can vary greatly from specimen to specimen. How did Hoffmann normalize his data to the number of monocytes in the cultures? Some variations could be explained by the variation of the adhesiveness of the cells even if the same number of lymphocytes and monocytes (Ficoll-Hypaque fractions) were added to wells at the beginning of the test.

HOFFMANN: We have not normalized our data to the actual number of monocytes. The data represent the response of monocytes that adhered to the culture dishes following culture of 5×10^6PBM/well.

WOOD: Igal Gery has shown that the ratio of intracellular and secreted IL-1 can vary significantly, depending on the conditions of macrophage stimulation. Has Hoffmann measured *intracellular* IL-1 to take this into account?

HOFFMANN: We have assessed intracellular and extracellular IL-1 activity in mouse macrophages, but not in human monocytes.

SANDRU: Has Hoffmann any explanation why IL-1 levels are low in the early stages of cancer patients and is at normal levels later on?

HOFFMANN: This matter is dealt with in the report of our work.

MECHANISMS INVOLVED IN THE LOW RESPONSIVENESS OF MICE IN THE IN VIVO RELEASE OF LYMPHOKINES

Ruth Neta

Department of Microbiology
University of Notre Dame
Notre Dame, Indiana 46556

Samuel B. Salvin

Department of Microbiology
University of Pittsburgh School of Medicine
Pittsburgh, Pennsylvania 15261

We have been conducting studies on the variations in the release of lymphokines into the circulation in inbred strains of mice (1-5). We have initiated this work with the hope of obtaining an insight into the role of lymphokines *in vivo* in the development of immunity. Like individual humans, inbred strains of mice vary in their resistance to infections (6). The question was, therefore, asked whether similar variations could be detected in the *in vivo* release of lymphokines in inbred mice.

To induce the release of lymphokines into the circulation, mice of various strains were sensitized intravenously with either one of two antigens: 300 μg cell walls of *Mycobacterium bovis* strain BCG in Drakeol-Tween 80 emulsion or 700 ng dead dried *Candida albicans* in Drakeol-Tween 80 emulsion. The sensitized mice and controls were both challenged three weeks later with the appropriate specific antigen, old tuberculin or cell filtrate of C. albicans, respectively, and bled 4 hours after challenge. The sera were assayed for migration inhibitory factor (MIF) and gamma-interferon (IFN-γ), as previously described (1).

Of the ten murine strains sensitized and challenged with antigens of BCG, five were high responders, namely C57Bl/6J, C57Bl/10J, C57Bl/10SnJ, C57Bl/KsJ, and Sec/1ReJ, and five were

ISBN 0-12-527540-4

TABLE I

In vivo release of lymphokines (MIF and IFN-γ) in inbred murine strains immunized with cell walls of BCG

Strain	Titer	
	MIF	IFN-γ
C57Bl/6J	256	500,000
C57Bl/10J	256	1,300,000
C57Bl/10SNJ	256	1,100,000
C57Bl/KsJ	256	1,300,000
Sec/1ReJ	64	200,000
CBA/CaJ	8-16	19,500
C3H/HeJ	4	850
AKR/J	4	1,100
RF/J	<4	<100
DBA/1J	<4	<100

Mice were sensitized intravenously with an emulsion of 300 μg BCG cell walls in Drakeol-Tween 80 and 3 weeks later were challenged intravenously with OT. In each experiment a pool of sera from 10 mice was used. Titers of MIF are given in reciprocal of serum dilution that caused 20% or more inhibition of migration of macrophages. Titers of interferon are given in reciprocal of serum dilution that produced a reduction of 0.5 log hemagglutination yield of EMC virus. (10 units equivalent to 1 unit of a reference mouse interferon standard obtained from the Antiviral Substances Program of NIAID, NIH).

---

low responders, namely CBA/CaJ, C3H/HeJ, AKR/J, RF/J, and DBA/1J (Table I).

Similarly, high- and low-responders were determined when the mice were sensitized and challenged with antigens of _C. albicans_. The high-responder strains to _C. albicans_ sensitization were the same as the high-responder strains to BCG sensitization, and the low-responder strains to _C. albicans_ sensitization were the low-responder strains to BCG sensitization (Table II).

To establish whether a correlation can be detected in the different strains between high- and low-responsiveness in the _in vivo_ release of lymphokines and the resistance or susceptibility to infections, some of the high- and low-responder strains were infected intravenously with $4x10^4$ - $1x10^5$ viable cells of _C. albicans_. The mice were sacrified at intervals after infection and their kidneys were cultured quantitatively

TABLE II

In vivo* release of lymphokines (MIF and IFN-γ) in inbred murine strains immunized with dead *Candida albicans

Strain	Titer	
	MIF	IFN-γ
C57Bl/6J	128	100,000
C57Bl/10SNJ	128	N.D.
C57Bl/KsJ	128	20,000
Sec/1ReJ	16	N.D.
CBA/CaJ	4	< 100
C3H/HeJ	< 4	< 100
AKR/J	< 4	1,000
RF/J	< 4	< 100

Mice were sensitized intravenously with an emulsion of 700 ng of dead C. albicans in Drakeol-Tween 80 and 3 weeks later challenged intravenously with a filtrate of C. albicans culture. Other details as in Table I.

for the presence of C. albicans colonies. Where the kidneys of the high-responder strains in the release of MIF and IFN-γ were free of the C. albicans organisms at 14 days following infection, the kidneys of the low-responder strains yielded increasing numbers of C. albicans. Most of the mice of these low-responder strains were dead or dying at 3 weeks after the infection. An apparent correlation was, therefore, observed between the ability of a strain to develop high titers of MIF and IFN-γ in the circulation and the ability to resist systemic infection with C. albicans. Inversely, the inability to respond with the presence of the high titers of the two lymphokines in the circulation was characteristic of mice susceptible to infection.

The release of lymphokines, as well as the resistance to C. albicans, is thought to depend on a normal function of thymus-derived lymphocytes. The maturation and proliferation of these cells has been shown to be affected by a thymic hormone - thymosin fraction 5 (7,8). The effect of treatment of mice with thymosin fraction 5 was, therefore, studied. Mice were treated intraperitoneally with an optimal dose, 5 ug of thymosin fraction 5, as experimentally established. Administration of this hormone before sensitization and again before challenge resulted in some of the low-responder strains, e.g. CBA/CaJ or C3H/HeJ, becoming highly responsive to challenge with specific antigen, C. albicans filtrate, with the titers of IFN-γ and MIF equal to those of the high-responder strains, i.e. >64 for MIF and >10,000 for interferon. However, other low-responder strains, such as RF/J, were not affected by treatment with thymosin, in that the titers of the two lymphokines remained very low, <100 for IFN-γ and <4 for MIF. When the high-responder strains C57Bl/10Sn and C57Bl/KsJ were treated similarly with thymosin fraction 5, the titers of the two lymphokines in the circulation were reduced at least four fold. Treatment with thymosin fraction 5, therefore, had different effects in different murine strains.

Footpad swelling 24-48 hours following administration of antigen into the footpads of sensitized mice is another expression of cell-mediated immune reaction that can be measured in vivo in the intact mouse. Following intravenous injection of mice with antigen from C. albicans, the high-responder resistant strain C57Bl/10Sn developed delayed hypersensitivity reaction after footpad challenge with the specific antigen, while low-responder susceptible mice of the C3H/HeJ strain did not show significant footpad swelling 24 hours after the footpad injection with specific antigen. However, while treatment with the thymic hormone caused a decline in the titers of circulatory lymphokines and a decline in resistance in a previously resistant high-responder strain, the delayed footpad reaction following administration of thymosin into this high-responder strain was greatly enhanced. In contrast, treatment with thymosine fraction 5, which resulted in greatly enhanced titers of interferon and MIF in circulation and in development of resistance in a previously susceptible low-responder strain, did not affect the development of delayed footpad hypersensitivity. This delayed reaction was not enhanced in the thymosin treated C3H/HeJ mice which developed high titers of interferon and MIF and from whose kidneys C. albicans cells were absent 14 days after intravenous infection (Figure 1).

In conclusion, inbred murine strains may be high- or low-responders in the in vivo release of IFN-γ and MIF. This high- and low-responsiveness of the different strains occured in parallel to antigens of two microorganisms, BCG and C. albicans.

Administration of thymosin fraction 5 to low-responder strains caused in some a marked increase in the titers of IFN-γ and MIF in vivo and increased their resistance to C. albicans, but did not affect the delayed hypersensitivity footpad reaction. In contrast, administration of this hormone into high-responder strains resulted in a decrease in the titers of IFN-γ and MIF, decreased resistance to C. albicans, but brought about a marked increase in delayed footpad reaction.

FIGURE 1

Effect of thymosin on immunologic responses to *Candida albicans*

Strain	MIF and IFN-γ	Resistance to *C. albicans*	DTH
C57Bl/10SNJ	↓	↓	↑
C3H/HeJ	↑	↑	O

The low-responder strain RF/J was not affected by thymosin treatment, indicating the multifactorial nature of the immunologic deficiencies leading to susceptibility to infections and low-responsiveness in *in vivo* lymphokine production.

ACKNOWLEDGEMENTS

This work was supported by grants AI-16064 and AM-2727 from NIH (SBS) and American Lung Association and IF-32CA06519 from NIH (RN).

REFERENCES

1. Neta, R., and Salvin, S. B., *Cell. Immunol.* *51*, 173 (1980).
2. Neta, R., Salvin, S. B., and Sabaawi, M., *Cell. Immunol.* *64*, 203 (1981).
3. Neta, R., *Cell. Immunol.* *60*, 100 (1981).
4. Neta, R., *J. Interferon Res.* *1*, 361 (1981).
5. Neta, R., and Salvin, S. B., *Infect. Immun.* *34*, 160 (1981).
6. Rosenstreich, D. L., Weinblatt, A. C., and O'Brien, A. D., *CRC Critical Rev. Immunol.* 263 (1982).
7. Schulof, R. S., and Goldstein, A. L., *Adv. Int. Med.* *22*, 121 (1977).
8. Low, T. L. K., Thurman, G. B., Chincarini, C., McClure, J. E., Marshall, G. D., Hu, S. K., and Goldstein, A. L., *Ann. N. Y. Acad. Sci.* *332*, 33 (1979).

DISCUSSION

FAHEY: My query regarding Neta's studies is directed to Wood. Neta's high and low responses are just the opposite of Wood's in the same mouse strains, e.g. C3H and C57. Would he comment on this.

WOOD: I have no explanation for the differences, but I can say that in addition to the strains which I reported on, CBA strains are also high responders to MDP. Accordingly, there would seem to be a perfect *negative* correlation between the results.

ORSON: What is Neta's view on a possible immunologic mechanism explaining the change in infection-susceptibility after thymosin treatment - especially as this did not correlate at all with DTH?

NETA: The explanation I favor is based on a number of situations in which DTH does *not* correlate with resistance (see Salvin et al., 1975). It is entirely possible that the reactive cells do not reach the skin test footpad site and are instead sequestered in lymphoid organs, - spleen, to give an example.

YOSHIDA: Does Neta have an explanation or postulated mechanism for the suppression of *in vivo* production of lymphokines after thymosin treatment?

NETA: I can think of two possibilities -- neither of them clearly convincing. I therefore pass on this query.

YOSHIDA: Can lymphocytes from the treated mice produce MIF normally *in vitro*?

NETA: Lymphocytes from the treated mice *all* produce MIF and IFNγ *in vitro,* but at very low levels.

ASHERSON: My query for Neta is: What is the effect of adult thymectomy on her phenomena. My reason for asking is that adult thymectomy prevents the appearance of T suppressor cells which act at the expression stage and the T acceptor (auxiliary) cell needed for their action. The underlying issue is whether the low reactors lack an effector system or whether they are inhibited by a suppressor system.

NETA: The system that Asherson's query brings to mind is that of Allen and Moore in which BCG-sensitized mice responded by production of granulomas. In their work the strains found responsive were the ones that proved to be high responders in our system. Subsequently Allen and Moore have shown that the high responder strains had increased suppressor cells. Accordingly, much of interest remains to be learned about these interactions.

DIAMANTSTEIN: I am rather surprised to learn that DTH-responses and lymphokine production are dissociated. A few months ago, Kaufmann and Hahn (Institute of Microbiology, Free University, Berlin) reported in J. Exp. Med. that a Ly^+ T cell-clone specific for Listeria was the T-effector cell for DTH and that the same cell produced the entire battery of lymphokines as well.

S. COHEN: It should not be surprising that there is a dissociation between MIF and DTH. There is good evidence, including data based on use of anti-lymphokine antibodies, studies of extracts of DTH sites, effects of anticoagulants *in vitro* and *in vivo*, and dissociation effects in desensitization models, that the major lymphokine driving the cutaneous reaction is MCF. MIF has very little to do with the skin reaction. Neta's results provide good evidence further supporting this analysis. I believe that the message is that infectious immunity and delayed hypersensitivity, though both are aspects of cellular immunity, do not necessarily depend on the same mediators.

LYMPHOKINES AS MEDIATORS OF EXTRINSIC IMMUNOREGULATORY CIRCUITS

Melanie Pulley, Dudley Dumonde

Department of Immunology
St Thomas' Hospital and Medical School
London, United Kingdom

Graham Carter, Bruce Muller

Department of Chemical Pathology
Charing Cross Hospital, London, U.K.

Barbara Southcott

Department of Radiotherapy
Charing Cross Hospital, London, U.K.

Frank den Hollander

Organon International B.V.
Oss, The Netherlands

Intravenous injection of lymphoid cell lymphokine (LCL-LK) in Man produces an acute phase reaction characterised by a dose related pyrexia and neutrophil leucocytosis as well as a lymphopenia and increases in plasma ACTH, cortisol and serum growth hormone, all maximal at 2-6hr. Plasma zinc and iron reaches a trough at 5-12hr and acute phase proteins (CRP and α_1-AG) peak later at 24-48hr. We suggest that the rise in plasma cortisol produces inhibitory effects on T-cell function which are sustained or modulated by the subsequent changes in plasma zinc, iron and acute phase proteins and that by stimulation of these extrinsic homeostatic mechanisms feedback inhibition of lymphokine production may be effected.

ISBN 0-12-527540-4

I. INTRODUCTION

It has been reported (1) that intravenous LCL-LK in advanced cancer patients results in an acute reaction characterised by pyrexia, neutrophil leucocytosis, lymphopenia and increases in plasma ACTH, cortisol and serum growth hormone all maximal at 2-6hr. This early phase is followed by plasma iron and zinc falling to a trough at 5-12hr at which time C-reactive protein (CRP) and α-acid glycoprotein (α_1-AG) start to rise to reach their peak at 24-48hr. Pyrexial stress did not account for other components of the reaction.

This sequence of events supports the view that lymphokines act as mediators of an immune-neuroendocrine network (2) by signalling the state of lymphoid cell activation. Indeed, increases of corticosterone have been found in rats during antigenic competition (3) at which time lymphokines can be detected in the blood. The effects on WBC compartmentation of pharmacological doses of glucocorticoids in Man are similar to those seen after LCL-LK injection. Single parenteral or oral doses produce a profound but transient circulating blood lymphopenia due to retrafficking of recirculating lymphocytes out of the intravascular space and into other lymphoid compartments (4); the effects on neutrophil kinetics are well documented (5-7).

There is abundant evidence that fever is mediated by endogenous pyrogen released from monocytes (8) which has been shown to be identical to lymphocyte activating factor (Interleukin-1) (9). Its production in rats is associated with decreases in both iron and zinc (10) which are thought to be implicated in reactive protein synthesis by the liver (11). Both CRP (12) and α_1-AG (13) are known to have suppressive effects on T-cell function.

In this study we have taken the opportunity to investigate possible homeostatic mechanisms extrinsic to the immune system which may regulate lymphokine production. As Man is extremely sensitive to bacterial endotoxin (14), it has been of paramount importance to show that the responses we describe could not be attributed to this.

II. MATERIALS, PATIENTS AND METHODS

A. Materials and Patients

1. Materials. Four LCL-LK batches (Organon International B.V.) were used, prepared from supernatants of the RPMI 1788 cell line as previously described (15). Molecular weights were in the range 5-300 KD, protein content (amino acid analysis) 14.4-20.0% and endotoxin content (Limulus lysate assay) 13.6-16.0 ng/mg dry weight. A wide variety of lymphokine activities were present (18) including LIF, SRF and MAF (Listeria clearance) (16). A small amount of interferon activity (625 u/mg dry weight)was detected (neutralisation with anti-α-interferon) in the batch designated LCL-LK 4882 but none in the batch used most frequently (LCL-LK 5792).

Control protein (CP) preparations were prepared in an identical way but from culture medium only; protein and endotoxin content per mg dry weight were at least as high as for LCL-LK. Pyrogen-free NaCl (9 gm/l) injections were also given.

2. *Patients*. Twenty-three with advanced cancer (12 breast, 5 prostate, 6 other tumours); aged 15-79. Disease was either unresponsive to other treatment modalities or recurrent.

B. Administration of LCL-LK and Assessments

Intravenous LCL-LK doses were escalated until a 'high systemic response' was obtained, ie a temperature rise of 2.0°C usually accompanied by rigors of 45 min duration, on which the patient was maintained. CP, containing at least as much endotoxin as in the LCL-LK the patient was about to have, was given prior to LCL-LK; injections were all at approx. 10am. 385 LCL-LK injections (2 μg-3.5 mg dry weight) and 23 control injections (12 CP; 11 NaCl) were given.

The fever index (FI) was determined by measuring the area under a fever curve (half-hourly temperatures). Full blood counts were at 0, 4 and 24hr. Time course studies with hourly estimations after CP and LCL-LK were done in six patients. (Additional estimations every 10-15 min during the first 90min in two patients). Endocrine, zinc, iron, CRP, α_1-AG, C3, C4 and C3d (every 10 min during first 60 min) estimations were at the same times.

III. RESULTS

A. Pyrexial response

The mean FI for all LCL-LK injections assessed, including doses that gave less than a 'high systemic response', was 155 compared to 4 (probably due to diurnal variation) for the control injections (0.001 > P). The FI after CP was no higher than after NaCl injections. Despite not standardising the many variables (eg frequency of preceding injections and clinical state) there was a strong dose relationship (0.001 > P for LCL-LK 4882). No correlation was found with the protein or the endotoxin content (all batches considered together).

B. *Effects on leucocyte kinetics*

The mean neutrophil increase at 4hr after all LCL-LK injections was 2.3 x 10^9/l (69%) compared to a mean slight decrease of 0.3 x 10^9/l (7%) after control injections (0.01 > P). The increment was dose related (0.05 > P for LCL-LK 5792); no correlation was found with either the protein or the endotoxin content.

The mean lymphocyte decrease at 4hr after all LCL-LK injections was 0.26 x 10^9/l (8%) compared to 0.04 x 10^9/l (1%) after the control injections. These differences were not significant due to the trough being missed as when the time course was studied the differences between the mean maximum percentage decreases after LCL-LK (65%) and CP (35%) were significant (0.05 > P). The time of both neutrophil and lymphocyte maximum responses varied between 2 and 6hr; counts returned to pre-injection values inside 12 hr. No neutropenia was seen to precede the leucocytosis and there were no significant differences between responses to CP and NaCl.

C. *Endocrine changes*

Mean maximal cortisol increases, attained at 2-4hr, after LCL-LK were 681 n mol/l (245%) compared to 147 n mol/l (42%) after CP (0.01 > P). ACTH levels were estimated in a single patient where there was an increase of 196 ng/l (370%) at 3hr, 1hr before the cortisol peak. There was no increase after CP.

There was a mean difference between the maximum increments of growth hormone after LCL-LK (all reached at 2hr) and those after CP of 8.9 mu/l (0.02 > P).

D. *Effects on iron, zinc and acute phase proteins*

The mean decrease in iron after LCL-LK was 8.2 μ mol/l (46%) and 4.5 μ mol/l (21%) after CP. The mean decrease in zinc after LCL-LK was 4.7 μ mol/l (36%) and 2.4 μ mol/l (20%) after CP. Iron and zinc troughs were both at 5-12hr.

CRP levels started to rise at 5-12hr after LCL-LK to peak at 24hr. The mean increment was 34 mg/l compared to 6 mg/l after CP (0.01 > P). α_1-AG levels commenced to rise at 12hr to peak at 24-48hr. The mean increment was 117 mg/l.

IV. DISCUSSION

Intravenous LCL-LK in Man produces changes extrinsic to the immune system which are likely to have an inhibitory effect on T-cell function. They are not due to endotoxin.

The increase in plasma ACTH indicates that the cortisol secretion is a consequence of hypothalamic stimulation; indeed endogenous pyrogen (interleukin-1) is thought to act on the hypothalamus, probably by the secondary mediation of prostaglandins (17). With knowledge of the *in vivo* effects of glucocorticoids our evidence suggests that LCL-LK-induced lymphopenia is mediated by the rise in plasma cortisol. There is a greater depletion of T-cells from the intravascular space after glucocorticoid administration (4); and furthermore, helper-cells (T_M) are depleted more than suppressor-cells (T_G) (18). Thus lymphokine-induced cortisol secretion may selectively modulate the immune response. The increases in growth hormone are also indicative of immunomodulatory effects (19).

Iron and zinc both have effects on T-cell function as well as being thought necessary for acute phase protein synthesis. Both CRP and α_1-AG are known to suppress T-cell function: CRP selectively binds to T-cells and inhibits their ability to form E-rosettes and their response in mixed lymphocyte culture. It also inhibits the induction of *in vitro* antibody formation to T-dependent antigens by generation of a T-suppressor population (20). α_1-AG suppresses mitogenic responses, antibody formation and mixed lymphocyte responses. LCL-LK-induced secretion of these proteins thus provides another route whereby feedback inhibition of lymphokine production may be effected.

The lack of any significant response to CP showed the reaction was not due to bacterial endotoxin and confirmed the Limulus lysate estimation that there was insufficient endotoxin in the preparations to be pyrogenic. Furthermore, the FI and neutrophil leucocytosis correlated with the dry weight of individual LCL-LK batches but not with the protein or endotoxin content of the preparations. No leucopenia was seen to precede the neutrophil leucocytosis unlike reports of endotoxin injection (16) and although endotoxin is a potent complement activator there was no evidence of this.

This acute response to LCL-LK illustrates mechanisms extrinsic to the immune system whereby lymphokine production can be controlled both in the shorter-term eg cortisol secretion and in the longer-term eg acute phase protein secretion. These homeostatic mechanisms warrant further dissection within the context of therapeutic investigation with a view to modulation of the immune response.

REFERENCES

1. Pulley, M. S., Dumonde, D. C., Carter, G., Muller, B., Fleck, A., Southcott, B. M. and den Hollander, F. C., *in* "Human Lymphokines: Biological Immune Response Modifiers" (A. Khan, ed.) Academic Press, New York (1982), in press.
2. Besedovsky, H., and Sorkin, E., *Clin. exp. Immunol.* 27, 1 (1977).
3. Besedovsky, H. O., del Rey, A., and Sorkin, E., *Clin. exp. Immunol.* 37, 106 (1979).
4. Fauci, A. S., and Dale, D. C., *J. Clin. Invest.* 53, 240 (1974).
5. Thompson, J., and van Furth, R., *J. Exp. Med.* 131, 429 (1970).
6. Athens, J. W., Haab, D. P., Raab, S. O., Mavez, A. M., Ashenbucker, H., Cartwright, G. E., and Wintrobe, M. M., *J. Clin. Invest.* 40, 989 (1961).
7. Dale, D. C., Fauci, A. S., and Wolff, S. M., *N. Engl. J. Med.* 291, 1154 (1974).
8. Atkins, E., Francis, L., and Bernheim, H. A., *Infect. Immun.* 21, 813 (1978).
9. Murphy, P. A., Simon, P. L., and Willoughby, W. F., *J. Immunol.* 124, 2498 (1980).
10. Kampschmidt, R. F., and Upchurch, H. F., *P.S.E.B.M.* 134, 1150 (1970).
11. Powanda, M. C., Cockerell, G. L., and Pekarek, R. S., *Am. J. Physiol.* 225, 399 (1973).
12. Mortensen, R. F., Osmand, A. P., and Gewurz, H., *J. Exp. Med.* 141, 821 (1975).
13. Bennett, M., and Schmid, K., *Proc. Natl. Acad. Sci. USA.* 77, 6109 (1980).
14. Joiner, K. A., and Wolff., *in* "Augmenting Agents in Cancer Therapy" (Hersh, E., M., ed.), p.125. Raven Press, New York. (1981).
15. den Hollander, F. C., van Lieshout, J. I., and Schuurs, A. H. W. M., *Immunopharmacol.* 3, 161 (1981).
16. Verheul, H. A. M., den Hollander, F. C., Dijkhuizen, D.M. Gijzen, J. W. L., Kessels, J. G. M. and Schuurs, A. H. W. M., *in* "Human Lymphokines: Biological Immune Response Modifiers" (A. Khan, ed.). Academic Press, New York (1982), in press.
17. Riveau, G., Masek, K., Parant, M., and Chedid, L., *J. Exp. Med.* 152, 869 (1980).
18. Haynes, B. F., and Fauci, A. S., *J. Clin. Invest.* 61, 703 (1978).
19. Saxena, Q. B., Saxena, R. K., and Adler, W. H., *Int. Archs Allergy appl. Immun.* 67, 169 (1982).
20. Mortensen, R. F., *Cell.Immunol.* 44, 270 (1979).

DISCUSSION

YOSHIDA: In guinea pigs, we showed long ago that intravenous injection of exogenous MIF-rich lymphokines brings about a drastic drop in the number of circulating monocytes. Did Pulley note any effect of her treatment on monocyte numbers?

PULLEY: No, we did not.

YOSHIDA: What kind of *in vitro* lymphokine activities were present in your preparation? Do such activities correlate with the effects that Pulley has seen *in vivo*?

PULLEY: The preparation is from unpurified supernatants of the 1788 line, containing a broad spectrum of lymphokine activities. Amongst these are LIF, active on both human and guinea pig leucocytes; skin inflammatory factors, active in both man and guinea pig, and MAF (Listeria clearance from mouse spleen). One batch (LCL-LK4882) contained a small amount of IFNγ. Negligible amounts of IL-2 were also found. The skin inflammatory activity in man appeared to be correlated with the extent of the acute clinical response.

S. COHEN: Pulley's findings are very interesting, but I am not sure it is fair to ascribe them entirely to lymphokine activity. Such supernatants contain many other components including products of cell metabolism, degradation products, etc. The culture medium controls validate the phenomenon--but do not exclude the participation of some of these toxic products. For that, one needs supernatants from antigen-activated lymphocytes - so that a proper control *is* available.

PULLEY: If one adheres to the designation of lymphokines as products of antigen-activated lymphocytes I accept that the phenomenon may be due to many things other than lymphokine activity. However, it should be noted that 1788 cell walls, when injected into animals, have proved nonpyrogenic.

BERTAGLIO: I would like to know the rationale for treating 23 patients with this heterogeneous preparation of ill-defined lymphokines?

PULLEY: Systemic administration was begun after we had encouraging results from pilot tests involving local injection into metastic breast modules. We thought the range of lymphokine activities present in the preparations might be beneficial for immunocompromised hosts. If one were to use a single, purified, well-defined lymphokine there would be no way of knowing which one would be the most helpful.

Although one will probably be giving inhibitory factors, as well, with heterogeneous preparations such as ours -- it may play a more physiological role *in vivo*, especially if the host is deficient in his capability to produce lymphokines.

KAMPSCHMIDT: What is Pulley's evidence that decreases in plasma iron and zinc affect acute phase protein synthesis?

PULLEY: I have no direct evidence that decreases in plasma iron and zinc affect acute phase protein synthesis. I can only say that the decreases seen are associated with the *subsequent* increases in acute phase proteins. Others have found iron and zinc levels to increase in the liver during an acute phase reaction - but the exact relationship between the changes await further biochemical analysis.

PHENOTYPE AND FUNCTIONAL ACTIVITIES OF IL-2 DEPENDENT CYTOTOXIC T-CELL CLONES

Michael A. Palladino
Margrit P. Scheid
Osias Stutman
Peter V Wussow
Herbert F. Oettgen

Memorial Sloan-Kettering Cancer Center
New York, New York

Il-2 dependent cloned cytotoxic T-cell lines derived from the peritoneal cavity or spleen of immunized mice were examined. By direct cytotoxicity tests and competitive inhibition tests, the lines were found to be specific for H-2D^{d}(CTLL-All) or a unique determinant on BALB/c leukemia RL♂1 (CTLL-R5 and CTLL-R9). Their cell surface phenotype, as determined by protein A-SRBC rosetting and indirect immunofluoresence was Lyt-2,3^{+}. By quantitative absorption, low levels of Lyt-1 was detectable on all cytotoxic T-cell clones. In vivo transfer of resistance to tumor grafts was accomplished locally with CTLL-R5 against RL♂1 (syngeneic graft) and systemically with CTLL-All, combined with IL-2 against sarcoma Meth A (allograft). After exposure to Sendai virus or Newcastle Disease Virus, CTLL-R5 and CTLL-All produced IFN-α,β. After stimulation with PMA and Con A, CTLL-R9 but not CTLL-R5 and CTLL-All produced both IFN-γ *and IL-2.*

INTRODUCTION

Advances in our understanding of the culture conditions that permit continued growth of T-cells have provided the

Supported by NCI grant CA-08748, ACS grants 310,162C and an ACS Junior Faculty Research Award (M.A.P.).

Interleukins, Lymphokines, and Cytokines

ISBN 0-12-527540-4

means of obtaining homogeneous T-cell populations (1). T-cell growth factor (TCGF), subsequently called interleukin-2 (IL-2), supports the proliferation of activated but not resting T-lymphocytes (2,3). IL-2 also permits continuous growth of antigen-specific T-lymphocytes. Some IL-2 dependent T-cell lines have retained specificity and reactivity in continuous culture for over three years without repeated exposure to antigen (4). We describe here some of the characteristics of three cloned cytotoxic T-cell lines recognizing H-2D^d or a unique leukemia specific antigen (4,5).

RESULTS AND DISCUSSION

1. Derivation and Specificity of Cloned Cytotoxic T-Cell Lines

C57BL/6 (B6) mice were immunized with BALB/c sarcoma Meth A. Nonadherent peritoneal exudate cells (PEC) from these mice were established in continuous culture in IL-2 conditioned growth medium obtained from CD rat spleen cell cultures stimulated with Con A for 48 hours. The cultured cells were subsequently cloned by limiting dilution at <0.5 cells/well and tested for cytotoxicity. Direct tests, competitive inhibition tests and antibody blocking tests with clone CTLL-A11 showed reactivity for an H-2D^d determinant on normal and malignant cells (Table 1).

BALB/c x C57BL/6 F_1 ($CB6F_1$) mice were immunized with BALB/c leukemia RL♂1. Spleen cells from these mice were established in continuous culture in IL-2, subsequently cloned by limiting dilution at <0.5 cells/well and tested for cytotoxicity. Direct tests and competitive inhibition tests with clone CTLL-R5 showed specificity for a unique antigen on leukemia RL♂1, probably identical with an antigen recognized by non-cultured cytotoxic T-cells (6). Preliminary studies suggest that clone CTLL-R9 has the same specificity.

2. Cell Surface Phenotype of Cloned Cytotoxic T-Cell Lines

Analysis of the cell surface phenotypes of the three cloned cytotoxic T-cell lines by the protein A-SRBC rosette assay, and by indirect immunofluorescence tests indicated that they maintain the characteristics of the T-cell subset they initially represented (Table 1). By quantitative absorption analyses, low levels of Lyt-1 antigens were detectable on all cytotoxic T-cell clones tested (5,7). As previously suggested (8), the apparent absence of Lyt-1 antigens from some cytotoxic T-cell populations probably indicates low expression rather than lack of expression. Monoclonal antisera directed against the

TABLE 1. Specificity and Cell Surface Phenotype of IL-2 Dependent Cytotoxic T-Cell Clones

Clone[a]	Source of lymphocytes	Strain of origin	Antigenic specificity	Cell surface phenotype[b]
CTLL-A11	Peritoneal cavity	B6	$H\text{-}2D^d$	$Lyt\text{-}2,3^+$
CTLL-R5	Spleen	$CB6F_1$	RL♂1, unique	$Lyt\text{-}2,3^+$
CTLL-R9	Spleen	$CB6F_1$	RL♂1, unique	$Lyt\text{-}2,3^+$

[a] For immunization against $H\text{-}2D^d$ determinants, B6 mice received one intradermal injection of 10^7 Meth A cells. Six days later, the mice were injected intraperitoneally with $5x10^5$ irradiated (10,000 rads) Meth A cells. Peritoneal exudate cells (PEC) were harvested three days later. Nonadherent cells were placed in IL-2 containing growth medium and cloned on macrophage feeder layers by limiting dilution (4). For immunization against leukemia RL♂1, $CB6F_1$ mice were injected intradermally with $5x10^5$ RL♂1 cells. Six weeks after the tumor had been rejected, spleen cells were mixed with irradiated (10,000 rads) RL♂1 cells (ratio 10:1) and cultured for six days. The spleen cells were placed in IL-2 containing growth medium and cloned on macrophage feeder layers by limiting dilution (5).

[b] Lyt-2/3 was demonstrated by protein A-SRBC rosette assays and indirect immunofluorescence tests. Low levels of Lyt-1 were detected by quantitative absorption analyses only.

Lyt-2.2 or 3.2 determinants blocked the cytotoxicity of CTLL-R5 and CTLL-A11 (5), while antisera directed against the Lyt-1 or Ly-5 determinants did not.

3. *Inhibition of Tumor Growth by Cloned Cytotoxic T-Cell Lines In Vivo*

Local transfer of resistance to BALB/c leukemia RL♂1 was accomplished with clone CTLL-R5 (Table 2). The growth of subcutaneously RL♂1 injected cells in BALB/c mice was inhibited when they were mixed with CTLL-R5 cells at a ratio of 1:10. By contrast, another BALB/c leukemia RV1 (radiation virus induced) was not inhibited when mixed with CTLL-R5 cells. These results are compatible with the unique specificity of CTLL-R5 in in vitro tests (9). Systemic transfer of resistance to tumor allografts was investigated with CTLL-A11 which had been shown to be specific for $H\text{-}2D^d$ determinants on BALB/c sarcoma Meth A and other cells (Table 3). While intraperitoneal injection of CTLL-A11 alone, 2 hours after subcutaneous injection of Meth A tumor cells into B6 mice, did not inhibit tumor growth, injection of CTLL-A11 with IL-2 (but not IL-2 alone) was inhibitory. Whether CTLL alone will be effective when injected in larger numbers, remains to be determined. However, recent studies have also demonstrated that exogenous IL-2 was needed for in vitro activity of cytotoxic T-cell lines (10). Whether the IL-2 is needed for the continued clonal expansion of the T-cell lines and/or to alter their migration patterns must also be determined (11).

TABLE 2. Local Transfer of Resistance to Syngeneic Tumor Grafts with Cloned Cytotoxic T-Cells

Injected tumor	*CTLL-R5 admixed*	*Number of mice with tumors/total number*		
		Day 7	*Day 10*	*Day 14*
BALB/c RL♂1	-	*2/9*	*6/9*	*8/9*
BALB/c RL♂1	+	*0/9*	*0/9*	*3/9*
BALB/c RV1	-	*0/5*	*3/5*	*5/5*
BALB/c RV1	+	*0/5*	*3/5*	*5/5*

10^5 tumor cells were mixed with 10^6 CTLL-R5 (washed free of IL-2) and injected intradermally into BALB/c recipients.

TABLE 3. Systemic Transfer of Resistance to Tumor Allografts with Cloned Cytotoxic T-Cells Combined with IL-2

BALB/c sarcoma Meth A	*CTLL-A11 IP*	*IL-2 IP*	*Tumor diameter (mm) on Day 7*	*Day 10*	*Day 14*
+	-	-	53	49	21
+	-	+	50	54	26
+	+	-	49	56	28
+	+	+	27	31	17

10^7 Meth A cells were injected subcutaneously into B6 mice. $5x10^6$ CTLL-A11 or 100 units of IL-2 (DEAE-Sephadex purified IL-2 from Con A stimulated rat spleen cells provided by Collaborative Research, Waltham, MA) were injected intraperitoneally two hrs later as indicated.

4. *Lymphokine Production by Cloned Cytotoxic T-Cell Lines*

After stimulation with Sendai virus or Newcastle Disease virus clones CTLL-A11 and CTLL-R5 produced IFN-α,β (>1000 units/ml). After stimulation with phorbol myristate acetate (PMA), also called tumor promoting agent (TPA), and Con A, clone CTLL-R9 produced both IFN-γ (>1000 units/ml) and IL-2 (>50 units/ml). Unstimulated CTLL-R9 cells, however, were absolutely dependent on exogenous IL-2 for continued proliferation. CTLL-A11 and CTLL-R9 was tested for production of IL-3, with negative results (Table 4). The relationship between the production of IFN-α,β, IFN-γ, IL-2, IL-3 and CSF by murine T-cell clones, T-cell lymphomas and human peripheral blood cells requires clarification and is currently under investigation (12).

Cloned cytotoxic T-cells maintained in IL-2 provide stable reagents for the investigation of T-cell cytotoxicity in vitro, inhibition of tumor growth by T-cells in vivo, and lymphokine production by T-cells.

TABLE 4. Lymphokine Production by Cytotoxic T-Cell Clones[a]

Clone	Lymphokine IFN-α,β	IFN-γ	IL-2	IL-3
CTLL-A11	+	-	-	-
CTLL-R5	+	-	-	ND
CTLL-R9	ND	+	+	-

[a]*T-cell clones (10^5 cells/ml) were stimulated with Sendai virus, Newcastle Disease virus, PHA or PMA and Con A. The levels of lymphokines were measured on days 1, 2 and 3 after induction. IFN's were measured by a CPE assay on L929 cells challenged with 10 PFU's VSV per cell. The IFN's were classified by neutralization with anti α,β antisera and crossreactivity on guinea pig fibroblasts. IL-2 was assayed on CTLL-A2, an IL-2 dependent cytotoxic T-cell line (5). IL-3 was assayed for its growth factor activity on a Mast cell line 320, clone 5 (R. A. Miller and O. Stutman, in preparation).*

ND - not done.

ACKNOWLEDGMENTS

We thank Dr. R. A. Miller for performing IL-3 assays, Mr. R. Geismar for excellent technical assistance.

REFERENCES

1. Gillis, S., and Smith, K.A. *Nature 268*, 154 (1977).
2. Morgan, D.A., Burscetti, F.W., and Gallo, R., *Science 193*, 1007 (1976).
3. Anderson, J., Gronvik, K.-O., Larsson, E.-L., and Coutinho, A., *Eur. J. Immunol. 9*, 581 (1979).
4. Palladino, M.A., Obata, Y., Stockert, E., Old, L.J., and Oettgen, H.F., *Fed. Proc. 40*, 957 (1981).
5. Palladino, M.A., Obata, Y., Stockert, E., and Oettgen, H.F. In press.
6. Nakayama, E.H., Shiku, H., Takahashi, T., Oettgen, H.F., and Old, L.J., *Proc. Natl. Acad. Sci. USA 76*, 3486 (1979).
7. Palladino, M.A., Carroll, A.M., deSousa, M., Gillis, S., Scheid, M.P., and Oettgen, H.F. In press.

8. Nakayama, E.H., Shiku, H., Stockert, E., Oettgen, H.F., and Old, L.J., *Proc. Natl. Acad. Sci. USA 76*, 1977 (1979).
9. Palladino, M.A., Old, L.J., and Oettgen, H.F., *Fed. Proc. 41*, 406 (1982).
10. Cheever, M.A., Greenberg, P.D., Fefer, A., and Gillis, S., *J. Exp. Med. 155*, 968 (1982).
11. Carroll, A.M., Palladino, M.A., Oettgen, H.F., and deSousa, M., *Fed. Proc. 41*, 303 (1982).
12. Palladino, M.A., Wussou, P.v., Stewart, W.E., and Oettgen, H.F., *Int. J. Immunopharmacol. 4*, 253 (1982).

DISCUSSION

LATTIME: Since the Rlol specific cytotoxic cell line plus exogenous IL-2, produces interferon, and since RLol is an NK-sensitive tumor, might not Pallidino's reduction in RLol growth be due to boosted NK activity.

PALLADINO: That may very well be taking place.

FAHEY: In the tumor treatment-system described by Palladino, in view of the rapid removal of IL-2 after injection, does more intensive IL-2 administration help anti-tumor activity?

PALLADINO: We have given IL-2 repeatedly (about 5 times) without further effect.

FAHEY: If the tumor is placed intraperitoneally does the cytotoxic T cell line prove to be more effective?

PALLADINO: We have not explored that possibility.

OPPENHEIM: Stadler and I have had a running argument for two years regarding the observation made by Palladino, e.g. whether an IL-2-dependent population can produce IL-2. Has he ascertained via use of ^{125}I-IL-2 whether all, or a proportion of his cells, express receptors for IL-2? A cell that both produces and reacts to IL-2 should theoretically be malignant. Is that the case?

PALLADINO: We have not examined the proposition whether an IL-2-dependent line can produce IL-2, nor do we know whether all or only a portion of our cells express IL-2 receptors. I can however tell Oppenheim that the IL-2 independent line is *not* malignant; it will not grow in animals.

OPPENHEIM: I did not know lymphocytes could produce βIFN. How did Palladino ascertain this?

PALLADINO: We used anti-αIFN and anti-βIFN antisera and also by crossreactivity on guinea pig fibroblasts. It has been confirmed by other groups at Sloan-Kettering that they make all three types of interferon. We have not separated the α and β so we do not know what concentrations of either is present in that preparation.

THE LYMPHOKINE-ACTIVATED KILLER CELL PHENOMENON: IN VITRO AND IN VIVO STUDIES

E. A. Grimm
E. Gorelik
M. M. Rosenstein
S. A. Rosenberg

Surgery Branch
National Cancer Institute
Bethesda, Maryland

I. ABSTRACT

In both mouse and man, incubation of fresh lymphocytes in partially purified and lectin free preparations of IL-2 results in the development of lytic activity against fresh autologous and allogeneic solid tumor cells[1,2] including those which are NK resistant[3]. Experimental evidence (3) has suggested that the human lymphokine activated-killer cells (LAK) represent a class of cytotoxic effector cells distinct from either the classical cytotoxic thymus derived lymphocytes or natural killer cells (NK). Lysis of tumor cells occurs in a 4 hour chromium release assay and is not blocked by staph protein A. While the LAK effectors display the serologic phenotype of CTL, the precursors remain undefined. In the human, the precursors are negative for the pan T markers of OKT3 and Leul, and also are negative for the NK marker, OKM1. In contrast, preliminary evidence in mice indicates that the LAK precursor is sensitive to monoclonal anti Thyl antibody plus complement.

Development of LAK cells in man is sensitive to irradiation and requires 2 - 3 days of culture, indicating that a proliferative or radiation sensitive differentiation signal is required. The signal(s) responsible for LAK cell activation copurifies with human IL-2 through ion exchange and molecular sieve chromatography and is not destroyed by acid

ISBN 0-12-527540-4

pH conditions that destroy the leukocyte interferon contained in our IL-2 preparations.

To test the role of LAK in vivo, highly concentrated supernatants containing murine IL-2 were injected IV into C57BL/6 mice prior to inoculation of syngeneic B16 melanoma or Lewis lung carcinoma cells. In both systems the IL-2 was highly effective in inhibiting the development of visible lung metastases ($p. \leqslant .01$). This effect was not ablated by administration of anti asialoGM_1 antibody, which alone caused a doubling of lung metastases. Thus LAK cells represent a highly effective anti tumor lytic system in vitro, and our preliminary data suggests a possible role in vivo, especially in control of blood-borne metastases.

II. INTRODUCTION

Interleukin-2 (IL-2) preparations have been shown to contain a signal(s) that plays several roles in the development and growth of cytotoxic lymphocytes. It has been shown in both mouse and human systems that IL-2 (a) can support long-term growth of cytotoxic lymphocytes, and (b) can function as the "second signal" in the differentiation of primary cytotoxic T lymphocytes (CTL). We have recently reported a third property of IL-2, the direct activation of killer cells (in the absence of exogeneous antigen) that have the unique property of efficiently lysing NK-resistant fresh solid tumor cells (3). All previous attempts using classical in vitro sensitization procedures to generate killer cells to fresh human tumors have been either unsuccessful or reproducible. Therefore, we believe that the ability of IL-2 to generate these lymphokine activated killer cells (LAK) provides a great potential for immunotherapy.

III. LYMPHOKINE-ACTIVATED KILLER CELLS LYSE AUTOLOGOUS NK RESISTANT FRESH SOLID TUMOR CELLS

Single-cell suspensions were prepared from a variety of human tumors using mechanical and enzymatic means (3). These tumor cells were immediately cryopreserved in human serum until used as targets in 4 hour chromium ^{57}Cr release assays.

Table I. Fresh Human Tumors are Insensitive to NK but Sensitive to LAK

	Lytic units per 10^7 effectors[a]		
Target Cells / Effector Cells	K562 / Fresh PBL	Auto Tu / Fresh PBL	Auto Tu / LAK
Patient			
Melanoma	190	0	1,560
Leiomyosarcoma	120	0	200
Synovial Cell Sarcoma	13	0	1,000
Liposarcoma	3	0	62
Synovial Cell Sarcoma	15	0	20
Rhabdomyosarcoma	16	0	154
Malig. Fibrous Histiocytoma	7	0	164
Sarcoma	147	0	60
Sarcoma	50	0	169
Sarcoma	50	0	2

[a]One lytic unit is defined the number of effector cells capable of causing 33% lysis of 5×10^3 tumor cells. Fresh PBL were tested at 100:1, effector to target ratios and lower.

Peripheral blood leukocytes (PBL) were collected from the tumor patients, and then simultaneously tested (a) to express classical natural killer activity (NK) by their ability to lyse K562 target cells; (b) to express NK activity by direct lysis of autologous fresh tumor or (c) by their ability to generate LAK cells capable of lysing autologous fresh tumor. LAK were generated as previously described (3) by incubation of PBL in partially purified and lectin free IL-2(4) for up to 7 days. Table 1 shows that in 10/10 combinations, the patients contain classical NK cells which lyse K562 targets, but do not contain cells which lyse fresh autologous tumor. In all cases the fresh tumors were lysable, since they were lysed by the LAK cells.

IV. IS IL-2 THE SIGNAL RESPONSIBLE FOR LAK ACTIVATION?

LAK activation by partially purified and lectin free IL-2 is radiosensitive and requires at least three days of culture (3). To determine the nature of the lymphokine responsible for LAK activation, we performed several experiments. Initially we reported that human IL-2 preparations depleted of all interferon retained stimulatory activity; also autologous IL-2 preparations were highly active (3). Purification via the scheme reported by Meir and Gallo (5) through saturated ammonium sulfate and DEAE ion exchange chromatography, revealed that the LAK stimulatory activity copurified with the IL-2 activity.

When our partially purified IL-2 preparations were absorbed with IL-2 dependent cell lines, the IL-2 and the LAK stimulatory activity were significantly diminished. While it remains to be tested whether human IL-2 purified to molecular homogeneity would activate LAK, all current evidence strongly supports the hypothesis that IL-2 is the lymphokine responsible.

V. SEROLOGIC PHENOTYPE OF LYMPHOCYTES REQUIRED FOR LAK DEVELOPMENT AND EXPRESSION

In both mouse and human, we have shown that LAK effectors display the classical CTL phenotypes of Thy +, OKT3, and OKT8+ respectively (2, 3). However, when these studies were extended to identify the serologic phenotype of the precursor, an anomaly became apparent: the pretreatment of the murine spleen precursor population with anti Thy plus complement eliminated the generator of LAK (Table II); while pretreatment of human PBL with either pan T cell reagent OKT3, or LEU 1 plus complement lysed up to 95% of the cells, but resulted in a dramatic augmentation of LAK generation (Table III). To show that our reagents were used at optimal concentrations, both experiments were performed in parallel with allogeneic CTL activation.

Table II. Murine LAK Precursors are Thy Sensitive

		Lytic Units Per 10^6 Cells	
TREATMENT	% Cells[a] lysed	102 Tumor[b] Targets IL-2 Culture	H-2d P815 Targets Post Allo Sens
None	15	2.5	100
C'	30	5.0	105
Anti Thy + C'	57	<.01	<.01

[a]C57BL/6 splenic cells were treated as shown and then used as responders for LAK by culture with murine lectin free IL-2, or for generation of CTL by culture with allogeneic DBA/2 spleen cells for 4 days.

[b]The 102 tumor target is an NK insensitive C57BL/6 methylcholanthrene induced tumor.

Table III also shows that the human LAK precursors are insensitive to the monoclonal antibody OKM1, reported to bind a majority of NK cells. In other tests (data not shown) we have been able to eliminate NK activity to K562 with OKM1, Leu 7 and with OKT 11, and still generate LAK cells. Therefore, this data suggests the human LAK precursor is neither a mature T cell, nor a classical NK cell. The sensitivity of murine LAK cells to anti Thy does support a T cell lineage, and perhaps the human reagents are not as broad in specificity for all human T cells as the monoclonal anti Thy reagent is for all murine T cells.

Table III. Human LAK Precursors are Insensitive to Anti T Cell Antibodies

TREATMENT[a],	% CELLS LYSED	LYTIC UNITS PER 10^6 CELLS: Tumor Targets — Activation with IL-2[b]	LYTIC UNITS PER 10^6 CELLS: Allo PBL — Activation with Allo PBL[b]
NONE	23	11.0	10.5
C'	32	-	8.7
OKT3 + C'	79	62.0	<1
OKM1 + C'	40	9.0	20
LEU1 + C'	86	66.0	<1

[a]Whole PBL were used after removal of plastic and nylon wool adherent cells

[b]Cultures were for 7 days with either IL-2 or Allo PBL. The IL-2 was generated from EL-4 cells cultured with phorbol myristate acetate (PMA, 10 ng/ml) for 48 hours in serum free conditions, then concentrated 100-fold.

VI. ANTIMETASTATIC EFFECT OF INTERLEUKIN-2 IN MICE

To test whether IL-2 would have an impact on tumor growth in vivo, possibly by generating LAK cells in situ, C57BL/6 mice were inoculated IV with various amounts of murine IL-2 for several days (Table IV). These mice were then injected with B16 melanoma cells. One group of animals received more IL-2 after the tumor injection. Fifteen days later mice were sacrificed and lung metastases were counted. A three to six fold decrease in the number of metastases was observed in the IL-2 treated mice.

TABLE IV. Inhibition of Metastases Formation in IL-2 Treated Mice Following IV Inoculation of B16 Melanoma Cells

Group	Treatment[a]	Median Number of Metastases in the lungs[b]	Range
1	Nontreated	48	27-76
2	IL-2. 64 U days -4 to -1	12 ($p<.01$)[c]	4-17
3	IL-2, 32 U days -4 to -1	11 ($p<.01$)	2-56
4	IL-2, 32 U days -4 to -1 and days +1 to +3	8 ($p<.01$)	1-25

[a]C57BL/6 mice were inoculated IV with 1.0 or 0.5 ml of IL-2 (64 U or 32 U) for 4 days. Twenty-four hours after the last treatment mice received 7.5 x 10^4 B16 melanoma cells IV; group 4 received an additional 3 daily injections of IL-2 (32 U) IP, starting one day after tumor cell inoculation.

[b]The number of metastases in the lungs was counted 15 days after tumor cell injection (9-10 mice/group).

[c]Significantly different from group 1, according to Mann-Whitney U test. There was no significant difference between groups 2, 3 and 4.

In other experiments (data not shown) treatment of mice with IL-2 only at time periods after tumor inoculation had no significant influence on the number of lung metastases. Pretreatment of mice with anti asialo GMI alone suppressed NK activity and caused a dramatic increase in B16 lung and liver metastases; however, the animals also receiving IL-2 generated no metastases in the liver, and a significantly reduced number in the lungs (to be published elsewhere). This information indicates that the IL-2 is not acting via an NK cell potentiation mechanism, since its effect is still apparent when all detectable NK activity is eliminated. Other experiments with the spontaneously metastasizing Lewis lung carcinoma have also shown encouraging results in control of metastases by IL-2. Efforts are currently in progress to determine whether conditions exist in which IL-2 can control the growth of established tumors, by continuous release of large quantities of purified IL-2 at the tumor site. VII.

CONCLUSIONS

Substantial in vitro lysis of both mouse and human solid tumors by autologous LAK cells has provided encouraging prospects for the use of IL-2 in the control of tumor growth. These LAK cells are distinct from NK cells and appear to express some T cell characteristics; however, in the human they also display properties distinct from classical CTL such as a rapid kinetics of activation, the lack of need for an antigenic stimulus, and precursor insensitivity to anti T cell antibodies and complement. Murine in vivo studies have shown that IL-2 can control blood borne metastases. It is quite possible that IL-2 will prove to be an efficient immunotherapeutic agent.

References

1. Lotze, M.T., Grimm, E.A., Mazumder, A., Strausser, J.L. and Rosenberg, S.A. Cancer Research. 41, 4420-4425 (1981).

2. Yron, I., Wood, T.A., Spiess, P.J. and Rosenberg, S.A. J. Immunol. 125, 238 (1980).

3. Grimm, E.A., Mazumder, A., Zhang, H.Z. and Rosenberg, S.A. J. Exp. Med. 155, 1823 (1982).

4. Grimm, E.A. and Rosenberg, S.A. in "Isolation, Characterization, and Utilization of T Lymphocyte Clones". (Fathman and Fitch, eds.) p. 57, Academic Press (1982).

5. Meir, J.W. and Gallo, R.C. J. Immunol. 128, 1122 (1982).

DISCUSSION

LATTIME: Does the LAK cell kill fresh solid tumors other than the syngeneic?

GRIMM: Yes, allogeneic as well as autologous fresh tumor cells are lysed, equally well.

LATTIME: Can Grimm separate the killing of the NK-sensitive targets from that of the syngeneic tumor by removing such populations with NK-reactive antisera?

GRIMM: Yes. Pretreatment with OKT-11, OKM-1, or OKH-1, Leu 7, with complement, did eliminate all detectable NK activity towards K562 from fresh PBL. After culture in IL-2 preparations, these remaining cells generated LAK as well as untreated cells and, in some cases, actually generated an augmented LAK activity.

ORTALDO: Has Grimm ruled out a suppressor cell since the LAK cell emerges in about 2-3 days?

GRIMM: No, this has not been tested.

ORTALDO: Does Grimm deplete K652 killing via NK depletion with growth HNK1 + OKM1) at day 2-3?

GRIMM: Yes, and I believe that K562 can be lysed by LAK simply because it *is* a tumor cell and LAK *can* lyse *all* tumor cells.

ORTALDO: The Daudi cell line is sensitive to purified NK cells.

GRIMM: I am only aware of reports to the opposite, and my Daudi line, which is grown in human AB serum, and is free of mycoplasma, is *not* sensitive to NK cells, even in whole PBL populations, at very high effector to target ratios of 100:1.

ORTALDO: One should be careful in concluding that LAK is a LGL (NK cell) or T cell -- as markers are often inconsistent regarding LGL.

GRIMM: Yes, I can only conclude that the LAK phenomenon is quite distinct from the classical NK or CTL phenomena.

LANDY: The choice of normal target cells as controls is, of course, of critical importance in studies such as the one Grimm has conducted. I have long felt that primary explants of normal tissues constitute the most rigorous kind of control in this regard. Has Grimm examined the effects of her LAK cells on *primary explants* of normal tissues as targets?

GRIMM: Yes, I have tested fresh single-cell preparations of normal lung, liver, and bowel from cancer patients. These patients' fresh tumor cells were lysed, but the autologous normal tissue cells remained intact.

SCHRADER: Do murine LAK's kill Con-A blast cells?

GRIMM: I do not know that this has been tested - but I rather doubt that they do.

VADAS: Which cell type is generated when IL-2 and antigens are present?

GRIMM: Both CTL and LAK are generated.

VADAS: Does Grimm find granules in her LAK cells?

GRIMM: Yes, a few perinuclear granules are evident. These granules appear to be colorless in our preliminary studies.

GERARD: Has Grimm tried to separate lymphocytes on percoll gradients to look for the precursors of LAK?

GRIMM: Yes, the LAK *do* sediment around the 47% percoll band.

GERARD: We too have done that experiment and looked for cytotoxicity against a melanoma cell line after 3 days. Surprisingly we found that the fraction, where LGL are sedimented, needs no added IL-2 to be active against the melanoma cell line; the other fractions become active *only* if IL-2 was added (the order of activity being in the range of increasing density). We have essentially the same results if the cells are cultured in presence of fetal calf serum or in serum-free medium (to avoid the inhibiting effect on NK activity of IgG contained in human serum).

GRIMM: We know that a variety of cytotoxic mechanisms are operative in the lysing of cultured cells. Therefore we have not performed the experiments that Gerard describes. It is of interest that we see no inhibiting effect by human AB serum, and can efficiently lyse fresh tumor cells as measured in a 4 hour chromium release assay. The cells we find in the 47% percoll band do become activated to lyse fresh tumor after 3 days in human serum and IL-2. The higher density fractions will develop LAK activity only if other stimuli, in addition to IL-2, are included -- such as allogeneic stimulator cells and macrophages.

CHAIRMEN'S COMMENTS

These presentations noted the effects of lymphokines administered in vivo both in animals and in man, as well as possible relationships to disease. *Wood* and associates reported that the muramyl dipeptide (MDP) has an adjuvant effect that is mediated through the stimulation of IL-1 release in vivo. Soluble amyloid A protein (SAA) elevation was used as an indirect measure of IL-1 effect. Mouse strain differences were noted in adjuvant effect of muramyl dipeptide on primary antibody response to BSA. *Leclerc* reported that production of IL-2 was reduced in some mouse strains after MDP administration. Initial studies on the possible mechanism failed to show a change in IL-2 receptor expression or T suppressor cell or PGE2 production that could account for this change. *Neta* also reported mouse strain differences in interferon and migration inhibitory factor production following tuberculin challenge to mice sensitized with BCG. Thymosine fraction 5 increased lymphokine production in some deficient strains when administered both at the time of sensitization as well as at the time of tuberculin challenge. *Dumonde* reported on initial studies with an in vitro system for measuring trans-endothelial leukocyte chemotaxis. Evidence was presented that lymphokines acting at the site of lymphocyte-endothelial cell interaction influence the induction and regulation of T cell immunity.

Cancer related studies were the focus of several reports. *Hoffman* noted that monocytes from early and late stages but not intermediate stages of cancer showed decrease of IL-1 production (in response to LPS stimulation). This was true for breast cancer, melanoma and colon cancer. *Pulley* reported on a Phase I clinical trial of lymphokine preparations from lymphoblastoid cell line RPMI 1788. Giving 250 micrograms intravenously daily or on alternate days to 23 cancer patients induced fever, and increase in several pituitary hormones and plasma cortisol by 2 to 4 hours, and a subsequent decrease in plasma zinc and iron, and rise in acute phase proteins. Immunoregulatory effects via the hypothalamus and pituitary were postulated. *Grimm* reported in vitro evidence that lymphokines can stimulate human and murine lymphokine activated killer cells (LAK), a distinct class of cytotoxic effector cells. The LAK cells are cytotoxic against fresh autologous and allogeneic solid tumor cells. Intravenous administration of IL-2 containing lymphokine preparations prior to innoculation of murine experimental tumors caused increased LAK activity and marked inhibition of visible lung metastases, indicating that such a system might be effective in controlling blood-borne metastases.

The discussion of these papers indicated great interest in the role of lymphokines in disease mechanisms and as possible modes of therapy. These will be lively and fruitful areas.

V. Paetkau
J. Fahey

SECTION X: OVERVIEWS AND CONCLUSIONS

LYMPHOKINES: PAST, PRESENT, AND FUTURE[1]

Stanley Cohen

Department of Pathology
University of Connecticut Health Center
Farmington, Connecticut

It is an almost impossible task to summarize a meeting such as this, in which so much new and exciting information has been presented. For this reason, I will confine my remarks to a consideration of where we have been and where we are going. Since this approach is based upon prediction, I will begin with some of my past predictions. Only after we evaluate my previous guesses will we attempt to predict the future. This will impose a sort of "confidence limit" or standard error" on my ramblings.

At the time of the Second International Lymphokine Workshop, I made a large number of predictions (1). First, I predicted that there would be a Third International Lymphokine Workshop. This was correct! Second, I predicted that answers would be available to a number of questions. For example, what are the restrictions in the subpopulations of cells that can make lymphokines? Also, is any given cell restricted to the production of a single lymphokine? Session 8 dealt, in one way or another, with issues related to this. The ability to construct hybridomas secreting specific sets of mediators is becoming almost routine. This provides exciting possibilities for ascribing functional behavior to specific macromolecules. We still can't really answer the second part of the question I posed using hybridomas, however, because of the problems of phenotypic restriction on the one hand, and uncertainty about monoclonality on the other. Experiments involving direct gene transcription will be required.

[1] *Concluding Remarks: Third International Lymphokine Workshop*

Interleukins, Lymphokines, and Cytokines

ISBN 0-12-527540-4

Another question I expected to be answered by this time related to the actual number of discrete lymphokines. The answer to that is easy--too many! We haven't even approached an estimate of the number of fundamental molecular entities (the well-known "Mother Factors" of Janicki) that comprise the set of mediators that we call lymphokines. A question that is related to this is which mediator do we blame for a given *in vivo* activity when many are present. This is of obvious importance for any attempt to understand pathogenesis both in experimental models and in human disease. The answer, in the past, has been based on fashion and fad, rather than scientific evidence. First it was complement, then Migration Inhibition Factor (MIF), then chemotactic factors, then interferon, and now Interleukin 2. Two years from now it will be Macromedin 3a, which hasn't even been discovered yet!

Another question I raised at the previous Workshop was how many molecules of mediator are required to activate a cell. The previous hints about single hit kinetics remain, but there is no new data on this point. Studies in other fields, such as those demonstrating molecular requirements for interaction between complement and immunoglobulin (2) show that this kind of question is answerable even in the absence of purified reagents.

What is the nature of the interaction of mediator and target? Here we are in much better shape. Since we are finally at the stage where pure mediators are becoming available, it is only a matter of time before we can use these as probes for the study of their cellular receptors. A related question deals with the nature of the internal activating events in the cell, both in effector lymphocyte and target cell. As was reported at this meeting (3), our laboratory has just begun to look at lymphocyte proliferation from this point of view. Other functions, such as those dependent upon or involving lymphokines, must await the availability of functionally and chemically pure mediators as probes.

Another major question at the time of the Second Workshop was the number and nature of the regulatory networks governing lymphokine production and the expression of lymphokine activity. It is now clear that the situation is much more complicated that that for antibody. Complex multicellular feedback and feedforward loops are involved (4). Only the broad features of these are known at present. However, rapid progress is being made in this area and I think that we can expect detailed answers by the time of the next Workshop.

I also indicated, three years ago, that it was likely that some of the mediators would be found to be released by the cell in precursor form. In addition to the examples of Eosinophil Chemotactic Factor (ECG_p) (5) and a subclass of murine MIF (6) that I used as illustrations at that time, we have heard in Session 7 of this meeting that chemotactic factors may exist in precursor forms that are activated by macrophage-dependent mechanisms (7). It is likely that Dr. Postlethwaite's studies along these lines will have important implications for other lymphokines as well. A related prediction, namely that we would know a lot about lymphokine subunit structure by now, was not quite as much on target as the one about precursors.

Finally, I emphasized that not only could different assay systems measure different aspects of the same mediator, they could, in theory, even measure opposite properties. I gave an example involving a well-defined synthetic tripeptide with both chemotactic and migration inhibitory properties. Another example from the present meeting involves the dual presence of endothelial chemokinesis activity as described by Dr. Watt (8) and endothelial cell migration inhibition as described by Dr. M. Cohen (9) in Sessions 4 and 7, respectively. Even if the supernatant sources and target cells were identical, which they were not, these results could be explained by a single factor, because one assay involved a single cell (gold track assay), and the other a cellular aggregate (agarose droplet and capillary tube assays). Moreover, in both cases, there are variable contributions of effects on locomotion per se and effects on cellular adhesiveness either to other cells or the supporting matrix. In the specific example cited, the endothelial cell likes to wander around looking for gold. When it is in the presence of a suitable partner or partners it likes to settle down and stick around for a while. Similar considerations hold for many other assay systems that may measure many more cellular properties than we expect them to. I have discussed this matter in detail because it illustrates a kind of biological Heisenberg uncertainty principle that we disregard at our peril.

So much for the past. My list of predictions for the future is much briefer and more optimistic. As indicated above, I expect a profound understanding of mediator-receptor interactions to emerge. This is analogous to the situation that already holds for many of the peptide hormones and many of the non-lymphokine growth factors such as Epidermal Growth Factor (EGF). This understanding will come about as a direct consequence of the availability of pure mediators based upon the application of hybridoma and recombinant technologies.

It will also reflect a growing capability for the construction of hybrid mediators with synthetic cellular binding sites and native active sites, and vice versa. I suspect that it will also prove feasible to generate on lymphocyte membranes well-defined receptors not ordinarily found there. If such receptors are specific for biologically active macromolecules and retain functional activity, then the possibility for novel kinds of therapeutic strategies emerges from such studies.

These studies, focussing not on biochemistry, but on the use of chemically defined probes to understand biology, should pave the way for the development of synthetic and semi-synthetic mediators and mediator analogues. These compounds will have obvious importance as therapeutic agents. With a little luck, we'll be able to treat cancer, the common cold, ant Type II Herpes. At the Second International Lymphokine Workshop, I indicated that the last few years have demonstrated that what we are studying is biologically relevant. I think that the next few years will show that it is clinically relevant as well.

REFERENCES

1. Cohen, S., *In* "Biochemical Characterization of Lymphokines" (A.L. deWeck, F. Kristensen, and M. Landy, eds.),p. 607. Academic Press, New York, (1979).
2. Cohen, S., *J. Immunol. 100*, 407 (1968).
3. Gutowski J., and Cohen, S., *In* "Interleukins, Lymphokines and Cytokines" (J. J. Oppenheim and S. Cohen, eds.) Academic Press, New York, in press.
4. Yoshida, T. and Cohen, S., *Fed. Proceedings Vol. 41*, 2480 (1982).
5. Torisu, M., Yoshida, T., Ward, P.A., and Cohen, S., *J. Immunol. 111*, 1450 (1973).
6. Adelman, N., Cohen, S., and Yoshida, T., *J. Immunol. 121*, 209 (1978).
7. Postlethwaite, A.E. and Kang, A. H., *In* "Interleukins, Lymphokines, and Cytokines" (J.J. Oppenheim and S. Cohen, eds.), Academic Press, New York, in press.
8. Watt, S. and Auerbach, R., Ibid, in press.
9. Cohen, M. D., D'Silva, H., Yoshida, T. and Cohen, S., Ibid, in press.

THE DEVELOPMENT OF A CONCEPTUAL FRAMEWORK FOR LYMPHOKINE FUNCTION

John J. Farrar[1]

Laboratory of Microbiology and Immunology
National Insitute for Dental Research
National Institutes of Health
Bethesda, Maryland

I realize I am now in a particularly unenviable position for two very specific reasons. The first, is that it is very difficult to add anything particularly new or profound at the end of the workshop. All that is profound has surely already been said by previous speakers. The second reason is that including Drs. Vilcek's and Cohen's summary comments, we have now been subjected to 91 ten minute presentations in four days. It is inevitable that such a gruelling schedule will sooner or later evoke some hostile behavior on the part of the audience and I would rather not be the speaker at that particular time.

However, at the risk of boring you all, I will attempt to summarize the progress I believe we have achieved since the last lymphokine workshop and at the same time raise a few criticisms which if acted upon may help guide us during the next three years.

We have witnessed a veritable explosion in the number of lymphokines or rather biological activities delineated in the past three years. These observations have, obviously made life for the lymphokinologist much more complicated.

[1]*Present address: Hoffmann-La Roche, Nutley, New Jersey*

ISBN 0-12-527540-4

However, at the same time we have been able to construct a conceptual framework in which to put most of these non-specific lymphokines. The conceptual framework that has been developed (at least with respect to lymphocyte activation) is centered on the idea that the activation of a precursor lymphocyte to proliferate and differentiate into a terminal effector B-cell or T-cell is dependent upon multiple signals each of which independently regulates a distinct phase of the response and is dependent on the application of a number of these signals in the appropriate sequence. This idea also apparently applies to the activation of macrophages, as well as T and B-cells. The data which attests to the veracity of this concept comes from the numerous examples of synergy between distinct factors in some particular lymphocyte activation process. For example, three years ago we were arguing whether or not T-cell replacing factor (TRF) and Interleukin-2 (IL-2) were identical because both factors seemed to be capable of enhancing antibody synthesis. Since that time we have been able to demonstrate the biochemical difference between TRF and IL-2 and, in fact, have extended the list of helper factors for antibody synthesis to an additional TRF, a B-cell growth factor, and two isotype switching factors or B-cell differentiation factors. From the data presented at this meeting, it is likely that all or many of these factors are functioning in a common lymphocyte activation pathway; each factor responsible for the initiation or activation of some distinct phase of the lymphocyte response. The data of Nakanishi et al., presented at the poster session two evenings ago, which demonstrated that four factors were required to activate B-cells to synthesize immunoglobulin is a good example of this phenomenon. Similarly, at this meeting we have also seen three or four examples from as many different labs of differentiation factors which appear to work in concert with the growth factor, IL-2 to affect the induction of the cytotoxic T-cell response. I believe that these observations and others represent slow but real progress in the development of our understanding of lymphokine function. These results may not be making life more simple for the lymphokinologist, but they are bringing us, I believe, to a clearer understanding of what is actually happening within these lymphocyte induction pathways.

The other important point to be made concerning the progress we have made in the past three years is that in the pursuit of our work, we have begun to employ some newer technologies which have recently become available. For example, whereas three years ago, one monoclonal antibody against a lymphokine was described, at the present meeting we have seen many examples of monoclonal antibodies against a number of factors and receptors for factors. Three years ago at the 2nd International Lymphokine Workshop just a few T-cell lines or hybridomas were utilized as sources of lymphokines. At this meeting, there have been multiple reports of T-cell lines and hybridomas being used either to produce very high titers of factors or selective factors not contaminated by other lymphokines. In addition, there are some other new and unique sources of lymphokines being exploited such as the human placenta by Flynn et al. as a source of Interleukin-1.

In terms of the purification techniques that have been employed in the pursuit of lymphokine purification and characterization we have witnessed considerable progress in the past three years. For example, there was but one report of the use of high pressure liquid chromatography by David Webb for the purification of lymphokines three years ago. At this meeting there were many more reports of the use of HPLC and other sophisticated techniques in purification such as affinity chromatography and immunoadsorption chromatography. Some really sterling examples of lymphokine purification have been presented at this meeting. One such example is the work of Aggarwal on the purification of lymphotoxin. This piece of work has gone a long way towards clarifying the complexity associated with the overwhelming degree of molecular heterogeneity that has hertofore been reported for this class of molecules.

Finally, we have witnessed the advent of the use of recombinant DNA technology in the field of lymphokine research. This technology is being applied not only in the laboratories of the bio-technology firms as exemplified by the work of David Goedel and Genentech but has also been shown to be within the realm of possibility for smaller labs as witnessed by the success of Richard Luben in his work with osteoclast activating factor.

So, in all, I believe we have made significant progress in the past three years although there is obviously some room for criticism. I began to think about negative aspects after a younger colleague of mine (who comes to lymphokines indirectly from the field of molecular biology) commented that, although he had learned a lot at the meeting, he felt that many investigators were simply doing the easy experiments, perhaps in order to obtain easy data, but that the critical, the really crucial experiments, were being avoided. After thinking about that criticism for a moment, I agreed. However, upon further consideration I proposed a less malevolent explanation for the phenomological approach often taken. Most of the investigators who have traditionally come into lymphokinology have had their training in classical cellular immunology and have not had extensive training in biochemistry, pharmacology, melecular biology, etc. which would allow them to perform these more critical experiments. Although this lack of the appropriate expertise within the field is changing, it is up to us, the cellular immunologist, to try to recruit as many of these trained scientists into this field. That we have failed to recruit into our field the investigators with the appropriate expertise may be a more appropriate criticism.

Finally, predictions for the future. As I don't have a crystal ball, I am not particularly well equipped to make very many predictions. Actually, there is one prediction I wish I could make but am afraid that I cannot. I wish that I could predict when we will have discovered and delineated the basic set of lymphokines. I wish I could predict when we will reach that point in time because I believe *that* point will mark the beginning of the third and most exciting phase of this research; that is when we begin to perform the critical mechanism-of-action studies on the 10, 20, or 80 lymphokines that comprise the immunoregulatory network.

The one prediction I can make with some degree of certainty, is that there will be an ever increasing role of industry; both of the biotechnology and pharmaceutical types, in lymphokines research. This fact has been bemoaned by many academic scientists but I think that we are wrong to begrudge the involvement of industry. There are obviously a lot of very complex and difficult issues to be resolved regarding this blending of academia and industry. These weighty problems were addressed in great detail last evening by Dr. George Poste. However, taken at its simplest level, the interest that industry is exhibiting in lymphokines represents a positive signal to us, the academicians. This interest is a signal from the real world, if you will, that a judgement has been made that what we have done and what we are doing is important and worthwhile and at some point in the future will have medical benefit. A fact of life that we must face is that in this capitalistic society in which we live, the one and only mechanism by which research achievements may be translated into medical benefit for the population as a whole is through the involvement of commercial interests. Therefore, in as much as I am very anxious for the day in which lymphokine studies are brought to fruition in a medically beneficial way, I welcome the interest of industrial concerns. I believe there remains plenty of room for the more basic scientific studies on lymphokines within the academic institutions.

INTERFERON IS A MODEL CYTOKINE

Jan Vilček

Department of Microbiology
New York University School of Medicine
New York, New York

There are many parallels between the development of the interferons and lymphokines. It took more than 20 years to obtain pure preparations of interferons. Yet, since the first successful purification of interferons in the late 1970s, a staggering amount of important new information has been accumulated about their nature and actions. The rate at which new information about interferons is accumulating has further accelarated after the breakthroughs in interferon gene cloning in 1980. Now that many purified lymphokines and better assays for them have become available it is easy to foresee that lymphokinology (pardon this clumsy expression) is on the verge of a similar information explosion. Given the attention of DNA cloners, how long can it be before we are swamped with information about lymphokine gene families, pseudogenes, introns and exons, flanking sequences and TATA boxes? We'd better get used to gazing at slides with illegible nucleotide sequences before we all meet again at the next lymphokine workshop.

It will be obvious to anyone willing to spend a half hour on a comparison of the Proceedings of the 1979 Lymphokine Workshop held at Ermatigen and the contents of the present volume that there has been enormous progress, stemming in a large measure from the recent successful isolation and physicochemical characterization of about half a dozen lymphokines and monokines. Work with defined molecules, in turn, made possible a preliminary characterization of an intricate network of immunoregulatory interactions, termed by aficionados "the lymphokine cascade". It now seems hard to believe that the first publications demonstrating the ability of IL-1

Interleukins, Lymphokines, and Cytokines

ISBN 0-12-527540-4

to promote IL-2 production appeared only two years ago. To lymphokinologists 1980 seems more like a decade ago.

It might be interesting to mention some of the factors which led to the ultimate fulfillment of all lymphokine purification studies -- the demonstration of a single Coomassie blue-stained band on SDS gels. The expertise and gadgetry of protein chemists are certainly indispensable, but even the fanciest HPLC equipment will not produce an isolated lymphokine peak from lousy starting materials. One cannot overemphasize the importance of defined cell systems and induction methods which result in preparations enriched for a particular lymphokine.

Chemical isolation of "classical" or "type I" interferons became a realistic task when methods had been devised allowing the preparation of starting materials containing 0.1 to 1% interferon, instead of only 0.001% or so. Purification of IL-2 became possible with the identification of murine and human transformed T cell lines producing high levels of the lymphokine on appropriate stimulation. Purification of human immune interferon (IFN-γ) was also preceded by the development of a highly effective induction protocol in cultures of PBL. It is interesting that the starting materials used for the purification of IL-2 and IFN-γ were prepared with the aid of the same inducer -- the tumor promoting phorbol ester, 12-O-tetradecanoylphorbol-13-acetate (TPA), also known as phorbol myristate acetate (PMA). Along with some related compounds, this phorbol ester comes close to being a universal potent lymphokine/ monokine inducer.

Parenthetically, a survey of the literature shows that immunologists favor the abbreviation PMA, while investigators involved in the study of tumor promotion and other biological effects of the same molecule generally use the acronym TPA. This situation can lead to amusing confusions. I was recently asked to review a manuscript authored by a distinguished group of cellular immunologists, describing the induction of a lymphokine by PMA. In the discussion the authors explained that observations similar to theirs were made by other investigators who used a different inducing agent termed TPA... The index of the Proceedings of the 1979 Lymphokine Workshop also lists phorbol myristate acetate separately from 12-O-tetradecanoylphorbol-13-acetate, conveying the impression that they are two different molecules. (If the person preparing the index for the current volume ever reads this text, a similar confusion might be avoided. TPA is PMA is TPA.)

At the 1979 Workshop there was a clear dividing line between research on interferons and the other lymphokines, partly because interferon research was more advanced, and partly because the two fields had developed from different bases. By 1982 this distinction has largely disappeared. Not only has the information gap narrowed considerably, but also, as shown in many papers presented at this Workshop, we now have several examples of interactions between the interferons and other lymphokines. IFN-γ is an integral part of the lymphokine cascade, with its production in T lymphocytes being promoted by IL-2. It now appears that IFN-γ can also have a positive feedback effect on the synthesis of interleukins. Many details of these interactions still need to be defined. Some effects seen on the administration of interferons to patients could be due to the stimulation of other lymphokines. One thought that comes to mind is that the pyrogenic reaction, universally observed in patients treated with various interferons, could be due to the induction of IL-1 (endogenous pyrogen). Some of the immunoregulatory activities observed with IFN-γ might be due to its ability to stimulate IL-2 production. Several papers presented at this workshop demonstrated that IFN- can cause macrophage activation. One of the unanswered questions is whether IFN-γ is in fact the only lymphokine mediating MAF activity or whether there are other structurally unrelated lymphokines exerting this function. In view of these many interactions and overlapping functions, interferonologists can no longer afford to ignore developments in the lymphokinologists' court and vice versa. In the last couple of years the two fields have virtually coalesced.

It is interesting that, although IFN-γ was among the first lymphokines to be purified to homogeneity, and together with the leukocyte interferons (IFN-α) it was the first one to be successfully "cloned" and completely sequenced, its major function is still not known. In view of their many known actions in different target cells, interferons appear to be more versatile than most other lymphokines. With progress continuing at the present rate we are likely to have a better understanding of the immunoregulatory functions of interferons when we all meet again at the Fourth International Lymphokine Workshop.

THE FOX AND THE RAVEN

Edgar Pick

Laboratory of Immunopharmacology
Department of Human Microbiology
Sackler School of Medicine
Tel-Aviv University
Tel-Aviv, Israel

In responding to the organizer's invitation to say a few words as we close this workshop, I remind you that I come from Israel, a country that has provided the world with a fair share of futurologists, better known to some as the prophets of doom.

I find myself leaving Haverford with an acute feeling of need to reassess where we stand. Communication, and especially oral communication between members of a closed group such as ours, has a two-edged effect: a *favorable* one, because it enables one to talk with colleagues who share his interests; but also a *detrimental* one, because excessive inbreeding tends to generate self-congratulation and reciprocal reassurance that what we are doing is good. When I analyze the reasons behind the slightly depressed feeling with which I leave this workshop, I can find several:

The first one is the fine presentation by molecular biologist, David Goeddel on the application of recombinant DNA technology to IFN production. I felt like a passenger on a provincial train pushed onto a sidetrack, while the luxury express thunders by at 200 miles an hour. Our little peaks on Sephadex columns (these days called Sephacryl or AcA34), our pitiful cpm of ^{3}H-thymidine ordinates and 1:2 to 1:16 dilution abscissae look so ridiculous! The best we can come up with is an HPLC tracing, the rationale behind which is that "this is the way we separate most polypeptides". And while most of us grew up knowing what a Sephadex G-100 or a CM-cellulose column is, who has some

Interleukins, Lymphokines, and Cytokines

ISBN 0-12-527540-4

comprehension of what an "C18-Ultra-Super-Bondalite" column is? And then look at *our* terminology! We are now in the process of adding numbers to the two ill-chosen letters which were supposed to save us from the darkness of the nomenclatural morass, but may yet lead us to our destruction.

Fellow lymphokinologists, have you ever had a nightmare in which you imagined that we all suddenly disappeared--or were never born? Think about the post-lymphokine world: Will textbooks of biochemistry, pathology, physiology or pharmacology be any different? Very little I fear! Sure, textbooks of immunology will be different--but remember, some of *us* have written them! What is our impact on the *mainstream* of biological thought? What would be different if we weren't around? At the practical level, how many patients would be worse off in our absence? We kneel humbly in front of the towering colossus of gene cloning and are trying to make him the offerings of our modest products, which he, benevolently, accepts. But beware! the day will come when gene cloners will be so busy that they will become selective. That day, eyebrows will be raised about the usefulness of the translated gene products, carefully labelled with two letter-one number or three or four letter acronyms, and these will be returned to us, with minimal thanks. I am afraid that we may have become victims of our vanity -- after working for years in the relative peace of academe, we are suddenly surrounded by representatives of "TECHS" with various prefixes who are telling us that they would love to clone our products, if only we could tell them which one to do first. Some of us, even, may be more than willing to exchange blue jeans for striped trousers. But, remember the fable of LaFontaine-the fox telling the raven that it has such a beautiful voice. The raven drops the cheese when it decides to demonstrate its vocal (scientific) prowess and that is the end of LaFontaine's story. But have we considered the possibility of an even more tragic ending such as the fox finding out that the piece of cheese doesn't really taste so good and deciding to abandon the poor raven, all alone up in the tree. What I am really saying is -- let's wait and do more work on the cell biology of lymphokine production and lymphokine action, and on the reality and usefulness of lymphokines *in vivo*. To be sure, IL-2 *is* a growth factor, but would any company have cloned the insulin gene just because insulin is a growth factor? Let us regain some of our lost modesty and innocence--the fact that we are clever enough to make one cell line produce a factor which makes another cell line make another factor which makes still another cell line grow or kill, will not provide us with a ticket to Paradise--and there is more to lymphokines than that.

I would end my friendly admonition by reminding you of a famous cytokine, and that is EGF. It has two wonderful qualities--it is present *in vivo*, and it is an eyelid opener. The first quality is something to look for--the second quality would make it a useful therapeutic agent for all of us.

STATE OF STANDARDIZATION, NOMENCLATURE, AND FUTURE PLANS

Joost J. Oppenheim

Laboratory of Microbiology and Immunology
National Institute of Dental Research
National Institutes of Health
Bethesda, Maryland

There is little that I can add to either the foregoing manic and depressive overviews or to the remarkable outpouring of new data and lively discussions between speakers and audience that filled the past four days. Despite this, there remain three relevant issues that were not considered during the workshop. The first of these issues is whether we should have organized a "wet" rather than "dry" workshop? In principle, a comparison of the effects of various lymphokine preparations at a workshop might have reduced certain complexities and controversies in this research area. In practice, however, the organizers felt that it was still premature to organize any formal testing of materials since most cytokines are neither available in sufficient purity nor in quantity to permit extensive testing by a panel of investigators. On the other hand, the format of the two previous international lymphokinesworkshops has succeeded in fostering relationships between investigators that has led to numerous collaborative projects and informal exchanges of information, cytokines, cell lines and antibodies. It is also apparent from this workshop that such interchanges have resulted in considerable agreement as the identity and property of cytokines and on the standardization of bioassays.

Both at the Human Lymphokine Conference held last year at Dallas, Texas and at this workshop, Dr. Dudley Dumonde has organized discussions concerning the issue of standardization. As co-chairman of the subcommittee of the IUIS on lymphokine standardization, he considers it timely to initiate a more organized exchange of mediators. Accordingly, at this meeting a feasibility study has been planned. Several

Interleukins, Lymphokines, and Cytokines

ISBN 0-12-527540-4

laboratories have agreed to make available large batches of partially purified Interleukin 2 and unseparated supernatants rich in Leucocyte Migration Inhibitory Factor (LIF), for testing by laboratories actively engaged in studies of these lymphokines. Although it can be argued that such an effort is warranted only when refined products of gene cloning become available, this pilot study will provide the organizational groundwork for the exchange of such materials. In the meantime such an effort may also result in greater uniformity in the bioassays of these lymphokines.

The second issue that rather surprisingly has not surfaced during these four days of torrential verbiage at this workshop has been the matter of terminology. Discussion of this issue was intentionally omitted from the formal program. Although there was likewise no organized discussion of this problem either at the preceding lymphokine workshop at Ermatingen, Switzerland in 1979, a number of the participants at that meeting agreed, that since a number of biological activities produced by macrophages and lymphocytes respectively could apparently be attributed to the same molecules, these activities should be renamed Interleukin 1 and 2, respectively. Fortuitously it turns out that these 2 factors can also act in sequence. This commendable approach to simplify terminology and improve communication, however, resulted in considerable controversy, since it broke with the tradition of investigators descriptive naming of factors on the basis of their biological activities. Paradoxically, this controversy has had considerable positive effect in focusing increased interest and galvanizing research on the subject of interleukins. Unfortunately, I did not detect any comparable synthesis developing at this meeting.

As is evident from the experience with the interleukin saga, collaborative identification of multipotent factors amplifies progress by simplifying communications and concepts. Lymphokine research has become too complex and has too many ramifications to be resolved by individual investigators. We have reached the point where the phenomenological approach of discovering and the naming of another factor by an investigator is nonproductive and merely enlarges the "Tower of Babel". It is imperative, to establish the relationship of such activities to already established cytokines. Until this can be achieved it remains operationally necessary to continue to name them on the basis of their apparent biological function. However, it behooves all of us to maximize exchange of moieties with multiple biological activities and thereby to further reduce the number of designated factors.

Finally, the concept of lymphokines and cytokines has been misconstrued in two different ways. First of all, these terms have been interpreted very broadly to include any product of activated cells. Such a definition is too broad since it includes metabolic waste products, leucotrienes, prostaglandins, RNA, transfer factor, etc. We should focus on studies of glycoprotein and lipoprotein products of activated cells. Insistence on a more regorous definition of this research area is in no way intended to restrict our research effort; collaboration with scientists using other technologies to study the other cell products should be encouraged. Second of all, investigators have if anything viewed the lymphokine and cytokine issues in too parachial a fashion. For example, there is a tendency on the part of immunologists to unduly limit their research to problems involving specificity, rather than perceiving their concerns as part of the larger problems of the role of macrophages and lymphocytes in the context of host defense, development and differentiation. Furthermore, since polypeptide products similar to lymphokines both are made by and can act on nonlymphoid cell types, we must study not only immunological competent cells, but the interdigitation of immunological mechanisms with inflammation, wound healing and other homeostatic systems. We should, therefore, not limit our considerations to the discipline of immunology, but we should welcome the opportunity to study the products of nonlymphoid cells and their effects on nonlymphoid cells. In this regard, I am delighted that many of the scientists who are interested in the endogenous pyrogens have participated in this workshop and feel that they can identify with us as a group with common interests. Furthermore, although investigators studying interferons (IFN) and colony stimulating factors (CSF) hold their own workshops, I hope that more of these colleagues can be encouraged to join us at our next meeting especially since the study of these cytokines has progressed more swiftly and we can learn from them, just as they will surely benefit by considering the interactions between IFN, CSF and other cytokines.

This brings me to the issue of future plans. The program committee discussed the date and site for the 4th International Lymphokine Workshop. It was decided to schedule the next workshop in two rather than three years for several reasons. As judged from the ever expanding literature and increases in the frequency of meetings, the momentum of lymphokine research continues its rapid acceleration. A number of smaller lymphokine workshops have been held between our meeting in Ermatingen and now in Haverford and we seek to avoid undue fragmentation of this field. In addition, in view of the impressive progress in molecular genetics, there was some anxiety that if we delayed the next workshop until 1985, we might find ourselves cloned out of business and totally redundant.

Fortunately, we have the offer of two possible sites in Europe for 1984 and we will let you know when the plans are finalized. I would like to finish by thanking all of you for making this an exciting meeting and for providing us with many stimulating papers and comments for the book.

INDEX

V

W